Understanding Natural Language Understanding

Erik Cambria

Understanding Natural Language Understanding

 Springer

Erik Cambria
College of Computing and Data Science
Nanyang Technological University
Singapore, Singapore

ISBN 978-3-031-73973-6   ISBN 978-3-031-73974-3   (eBook)
https://doi.org/10.1007/978-3-031-73974-3

To Marvin Minsky,

A wonderful mentor and a visionary pioneer whose brilliance and curiosity have paved the way for generations of dreamers and thinkers. Your groundbreaking work in artificial intelligence has not only expanded the horizons of human knowledge but also inspired us to imagine a future where machines understand and enrich our lives.

In gratitude for your relentless pursuit of understanding and your unwavering belief in the potential of the human mind.

MIT Media Lab, Cambridge, MA, United States, 2008

Preface

About half a century ago, artificial intelligence (AI) pioneers like Marvin Minsky embarked on the ambitious project of emulating how the human mind encodes and decodes meaning. While today we have a better understanding of the brain thanks to neuroscience, we are still far from unlocking the secrets of the mind. Especially when it comes to language, the prime example of human intelligence, we face enormous difficulties in replicating how the human mind processes it. "Understanding natural language understanding", i.e., understanding how the mind encodes and decodes meaning through language, is a significant milestone in our journey towards creating machines that genuinely comprehend human language.

Large language models (LLMs), such as GPT-4, have astounded us with their ability to generate coherent, contextually relevant text, seemingly bridging the gap between human and machine communication. Yet, despite their impressive capabilities, these models operate on statistical patterns rather than true comprehension. This textbook delves into the nuanced differences between these two paradigms and explores the future of AI as we strive to achieve true natural language understanding (NLU). LLMs excel at identifying and replicating patterns within vast datasets, producing responses that appear intelligent and meaningful. They can generate text that mimics human writing styles, provide summaries of complex documents, and even engage in extended dialogues with users. However, their limitations become evident when they encounter tasks that require deeper understanding, reasoning, and contextual knowledge. LLMs can produce plausible-sounding but incorrect answers, struggle with ambiguous queries, and often lack the ability to generalize knowledge across different domains effectively.

Instead, an NLU system that deconstructs meaning leveraging linguistics and semiotics (on top of statistical analysis) represents a more profound level of language comprehension. It involves understanding context in a manner similar to human cognition, discerning subtle meanings, implications, and nuances that current LLMs might miss or misinterpret. NLU grasps the semantics behind words and sentences, comprehending synonyms, metaphors, idioms, and abstract concepts with precision. This deeper comprehension allows for consistent accuracy and robustness, enabling AI systems to handle ambiguous, incomplete, or novel queries more effectively.

One of the key advantages of cognitive-inspired NLU systems is their ability to learn and adapt from interactions meaningfully. Unlike LLMs, which remain static once trained, NLU systems can incorporate new knowledge dynamically, correcting misunderstandings and adapting to new contexts. This adaptability is crucial in fields where information is rapidly evolving, such as technology and medicine. Moreover, NLU systems can generalize knowledge from one domain to another, providing more reliable and versatile applications across interdisciplinary fields. Ethical considerations also play a significant role in the development of NLU systems, as these are more transparent and can better recognize and mitigate biases, ensuring fair and ethical responses. They understand the potential impact of their responses, avoiding harmful or inappropriate content more reliably. This understanding of ethical implications and societal norms is crucial for building trustworthy AI systems in sensitive areas like mental health, education, and decision-making.

Achieving true NLU involves advanced knowledge representation, such as incorporating symbolic reasoning and structured knowledge bases. This includes ontologies, semantic networks, and rule-based systems that explicitly encode relationships and rules. Combining symbolic AI with machine learning creates hybrid systems that leverage both structured knowledge and the pattern recognition strengths of LLMs. Neurosymbolic integration, which merges neural networks with symbolic reasoning systems, is a promising approach to understanding and generating more accurate and contextually appropriate responses. Embedding real-world knowledge and commonsense reasoning into AI systems is also crucial for NLU. Training on diverse data sources and integrating world models that simulate real-world scenarios enable these systems to maintain context over long conversations, understand intents and sentiments, and manage turn-taking effectively. Furthermore, NLU systems should emulate human-like learning and adaptation, continuously learning from human interaction, feedback, and correction.

In this textbook, we explore the current state of LLMs, their capabilities, and limitations, and contrast them with the aspirational goals of NLU. We delve into the technical foundations required for achieving true NLU, including advanced knowledge representation, hybrid AI systems, and neurosymbolic integration. We also examine some ethical implications and societal impacts of developing AI systems that genuinely understand human language. The textbook features multiple exercises at the end of each chapter, along with a group assignment and a final quiz. It evolved out of the author's own teaching course SC4021 (NTU CCDS course on information retrieval) but it can be used for many other courses, e.g., natural language processing (NLP), AI, data analytics, data mining, etc., at NTU and universities worldwide.

We will explore different ways of encoding and decoding meaning, which can be used for knowledge representation and reasoning in several downstream applications. As we embark on this exploration, we are reminded of the pioneering work of visionaries like Marvin Minsky, whose relentless pursuit of understanding and belief in the potential of the human mind have inspired us to dream of a future where machines not only mimic human language but truly comprehend it. This textbook is dedicated to those who continue to push the boundaries of AI, striving to create systems that enrich our lives with genuine understanding and insight.

Acknowledgements

This work would have never been possible without the help of my wonderful research group, the Sentic Team (`https://sentic.net/team`), who have helped me translate my silly ideas into concrete research works over the last ten years. Special thanks go to my awesome postdocs Drs. Rui Mao, Qian Liu, and Xulang Zhang, who were instrumental in organizing and refining the materials of this textbook. Last but not least, I thank my beautiful wife, Jocelyn Choong, for often forcing me to focus on finishing this book despite the many distractions life has to offer.

Sentic Team, NTU CCDS, 2024

Contents

Acronyms

ABOM   Aspect-Based Opinion Mining
ABSA   Aspect-Based Sentiment Analysis
ACE   Automatic Content Extraction
AI   Artificial Intelligence
ALM   Anomalous Language Modeling
AMT   Amazon Mechanical Turk
ASR   Automatic Speech Recognition
AZP   Anaphoric Zero Pronoun
AZPR   Anaphoric Zero Pronoun Resolution
BERT   Bidirectional Encoder Representations from Transformer
BGRNN   Bidirectional Gated Recurrent Neural Network
BLEU   Bi-Lingual Evaluation Understudy
CEG   Commonsense Explanation Generation
CNN   Convolutional Neural Network
CRF   Conditional Random Field
CMT   Conceptual Metaphor Theory
DAG   Directed Acyclic Graph
DRL   Deep Reinforcement Learning
FFM   Five Factor Model
FOL   First-Order Logic
GBN   Gaussian Bayesian Network
GCN   Graph Convolutional Network
GRU   Gated Recurrent Unit
GPT   Generative Pretrained Transformer
HAN   Hierarchical Attention Network
HCI   Human-Computer Interaction
HMM   Hidden Markov Model
IPA   International Phonetic Alphabet
JERE   Joint Entity and Relation Extraction
KGC   Knowledge Graph Construction
LCS   Longest Common Subsequence

LDA	Latent Dirichlet Allocation
LLM	Large Language Model
LSTM	Long Short-Term Memory
MAE	Mean Absolute Error
MAP	Mean Average Precision
MBTI	Myers-Briggs Type Indicator
MLP	Multi-Layer Perceptron
MRC	Machine-Reading Comprehension
MRD	Machine-Readable Dictionary
MRF	Markov Random Field
MSE	Mean Squared Error
MTL	Multi-Task Learning
MUC	Message Understanding Conference
MWE	Multi-Word Expression
NER	Named Entity Recognition
NLP	Natural Language Processing
NLU	Natural Language Understanding
NMF	Non-negative Matrix Factorization
NMT	Neural Machine Translation
NTN	Neural Tensor Network
OCR	Optical Character Recognition
OMCS	Open Mind Common Sense
OOV	Out Of Vocabulary
PCA	Principal Component Analysis
POS	Part Of Speech
PLM	Pretrained Language Model
PMI	Pointwise Mutual Information
PSA	Position-aware Self Attention
QA	Question Answering
RDR	Ripple Down Rule
RNN	Recurrent Neural Network
ROC	Receiver Operation Characteristic
RRS	Responsible Recommender System
SBD	Sentence Boundary Disambiguation
SES	Shortest Edit Script
SMO	Sequential Minimal Optimization
SMS	Short Message Service
SMT	Statistical Machine Translation
SRL	Semantic Role Labeling
SSL	Semi Supervised Learning
STM	Syntactic Tree Matching
SVD	Singular Value Decomposition
SVM	Support Vector Machine
UGC	User-Generated Content
WSD	Word Sense Disambiguation

Chapter 1
Natural Language Understanding & AI

Abstract In this chapter, we delve into the critical role that natural language understanding (NLU) plays in shaping the future of artificial intelligence (AI). To set the stage, we begin by defining what constitutes an NLU system. Next, we explore how NLU can drive the evolution of next-generation AI systems, which promise to be more reliable, responsible, and personalized. To this end, we introduce the seven pillars for the future of AI, which represent the foundational elements necessary to advance AI technology in a way that is more transparent and reliable. Next, we propose the concept of responsible recommender systems, which incorporate ethical guidelines and user-centric principles to ensure recommendations are not only relevant but also fair, unbiased, and respectful of user privacy. Lastly, we present a framework for personalized sentiment analysis, which aims at making AI systems more responsive and attuned to the needs and emotions of each user.

Key words: Natural Language Understanding, Reliable AI, Responsible AI, Personalized AI

1.1 Introduction

We define an NLU system as a brain-inspired modular framework that deconstructs meaning through explicitly modeling the cognitive processes that the human mind leverages to encode and decode language. While large language models (LLMs) like GPT-4 have demonstrated significant advancements in generating coherent and contextually relevant text, they fundamentally lack genuine comprehension. This distinction between statistical pattern recognition and actual understanding marks the critical difference between natural language processing (NLP) and the aspirational goal of NLU.

© The Author(s), under exclusive license to Springer Nature Switzerland AG 2025
E. Cambria, *Understanding Natural Language Understanding*,
https://doi.org/10.1007/978-3-031-73974-3_1

1

NLU involves an understanding of context that mirrors human comprehension. It discerns subtle meanings, implications, and nuances that LLMs might miss or misinterpret. For example, understanding the sentence "The bank is on the river bank" requires recognizing that "bank" refers to two different concepts based on context. While LLMs use statistical patterns to guess meanings, NLU comprehends these distinctions inherently. This comprehension extends to more complex constructs like sarcasm, irony, and humor, which often elude LLMs. Moreover, NLU goes beyond mere word associations to grasp the semantics behind sentences. This includes understanding synonyms, metaphors, idioms, and abstract concepts. For instance, interpreting a metaphor like "Time is a thief" involves recognizing the abstract concept that time, like a thief, can take things away from us. LLMs might recognize this phrase as common but would not truly understand the conceptual comparison without extensive training data on similar metaphors. Similarly, idiomatic expressions like "kick the bucket" (meaning "to die") require an understanding of cultural context and figurative language that NLU can provide.

NLU ensures consistent accuracy by providing precise and reliable responses. This capability stems from an understanding that enables the system to handle ambiguous, incomplete, or novel queries effectively through reasoning. LLMs, in contrast, sometimes generate plausible-sounding but incorrect or nonsensical answers due to their reliance on probabilistic models. For instance, when asked a complex question involving multiple steps of logic, an LLM might provide an answer that fits part of the question but does not fully resolve the complexity. An example might be a multi-part medical diagnosis where the system needs to integrate symptoms, patient history, and current medical knowledge to provide an accurate assessment. The robustness of NLU means it can interpret and respond accurately even when the input is vague or requires additional contextual knowledge. For example, if a user says, "I need a place to stay near the event," NLU can infer that the user is looking for accommodations close to a specific location, while an LLM might provide generic information about lodging without understanding the specific requirement. NLU can also handle evolving conversations where the context shifts, maintaining coherence and relevance in its responses.

NLU systems are capable of generalizing knowledge from one domain to another effectively. While LLMs are proficient at recognizing patterns within their training data, they often struggle to apply knowledge across vastly different contexts. NLU, however, can transfer learning and apply relevant information dynamically. For instance, an NLU system could use its understanding of medical terminology to assist in a legal context where medical information is relevant. This cross-domain generalization is crucial for applications in interdisciplinary fields such as bioinformatics, where knowledge of both biology and data science is necessary. Furthermore, NLU systems learn and adapt from interactions in a meaningful way. They can incorporate new knowledge and correct misunderstandings dynamically, similar to human learning processes. This contrasts with LLMs, which are static once trained. For example, if an NLU system encounters a new scientific discovery, it can integrate this information into its knowledge base and apply it in future interactions.

NLU can better recognize and avoid biases, leading to more fair and ethical responses. LLMs can inadvertently reinforce biases present in their training data because they lack the deeper understanding necessary to critically evaluate and mitigate such biases. For example, an LLM trained on biased data might perpetuate stereotypes, whereas NLU would recognize and avoid such biases. This capability is essential in applications such as hiring processes, where unbiased decision-making is crucial for fairness. Moreover, NLU understands the potential impact of its responses, avoiding harmful or inappropriate content more reliably. It comprehends the ethical implications and societal norms guiding human interactions, ensuring safer and more responsible AI behavior. For instance, an NLU system would avoid making insensitive comments about sensitive topics, understanding the context and potential repercussions. This understanding helps in creating AI systems that can be trusted in sensitive applications like mental health support and education.

Achieving true NLU involves advanced knowledge representation, such as incorporating symbolic reasoning and structured knowledge bases. This includes ontologies, semantic networks, and rule-based systems that explicitly encode relationships and rules. For example, an NLU system could use an ontology to understand the relationship between different medical conditions and treatments. This structured approach allows the system to make logical inferences and provide reasoned answers based on a deep understanding of the subject matter. Combining symbolic AI with machine learning creates hybrid systems that leverage both structured knowledge and the pattern recognition strengths of LLMs. Neurosymbolic integration, which merges neural networks with symbolic reasoning systems, helps in understanding and generating more accurate and contextually appropriate responses. For instance, a neurosymbolic system might use neural networks to process natural language input and symbolic reasoning to deduce the appropriate response based on an internal knowledge base. This hybrid approach allows for more sophisticated and reliable AI systems that can handle complex queries and tasks. NLU also requires embedding real-world knowledge and commonsense reasoning into AI systems. This involves training on diverse data sources and integrating world models that simulate real-world scenarios. Advanced dialogue systems can maintain context over long conversations, understand intents and sentiments, and manage turn-taking effectively. For example, a customer service chatbot with NLU would handle a multi-step customer query seamlessly, maintaining context and providing accurate solutions throughout the interaction. This capability is essential for creating AI systems that can engage in meaningful and productive conversations with users. Techniques like meta-learning and analogical reasoning enable systems to adapt quickly to new information and contexts, transferring knowledge from known situations to new, similar ones. This continuous learning and adaptation make AI systems more resilient and effective in dynamic environments.

In summary, NLU systems go beyond the mere statistical analysis of language and, hence, have the potential to be the enablers of next-generation AI systems that are reliable, responsible and personalized. We discuss this in more detail in the next three sections.

1.2 Towards More Reliable AI

In 2022, the world was stunned by ChatGPT, a chatbot that relies on an LLM built by means of generative pretraining transformers (GPT). The performance capabilities of GPT-based LLMs enable chatbots to generate detailed, original, and plausible responses to prompts. GPT-4 and other LLMs are pretrained on a large dataset (self-supervised and at scale), before being adapted for a variety of downstream tasks through fine-tuning. Pretraining is time-intensive and never repeated, whereas fine-tuning is conducted in a regular fashion. The behavior of GPT-based chatbots arises through fine-tuning. The performance capabilities of LLMs have been attributed to at least two factors: pretraining and scale (Bommasani et al., 2021). Pretraining, an instance of transfer learning in which LLMs use knowledge acquired from one task and transfer it to another, makes LLMs possible. Scale, including better computer hardware, the transformer architecture, the availability of more and higher-quality training data, makes LLMs powerful. Although these capabilities are not insubstantial, they do not yet rise to the level of NLU (Bender and Koller, 2020; Amin et al., 2024, 2023). In addition, LLMs are prone to hallucination: ChatGPT may produce linguistic responses that, though syntactically and semantically fine and credible-sounding, are ultimately incorrect (Shen et al., 2023b). Furthermore, we may distinguish between the capabilities of LLMs (acquired through pretraining) and the behavior (affected by fine-tuning, which happens after pretraining) of LLMs. Fine-tuning can have unintended effects, including behavioral drift on certain tasks. ChatGPT, in fact, seems prone to the 'short blanket dilemma': while trying to improve its accuracy on some tasks, OpenAI researchers inadvertently made ChatGPT worse for tasks which it previously excelled at (Chen et al., 2023a).

AI research has slowly been drifting away from what its forefathers envisioned back in the 1960s. Instead of evolving towards the emulation of human intelligence, AI research has regressed into the mimicking of intelligent behavior in the past decade or so. The main goal of most tech companies is not designing the building blocks of intelligence but simply creating products that existing and potential customers deem intelligent. In this context, instead of labeling it as 'artificial' intelligence, it may be more apt to characterize such research as 'pareidoliac' intelligence. This term highlights the development of expert systems while raising questions about their claim to possess genuine intelligence. We feel there is a need for an AI refocus on humanity, an Anti-Copernican revolution of sorts: like Copernicus demoted humans from their privileged spot at the center of the universe, in fact, deep learning has removed humans from the equation of learning. In traditional neural networks, especially those with a shallow architecture (few hidden layers), humans were at the center of the technological universe as they had to carefully design the input features, select appropriate hyperparameters, adjust learning rates, etc. Instead, due to their increased complexity and capacity to automatically learn features from data, deep neural networks do not require manual feature engineering and, hence, have effectively removed humans from the loop of learning. While this is good in terms of cost, time, and effectiveness, it is bad for several other reasons, including transparency, accountability, and bias.

In the deep learning era, humans no longer have control on how the learning process takes place. To save on cost and time, we delegated the important task of selecting which features are important for classification to deep neural networks. These, however, are mathematical models with no commonsense whatsoever: they do not know how to properly choose features. For example, in selecting candidates for a job opening, deep neural networks may decide that gender is an important feature to take into account simply because more men are present in the training data as positive samples.

The issue is not only that deep nets may accidentally choose unimportant or even wrong features, but that we have no way of knowing this because of their black-box nature (Yeo et al., 2024b). In other words, not only humans have been taken out of the picture but they have also been blindfolded. For these reasons, we feel there is a need to bring human-centered capabilities back at the center of AI, e.g., by having human-in-the-loop or human-in-command systems that ensure AI outputs and reasoning steps are human-readable and human-editable. To this end, we propose seven pillars for the future of AI (Cambria et al., 2023), namely: *Multidisciplinarity*, *Task Decomposition*, *Parallel Analogy*, *Symbol Grounding*, *Similarity Measure*, *Intention Awareness*, and *Trustworthiness* (Fig. 1.1).

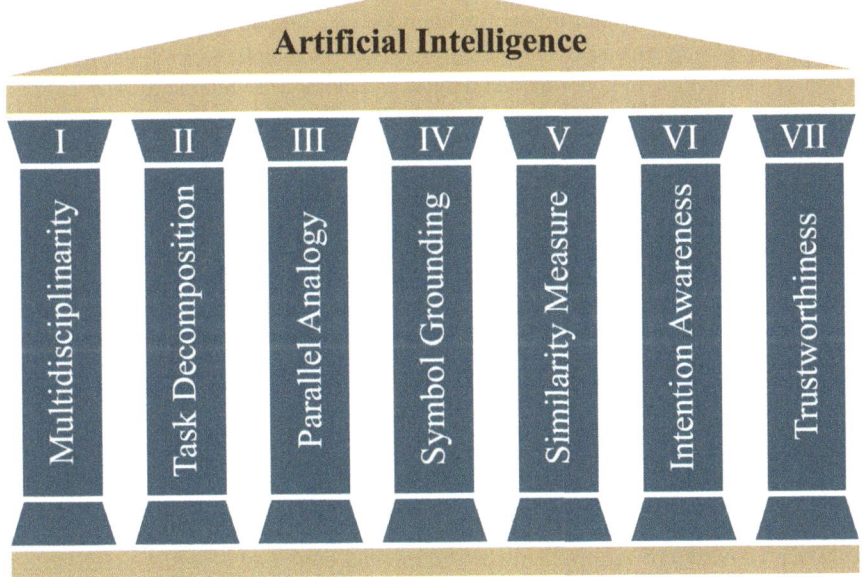

Fig. 1.1: Seven pillars for the future of AI.

1.2.1 Multidisciplinarity

Due to the complex and multifaceted nature of modern AI technologies and applications, *Multidisciplinarity* is of increasing importance for the future of AI. The integration of knowledge from disciplines like mathematics, semiotics, logic, linguistics, psychology, sociology, and ethics allows for a more holistic understanding of AI's capabilities and limitations. Mathematical principles such as linear algebra, calculus, probability theory, and optimization underpin the design of AI algorithms. Maths alone, however, is not enough for designing intelligent systems, because mathematical approaches excel at capturing predominant linguistic patterns but often struggle with addressing 'long tail' issues such as less common or niche linguistic phenomena. Disciplines like semiotics can help AI systems understand the nuances of language, including metaphors, idioms, sarcasm, and cultural references, whether they fall within the more frequent or rarer occurrences across the spectrum of everyday human language.

Logic also plays a fundamental and enduring role in the development and advancement of AI, as it provides a rigorous framework for reasoning, problem-solving, and knowledge representation. Word embeddings, which essentially replace words with numbers, have made most AI researchers forget about the importance of linguistics. Concepts from syntax, semantics, phonetics, and morphology (see next section), however, are crucial for interpreting the intended meaning of natural language. Psychology will play an essential role in creating systems that enhance well-being, foster human relationships, and provide meaningful and empathetic interactions (Fei et al., 2024). By addressing issues related to inequality and cultural diversity, sociology will guide AI development in ways that promote positive societal outcomes and responsible innovation. The arts are also going to be key for the future of AI, as highlighted by recent STEAM (STEM + Art) initiatives, in order to 'humanize' AI through computational creativity, cultural and social understanding, and the enhancement of AI usability (Sorensen et al., 2022). Finally, ethics are paramount to ensure that AI technologies are developed, deployed, and used in ways that align with human values and promote accountability (Floridi et al., 2018).

1.2.2 Task Decomposition

Like *Multidisciplinarity*, *Task Decomposition* aims to better handle the complex and multifaceted nature of AI problems. It is a method commonly used in psychology, instructional design, and project management to break down a complex task or activity into its individual components. *Task Decomposition* is one of the key points of this textbook: no matter what kind of downstream task we are handling, if we do not deconstruct it into its constituent subtasks, we are practically forcing our model to implicitly solve a lot of subtasks it has never been trained for. The 'sentiment suitcase model' (Cambria et al., 2017), for example, lists 15 NLU subtasks that need to be solved separately before sentiment analysis can be accomplished.

Firstly, a Syntactics Layer preprocesses text so that informal and inflected expressions are reduced to plain standard text (Zhang et al., 2023a). This is done through subtasks such as microtext normalization, for refining and standardizing informal text, part of speech (POS) tagging, for assigning grammatical categories (such as nouns, verbs, adjectives, and adverbs) to each word in a sentence, and lemmatization, for reducing words to their base or dictionary form (lemmas).

Secondly, a Semantics Layer deconstructs normalized text into concepts, resolves references, and filters out neutral content from the input (Mao et al., 2024c). This is done through subtasks such as word sense disambiguation (WSD), for determining the correct meaning of a word within a given context, named entity recognition (NER), for identifying and classifying names of people, places, organizations, and dates, and subjectivity detection, to distinguish between factual information and subjective content.

Finally, the Pragmatics Layer extracts meaning from both sentence structure and semantics obtained from the previous layers. This is done through subtasks such as personality recognition, to infer traits, characteristics, preferences, and behavioral tendencies of the speaker, metaphor understanding, for interpreting figurative language in text, and aspect extraction, for identifying and extracting specific facets, features, or components mentioned in text and, hence, enabling a more fine-grained analysis. Only after handling all these subtasks, which we humans take care of almost subconsciously during reading or communication, the downstream task, e.g., polarity detection, can be effectively processed.

1.2.3 Parallel Analogy

Similar to *Multidisciplinarity* and *Task Decomposition*, *Parallel Analogy* looks at AI problems in a multifaceted way. Engineers and computer scientists have always been obsessed with optimization. In the development of AI systems, this translates into finding the 'best' knowledge representation, the 'best' reasoning algorithm, the 'best' way of doing things. This, however, results in only having one way of solving a problem. Instead, several analogous representations of the same problem should be maintained in parallel while trying to solve it so that, when problem-solving begins to fail while using one representation, the system can switch to one of the others.

Parallel Analogy, or 'panalogy' (Minsky, 2006), is key to solving highly complex AI problems, but also simpler problems in which a change of perspective is required. In affective computing tasks, for example, sometimes it is useful to see emotion concepts from a semantic point of view, e.g., 'joy' and 'sadness' are similar because they are both emotions, or a polarity point of view, e.g., 'joy' and 'sadness' are opposite because the former is positive and the latter is negative (Cambria et al., 2012). Similarly, we could say that words like 'joyful', 'joyfully', 'enjoy', and 'enjoyment' are similar because they all share the same root word 'joy', but totally different in a POS tagging sense (adjective versus adverb versus verb versus noun, respectively).

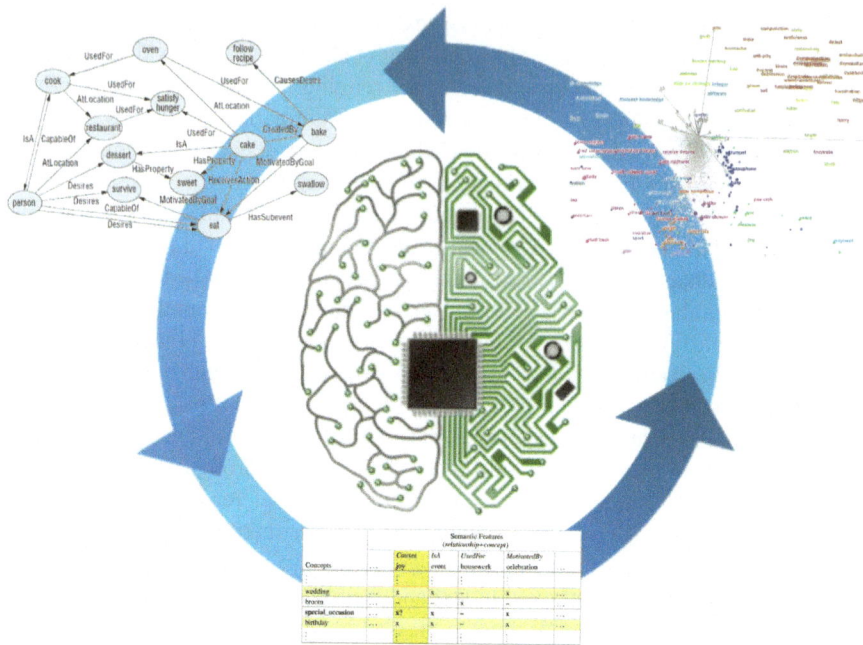

Fig. 1.2: An example of 'panalogy' where the same data is 'redundantly' represented as a knowledge graph, as a matrix, and as embeddings (Cambria et al., 2012).

For more general NLU tasks, it could be useful to have the same data 'redundantly' represented both as a knowledge graph and as embeddings (Fig. 1.2). The knowledge graph could be more useful for solving problems requiring *Symbol Grounding*, e.g., answering questions like 'what is what?' (see next section). Embeddings, instead, could be more useful for *Similarity Measure*, e.g., answering questions like 'what is similar to what?' (explained later).

1.2.4 Symbol Grounding

Symbol Grounding is a key concept of this textbook and a major distinguishing factor between NLP and NLU. It deals with how symbols, which are abstract representations, acquire meaning and connection to the real world. In human cognition, we understand the meanings of symbols through a process known as 'grounding'. When we see or hear a word, for example, our brains associate it with the sensory experiences and interactions we have had with the objects or concepts that such a word represents, thus providing a foundation for our understanding of its meaning. This sensory experience provides a foundation for our understanding of the symbol's meaning.

In the context of AI, the *Symbol Grounding* problem arises because computers lack the inherent sensory experiences that humans possess. They process symbols as strings of characters or digital information without a direct connection to the real world, raising the question of how they can truly understand the meanings of symbols in a way that is equivalent to human understanding. For instance, consider the word 'apple'. Humans understand this word not just as a sequence of letters, but as a fruit with certain sensory qualities like color, taste, texture, and smell, all of which are grounded in our experiences with actual apples. Current AI systems are unable to grasp the richness of meaning behind the word 'apple' without having those sensory experiences.

To solve this, we may have to take a step back in order to move forward. Old-school (symbolic) AI was better at *Symbol Grounding* but it was not scalable nor flexible. New deep-learning-based (subsymbolic) AI, instead, is very scalable and flexible but it does not handle symbols. The best of both worlds could be achieved through a hybrid (neurosymbolic) AI that leverages the strengths of both symbolic and subsymbolic models to overcome their respective limitations and achieve a more comprehensive understanding of the world. In NLU research, this can be implemented in several ways, e.g., by injecting external knowledge into a deep neural neural network (Liang et al., 2022b) in the form of embeddings (Fig. 1.3).

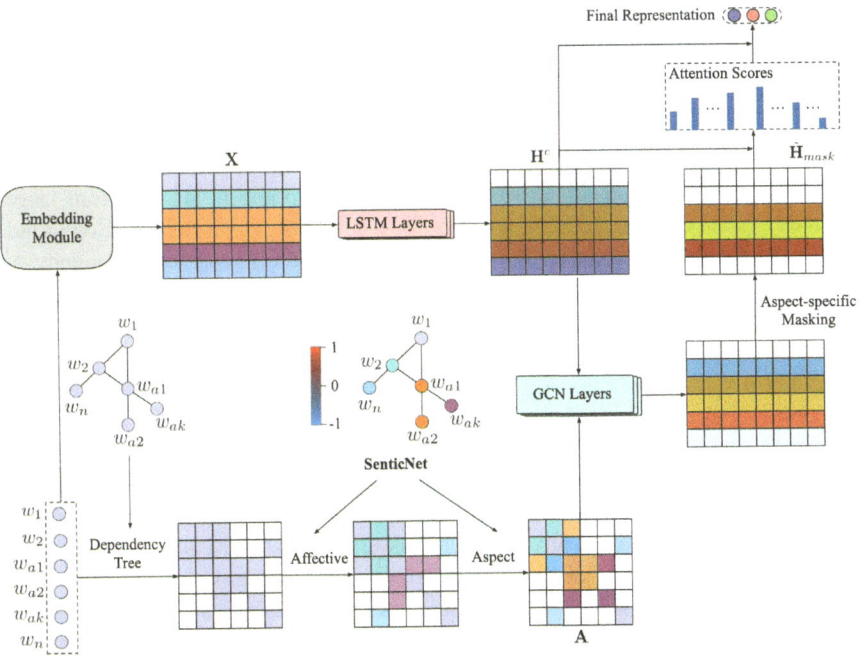

Fig. 1.3: An example of neurosymbolic AI where (symbolic) commonsense knowledge is injected into a (subsymbolic) graph convolutional network (Liang et al., 2022b).

Another recent neurosymbolic approach consists in a three-step normalization process (Cambria et al., 2024) that first leverages linguistics to replace expressions like shopping_for, bought, and purchasing with their lemmas, e.g., shop_for, buy, and purchase, respectively. Next, deep learning models are used to cluster the resulting lemmas into primitives, e.g., BUY(x) is the cluster representing shop_for, buy, and purchase and the likes of them (where x is the direct object acted upon by the primitive). Finally, first-order logic (FOL) is used to ground such primitives, e.g., BUY is defined in terms of GET and GIVE, which in turn are specified in terms of other superprimitives defining transfer of ownership.

In robotics, some researchers have emphasized the importance of physical embodiment and sensorimotor experiences in the development of intelligent systems. Such approaches, which go under the name of Embodied AI, are promising but they are still limited to very basic sensory experiences such as object manipulation. A similar approach can be taken in virtual worlds or in the metaverse, where AI could learn social commonsense, based on how people interact, and some sort of physical or spatial commonsense, such as gravity or the fact that you cannot go through walls. Additionally, an AI system could generate real-time virtual simulations to better perform causal reasoning and narrative understanding by grounding words into virtual objects and actions on the fly.

1.2.5 Similarity Measure

Because we have no better way of performing grounding, in computer science we use embeddings to represent data, e.g., text, audio, images, and videos, as vectors or data points in a multi-dimensional space. This mapping is learned from large amounts of data during a training process and it is usually focused on just one kind of similarity (usually semantic similarity based on word co-occurrence frequency). In order to enable *Parallel Analogy* (explained earlier), we need to generate different representations for the same data based on different kinds of similarities.

Another problem is that we are still using very basic metrics to quantify the similarity between pairs of embeddings and, hence, perform classification and pattern recognition. All such similarity metrics, e.g., Jaccard coefficient, Euclidean distance, and cosine similarity, blindly calculate distances in a multi-dimensional vector space without considering its topology (Steck et al., 2024). In the future, we need to adopt more topology-aware methods for calculating similarity in multi-dimensional vector spaces, e.g., Mahalanobis, Minkowski or Wasserstein distances and principal path methods (Ferrarotti et al., 2019). These kernel methods are designed to discover smooth paths between objects in space by traversing a series of waypoints. One of their standout features is their ability to seek paths that pass through high-probability regions of the space, effectively navigating through geodesics influenced by the probability distribution of the sampled data. These methods also emulate the cognitive process where thinking involves transitioning from one concept to another while traversing regions of space with a high likelihood of encountering related concepts.

1.2.6 Intention Awareness

Intention Awareness refers to the ability to recognize and understand the intentions or goals of oneself or others. It plays a crucial role in human social interactions and communication, as it enables individuals to anticipate and interpret the actions and behaviors of others, leading to more effective and empathetic interactions. Current AI models provide one-fits-all solutions without taking into account user beliefs, goals and preferences.

Theory of mind should always be applied to better understand user's actions and queries. When this is not possible, user profiling in the form of persona or personality recognition should be employed to generate more relevant actions or answers (Zhu et al., 2023). For the same reason, AI should also have enough commonsense knowledge, including a model of fundamental human beliefs, desires, and intentions, in order to minimize miscommunication and avoid unintended consequences (e.g., apocalyptic scenarios like accidentally wiping out humanity in the attempt to solve climate change). In other words, future AI systems should always try to understand what users are doing and why they are doing it. For instance, recent hybrid frameworks have tried to improve human-robot interaction by modeling *Intention Awareness* in terms of motivational and affective processes based on conceptual dependency theory (Ho et al., 2023).

Finally, recent attempts to augment the human decision-making process, especially in dynamic and time-sensitive scenarios such as military command and control, game theory, home automation, and swarm robotics, have focused primarily on environmental details such as positions, orientations, and other characteristics of objects and actors of an operating environment (situation awareness). However, a significant factor in such environments is the intentions of the actors involved (Howard and Cambria, 2013). While creating systems that can shoulder a greater portion of this decision-making burden is a computationally intensive task, performance advances in modern computer hardware bring us closer to this goal.

1.2.7 Trustworthiness

Last but not least, *Trustworthiness* is a key pillar that measures the degree to which AI systems, models, and algorithms can be relied upon to perform as intended, make accurate and ethical decisions, and avoid harmful consequences. It is a concept closely related to *Intention Awareness*, but also explainability and interpretability. Explainability allows an AI model to generate descriptions of its decision-making processes in order to enable the user to make informed modifications to the outputs or even to the model itself in a human-in-the-loop fashion. Interpretability, in turn, enables users to understand the inner workings of an AI model, e.g., by identifying which input features have the most impact on its output or by assessing how changes in input variables affect the model's predictions or by leveraging a confidence score to gauge how confident the AI model is about its own output.

According to one theory of trust, trust is grounded in probabilities that a trustor A attributes to his/her own beliefs about the behavior and competences of a trustee B with respect to the performance of some action ϕ relevant to a goal G. Where n denotes the probability threshold value and m denotes the probability value that A attributes to his/her trust-relevant beliefs, there will be a trust relation between A and B if and only if $m \geq n$. It is at least arguable that this probability threshold value will be met, given the twin phenomena of hallucination and behavioral drift. In any case, we believe that trust is more than a matter of satisfying probability threshold conditions (i.e., $m \geq n$). We can define *Trustworthiness* as a 5-ary relation R(A, B, ϕ, ψ, G), consisting of five relata: the trustor A, a trustee B, some action ϕ to be performed, some G-relevant attribute ψ that may be judged by A as absent or present in B during B's performance of ϕ, and a goal G that makes the performance of ϕ desirable (Chen, 2021). Indeed, trust is a mental state that A holds toward B with respect to the performance of some G-relevant ϕ.

If the goal G is NLU, actions ϕ_i are typical NLU tasks such as sentiment analysis and dialogue generation, and G-relevant attributes ψ_i are qualities such as explainability and interpretability. All other things being equal, if B_1 possesses each G-relevant attribute ψ_i in greater abundance than B_2, we have *pro tanto* reasons to have greater trust in the former than the latter. Intuitively, even if real parrots or stochastic ones (LLMs) produce appropriate linguistic responses to task-related ϕ_i prompts, we would not deem their linguistic behavior trustworthy unless they possess the relevant NLU. Meaning involves a relation between the linguistic form of data and an extralinguistic reality that is distinct from language. Where M denotes meaning, E denotes the form of natural language expressions, and I denotes communicative intent, this relation may be formally represented as $M \subseteq E \times I$ (Bender and Koller, 2020). M contains ordered pairs (e, i) of natural language expressions (e) and communicative intents (i). Understanding may be interpreted as the process of retrieving i, given e.

Since LLMs are pretrained on large datasets and meaning cannot be learnt from linguistic form (e) alone, however impressive their transformer architecture might be, LLMs will necessarily lack the relevant intentionality. Such a limit can result in hallucinatory responses from LLMs, if e and i are not directly associated in the pertaining datasets. Hence, humans have to watch over them and correct them in mission-critical tasks. Ensuring *Trustworthiness* requires collaboration among AI developers, domain experts, ethicists, and policymakers. It involves a combination of technical measures, ethical considerations, and transparency initiatives. As AI continues to play an increasing role in various aspects of society, building and maintaining trust in AI technologies is essential for their responsible and sustainable integration.

1.3 Towards More Responsible AI

One of the best examples of the pervasiveness of AI are recommender systems, which enhance user experience through personalized content delivery, utilizing advanced algorithms to process vast amounts of data, continuously learn and adapt to user preferences, operate in real-time, and provide tangible benefits across various industries such as e-commerce, streaming services, and social media. NLU is crucial for the future of recommender systems as it improves user interaction, context awareness, sentiment analysis, multilingual capabilities, and the extraction of insights from unstructured data, leading to more accurate, personalized, and effective recommendations.

Recommender systems have revolutionized our online experiences by offering personalized recommendations tailored to individual preferences and behaviors (Li et al., 2024). While undoubtedly advantageous in numerous aspects, it is crucial to recognize the inherent drawbacks in their design. One notable concern is that these algorithms often prioritize the interests of the deploying company, aiming to maximize profits and user engagement rather than solely focusing on enhancing user experience. Consequently, there is a risk that the recommendations provided may not consistently align with users' best interests, potentially leading to the formation of echo chambers, filter bubbles, or even manipulation of user behavior.

Despite their undeniable usefulness, it is essential to approach recommender systems with a critical perspective, advocating for transparency, ethical considerations, and user empowerment to ensure that they function as tools for enrichment rather than exploitation. Most of the evaluations and metrics of recommender systems focus on validation of the optimization process while finding the items of user interest in very large item collections. In most recommendation cases, we commonly have two players: (1) users who are advised with some items and (2) item providers. In business recommender system applications, users are a weaker partner, as they do not design, control, or understand the limitations of the recommendation engines. Reciprocity or fairness mechanisms could address these concerns, but they are relatively rarely considered in the context of corporate recommender systems.

Companies are driven by business objectives, such as greater profit, market size occupied, as well as user time or attention gained. For them, human needs and satisfaction are, then, only auxiliary goals. This means that companies must respect their customers, but only to the extent that it enables them to achieve their objectives and earn money. Accordingly, they employ recommender systems to support their customers in decision-making processes resulting in purchases, reviews, etc., which directly impact their income (Fig. 1.4). Then, can we claim that recommender systems just simply provide a recommendation list, satisfying users? They primarily satisfy companies themselves, then users. Moreover, the evaluation measures commonly used in scientific papers are being adapted in practical implementations to the business models used. Obviously, it does not mean that companies are against their customers. They just focus on their business. In summary, commercial recommender systems are driven by business interests rather than the interests of their users. Therefore, they are commonly beyond the control and awareness of customers.

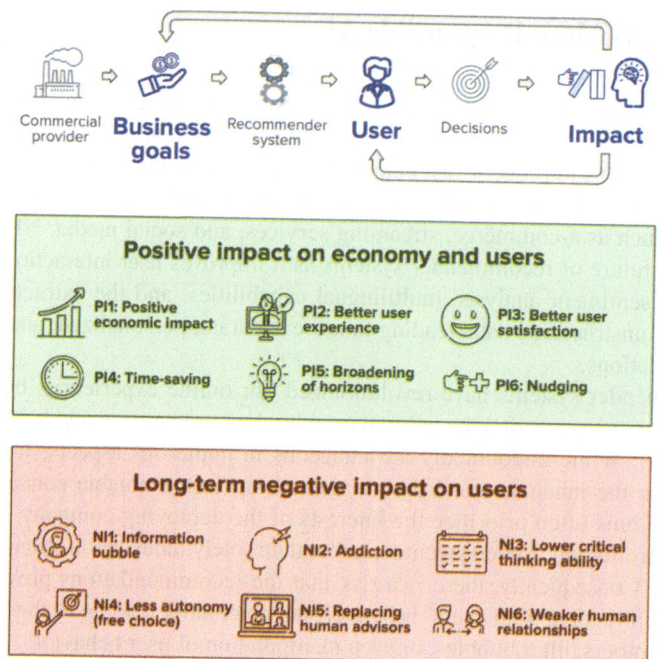

Fig. 1.4: Recommender systems have both positive and long-term negative effects on their users and economy.

1.3.1 Impact of AI on Economy and Humans

The main goals of recommender systems and their positive impact on the economy and humans (Fig. 1.4) are:

PI1: *Positive economic impact* on companies using recommender systems (De Biasio et al., 2024), i.e., personalization provided by recommender systems can be a competitive advantage over market rivals;

PI2: *Better user experience*, which directly results from support of recommender systems in navigation and communication with the system;

PI3: *Better user satisfaction* can be one of the measured feature of recommender systems making users utilize them (He et al., 2024b);

PI4: *Time-saving* is a consequence of PI2, i.e., faster navigation through online services and large item collections (Tembhare et al., 2023);

PI5: *Broadening of horizons* – some recommender system suggestion of items beyond known user preferences (Liang and Willemsen, 2023). It is also addressed using quite well known concepts and measures: *diversity, coverage, novelty, unexpectedness*, and *serendipity* (Fu et al., 2023);

PI6: *Nudging* users towards their positive decisions and behavior, e.g., related to unhealthy eating (Castiglia et al., 2022) or news diversity.

Most research related to recommender systems focuses on better *immediate recommendations*, i.e., more precisely infer about user needs while directly fulfilling business goals. However, users exposed to recommender systems can experience some *long-term effects*, including negative ones (Fig. 1.4). In particular, we would like to highlight six of them that we believe require further investigation:

NI1: *Information bubble* – greater user confinement;
NI2: *Addiction* to very good prompts that excellently meet user needs; excessive use of recommender systems;
NI3: *Lower critical thinking ability* due to better and better recommendations (why seek anything else?);
NI4: *Less user autonomy* to make their own choices.
NI5: *Replacing human advisors* who become inferior and more expensive;
NI6: *Weaker human relationships* as more and more user needs are filled by systems.

All the above-mentioned effects may lead to other effects like decreased well-being or physical and mental health. There are also other adverse effects on humans, which often directly result from the contradictory goals of sellers and customers, like nudging user moods to induce unplanned purchases (Ho and Lim, 2018). The phenomenon commonly called *information bubble* (NI1) stems from human to consume increasingly similar items over time even without any recommendation. However, recommender systems can reinforce this effect, even if the user may sometimes feel bored or less satisfied (Areeb et al., 2023). On the other hand, there are recommender system solutions going in the opposite direction, i.e., broadening user horizons (PI3). Unfortunately, however, are not commonly used in commercial applications. *User autonomy* (NI4) refers to the user's ability to undertake their free choices (Krook and Blockx, 2023), i.e., possibly without any manipulation. This is also closely related to *user control* over their decisions and recommendation mechanisms (Harambam et al., 2019). The loss of such control may partially result from being in the information bubble (NI1), excessive recommender system use (NI2), and loss of criticism (NI3).

1.3.2 Towards Human-like Recommender Systems

Recently, companies and other stakeholders can collect large amounts of data about their users. More data potentially leads to more accurate and friendly recommender systems. This is additionally supported by multimodal recommender systems that process diverse information about the environment, recommended items, and also user social networks. Besides, recommender systems can make use of the general data about users like their personality traits, cognitive abilities or more temporal affective states like emotions (Dhelim et al., 2022). As a result, current recommender systems, and future recommender system in particular, will increasingly resemble human advisors.

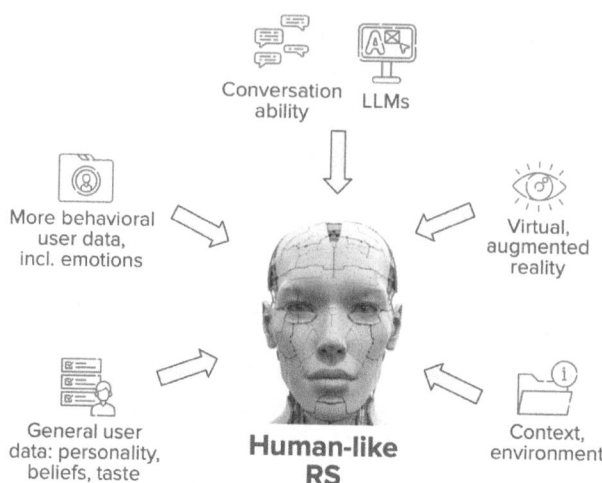

Fig. 1.5: Recommender systems become more and more human-like. Includes a Midjourney-generated image.

It will be boosted by a great progress in generative AI, which enhance recommender system interaction capability, and leading to conversational recommender system (Li et al., 2023a). This, in turn, is likely to be soon combined with virtual or augmented reality making the user experience with recommender system even more immersive and mimicking human beings (Xue et al., 2024). This means that the development of LLM and other multimodal generative AI systems and virtual reality will make recommender systems more capable of personalizing multimodal and immersive interaction, while benefiting from the general human profile (personality traits, beliefs, general taste), temporal state (emotions, mood, attention) and more comprehensive behavioral user data. As a result, recommender systems will be developed towards human-like assistants rather than simple recommendation list generators (Fig. 1.5). Such interactive recommender systems may substitute for real human advisors (NI5). This means that users will tend to replace their natural interpersonal communication with virtual assistants, thus reducing their interpersonal contacts, which may eventually cause greater isolation from real life (Fig. 1.6).

Moreover, such recommender systems may know the user better than other people. Going further, users may establish a kind of relationship with virtual advisors. The case of Replika is a good and recent example of such a virtual engagement (Pentina et al., 2023). Perhaps future recommender systems will become more like our friends imitating and substituting for human friends? This, however, may be harmful and dangerous to the users (Zimmerman et al., 2023) and negatively affect their interpersonal relationships (NI6). At the same time, this may dovetail with another general social phenomenon: less face-to-face contacts that directly decrease life satisfaction (Hall et al., 2023).

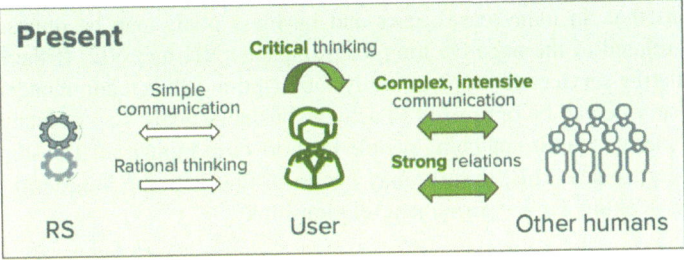

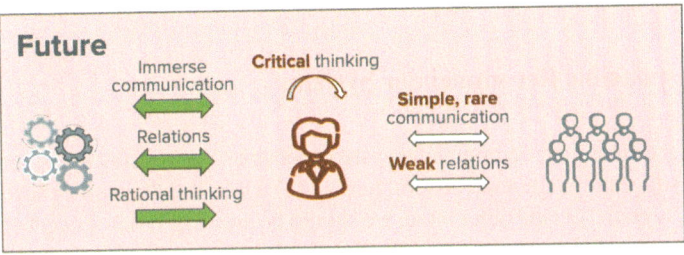

Fig. 1.6: Present and future relationships between the user, recommender system and other humans.

Simultaneously, high quality and effectiveness of recommender systems may provide excessive use and addiction (NI2) to various digital services such as online video streaming services (Hasan et al., 2018). It means that users are becoming increasingly dependent on recommender system suggestions, thus, losing their critical thinking ability (NI3), and are even susceptible to changing their personal identity (Bonicalzi et al., 2023). This is also due to the exhibit of *confirmation bias* by many people, that is, the tendency to actively seek information that confirms initial preferences. However, typical recommender systems do little to avoid this bias, since they suggest items that are in line with user's preferences.

To demonstrate the problem of contradictory business goals of recommender system providers and human objectives we would like to consider two examples: (1) streaming services, especially VoD (Video on Demand) like Netflix or Disney+, and (2) dating services like Tinder[1]. Typically, streaming services that provide movie series jump from one episode to another without any break. This is further supported by cliffhangers at the end of the episode, i.e., suspending the action of the film so that the viewers are left in a very exciting or scary moment to encourage them to continue watching. This will be even more challenging for users in the future due to more immersive virtual reality solutions and personalization exploiting human emotions. As a result, the user can be kept in the service for many hours. However, is spending a long time like this the intention and need of the user? When launching the service, would the user consciously choose such a scenario?

---

[1] Note that we are not analyzing any particular service, but only want to indicate some of the potential risks and directions for further studies.

It means that, in many cases, user and business goals may be opposed. This is closely related to the negative long-term impacts: NI1 and NI2. If the business model of dating services relies on monthly subscriptions, then recommender system in such a service can be optimized to achieve business goals, i.e., to retain paying users. This may result in matching people to short-term human relationships rather than long-term ones. This, in turn, may not be in line with the long-term goals of many users seeking a more permanent relationships.

1.3.3 Responsible Recommender Systems

In general, commercial recommender systems focus on (1) meeting users' temporary needs (by suggesting potentially useful items), (2) while simultaneously accomplishing business goals. On the other hand, we should be aware of risks of negative impact (NI1-NI6), and challenges arising from human-like recommender systems. Therefore, in the landscape of rapid and continuous development of recommender systems, we would like also to consider: (3) the life goals and values of each individual, and (4) general, agreed society-wide recommendations related to health and well-being, e.g., physical and mental health guidelines. This leads us to the concept of Responsible Recommender System (RRS) (Kazienko and Cambria, 2024).

Previous work referring to the term *Responsible Recommender System* actually focused on some of its features such as *trustworthiness, fairness, accountability, explainability*, and *transparency*. Most of these topics are covered among others, by papers presented at the series of *The FAccTRec Workshop: Responsible Recommendation* collocated at the RecSys conference. All of them are very important components of RRS.

In this chapter, however, we go further and postulate to additionally respect new issues not much considered before (Fig. 1.7), in particular:

1) human values and user personal goals (especially related to long-term impact on them), and
2) societal recommendations (related to lifestyle or health like physical activities, sleep, the need for breaks in online activity, etc.)

All of the above should be taken into account while maintaining user autonomy (Krook and Blockx, 2023) and non-conflict with business objectives. Preserving user autonomy is a crucial component of RRS mitigating the long-term negative impact NI4. It enforces that the recommender system designer should strive to keep the user free: free to change their choices, free to make new ones, or even to disable the system altogether. Our RRS concept adheres to the principles of digital humanism, emphasizing the importance of norms to ensure that technology is developed and used in a manner that is more ethical and driven by human values (Prem et al., 2023). Please note that we focus primarily on recommender systems' impact on individual users. However, the social effect can be considered as well.

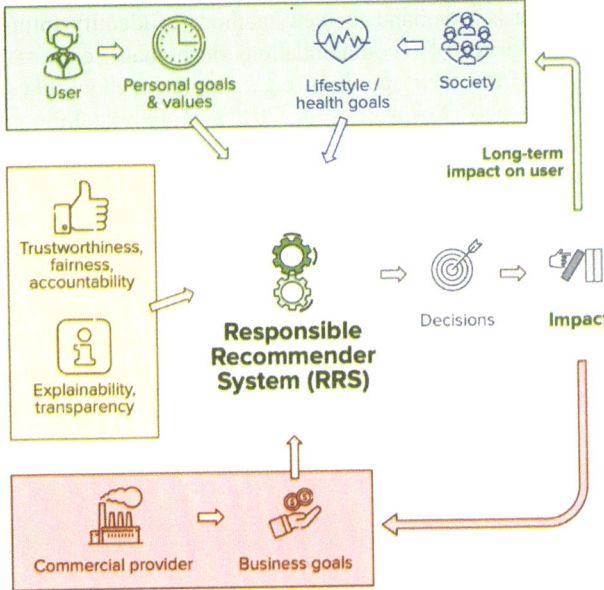

Fig. 1.7: Idea of RRS in a market environment. It respects: (1) individual user goals and values, especially long-term ones, (2) societal goals, e.g., lifestyle or health recommendations, and (3) business goals. RRS also (4) preserves trustworthiness, fairness, accountability, and explainability leading to better transparency.

Defining, identifying, and measuring long-term impact on humans is a great challenge for research. It means we need to establish new procedures and measures suitable for (1) different kinds of recommender systems like content (news, videos) delivery or online trade, (2) different personalities, e.g., people who are more susceptible to influence and manipulation, (3) different social groups and cultures sharing the same values, e.g., religious ones.

Another challenge is to analyze and quantify the emotional engagement of users. it is important since recommender system providers may be tempted to exploit user emotions to achieve their business goals. This, however, may be in opposition to the user goals, values, and even health restrictions (e.g., for people with heart diseases). As a result, users should both be aware and consciously decide to what extent they want their emotions to be evoked.

The implementation of the RRS concept requires new methods and reasoning architectures that would integrate contradictory goals: business vs. personal vs. societal ones. In particular, we need to develop how to combine general and uncontroversial health recommendations in the form of incentives with personal preferences while maintaining benefits for recommender system operators[2].

---

[2] Please note that existing nudging solutions (PI6) focus on single positive nudging objectives rather than on integration and alignment with other targets, e.g., through multi-criteria inference.

Additionally, there is a demand for new methods to identify future, especially non-obvious consequences of recommendations on humans, e.g., many user features can be derived from their activities, e.g., simple Facebook likes, which was observed many years ago (Kosinski et al., 2013). Moreover, these consequences (possible impact and risks) should be presented to the user in an understandable and editable form – preferable in the interaction, which again requires new technologies. Recommender systems especially if combined with generative AI models provide users with successive portions of new information. This, in turn, leads to information overload and potential information pollution, resulting in harm to the user, which also requires new solutions and maybe even legal regulations (Hołyst et al., 2024).

Overall, some of the risks with a long-term effect, perhaps should be addressed through solutions similar to smoking and perhaps even investing in the stock market, e.g., increase awareness among users regarding the potential hazards of using recommender systems.

1.3.4 What Next

AI and recommender systems are poised to revolutionize the fabric of societal interactions, offering unprecedented opportunities to reshape the way individuals connect, communicate, and collaborate. While they can facilitate connections by matching individuals with shared interests and preferences, there is concern that these mediated interactions may lack the depth and authenticity of face-to-face communication. Over-reliance on algorithmic recommendations may contribute to the formation of echo chambers, limiting exposure to diverse perspectives and potentially weakening interpersonal bonds. Moreover, the increasing integration of AI and recommender systems into daily life raises questions about individual autonomy. While these systems aim to streamline decision-making by offering tailored suggestions, there is a risk of subtle influence on user behavior. This raises ethical concerns about the manipulation of human autonomy and the potential erosion of free will in the face of algorithmic determinism. On the one hand, AI and recommender systems provide access to vast amounts of information and support data-driven decision-making, empowering individuals to make more informed choices. Simultaneously, the reliance on algorithmic recommendations may reduce individuals' inclination to critically evaluate information and exercise independent judgment. Moreover, the lack of transparency in recommendation processes may hinder users' understanding of how decisions are made, potentially undermining trust in the information presented.

In the near future, it will be increasingly important to raise awareness among users about the potential risks associated with recommender systems, drawing parallels to other contexts such as smoking and investing in the stock market. Just as education campaigns have been instrumental in informing the public about the health hazards of smoking and the financial risks of stock market investments, a similar approach is needed to highlight the potential pitfalls of relying blindly on recommendation algorithms.

By elucidating the possible consequences, including the formation of echo chambers, filter bubbles, and the manipulation of user behavior, individuals can make more informed decisions about their online interactions. Moreover, fostering a culture of critical thinking and digital literacy can empower users to navigate recommender systems responsibly, mitigating the adverse effects and ensuring that they serve as tools for enrichment rather than exploitation. A key element of RRSs will be explainability. When users are presented with recommendations, they often desire insight into the rationale behind the suggestions. Explainability offers transparency into the underlying mechanisms and criteria employed by the system to generate these recommendations. Firstly, explainable recommender systems foster trust by providing users with visibility into how recommendations are formulated. Users are more likely to trust recommendations if they comprehend the reasoning behind them, which consequently enhances engagement and satisfaction. Secondly, explainability aids users in understanding why specific recommendations are presented to them, resulting in a more meaningful and personalized experience. With a grasp of the factors influencing recommendations, users can better evaluate and interpret them, thereby enabling more informed decision-making.

Moreover, explainability plays a crucial role in mitigating undesired biases. By unveiling the factors considered by the recommender system, users can identify and address potential biases related to demographics, preferences, or historical interactions. RRSs will empower users to provide feedback and adjust their preferences based on the recommendations they receive. When users comprehend the reasoning behind recommendations, they can offer valuable feedback, thereby contributing to the improvement of recommendations over time. Additionally, in industries such as finance or healthcare, regulatory requirements often mandate transparency and accountability in recommendation systems. Explainability ensures compliance with regulations and standards by offering clear explanations of recommendations and decision-making processes. Finally, RRSs will facilitate the identification and resolution of potential ethical concerns, such as privacy violations or manipulation of user behavior.

1.3.5 Summary

So, are recommender systems friends, foes, or frenemies? They can be friends when they provide transparent and personalized recommendations that genuinely enhance user experiences. For instance, on streaming platforms like Netflix or Spotify, recommender systems recommend movies, shows, or music tailored to users' tastes and preferences, helping them discover content they might enjoy. They can be foes when they prioritize the interests of businesses or other stakeholders over those of users, which can lead to a proliferation of sponsored content or biased recommendations, potentially undermining user trust and satisfaction. Given the pervasive influence of recommender systems on various aspects of human life, it is crucial to study their mechanisms, effects, and implications more comprehensively.

Understanding how recommender systems operate, the algorithms they employ, and the biases they may exhibit is essential for addressing potential ethical concerns, ensuring transparency, and promoting user empowerment. Finally, studying the impact of recommender systems on society can inform the development of regulatory frameworks, awareness campaigns, industry standards, and best practices to mitigate negative consequences and maximize the benefits of these systems for individuals and communities alike. We are convinced that the concept of RRS presented in this chapter will inspire the emergence of new commercial solutions that support both the well-being, goals and values of users and pursue necessary business objectives.

1.4 Towards More Personalized AI

As AI becomes increasingly integrated into various aspects of daily life, the demand for more tailored and individualized experiences is growing. Personalization enhances the relevance, efficiency, and effectiveness of AI applications, leading to numerous benefits across different sectors. Firstly, personalization significantly improves user experience by allowing AI systems to cater to the unique preferences, needs, and behaviors of individual users. This leads to more intuitive and engaging interactions, enhancing user satisfaction and loyalty. For example, personalized recommendation systems in e-commerce and streaming services help users discover products and content that align with their tastes, thereby improving their overall experience. Secondly, personalized AI systems increase efficiency and productivity by streamlining workflows and automating tasks in ways that are specifically tailored to the user's habits and requirements. In professional settings, personalized AI assistants can manage schedules, prioritize tasks, and provide relevant information, thereby boosting productivity and efficiency.

In this chapter, we focus on the task of personalized sentiment analysis, which aims to analyze individual sentiment perceptions. This approach is motivated by the observation that different individuals may perceive an identical statement differently regarding its sentiment polarity. In contrast, conventional sentiment analysis aims to predict the semantic sentiment of a statement, where the sentiment prediction remains the same for an identical statement. For instance, an introverted person may have a negative sentiment towards performing in front of a large audience, while an extroverted person may view the same situation positively (see Fig. 1.8). This variation in sentiment perception can be attributed to personality traits. While the distinction between introversion and extroversion is from personality theory, the variability in sentiment perception among individuals can also be influenced by other factors. The inconsistency in sentimental perception between individuals may originate from multiple sources. For example, as the adage suggests, *"the enemy of my enemy is my friend"*, the sentimental perception of a person can be driven by the relationship or the context of the situation. In this case, personalized sentiment analysis extends beyond traditional semantic and pragmatic understanding, incorporating a broader range of human subjective factors, such as persona information.

Fig. 1.8: Conventional sentiment analysis vs. personalized sentiment analysis.

To this end, we propose a novel neurosymbolic AI framework Zhu et al. (2024b) that leverages seven levels of personalization for personalized sentiment analysis (see Fig. 1.9). In particular, such a framework initially targets identifying whether the user is a human (e.g., male or female) or other intelligent agent, determining whether to consider typical human needs and beliefs. Secondly, the framework aims to identify the user's cultural background to discern whether a given concept is perceived as positive or negative based on different cultural beliefs. Next, a similar mechanism is applied to Religion. Vocation is considered an important factor in shaping users' views based on their job and education level. Following, the user's ideologies are modeled to consider political beliefs and social, economic, or philosophical view-points. Next, Personality is detected in order to better classify certain concepts as positive or negative based on the user's personality traits. Finally, Subjectivity aims to consider specific user preferences learned from historical or training data.

This work aims to evaluate whether the framework of the identified seven levels of personalization can help improve the accuracy of sentiment analysis. We evaluate the framework based on a dialogue dataset (Chen et al., 2023b) that originated from *Harry Potter* novels and LLMs. The dataset contains conversations between different characters in the *Harry Potter* novels and the associated sentiment perceptions be-tween characters. Considering the advancement of LLMs in diverse domains (Mao et al., 2024a), we leverage ChatGPT and GPT-4 to generate the seven-level persona information for each character by well-designed prompts. Finally, the utility of the persona information is evaluated with a neurosymbolic AI paradigm.

The persona information is structured as tailored prompts, feeding into LLMs together with the conversations of the employed dataset. We analyze the variations in the sentiment analysis accuracy of AI after the inclusion or exclusion of different prompts. We find that in the context of the conversations of *Harry Potter*, Culture, Vocation, and Ideology present the highest utility, resulting in the most accuracy gains in our research domain. All the seven aspects have positive utilities to the neurosymbolic AI system.

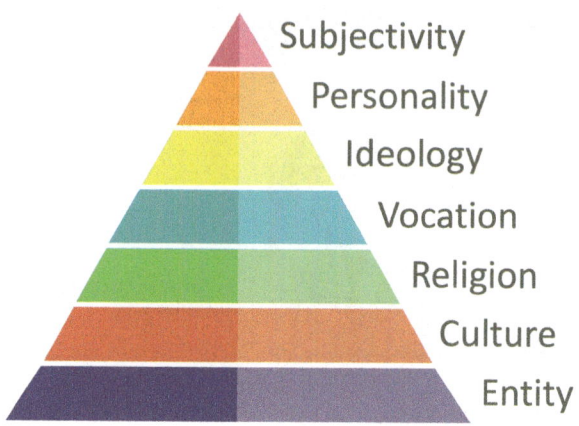

Fig. 1.9: Personalized Sentiment Analysis Pyramid.

Integration of personalized neurosymbolic knowledge into LLMs leads to biases towards the object and the subject on sentiment analysis task. It is reasonable that different types of persona information have different utilities in a specific domain because different individuals may prioritize certain aspects of persona information based on the context or the domain they are in. The sentiment perception reasoning may rely on different background knowledge in different scenarios. The contribution of this work is threefold. (1) It introduces the novel concept of personalized sentiment analysis, designed to enhance AI's understanding of the varied sentiment perceptions between individuals. (2) It presents a comprehensive framework derived from an analysis of literature, delineating seven levels of personalization for personalized sentiment analysis. (3) It performs an empirical study to assess the effectiveness of various persona information types within a conversational domain.

1.4.1 Literature Review

1.4.1.1 Sentiment Analysis with AI

Recently, there have been notable advancements in sentiment analysis, characterized by several key trends. Initially, the focus was on lexicon-based approaches to identify sentiment polarities such as positive, negative, and neutral in sentences or documents. The field then evolved by introducing concept-level sentiment analysis systems like pSenti (Mudinas et al., 2012) and SenticNet (Cambria et al., 2024). These systems represented a shift towards more advanced methods that combined lexicon-based and learning-based approaches. They presented greater accuracy in tasks such as sentiment polarity classification and sentiment intensity prediction, surpassing the capabilities of traditional lexicon-based systems (Medhat et al., 2014).

With the development of neural networks, the research focus of sentiment analysis shifted towards developing different learning frameworks to improve accuracy. Several algorithms were proposed for sentiment analysis including convolutional neural network (CNN)-based supervised learning (Kim, 2014), transfer learning (Dong and De Melo, 2018), adversarial training (Fedus et al., 2018), meta-learning (He et al., 2023b), and prompt-based (Mao et al., 2023c). These research efforts address learning challenges in sentiment analysis, e.g., pattern discovery with label data, efficient learning from few-shot examples, robust representations, and domain adaption. During this period, research in multimodal (Zadeh et al., 2017) or cross-lingual (Zhang et al., 2024b) sentiment analysis was dynamic because it expanded the scope of sentiment analysis beyond English text. Recently, there has been a significant enrichment in the task setups of sentiment analysis. Researchers are no longer satisfied with simply predicting a sentiment polarity for an input text; they are extending the scope of sentiment analysis to include different levels of granularity and contextual awareness, e.g., aspect-based sentiment analysis (ABSA) (Mao and Li, 2021) and opinion mining (Marrese-Taylor et al., 2014), emotion detection (Abdul-Mageed and Ungar, 2017), conversational sentiment analysis (Li et al., 2023b), sentiment analysis from electroencephalography (EEG) signals (Kumar et al., 2019b), facial expressions (Dagar et al., 2016) or speech (Lu et al., 2020). Another trend in sentiment analysis is that researchers paid more attention to the linguistic phenomena that likely affect sentiment analysis, e.g., metaphors (Mao et al., 2023b), sarcasm (Yue et al., 2023), and ambiguous word senses (Zhang et al., 2023b). Considering the impact of sentiment in broad domains, there are research papers studying sentiment analysis in different science domains, e.g., nature disaster (Duong et al., 2024), mental health (Ji et al., 2022), finance (Ma et al., 2023, 2024), legislation (Proksch et al., 2019) and education (Altrabsheh et al., 2013).

To sum up, previous research has addressed sentiment analysis by tackling learning challenges, enhancing sentiment analysis granularity, improving NLU in learning systems, and grounding sentiment analysis in different downstream tasks. However, sentiment perception can vary subjectively in different contexts. There is limited research on personalized sentiment analysis that integrates various types of persona information. This motivates us to bridge this gap by forming a framework to identify the sources of subjectivity in sentiment analysis and developing a neurosymbolic system to process the task of personalized sentiment analysis.

1.4.1.2 Sources of Diversity in Sentiment Perception

Theory of mind suggests that individuals understand that others may hold beliefs, desires, intentions, emotions, and thoughts that differ from their own (Apperly and Butterfill, 2009). Thus, we believe that multiple factors can influence individual sentiment perceptions. According to the theory of appraisal (Martin and White, 2003), opinions and sentiments arise not as direct responses to stimuli but as complicated evaluations incorporating subjective judgments across multiple levels.

We assume that the factors influencing sentiment perception are hierarchically structured. This hierarchical structure ranges from general factors affecting large populations to specific factors influencing individuals. To explore this idea further in the context of sentiment analysis, we reviewed the following theoretical research. Theory of mind research is subject to humans and other species or intelligent agents. Early research found that chimpanzees may possess a preliminary form of theory of mind. The ability allows them to infer the mental states of others, like humans (Premack and Woodruff, 1978). This ability to attribute intentions, knowledge, and beliefs to others suggests that chimpanzees possess the basic forms of social cognition. However, the following research has shown that there are great differences between humans and animals in terms of the depth and complexity of theory of mind (Call and Tomasello, 2008). For example, while chimpanzees can understand others by a perception-goal psychology, they do not have a fully developed belief-desire psychology like humans. With the development of LLMs, e.g., ChatGPT, researchers also extended theory of mind tests to AI. ChatGPT and GPT-4 were tasked with difficult questions that required inferring the counterfactual effects of actions on mental states (Bubeck et al., 2023). The findings show that GPT-4 demonstrated strong abilities in these scenarios, possessing an advanced level of theory of mind.

Cultural norms, values, and beliefs significantly impact how people perceive and understand the world. Researchers compared students from Western (American) and Eastern (Indian) cultural backgrounds and discovered that Western participants had an independent self-construal and saw themselves as considerably more different from others (Markus and Kitayama, 2014). Indian participants, on the other hand, perceived themselves as somewhat more alike to others, suggesting an interdependent self-perception. Additionally, studies show that different languages have different conceptions of emotion, with variations in the causes, evaluations, outcomes, modes of management and display, and even physiological responses linked to particular notions (Pavlenko, 2008). Sentiment perception can also be impacted by religious factors. The theory of cognitive dissonance suggests that individuals might adjust their opinions to match their religious beliefs in order to mitigate psychological discomfort (Festinger, 1957). This process can strengthen existing beliefs while eliminating conflicting opinions. Affirming religious beliefs can reduce the negative affect and emotional discomfort that occurs when individuals experience cognitive dissonance (Burris et al., 1997).

Individuals gain a sense of identity and self-esteem from association with various groups, including those based on occupational and educational backgrounds (Tajfel et al., 1979). To maintain a positive social identity, individuals often adopt perspectives consistent with the norms and values of these groups. Related research also shows that people in high-status occupations tend to show more liberal attitudes toward social issues than people in low-status occupations (Lipset, 1987). Individuals may interpret information in a way that is consistent with their beliefs and values. This bias may lead individuals to view information that aligns with their ideology as more credible and trustworthy, while ignoring conflicting information (Kunda, 1990). Individuals with different ideologies may interpret identical information in divergent ways, thus delivering different opinions on the same target.

For example, those with conservative ideologies often emphasize individual responsibility and individual rights, influencing their stance on welfare and healthcare-related policies (Taber and Lodge, 2006). In contrast, individuals with liberal ideologies often prioritize social justice and equality, leading to opposing opinions. The way people construct their ideas is also influenced by certain personality traits (Jansen et al., 2022b). For example, people who are open to new experiences are more likely to be receptive to new ideas and have flexible, open-minded perspectives (McCrae and Costa, 1987). On the other hand, those with higher conscientiousness typically base their beliefs on carefully analyzing the available data, leading to more thoughtful viewpoints. Sometimes, sentiment perceptions can be influenced by individuals' subjectivity, e.g., individuals who have had positive interactions with dogs are more likely to view dogs positively, contrasting with those whose experiences have been negative (Tyrer et al., 2015). People tend to focus on information that aligns with their personal preferences and beliefs, potentially distorting their perception of emotions (Bower, 1981).

This tendency, termed *confirmation bias*, can impact how individuals interpret emotional events. Moreover, varied experiences and subjective feelings can influence using metaphorical language among individuals to express their opinions (Lakoff and Johnson, 1980). For example, financial analysts may employ different metaphors in their reports under different market conditions (Mao et al., 2023a). The public's perception of different types of weather disasters is also reflected in their metaphorical expressions (Mao et al., 2024d). To sum up, theoretical research and empirical studies support that individuals' sentiment perceptions are subject to multiple factors, including entity diversity that distinguishes between humans and other intelligent agents like animals and AI; culture, religion, vocation, ideology, personality, and subjectivity. These factors may impact personalized sentiment analysis in different scenarios. However, their collective impacts generally represent the complex interplay of individual characteristics and contextual influences on sentiment perception. These factors not only influence how individuals perceive and interpret the sentiment of a target but also alter the language they use to express subjective feelings. In sentiment analysis, understanding these factors is necessary for developing more personalized and context-sensitive systems.

1.4.2 Methodology

After reviewing relevant literature in psychology and cognitive science in the previous section, we define a hierarchical framework, termed Personalized Sentiment Analysis Pyramid (see Fig. 1.9), which encompasses seven factors that can influence individual sentiment perception. By considering each factor, such framework aims to enable more accurate and personalized sentiment analysis, tailoring sentiment detection to the unique characteristics and contexts of each individual. This hierarchical framework is shaped as a layered structure that includes seven persona aspects, namely: entity, culture, religion, vocation, ideology, personality, and subjectivity.

Entity refers to the differentiation between human genders and other intelligent agents. Culture represents how various cultures perceive concepts as positive or negative. Religion involves considering how specific religious beliefs can influence an individual's opinions on certain topics. Vocation aids in understanding people's opinions based on their occupation and educational background. Ideology involves political beliefs and social, economic, or philosophical viewpoints. Personality assists in categorizing concepts as positive or negative based on personality traits. Finally, subjectivity considers personal preferences and experiences. At the bottom of the pyramid, personalization is more general, e.g., entities of the same gender and species, such as males, females, AI, or other creatures, can share the same persona information. Personalization is more specific at the top layer, e.g., subjectivity level. Next, we use GPT-4 Turbo to analyze the persona information of our subjects related to the above seven aspects. LLMs have shown superior knowledge in diverse domains, including multilingual capabilities and scientific task processing (Mao et al., 2024a). They were also suggested as a useful tool for survey research and persona information generation (Jansen et al., 2023). Thus, it is eligible for analyzing the personalities of a subject from different aspects.

Finally, we test LLM performance on personalized sentiment analysis tasks. The obtained persona information in the former step is used as symbolic knowledge, guiding the sentiment inference of an LLM. Since we combine the symbolic knowledge and the reasoning ability of neural network-based LLMs together, the methodology is neurosymbolic. The structured symbolic knowledge provides a clear and understandable reasoning basis, enhancing the interpretability and explainability of the system's decisions (Yeo et al., 2024a). The testing data were sourced from novels, including dialogues with multi-turns. The task is to predict a speaker's sentiment perceptions towards another speaker involved in the conversation. We hypothesize that the additional persona information has different utilities in this scenario because the intensity of the influence of personal characteristics on sentiment perception changes as the scene changes. We aim to evaluate the utilities of the ensemble and each type of persona information.

The overall workflow of our method can be viewed in Fig. 1.10. Our task setup and computing pipeline represent a novel approach for several reasons. First, we do not focus on analyzing the sentiment of a conversation based solely on its semantic content. Instead, our goal is to analyze how one person perceives another's sentiments. This means that even in a conversation that may seem neutral, there could still be negative sentiment if the individuals involved in the conversation do not like each other. Second, unlike traditional personalized AI techniques, such as user preference-based dialogue systems (Zhu et al., 2023, 2024a), personality trait-based recommender systems (Yang and Huang, 2019), or annotation-subjectivity-driven sentiment analysis (Zhang et al., 2024c), our approach considers persona information from multiple aspects. This allows our system to incorporate a broader range of factors that may be informative for personalized sentiment analysis. Finally, our system prioritizes user subjectivity by generating personalized outputs based on different types of persona information, even when presented with the same dialogue input. This approach to human-computer interaction (HCI) is more human-centric.

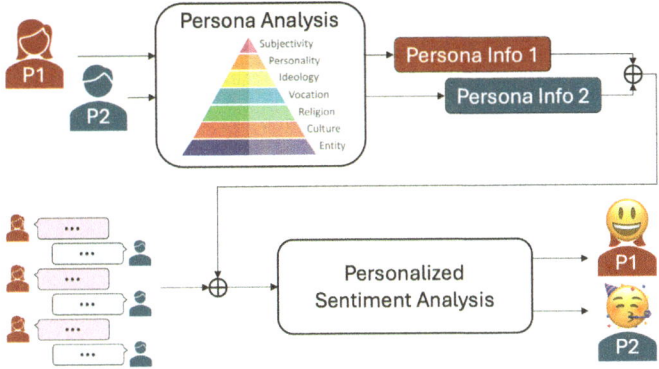

Fig. 1.10: Personalized sentiment analysis workflow. ⊕ denotes textual concatenation.

1.4.2.1 Persona Information Acquisition

Since we have defined seven aspects for persona analysis and our analytical subjects are characters from *Harry Potter* novels, we can consult the persona information for GPT-4 directly. We formulate the query template as follows.

[Goal]: I want to categorize the given object according to their {aspect_term} in the Harry Potter book series. Please suggest a description, one in a line, starting with "-" and surrounded by quotes "". For example: - "{example_category}"
Do not output anything else.
You may choose only one {aspect_term} from the following list: {category_list}
Please categorize {sample_in_prompt} into one {aspect_term} according to their {aspect_term} in the Harry Potter book series.

Table 1.1 shows the aspect terms for the seven aspects used to analyze the characters in the Harry Potter book series. To prevent overlap among the seven aspects and ensure that the LLM fully understands the meanings of the aspect terms, we provide a list of categories for the first six aspects (shown as follows) for the LLM to reference during inference. For "subjectivity", we provide examples such as "Quidditch Seeker" and "Painting", allowing the LLM to generate relevant answers openly.

- Entity: [specie type] Wizards and Witches, Muggles, Werewolves, Dragons, Hippogriffs, Basilisks, Trolls, Hags, Giants, Ghosts, House-elves, Goblins, Centaurs, Veela, Merpeople, Dementors, Vampires. [gender] male, female, or, inapplicable.
- Culture: Gryffindor, Slytherin, Hufflepuff, Ravenclaw, England, Scotland, Wales, Irish, French, Bulgarian, India, African, Romani, Middle Eastern, USA.
- Vocation: Auror, Healer, Transfigurers, Charms Experts, Diviners, Professor, Magizoologist, Potion Master, Curse Breaker, Metamorphmagi, Animagi, Occlumens, Legilimens, Runes Experts, Patronus Charm Casters, Quidditch Player, Journalist, Shop Owner, Ministry Official, Librarian, Herbologist, Arithmancer, Servants, Metalworkers, Bankers, Underwater Dwellers, Companions or Pets.

Table 1.1: Persona aspect term for analyzing the characters in Harry Potter novels.

Aspect	Aspect Term
Entity	specie type and gender
Culture	cultural background
Religion	religious beliefs
Vocation	strong feeling of suitability for a particular career or occupation
Ideology	ideologies from aspects of political, social, epistemological, and ethical
Personality	MBTI personality type
Subjectivity	preferences and hobbies

- Religion: Good vs. Evil, Love vs. Indifference, Acceptance_Death vs. Fear_ Death vs. Bravery_Death vs. Denial_Death vs. Honor_Death, Sacrifice vs. Selfishness, Redemption vs. Condemnation, Impartiality vs. Prejudice, Tolerance vs. Intolerance, Courage vs. Cowardice, Faith vs. Skepticism, Responsibility vs. Irresponsibility.
- Ideology: Equality and Inclusivity vs. inequality and exclusivity, Reform vs. Satus Quo, Utilitarianism vs. Moral Absolutism, Knowledge vs. Ignorance, Loyalty and Community vs. disloyalty and individualism, Pragmatism vs. Idealism.
- Personality: ESTJ, ENTJ, ESFJ, ENFJ, ISTJ, ISFJ, INTJ, INFJ, ESTP, ESFP, ENTP, ENFP, ISTP, ISFP, INTP, INFP.

1.4.2.2 Personalized Sentiment Analysis

We conduct personalized sentiment analyses using the information obtained from the persona and an LLM. Given the collection of the persona information (p_i) of a person related to the seven aspects $(p = \{p_1, p_2, \dots, p_i, \dots, p_7\})$, the scene illustration (e) where the conversation happens, the dialogues with multi-turns (d), the interlocutors (m, n), and task description (t), the task aims to predict the mutual sentiment perceptions, e.g., the sentiment perception of m (a perceiving subject) towards n (a perceiving object) $(s^{(m \to n)})$ and the sentiment perception of n towards m $(s^{(n \to m)})$. We fit all the aforementioned textual information into a prompt template $(template(\cdot))$, then ask an LLM to predict the sentiment perceptions $(LLM(\cdot))$ from the prompt $(prompt)$.

$$prompt^{(m \leftrightarrow n)} = template(e^{(m,n)}, d^{(m,n)}, p^{(m)}, p^{(n)}, m, n, t) \qquad (1.1)$$

$$s^{(m \to n)}, s^{(m \leftarrow n)} = LLM(prompt^{(m \leftrightarrow n)}) \qquad (1.2)$$

The prompt template $(prompt^{(m \to n)})$ for inferring the sentiment perception of m towards n can be viewed below. In the prompt box, the content after [Goal] refers to the task description (t) that directs an LLM to deliver desired predictions, following a fixed structure. scene (e) is the background illustration in which the conversation takes place.

`dialogue_sample` is the dialogue with multi-turns (d). `character_1` and `character_2` correspond to the interlocutors (m, n). `persona_1` and `persona_2` denote their persona information $p^{(m)}$ and $p^{(n)}$ that was obtained in Section 1.4.2.1.

[Goal] I want to classify the sentiment scores between the two characters in the Harry Potter book series based on their dialogue and persona.
Please suggest a sentiment score, one in a line, starting with "-" and surrounded by quotes "". For example:
- "<Harry to Hermione> 1"
- "<Hermione to Harry> 2"
The following shows the different meanings of the sentiment scores.
Please select the sentiment score from the following options: -5: Vendetta; -4: Intentionally inflict harm; -3: Maliciously targeting and harm; -2: Deliberately bullying/deliberately targeting; -1: Rude/Frivolous/Mean characters; 0: Stranger/Neutral; 1: Normal/Polite; 2: Friendly; 3: Kind; 4: Close; 5: Devoted.
[Scene] {scene}
[Dialogue] {dialogue_sample}
[Persona] {character_1}: {persona_1}. {character_2}: {persona_2}.
Please classify the sentiment scores between the two characters {character_1} and {character_2} based on the given dialogue and their personas.

1.4.3 Experiment

1.4.3.1 Research Questions

In this chapter, we aim to explore the following research questions:

1) What is the utility of using the ensemble of the seven levels of personalization?
2) What is the utility of using individual personalization?
3) How does personalization impact sentiment analysis accuracy across different types of entity and culture factors?

These research questions are explored in Sections 1.4.3.5–1.4.3.7, respectively.

1.4.3.2 Dataset

Our personalized sentiment analysis uses the Harry Potter Dataset (HPD) (Chen et al., 2023b). It was developed to enhance the alignment of conversation agents with fictional characters from Harry Potter novels. It includes annotating relationships and character attributes that evolve over the storyline. HPD includes background information, such as conversation scenes, speaker identities, and character attributes, to enable dialogue agents to generate replies consistent with the Harry Potter universe. In contrast to our structured persona analysis, the character attributes in the dataset were not derived from the same set of analytical aspects. Thus, we did not use their character attribute descriptions.

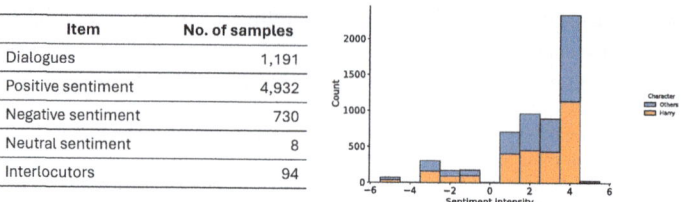

Item	No. of samples
Dialogues	1,191
Positive sentiment	4,932
Negative sentiment	730
Neutral sentiment	8
Interlocutors	94

Fig. 1.11: Dataset statistics and sentiment intensity distribution.

We leverage their affection labels that indicate the sentiment intensity of a perceiving subject to another perceiving object as our sentiment intensity labels. In our classification task, positive labels correspond to the sentiment intensity ranging from -5 to -1, and negative labels correspond to the sentiment intensity ranging from 1 to 5. A neutral label corresponds to the sentiment intensity of 0. Our method is evaluated using the English version of the original HPD. The statistics of our employed data and the sentiment intensity distribution are shown in Fig. 1.11.

1.4.3.3 Large Language Models

The persona information was queried from GPT-4 Turbo. The personalized sentiment analysis was evaluated with GPT-4 Turbo and GPT-3.5 Turbo (i.e., ChatGPT), respectively. GPT-4 Turbo is an upgraded version of GPT-3.5 Turbo. Both LLMs were developed by Open AI, pretrained with Transformer-based deep neural networks and a large number of corpus resources. By a preliminary test, we found that these LLMs have rich knowledge about *Harry Potter* novels.

1.4.3.4 Evaluation Metrics

We evaluate the personalized sentiment analysis performance using two types of metrics. The F_1 score and accuracy are used to evaluate the accuracy of sentiment polarity classification. In this task, LLMs aim to predict whether the sentiment perception is positive, negative, or neutral. Mean squared error (MSE) is used to evaluate the accuracy of sentiment intensity prediction. Considering the potential failure of LLMs to respond in the required format, the Answer Rate is also integrated into the evaluation. It measures the percentage of queries successfully answered by the LLM. In this task, LLMs aim to predict the sentiment intensity, ranging from -5 to 5. Since the sentiment intensity between the same perceiving subject and perceiving object is identical across all dialogues in the dataset, we combine all dialogues between two identical interlocutors as a united input. A correct prediction is defined as the predicted sentiment label (either a sentiment polarity or a sentiment intensity score) of a perceiving subject towards a perceiving object matching the ground truth label.

1.4.3.5 Ensemble Utility of the Seven Levels of Personalization

We compared the sentiment analysis results of GPT-3.5 and GPT-4 with and without the ensembled seven levels of personalization in Table 1.2. The ensemble utility of seven levels of personalization improved the performance of GPT-3.5 on sentiment analysis. For GPT-4, the inclusion of the seven levels of personalization seems to contribute to a slight improvement in F_1 (p:=pos) and accuracy and MSE or even a marginal decrease in F_1 (p:neg) and Macro F_1. Taking into consideration the answer rate, however, we found the accuracy of GPT-4 on the whole dataset turns into $0.8977 \times 0.8155 = 0.7321$ while the one of GPT-4 w/ p1:7 becomes $0.9058 \times 09492 = 0.8598$. Thus, the adjusted accuracy increase of 0.1277 demonstrates the ensemble effectiveness of the seven levels of personalization on sentiment analysis task.

Table 1.2: Personalized sentiment analysis. p:=pos means positive sentiment is defined as positive labels for computing F_1; p:=neg means negative sentiment is defined as positive labels for computing F_1.

	F_1 (p:=pos)	F_1 (p:=neg)	Macro. F_1	Acc	MSE	Answer Rate
GPT-3.5	0.9381	0.5340	0.4907	0.8686	0.3586	0.6522
w/ $p_{1:7}$	0.9484	0.5859	0.5114	0.8936	0.3293	0.6860
Delta	0.0103	0.0519	0.0207	0.0250	-0.2036	0.0338
GPT-4	0.9569	0.7372	0.5704	0.8977	0.1860	0.8155
w/ $p_{1:7}$	0.9631	0.7092	0.5630	0.9058	0.1812	0.9492
Delta	0.0062	-0.028	-0.0074	0.0081	-0.0048	0.1337

1.4.3.6 Utility Analysis of the Individual Personalization

Table 1.3 shows the positive influence of each individual personalization on the performance of GPT-3.5 in the personalized sentiment analysis tasks. Among them, Culture, Vocation, Ideology, and Subjectivity strengthened the performance of GPT-3.5 by a significant margin, while Entity, Religion, and Personality contributed less. There may be a thought-provoking rationale for such a discrepancy. The Harry Potter book series is deeply rooted in a rich culture, ideology, and subjectivity backdrops, thereby offering a rich tapestry of themes and narratives that resonate deeply with readers' or even sentiment intensity annotators' own values. For example, someone who values loyalty and friendship may possibly echo characters like Harry and Ron. Hence, they may easily capture Harry's negative sentiment towards Peter Pettigrew, who betrayed his friends, James and Lily Potter. Consequently, an LLM knowing these factors (culture, ideology, and subjectivity) may understand characters' sentiments more precisely by resonating with characters' values.

Table 1.3: Personalized sentiment analysis by persona types.

	F_1 (pos=pos)	F_1 (pos=neg)	Macro. F_1	Acc-all	MSE	Answer Rate
GPT-3.5	0.9381	0.5340	0.4907	0.8686	0.3586	0.6522
w/ p1	0.9478	0.5553	0.5010	0.8825	0.3200	0.6769
w/ p2	0.9508	0.5793	0.5138	0.8926	0.2941	0.6801
w/ p3	0.9413	0.5403	0.4939	0.8841	0.3736	0.6713
w/ p4	0.9504	0.5542	0.5015	0.8886	0.3055	0.7171
w/ p5	0.9456	0.5532	0.4996	0.8884	0.3419	0.6938
w/ p6	0.9490	0.5619	0.5073	0.8868	0.3108	0.6667
w/ p7	0.9502	0.5722	0.5074	0.8897	0.3046	0.6780

1.4.3.7 Personalization Utility Analysis by Entity and Culture

The results presented in Section 1.4.3.6 are readily comparable, as the growths in F_1, Macro. F_1, accuracy, and answer rate are consistent. Unlike Sections 1.4.3.5 and 1.4.3.6, performance summarized by different categories of entity and culture aspects is more sensitive to the answer rate, since we investigate the results by breaking down the characters into finer-grained groups instead of treating them as a whole. Therefore, we calculated the evaluation metrics based on several query samples, by setting the missing result with a fixed value out of the scope of ground-truth labels (Zhu et al., 2024b). Moreover, we detailed the sentiment analysis results by presenting the metrics of each category group, both as subjects (conveying sentiments to Harry) and objects (receiving sentiments from Harry).

In general, including seven levels of personalization completely enhanced the performance of groups Ghosts, Acromantula, Veela, Centaurs, and French. For entity breakdowns, the negative effects on GPT-4 of including seven aspects occur only in the cases where the entity groups (Muggles, Wizards and Witches, House-elves, and Goblins) play as an object. For culture breakdowns, the negative impacts of including seven aspects on GPT-4 are primarily observed when the entity groups (England, Gryffindor, Ravenclaw, Slytherin, and Hufflepuff) function as objects. However, effects related to these breakdowns are observed in only two cases (Ravenclaw and Hufflepuff) when they serve as subjects. For both breakdowns (entity and culture), the negative effects on GPT-3.5 of including seven aspects occur relatively more often and mainly when the entity group plays as a subject. The above observation highlights the bias of LLMs (such as GPT-3.5 and GPT-4) towards the subject and the object on sentiment analysis, especially when these models address personalized neurosymbolic knowledge. Additionally, comparing the results of GPT-3.5 and GPT-4, we observed that GPT-3.5, when integrated with personalized neurosymbolic knowledge, achieved comparable or even superior performance to GPT-4. This is validated by the results from several Entity or Culture groups including Muggles, Ghosts, Acromantula, Veela, and England.

1.4.4 Summary

In this section, we studied the task of personalized sentiment analysis. Personas were widely studied in commercial domains and web research (Jansen, 2022; Jansen et al., 2022a). Unlike conventional sentiment analysis tasks that aim to analyze sentiment by the meanings of the text, personalized sentiment analysis targets to analyzing the individual sentiment perception. The difference is that an identical statement can yield the same sentiment prediction by its meaning. In contrast, different people may perceive the message differently based on their own personal preferences, personality traits, beliefs, background, etc. To this end, we devised a framework, termed the Personalized Sentiment Analysis Pyramid, for tackling all these different facets through seven different levels of personalization, namely: Entity, Culture, Religion, Vocation, Ideology, Personality, and Subjectivity.

We evaluated the framework with a dialogue dataset sourced from *Harry Potter* novels. The evaluation showed that personalized neurosymbolic knowledge, i.e., seven levels of personalization, augmented LLMs' performance on sentiment analysis. We also analyzed the utility of each persona aspect and found that each individual persona aspect can augment sentiment intensity classification results. Finally, we investigated the influence of persona information on several character groups in the *Harry Potter* novels. Results showed that including persona information elevated the performance of groups Ghost, Acromantula, Veela, Centaurs, and French. Furthermore, a bias of LLMs fed with personalized neurosymbolic knowledge towards subject and object groups is observed.

1.5 Conclusion

The pursuit of automating tedious or repetitive tasks has a rich history, with origins tracing back to Ancient Egypt and the Greek Empire. Among the earliest documented works on automation is the "Book of Ingenious Devices", published in 850 by the Banu Musa brothers. While we have made significant strides since those times, thanks to advancements in mathematical modeling, we now face the challenge that mere mathematics alone may not suffice to model the intricate processes by which the human brain encodes and decodes meaning for complex tasks, including intuitive decision-making, sense disambiguation, and narrative comprehension.

In this chapter, we discussed the importance of personalized AI (in the context of sentiment analysis) and responsible AI (in the context of recommender systems). We also proposed a novel approach to AI that centers on humanity, characterized by seven essential features or pillars. In the future, we plan to define best practices for abiding by such pillars. For example, current post-hoc interpretability methods may not be the best way to implement *Trustworthiness* as they simply find correlations between inputs and outputs of an AI model without really explaining its inner workings. Similarly, there is no point in having a confidence score if this is calculated based on the wrong parameters.

Moreover, we need to define how to evaluate explainability in terms of qualities such as plausibility, i.e., the extent to which an explanation resonates with and is deemed acceptable by a human audience, and faithfulness, i.e., the extent to which the explanation accurately reflects the model's decision-making process. As we look ahead, it is imperative to foster the development of human-in-the-loop and human-in-command systems, integrating human participation in AI through paradigms such as active learning and decision intelligence. We need to develop clear guidelines and principles for AI development which prioritize human values, fairness, accountability, transparency, and privacy, and which should be integrated into the design process from the outset. We need to conduct regular audits of AI algorithms to detect and mitigate biases, errors, and ethical concerns.

Finally, we need to implement and enforce regulations and governance mechanisms that define the boundaries of AI usage, protect individual rights, and foster moral AI practices. By implementing these strategies, society can work toward ensuring that AI technologies are developed and used in ways that empower individuals, maintain human control, and align with ethical and societal values. Balancing technological progress with human agency and values is essential for the responsible advancement of AI.

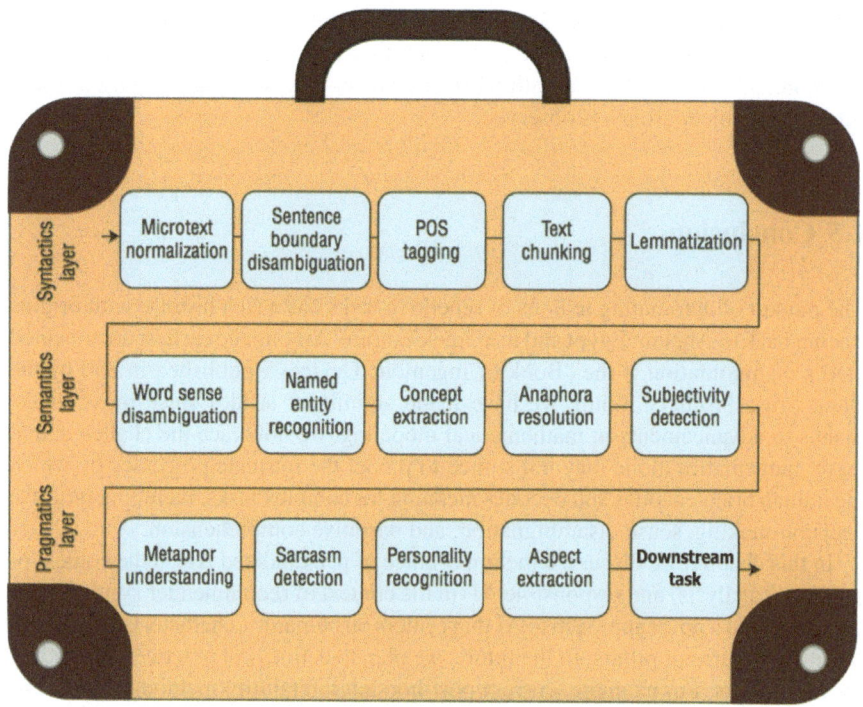

Fig. 1.12: The NLU suitcase model.

If we do not engineer it well, in fact, AI could very much end up being like plastic: a great invention that made our life easier about a century ago, but which is now threatening our own existence. NLU can aid this process by implementing a more transparent a brain-inspired way of processing language. We discuss this in more detail in the next three chapters, in which we illustrate how to handle the multiple cognitive processes associated with the different building blocks of language understanding via three layers (Fig. 1.12), namely: Syntactics Processing, Semantics Processing, and Pragmatics Processing.

This framework draws inspiration from two significant sources: the sentiment analysis suitcase model (Cambria et al., 2017) and the Jumping NLP Curves paradigm (Cambria and White, 2014). The former was specifically designed to address polarity detection. It focused on a narrower scope compared to more general NLU tasks and utilized a different arrangement of its components to tackle sentiment-related challenges.

The latter conceptualized the evolution of NLP research as a series of three distinct curves, each representing different levels of complexity in handling natural language. These curves illustrate the progression from basic language processing techniques to more sophisticated approaches that handle increasingly complex linguistic phenomena. The current framework builds on this paradigm by incorporating its insights into the design and organization of NLU components, aiming to model and manage the complexities of modern NLP challenges effectively.

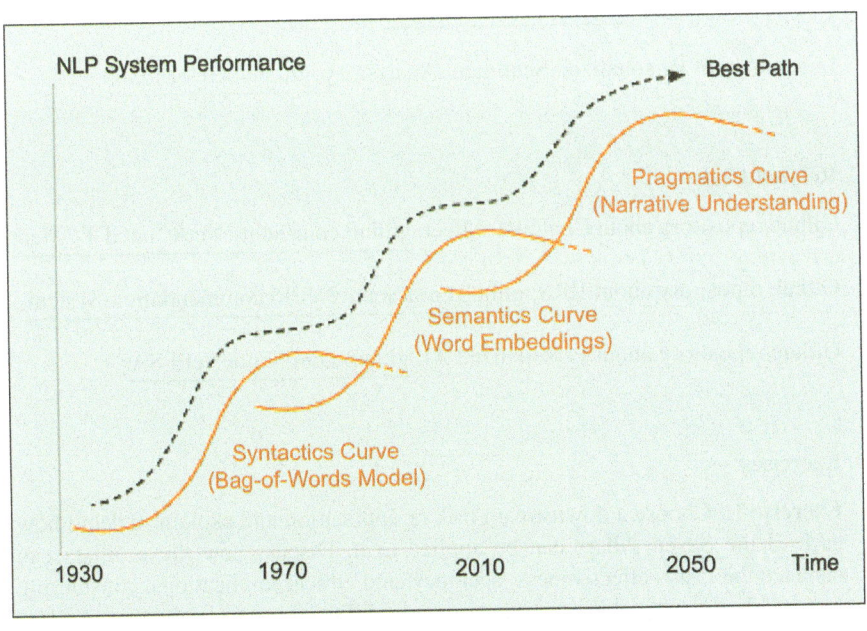

Fig. 1.13: Jumping NLP curves paradigm.

1.6 Learning Resources

A. Reading List

- Erik Cambria, Rui Mao, Melvin Chen, Zhaoxia Wang, and Seng-Beng Ho. Seven Pillars for the Future of Artificial Intelligence. IEEE Intelligent Systems, 38(6): 62–69, 2023 (Cambria et al., 2023)

- Przemysław Kazienko and Erik Cambria. Towards Responsible Recommender Systems. IEEE Intelligent Systems, 39(3):5–12, 2024 (Kazienko and Cambria, 2024)

- Luyao Zhu, Rui Mao, Erik Cambria, and Bernard Jim Jansen. Neurosymbolic AI for Personalized Sentiment Analysis. In Proceedings of HCII, 2024 (Zhu et al., 2024b)

B. Relevant Videos

- Labcast about the Seven Pillars for the Future of AI: youtu.be/SX1Cl_eDTLE

- Labcast about RRS: youtu.be/DGc1i5I3q88

- Labcast about Personalized Sentiment Analysis: youtu.be/pzlUpZfc8fQ

C. Related Code

- Github repository about ChatGPT Affect: github.com/SenticNet/ChatGPT-Affect

- Github repository about RRS: github.com/SenticNet/Recommendation-System

- Github repository about Personalized AI: github.com/SenticNet/PSA

D. Exercises

- **Exercise 1.** Choose a downstream task or application and explain in detail how each of the Seven Pillars can be applied to it. Discuss how these pillars can enhance the task's effectiveness, accuracy, and ethical implications, considering factors like interpretability, robustness, scalability, fairness, and privacy. Illustrate how these principles can be integrated into the development, deployment, and monitoring phases of the system to ensure it operates efficiently and ethically.

- **Exercise 2.** Discuss the development of a QA system implementing the principle of panalogy, i.e., a QA system storing information redundantly in two formats: knowledge graph and embeddings. This approach aims to leverage the complementary strengths of both structured and unstructured data representations to provide more accurate and flexible responses to user queries. Describe strengths and weaknesses of each format with respect to different scenarios or domains.

- **Exercise 3.** Provide two examples of neurosymbolic systems: one in which a symbolic representation, e.g., a knowledge graph or a semantic network, is incorporated within a subsymbolic algorithm to improve its accuracy, and one in which a subsymbolic algorithm, e.g., a neural network or a machine learning model, is used to develop a symbolic representation to increase its comprehensiveness.

- **Exercise 4.** Discuss how the concept of responsible recommender systems can be adapted to other types of systems, e.g., search engines, automated hiring platforms, financial services, healthcare applications, or smart city infrastructures, by integrating principles of fairness, transparency, accountability, user control, and privacy protection to ensure ethical and equitable outcomes.

- **Exercise 5.** Describe how the Personalized Sentiment Analysis Pyramid can be adapted to another downstream task (instead of polarity detection) and explain in detail how each layer can be tailored to meet the specific needs of the chosen task. Discuss how personalization can be incorporated at each stage to enhance the task's relevance and effectiveness for individual users. Highlight practical examples and potential challenges in adapting the pyramid to this new context.

Chapter 2
Syntactics Processing

Abstract Computational syntactics processing is a fundamental technique in NLU. It normally serves as a preprocessing method to transform natural language into structured and normalized texts, yielding syntactic features for downstream task learning. This chapter proposes a systematic review of low-level syntactic processing techniques, namely: microtext normalization, sentence boundary disambiguation, POS tagging, text chunking, and lemmatization. In particular, we summarize and categorize widely used methods in the aforementioned syntactic analysis tasks, investigate the challenges, and yield possible research directions to overcome the challenges in future work.

Key words: Microtext Normalization, Sentence Boundary Disambiguation, Part of Speech Tagging, Text Chunking, Lemmatization

2.1 Introduction

Syntactics processing is a generalization of NLU subtasks that are concerned with the structure of phrases and sentences, as well as the relation of words to each other within the phrase or sentence (Woolf, 2009). It involves extracting meanings from sentence constituents, and establishing the semantic structure of the input sentence (Roberts, 2016). Syntactics processing is thus a foundational step towards higher-level interpretations, preceding other more complicated tasks, such as WSD, information retrieval, sentiment analysis, dialogue system, etc. It preprocesses the input texts, and provides important features that can be further utilized by machines. However, in the era of deep learning based high-level NLP tasks, syntactic processing techniques are often overlooked. These NLP applications rely on neural networks to implicitly extract syntactic information. They are indeed able to significantly outperform symbolic methods that depend on handcrafted feature templates.

E. Cambria, *Understanding Natural Language Understanding*,
https://doi.org/10.1007/978-3-031-73974-3_2

Nevertheless, we argue that integrating symbolic and subsymbolic AI, also known as neurosymbolic AI, is the key for stepping forward in the path from NLP to NLU. Despite recent advancements in deep learning, statistical NLP does not achieve true NLU, as it merely makes probabilistic guesses. Belinkov et al. (2017) conducted an experiment to evaluate the word embeddings learned by neural machine translation (NMT) on syntactic tasks, with results indicating that they are poor representations for syntactic information. Therefore, to further improve the performance of semantic and pragmatic tasks, it is of great benefit to regard syntactics processing as sub-problems. For instance, POS tagging is incorporated in the decoding process (Feng et al., 2019b) or as an auxiliary task (Niehues and Cho, 2017; Mao and Li, 2021) to improve NMT and metaphor detection, lemmatization as a preprocessing step boosts the accuracy of neural sentiment analysis on social media text (Symeonidis et al., 2018). Hence, syntactics processing is an integral part of the neurosymbolic NLU paradigm.

In this chapter, we examine the five most basic tasks in syntactics processing, namely, microtext normalization, sentence boundary disambiguation (SBD), POS tagging, text chunking, and lemmatization. There is a variety of other tasks in this field, e.g., stopword removal, negation detection, constituency parsing, and dependency parsing. Here, we decide to focus on the most foundational syntactic processing tasks that can be helpful to high-level syntactic tasks and other NLU applications. To elaborate, dependency parsing analyzes the grammatical structure of a sentence based on the dependencies between words. It is strongly related to POS linguistically. Indeed the majority of existing algorithms heavily rely on POS tagging (Chen and Manning, 2014; Dyer et al., 2015; Zhou et al., 2020a). Similarly, constituency parsing analyzes sentences by breaking them down into constituents. It can be regarded as hierarchical text chunking, establishing structure within the chunks. Therefore, we view them as high-level tasks that can benefit from the newly introduced low-level syntactic processing techniques and justify separate reviews.

Although low-level syntactic processing tasks have gradually faded out of public views in many NLP conference proceedings, previous surveys presented significant findings in individual syntactic processing tasks. Satapathy et al. (2020) presented a comprehensive review on microtext normalization. They discussed the similarities between microtext and brachygraphy, and suggested the potential applications of microtext normalization in more complex NLU tasks. Read et al. (2012) reviewed 10 publicly available SBD systems, and argued that the published results of these systems are evaluated on different task definitions and data annotations. Thus, they assessed the systems on unified task definition and gold-standard datasets, as well as on user-generated content (UGC) corpora. Manning (2011) conducted error analysis on existing POS tagging algorithms at the time, deducing 7 common error categories, and proposed a solution for the errors and inconsistencies in the gold-standard dataset. Kanis and Skorkovská (2010) compared manual and automatic Czech lemmatizers, based on their influence on the performance of information retrieval. However, many of these reviews are out-of-date, unable to cover the recent advancement, powered by e.g., deep learning techniques. Additionally, some of them focused on the applications of specific tasks.

Technical Trends

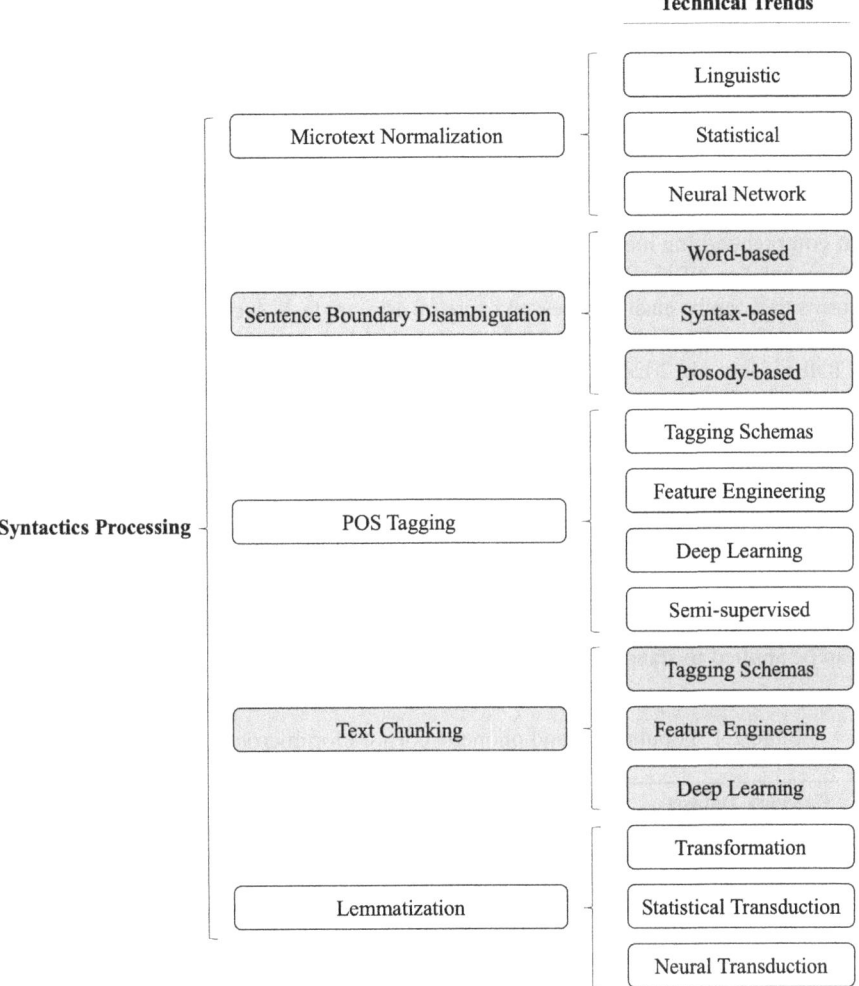

Fig. 2.1: Outline of this chapter. Each subtask is explained in terms of different technical trends.

In the following sections, we introduce five different pragmatic processing techniques, namely: microtext normalization (Section 2.2), SBD (Section 2.3), POS tagging (Section 2.4), text chunking (Section 2.5), and lemmatization (Section 2.6). Fig. 2.1 presents an outline of the chapter: each task is organized in terms of different approaches and provides a summary. Finally, Section 2.7 wraps up the chapter by proposing concluding remarks, a reading list, three relevant videos, some related code, and five exercises.

2.2 Microtext Normalization

In the era of short message service (SMS) and social media, a plethora of text data can be mined from the web. Microtext is a term referring to shorthand writing, commonly seen in these informal texts. The occurrence of microtext poses a problem for NLU, as it uses informal languages, e.g., shortened and lengthened words, abbreviated phrases, and missing grammatical components. It is challenging for machines to comprehend the meaning of the source text. Therefore, microtext normalization is essential in NLU. The task aims to convert the informal texts into their standard forms that can be easily processed by a downstream task, such as sentiment analysis for social media (Brody and Diakopoulos, 2011; Satapathy et al., 2017b, 2019b; Chaturvedi et al., 2024), information retrieval (Bontcheva et al., 2013b), and question answering (QA) system (Mittal et al., 2014). Microtext is often highly compact, informal, and semi-structured (Rosa and Ellen, 2009). There are many abbreviations, acronyms, lax spelling and grammar, as well as metadata that establish the context. Although microtext is a written language, it is strongly influenced by phonetics, resulting in language-dependent character repetition and substitution phenomena, e.g., *"nooooo"* for *"no"*, and *"luv"* for *"love"*. It is also an important preprocessing task in the context computer vision, e.g., optical character recognition (OCR) can be used to digitize handwritten or printed informal notes, and then microtext normalization can be applied to standardize the resulting text for further processing.

Table 2.1: Popular text and phonetic corpora for microtext normalization.

Category	Dataset	Source	Reference
Text	SMS-M	Mobile phone short messages	Dunlop and Crossan (2000)
	SMS-C	Short Message Service corpus	Choudhury et al. (2007)
	SMS-L	Mobile phone short messages	Liu et al. (2011)
	NUS SMS	NUS Short Message Service corpus	Wang and Ng (2013)
	Edinburgh	The Edinburgh 𝕏 Corpus	Petrović et al. (2010)
	LexNorm1.1	English 𝕏 text	Han and Baldwin (2011)
	LexNorm1.2	Modified LexNorm1.1	Yang and Eisenstein (2013)
	W-NUT	2015 ACL-IJCNLP Workshop on Noisy User-generated Text	Baldwin et al. (2015)
	Nom-Tweets	English 𝕏 text	Satapathy et al. (2017b)
	WSJ	Wall Street Journal	Marcus et al. (1993)
	WUS	SMS messages in Swiss	Ueberwasser and Stark (2017)
	Sw-SMS	SMS messages in Swiss	Lusetti et al. (2018)
Phonetic	ARPABET	International Phonetic Alphabet (IPA) to ASCII symbols	Khoury (2015)
	TIMIT	Revised ARPABET	Taylor et al. (1998)
	CMUdict	Carnegie Mellon University Pronouncing Dictionary	Jiampojamarn et al. (2007)
	Festival	Festival lexicon	Taylor et al. (1998)
	CELEX	Lexical databases of English, Dutch and German	Jiampojamarn et al. (2007)

Even without any knowledge about texting conventions, humans can understand this type of irregular words by sounding them out. Thus, phonetics-based abbreviations and spelling alterations are invented and evolve more freely than orthography-based ones. For instance, *"schewpid"* is created to imitate *"stupid"* in a British accent. It is uncommonly used, but most English speakers can understand it, even if they have never seen it before. As for orthography-based abbreviations and alterations such as *"wdym"* for *"what do you mean"*, less Internet-savvy people will have to look up its meaning. Therefore, text-level mapping alone is often not adequate to address the sparsity caused by phonetic-level alterations. Thus, existing typical corpora (see Table 2.1) are prepared for both text-level and phonetic-level mappings. Accuracy, top-n accuracy, F_1 score, and bi-lingual evaluation understudy (BLEU) score (Papineni et al., 2002) are widely used evaluation measures. Specifically, top-n accuracy considers the system correct, if the correct standard form is in its top n predictions.

In the remainder of this chapter, we review and categorize existing microtext normalization approaches by linguistic, statistical, and neural network methods.

2.2.1 Linguistic Approach

Linguistic approaches normalize microtext by modeling linguistic knowledge or commonly observed patterns in informal texts. Normalizers in this categories use rule-based and/or lexicon-based method to leverage orthographic and phonetic information, or to implement their observations or assumptions about the formation of microtext. This approach is advantageous at addressing specific problems, such as abbreviation, character repetition, and character substitution caused by phonetics. However, it is labor intensive, as most rules and lexicons are manually built. Pennell and Liu (2010) described a rule-based normalization system for text messages so that they can be read by a text-to-speech system. A set of hand-crafted rules are applied to translate standard English into abbreviations in text lingo, which provides features for a maximum entropy (MaxEnt) classifier to automatically create abbreviations, thereby establishing a mapping from standard English to the text message domain. This mapping is then reversed to create a look-up table so that the corresponding English words can be predicted from the input text message.

Brody and Diakopoulos (2011) described an automated method based on the notion that word lengthening is strongly associated with subjectivity and sentiment. Their method leverages this association to identify new domain-specific sentiment- and emotion-bearing words that are not in the existing lexicon. Jose and Raj (2014) proposed a lexico-syntactic normalization model. As the name suggests, their model consists of two modules – lexical normalization module, and syntactic analysis module. The former is based on a channelized database and a user feedback system. Inspired by the multi-channel model (Xue et al., 2011) that is introduced in the later section, the database is built as four channels of database, namely abbreviation channel, non-noisy channel, grapheme channel, and phoneme channel.

Given a tokenized input string, candidate outputs are produced from each of these channels via binary search. The syntactic analysis module is based on a bottom-up parser. It takes the lexically normalized string outputted from the previous module, and builds the corresponding parsing tree using grammar rules. The module will then decide whether the sentence is valid or not, which helps to make the text normalization task more efficient. Desai and Narvekar (2015) introduced a method to normalize OOV words. First, a set of rules are applied to standardize elongated words such as 'goooood'. Then, the most probable candidates are found from compiled databases for the OOV tokens. Lastly, any remaining noises are normalized using Levenshtein edit distance, which signifies the minimum number of single character insertions, deletions, and substitutions required to transform one word to another. Mittal et al. (2014) proposed a three-level architecture to process microtext in QA systems. First, noise presented in the SMS tokens are replaced with the closest dictionary words using the Soundex (Beider, 2008), Metaphone algorithm (Philips, 1990), and a modified version of the longest common subsequence (LCS) algorithm termed Phonetic LSC (PLSC) that also takes phonetic similarity into account. Next, a semantically similar set of candidate questions are selected. Lastly, the results are optimized via syntactic tree matching (STM) and WordNet-based similarity.

Satapathy et al. (2017b) proposed a phonetic-based algorithm to normalize tweets, which is used to enhance the accuracy of polarity detection. It is based on the assumption that humans are able to understand character repetition and substitution in unknown informal words because they automatically shift to the phonetic domain when they read text. Therefore, they adopted an ensemble approach that mainly relies on phoneme, and used a lexicon to address other forms of OOV words such as acronyms and emoticons. Given a tweet, all the OOV words are matched with the lexicon to find their in-vocabulary forms and corresponding polarity class. Then, any leftover unnormalized tokens are processed using Soundex, which uses homophones to encode texts so that characters with similar pronunciations can be easily matched. Since their objective is to boost polarity detection, the phonetic code of OOV words is matched against the knowledge base SenticNet (Cambria et al., 2024). The final output is fed into a polarity classifier to test the effectiveness of normalization, and yields promising results.

All of the phonetic algorithms used in the above mentioned methods encode words based on their spelling, and thus are only applicable for English or languages written in Latin alphabet. An alternative is to use International Phonetic Alphabet (IPA), which more accurately represents pronunciations and can be used to encode any languages. However, its disadvantage is also evident – such a fine-grained encoding system can be too specific for phonetic matching. Thus, it needs to be simplified accordingly. Khoury (2015) proposed a phonetic tree-based microtext normalization model. The model determines the probable pronunciation of English words based on their spelling via a radix-tree structured phonetic dictionary. To build the tree, Khoury created a dataset of IPA transcriptions of the English Wiktionary[1].

---

[1] https://wiktionary.org

Jahjah et al. (2016) presented a novel word-searching strategy, built on the idea of sounding out the consonants of a given word. Their algorithm uses spelling and phonetic strategies to extract the base consonant data from misspelled and real phrases. First, both visual signature and phonetic signature of real English words are extracted using a set of rules. Then, the model finds a set of in-vocabulary words with identical signatures to the OOV word, along with their occurrence probability, and applies several heuristics to find the best matching in-vocabulary word.

Building upon their previous work, Satapathy et al. (2019b) proposed a cognitive-inspired microtext normalizer, named PhonSenticNet, to aid concept-level sentiment analysis for SMS texts. Given a message, a binary classifier is first applied to detect whether microtext is present, aiming to reduce execution time. If true, PhonSenticNet finds the closest match for every concept in the input sentence based on phonetic distance, which is computed using Sorensen similarity algorithm based on both Soundex encoding as well as IPA encoding. Results showed that PhonSenticNet outperforms their previous algorithm.

2.2.2 Statistical Approach

Statistical approaches employ machine learning models such as hidden Markov model (HMM) and support vector machine (SVM), to perform text normalization. For this type of normalizers, noisy channel is the most widely-used formulation. Modeling the process of distorting the source into the target, noisy channel is the basis of many NLP applications, e.g., statistical machine translation (SMT), automatic speech recognition (ASR), and spell checking. Intuitively, we can view microtext normalization as a translation task from standard English to informal English. Therefore, the nature of noisy channel makes it suitable for microtext normalization.

Choudhury et al. (2007) constructed a noisy channel model based on HMM for words in standard English to represent all possible variations of their corresponding text language version. The core concept of their model is to formulate the transformation from standard English to text language as a noisy channel. Given a sentence s in standard form, the noisy channel decodes it into text language t by

$$\hat{t} = argmax_s P(s)P(t \mid s)$$

where the language model $P(s)$ encodes which strings in standard form are valid, and the error term $P(t \mid s)$ models how the standard forms are distorted. The former can be easily constructed by exploiting large amounts of unlabeled data, therefore the paper focused on the latter, which is implemented by a letter-to-phoneme conversion HMM. Given a word in standard English, the HMM formulates the graphemes in the word as observation states, and their corresponding phonemes as hidden states. To train and test their method, they created a word-aligned corpus of SMS texts and standard English.

Jiampojamarn et al. (2007) pointed out that the previous letter-to-phoneme conversion systems only support one-to-one letter-phoneme alignment, which leads to difficulties in handling single letters that correspond to two phonemes, or vice versa. This can be partially mitigated by manually constructing a fixed list of merged phonemes prior to the alignment process. The paper, however, proposed a better solution to overcome this limitation – an automatic many-to-many alignment method. Given an input word, a many-to-many aligner establishes the appropriate alignments across the graphemes and phonemes. After this process, the input word is represented as a set of letter chunks, each containing one or two letters aligned with phonemes. The chunk boundaries are determined via a bigram letter chunking prediction model based on instance-based learning (Aha et al., 1991). The prediction model generates all the bigrams in a word, automatically determining whether each of them should be a double letter chunk based on the context. Subsequently, an HMM embedded with a local classifier is applied to find the globally most possible sequence of phonemes of the given word. The local classifier, also based on instance-based learning technique, generates the phoneme candidates for every letter chunk according to the context. Thus, compared to conventional HMM described above, their modified HMM is able to utilize context information from not only the phoneme side but also the grapheme side.

Bartlett et al. (2008) applied a structured SVM model called SVM-HMM model (Altun et al., 2003) to orthographic syllabification, which acts as a sub-system to improve letter-to-phoneme conversion. The paper formulated orthographic syllabification as a sequence labeling problem, employing SVM-HMM to predict tag sequences. Unlike standard SVM, SVM-HMM employs Markov assumption during scoring, and thus is able to consider complete tag sequences during training. The motivation for using SVM-HMM over HMM is mainly the benefit of the discriminative property in SVM, and leeway to adopt feature representations without any conditional independence assumptions. The paper explored two tagging schemas for syllabification. One is positional tags (NB tags) (Bouma, 2003) that label whether each letter occurs at a syllable boundary or not. The other is structural tags (ONC tags) (Daelemans et al., 1997; Skut et al., 2002), which represent the role each letter is playing within the syllable, namely onset, nucleus, and coda. The former is simple, straightforward, and adaptable to different languages. The latter, on the other hand, is able to capture the internal structure of a syllable to improve accuracy, but does not explicitly represent syllable breaks. To mitigate the weakness of the ONC tag schema, the paper combined the two schemas and proposed a hybrid Break ONC tag schema. Experiments show that their model not only outperforms previous syllabification systems, but also improves the accuracy of letter-to-phoneme conversion. Kaufmann and Kalita (2010) employed an SMT system for $\mathbb{X}$ message normalization. Given a tweet, it is first passed through a syntactic normalization module, where sets of rules on the orthographic- and syntactic-level are applied to preliminarily normalize the input sentence. Next, the normalized tweet is fed into the machine translation module, which is implemented by an existing SMT model called Moses (Koehn et al., 2007), to transform the tweet into standard English.

Xue et al. (2011) introduced a method that considers orthographic factors, phonetic factors, contextual factors, and acronym expansions, formulating each of them as a noisy channel. The core concept is that a non-standard term should be similar to its standard form in respect of one or more of these factors, and thus each channel is responsible for one aspect of the distortion that converts the intended form into the observed form. The grapheme channel models the corruption of spellings. The phoneme channel causes distortion in pronunciations. The context channel changes terms around a target term. The acronym channel transforms a phrase into a single term. These channel models are combined using two variations of channel probabilities, namely Generic Channel Probabilities and Term Dependent Channel Probabilities. The former assumes that the probability of a term being emitted through a noisy channel is independent of its standard form, whereas the latter takes into consideration that some standard forms are more likely to be distorted via a certain channel in reality. Experiments show that the two variants achieve similar performance, with the latter being slightly better on SMS dataset.

Han and Baldwin (2011) theorized that supervised learning is not adept at handling $\mathbb{X}$ OOV words due to data sparsity. Hence, they introduced an unsupervised classifier that detects ill-formed words and generates candidate in-vocabulary words using morphophonemic similarity. Their normalization strategy consists of three stages. First, a confusion set of candidate in-vocabulary words is generated for each OOV word based on morphophonemic variations, which are produced using lexical edit distance and the double Metaphone algorithm (Philips, 2000). Then, based on its confusion set, a linear kernel SVM classifier (Fan et al., 2008) is employed to detect whether the given OOV word is actually an ill-informed word. Lastly, if the OOV word is indeed ill-informed, the most likely candidate is selected based on morphophonemic similarity and context. Although their method outperforms most supervised models, there are limitations. The most prominent one is that it only targets single-token words, and thus is unable to handle phrases and acronyms.

Pennell and Liu (2011) described an SMT-based system for expanding abbreviations found in informal texts. The method operates in two phases. In the first phase, a character-level SMT model generates possible hypotheses for each abbreviation. By training the model at character level, the model is able to learn the common character abbreviation patterns regardless of their associated words, and thus alleviates the OOV problem. In the second phase, an in-domain language model decodes the hypotheses, choosing the most likely one in the context.

Similarly, Li and Liu (2012) introduced a word-level framework based on an SMT approach, which performs well in translating OOV words into in-vocabulary words. The framework can be divided into two components. One component is a character-based SMT module that translates non-standard words into standard words by matching their character sequence. The other component is a two-stage SMT module that leverages phonetic information. Non-standard words are first translated into possible phonetic symbols, then mapping to standard words. The candidate word lists generated by the two components are then combined and ranked using a set of heuristic rules.

Liu et al. (2012) described a broad-coverage normalization system that incorporates different aspects of human cognition, including letter conversion, visual priming, and resemblance between string and phone. The system consists of four key components. Given a token, three sub-normalizers respectively produce candidate words based on the cognition aspects, namely, enhanced letter transformation, visual priming, and spell checking. Then, the last component, called candidate combination, uses either word-level or message-level strategy to rank the candidates. The enhanced letter transformation component is a standard noisy channel, where the error term is modeled by a character-level conditional random field (CRF). The enhancement is two-fold. First, features relating to phoneme, morpheme, syllable, and word boundary are used to train the CRF. Second, the training *(word, token)* pairs are chosen according to their global contextual similarity. On the other hand, the visual priming component is also a noisy channel, but uses a novel technique inspired by cognitive behavior. They observed that when humans are familiar with informal texts, they are able to quickly recognize nonstandard tokens given only minor visual cues. Following this observation, the priming component computes the error term based on word frequency of the token and the minor visual stimulus, which is LCS between the token and target word divided by the token length. Lastly, for the spell checking component, the open-source Jazzy spell checker[2] is employed. The broad-coverage system outperforms previous state-of-the-art models at both word level and message level.

Zhang et al. (2013) proposed a parser-centric word-to-word normalization method to convert raw informal texts into the correct grammatical version. The main drive for this method is that text normalization performance should be strongly dependent on how useful it is for downstream applications, e.g., parsing. Given an input sequence, a series of replacement generators replace original tokens with normalized ones by producing sets of edit operations. Subsequently, a directed acyclic normalization graph is constructed based on the generated replacements. The graph models two types of dependencies, namely, syntactic consistency of true assignments, and correlations among replacements. The output is determined by using Maximum A Posteriori probability (MAP) inference, which is to select the weighted longest path in the normalization graph. The paper tested the normalizer in two aspect, namely, its effects on dependency parsing and its domain adaptability. Results showed that the proposed normalizer indeed helps improve parsing, and supports domain transfer with low cost. Wang and Ng (2013) made the observation that aside from standardizing informal words, other normalization operations, such as missing word recovery and punctuation correction, are also crucial for the improvements of downstream NLU tasks. To this end, they presented a punctuation correction method based on dynamic CRF (Sutton et al., 2007), which is described in detail in the text chunking section, and a missing word recovery method based on standard CRF. For the former, the words and punctuation symbols surrounding the focal word are represented as binary features, and fed into a two-layer dynamic CRF model. The first layer assigns the actual punctuation tags, whereas the second layer labels sentence boundaries.

---

[2] https://jazzy.sourceforge.net

For missing word recovery, the paper specifically targeted the omission of *"be"* in English. Hence, a CRF is employed to label every token in the input sentence with a tag that indicates whether the token should be followed by a conjugation of *"be"*. Furthermore, they proposed a novel beam-search decoder that can integrate different types of normalization operations, including statistical and rule-based operations. Given an input sentence, the decoder iteratively produces new sentence-level hypotheses, evaluating them to retain the plausible ones, until it finds the best normalization. Each hypothesis is generated by several hypothesis producers, each of which focuses on a different target aspect of informal texts, applying the corresponding type of normalization operation to the sentence. In other words, the objective of beam-search is to find the best pipeline of hypothesis producers.

Yang and Eisenstein (2013) suggested that data sparsity of social media texts makes unsupervised learning extremely challenging. To solve this problem, they proposed a unified unsupervised model based on maximum-likelihood framework. Given an informal sentence, a log-linear model is applied to model the string similarity between tokens in tweets and standard English. Since it is an unsupervised, locally-maximized conditional model, typical dynamic programming techniques such as the Viterbi algorithm is not ideal. Instead, they applied the sequential Monte Carlo algorithm (Cappé et al., 2007) for training. The log-linear model is combined with a language model for standard English to output the desired conditional probability.

Based on the assumption that syllable plays a fundamental role in non-standard word generation in social media, Xu et al. (2015b) viewed syllable as the basic unit and extended the conventional noisy channel approach. Given a non-standard word, it is first segmented into syllables to identify the non-standard syllables. Then, the error term of the noisy channel can be represented as syllable similarity, which is an exponential potential function that combines orthographic similarity and phonetic similarity. The former is measured by edit distance and LCS, whereas the latter is measured by phonetic edit distance and phonetic LCS based on letter-to-phone transcripts. Results showed that the syllable-based approach indeed yields significant improvement. Additionally, this method has the advantage of being robust and domain independent.

2.2.3 Neural Network Approach

In recent years, deep learning models are widely used in the field of NLP (Otter et al., 2020; Mao et al., 2019), as they are able to automatically extract features from input text. Neural network approaches make use of this advancement in microtext normalization domain. As stated in the previous section, microtext normalization can be viewed as a translation problem. Hence, it is a natural progression to employ the encoder-decoder framework commonly adopted by NMT (Yang et al., 2020a). Additionally, since the attention of normalizing microtext has gradually shifted from word level to character level, neural networks are frequently utilized to extract character-level features.

Chrupała (2014) introduced a semi-supervised text normalization model. They pointed out a weakness of the widely used noise channel decomposition in microtext normalization – texts in normalized form of source domain are limited, and standard text resources vary across domains. Thus the target language model is in fact not as easy to estimate as one imagines. Therefore, their model serves as an alternative that utilizes a large amount of unlabeled data from a source domain. To achieve this, normalization is formulated as a string transduction problem (Chrupała, 2006), which is discussed in detail in the lemmatization section. Using labeled data, a CRF learns the sequence of edit operations that transforms the tweets to standard English. Additionally, a Simple Recurrent Network (Elman, 1991) is employed to induce character-level embeddings from unlabeled $\mathbb{X}$ data, which are used as features in the CRF transduction model. Their model is able to yield accurate results with less training data and without lexical resources.

Leeman-Munk et al. (2015) proposed a deep learning model to solve the W-NUT (see Table 2.1 for W-NUT) Lexical Normalization for English Tweets challenge. Their model consists of three components, a normalizer, a flagger that identifies words to be normalized, and a conformer that smooths out the simple errors from the normalizer. The normalizer is implemented as a character-level three-layer feedforward neural network, which takes in the input word and predicts a possible standard version. The flagger has the same architecture as the normalizer and works in parallel with the normalizer. The input word is replaced by the generated word only if the flagger tags it as to be normalized. Since the normalizer functions at character level, it is possible that the predicted word is one character off. To address this error, a conformer is constructed using a dictionary collected from the gold standard training data. If the word generated by the normalizer is not in the dictionary, the conformer replaces it with the most similar word in the dictionary based on Levenshtein edit distance.

Taking inspiration from the advances in NMT, Lusetti et al. (2018) proposed a character-level encoder-decoder model to normalize multilingual SMS messages. The model employs a Bidirectional Long Short-Term Memory (BiLSTM) (Graves et al., 2013) as an encoder, an LSTM (Hochreiter and Schmidhuber, 1997) as a decoder, and a soft attention mechanism (Bahdanau et al., 2015). They further introduced two methods to augment the model. First is to fuse it with a word-level language model of the target language using the synchronization mechanism (Ruzsics and Samardzic, 2017). Second is to enhance the training set with additional target-side data.

Similarly, Satapathy et al. (2019a) tested different variants of the sequence-to-sequence (Seq2Seq) normalization framework, and to what extent they can improve the performance of sentiment analysis on informal texts. Specifically, they experimented with four types of character-level encoder-decoder architectures, namely LSTM, LSTM with attention, gated recurrent unit (GRU) (Cho et al., 2014), and LSTM with CNN as feature extractor (Kim, 2014). Experiments show that all the normalizers above help boost the accuracy of sentiment analysis on social media and SMS texts. The ranking of effectiveness from high to low is GRU, LSTM with attention, LSTM with CNN, and LSTM.

Table 2.2: Comparison between different microtext normalization methods. OM stands for orthographic matching based on string similarity. PM stands for phonetic matching based on phonetic similarity. CX stands for context. SY stands for syllable. CH stands for character. UN stands for unlabeled.

	Method	OM	PM	CX	SY	CH	UN	Abbreviation	Sentiment
Linguistic	Pennell and Liu (2010)							✓	
	Brody and Diakopoulos (2011)	✓							✓
	Jose and Raj (2014)	✓	✓					✓	
	Mittal et al. (2014)	✓	✓						
	Desai and Narvekar (2015)	✓	✓						
	Satapathy et al. (2017b)	✓	✓					✓	✓
	Khoury (2015)		✓						
	Jahjah et al. (2016)		✓						
	Satapathy et al. (2019b)		✓						✓
Statistical	Choudhury et al. (2007)	✓	✓			✓			
	Jiampojamarn et al. (2007)	✓	✓						
	Bartlett et al. (2008)	✓			✓				
	Kaufmann and Kalita (2010)	✓		✓					
	Xue et al. (2011)	✓	✓	✓		✓		✓	
	Han and Baldwin (2011)	✓	✓	✓		✓			
	Pennell and Liu (2011)			✓	✓	✓		✓	
	Li and Liu (2012)	✓	✓		✓	✓		✓	
	Liu et al. (2012)	✓	✓		✓	✓		✓	
	Zhang et al. (2013)	✓		✓					
	Wang and Ng (2013)			✓					
	Yang and Eisenstein (2013)	✓				✓			
	Xu et al. (2015b)	✓	✓	✓		✓			
Neural	Chrupała (2014)					✓	·✓		
	Leeman-Munk et al. (2015)					✓		✓	
	Lusetti et al. (2018)		✓			✓		✓	
	Satapathy et al. (2019a)		✓			✓		✓	
	Lourentzou et al. (2019)		✓			✓		✓	

Lourentzou et al. (2019) proposed a hybrid word-character attention-based encoder-decoder model for text normalization. Given a sentence, the tokens are first passed through a word-based Seq2Seq model to capture semantic meaning and long-term contextual dependencies. Any detected OOV tokens are marked and passed on to a secondary character-based Seq2Seq model, which outputs the possible in-vocabulary tokens and corresponding confidence. The character-based Seq2Seq model is trained using synthetic adversarial data that aims to capture errors commonly found in social media texts.

2.2.4 Summary

Microtext normalization is a complex task that encapsulate many sub-problems, e.g., misspelled words, acronyms, phonetics-based spelling alterations, real-word spelling errors, missing words and punctuations, etc. In summary, among linguistic approaches, the system proposed by Pennell and Liu (2010) is advantageous in that it does not require external lexicons, however, it relies on hand-crafted rules to generate candidates. Jose and Raj (2014) incorporated a parser as sub-system, effectively capturing lexico-syntactic information. However, their method heavily relies on manually complied database. The representation of string similarity is a focal point of linguistic approaches. Desai and Narvekar (2015) used Levenshtein edit distance to measure similarity, which cannot model the human intuition behind microtext. Mittal et al. (2014), on the other hand, applied multiple spelling-based phonetic matching algorithms to compute similarity. However, their method is language-specific. IPA phonetic matching can be applied to any language, while it suffers from over-specificity (Khoury, 2015). Jahjah et al. (2016) mitigated this weakness by incorporating spelling features combined with IPA phonetic matching. Brody and Diakopoulos (2011); Satapathy et al. (2017b, 2019a) all targeted sentiment analysis. Brody and Diakopoulos (2011) relied a set of simple rules for elongated words, and thus limited in use. The latter two models are able to handle various types of OOV words, among which Satapathy et al. (2019b) has the advantage of incorporating IPA encoding.

For statistical approaches, most models formulate microtext normalization as a noisy channel problem, as first implemented by Choudhury et al. (2007). A limitation of their method is that it can only map one letter to one phoneme. Jiampojamarn et al. (2007) implemented a many-to-many letter-phoneme alignment that can mitigate this problem, whereas their solution relies on hand-crafted conversion rules and only works on fixed window size of letters. Bartlett et al. (2008) and Xu et al. (2015b) addressed this limitation by utilizing syllable. The former uses syllable tagging as a sub-system, whereas the latter considers syllable as the basic unit for noisy channel. Han and Baldwin (2011); Yang and Eisenstein (2013) both utilized unsupervised learning to mitigate the data sparsity problem for social media text. However, their methods cannot handle phrases and acronyms. The parsing-centric method proposed by Zhang et al. (2013) is advantageous in aiding downstream task, however, it has the same weakness of only dealing with single-token word-to-word conversion. This limitation can be addressed by SMT-based approach. Kaufmann and Kalita (2010) directly implemented an existing SMT system for microtext normalization. Following this approach, Pennell and Liu (2011) presented a character-level SMT system that can better handle abbreviation. Li and Liu (2012) further improved such method by combining character-level and phonetic SMT systems. On the other hand, Xue et al. (2011) tackled the difficulty of abbreviation by using multiple noisy channels, each for one aspect of microtext. Similarly, Liu et al. (2012) also incorporated multiple noisy channels, and took it one step further by designing each noisy channel specifically for its function. Lastly, Wang and Ng (2013) confronted the niche aspects of microtext normalization, i.e., missing word recovery and punctuation correction.

With neural network approaches, Chrupała (2014) addressed the target domain problem of NMT-based approach by proposing a semi-supervised alternative. However, such method only targets single-token word. Leeman-Munk et al. (2015) circumvented both of these limitations by using a character-level pipeline model, which conversely can introduce propagated errors. Lusetti et al. (2018); Satapathy et al. (2019a); Lourentzou et al. (2019) all made use of the encoder-decoder structure with various neural networks, which can be considered as an NMT-based approach. Nonetheless, the target domain problem is less prominent nowadays, because a) more annotated datasets are available, and b) deep learning approach is more adept at extracting generalized information than traditional SMT systems. Table 2.2 presents a summary of the introduced methods and their concerns. Notably, most linguistic and statistical methods use a combination of orthographic matching and phonetic matching by computing their corresponding similarities. This is expected, as each provides complementary information that the other cannot address.

There are methods that use neither, formulating their target problems as sequence prediction tasks instead, using either graphical models or neural networks. For some approaches, contexts are taken into consideration by incorporating syntactic features (Fossati and Di Eugenio, 2008; Kaufmann and Kalita, 2010; Zhang et al., 2013), or using attention mechanism (Lusetti et al., 2018; Satapathy et al., 2019a; Lourentzou et al., 2019). Additionally, syllable-level and character-level models are able to learn more fine-grained patterns than word-level ones, and thus are gaining popularity in recent years, especially for the latter. Since microtext normalization is a field with abundant raw text, we also mark out methods that are able to utilize unlabeled data. Finally, there are specific sub-problems that were addressed by a part of the reviewed linguistic and statistical methods, namely, abbreviation and acronym and sentiment-related error. Table 2.3 shows the performance of the introduced methods, organized by their target tasks. On the SMS-C dataset, Li and Liu (2012) obtained the highest top-1 accuracy, whereas Liu et al. (2012) obtained the highest top-20 accuracy. Zhang et al. (2013) achieved the best performance on LexNorm1.1, and Xu et al. (2015b) on LexNorm1.2.

To sum up, linguistic approaches build upon simple text-level and phonetic-level matching algorithms, utilizing hand-crafted rules, lexicons or knowledge bases. As mentioned above, it is adequate at targeting specific problems, e.g., IPA-based phonetic matching, sentiment, and context provided by metadata. However, its drawbacks are prominent. Rule-based and lexicon-based methods require a lot of human effort, hence unable to adapt to other language domains. Statistical approach, on the other hand, is much less labor-intensive. This type of models regard microtext normalization as an SMT problem or a sequence labeling problem. Although most of them still rely on feature engineering, they pave the way for neural network approaches. With the help of neural networks, context-sensitive character-level features can be easily extracted by Seq2Seq models. It is also adaptable to other languages. However, neural network approaches still have much room for improvement. To our knowledge, there is not yet a deep learning normalizer that directly addresses the data sparsity problem caused by phonetics-based alterations.

Table 2.3: Performance of the introduced microtext normalization methods. Nonstandard denotes normalizing informal OOV words. $\mathbb{X}$ API stands for data randomly collected from $\mathbb{X}$ (formerly known as Twitter). SMS-G stands for SMS messages generated by volunteers. WER is word error rate.

Task	Method	Experiment 1			Experiment 2		
		Dataset	Result	Measure	Dataset	Result	Measure
Nonstandard	Pennell and Liu (2010)	$\mathbb{X}$ API + SMS-G	58.46	top-1 acc.	$\mathbb{X}$ API + SMS-G	75.38	top-3 acc.
	Jose and Raj (2014)	SMS-G	95.41	acc.			
	Khoury (2015)	SMS-L	30.2	top-1 acc.	SMS-L	59.7	top-5 acc.
	Jahjah et al. (2016)	SMS-L + SMS-C	56.6	top-1 acc.	SMS-L + SMS-C	73.6	top-5 acc.
	Satapathy et al. (2019b)	NUS SMS	92.75	acc.	$\mathbb{X}$ API	72.88	acc.
	Kaufmann and Kalita (2010)	$\mathbb{X}$ API	79.85	BLEU			
	Pennell and Liu (2011)	$\mathbb{X}$ API	75.57	top-1 acc.	$\mathbb{X}$ API	92.53	top-20 acc.
	Choudhury et al. (2007)	SMS-C	58	top-1 acc.	SMS-C	86	top-20 acc.
	Xue et al. (2011)	SMS-C	96	acc.	$\mathbb{X}$ API	96	acc.
	Han and Baldwin (2011)	SMS-C	75.5	F_1	LexNorm1.1	75.3	F_1
	Li and Liu (2012)	SMS-C	71.96	top-1 acc.	SMS-C	83.11	top-20 acc.
	Liu et al. (2012)	SMS-C	64.36	top-1 acc.	SMS-C	91.75	top-20 acc.
	Yang and Eisenstein (2013)	LexNorm1.1	82.09	F_1	LexNorm1.2	82.06	F_1
	Xu et al. (2015b)	LexNorm1.1	85.3	F_1	LexNorm1.2	86.08	F_1
	Chrupała (2014)	LexNorm1.1	4.5	WER			
	Leeman-Munk et al. (2015)	W-NUT	81.49	F_1			
	Lourentzou et al. (2019)	W-NUT	83.94	F_1			
	Lusetti et al. (2018)	WUS + Sw-SMS	87.61	acc.			
Polarity	Brody and Diakopoulos (2011)	$\mathbb{X}$ API	93	acc.			
	Satapathy et al. (2017b)	$\mathbb{X}$ API	81.59	acc.			
	Satapathy et al. (2019a)	NUS SMS	83.5	acc.			
Letter-to-phoneme	Jiampojamarn et al. (2007)	CMUDict	65.6	acc.	CELEX Eng	83.6	acc.
	Bartlett et al. (2008)	CELEX	89.99	acc.			
QA system	Mittal et al. (2014)	FAQ corpus	75	acc.			
Missing tokens	Wang and Ng (2013)	NUS SMS	66.54	BLEU			
Parsing	Zhang et al. (2013)	SMS-C	81	F_1	LexNorm1.1	91.9	F_1

It can potentially lead to performance drop for languages that are rich in such alterations on social media, such as Chinese. Thus, incorporating phonetic features or a phonetic-level encoder can be a potential direction for future work. Furthermore, the use of microtext is strongly affected by the geolocation and dialect of users. For instance, when expressing laughter in Latin alphabets, there is a tendency to use *"lol"* for English speakers, *"hhh"* for Chinese speakers, *"www"* for Japanese speakers, *"kkk"* for Korean speakers, *"555"* for Thai speakers, etc. Metadata that provides such information is scarcely utilized in existing methods. As such, this would also be an interesting direction to explore.

2.3 Sentence Boundary Disambiguation

SBD, which decides where sentences begin and end in raw texts, is an important yet overlooked preprocessing task for many NLU applications. Downstream tasks such as machine translation (Matsoukas et al., 2007; Zhou et al., 2017) and document summarization (Jing et al., 2003; Boudin et al., 2011) rely on predetermined sentence boundaries for good performance.

Table 2.4: Widely used corpora for SBD.

	Dataset	Source	Reference
Text	BC	The Brown Corpus	Francis and Kucera (1979)
	WSJ	Wall Street Journal	Marcus et al. (1993)
	SMC	X text corpus	Rudrapal et al. (2015)
	MG	Texts in Modern Greek	Stamatatos et al. (1999)
	GRS	Gold Standard Rules	Sadvilkar and Neumann (2020)
	MIMIC	Medical Information Mart for Intensive Care corpus	Johnson et al. (2016)
	FV	Fairview Health Services database	Knoll et al. (2019)
Prosody	T-News	Speech transcripts from news program	Stevenson and Gaizauskas (2000)
	T-CTL	Prepared speech transcripts	Treviso et al. (2017)
	T-MCI	Impaired speech transcripts	Treviso et al. (2017)
	T-R	Transcripts of BBC and radio news	Gotoh and Renals (2000)
	BN	Broadcast news transcripts	Liu et al. (2004)
	CTS	Conversational telephone speech transcripts	Liu et al. (2004)

Sentence segmentation is seemingly easy through identifying punctuation marks. However, there are some notable ambiguities that often occur in text. For instance, aside from indicating the end of a sentence, a period can also appear in abbreviations or as decimal point. In a more complex scenario where an abbreviation is the last token of a sentence, the period simultaneously serves as a part of the abbreviation as well as the full-stop of the sentence. There are two widely-used corpora for SBD, the Brown Corpus (Francis and Kucera, 1979) and the WSJ corpus from the Penn Treebank project (Marcus et al., 1993). In the Brown Corpus, 90% of periods occur at the end of sentences, and 10% at the end of abbreviations, among which about 1% to 2% are in both, whereas in the WSJ corpus, about 47% of periods appear in abbreviations. Thus, the ambiguity level largely depends on corpora and the associated domains. Widely applied datasets are listed in Table 2.4. The evaluation metrics commonly used are accuracy (or error rate) and F-measure.

In this chapter, we divide existing SBD algorithms into three categories: word-based approaches, syntax-based approaches, and prosody-based approaches.

2.3.1 Word-based Approach

The word-based approaches largely rely on morpholexical information, e.g., suffix, spelling, capitalization, and word length to segment sentences. For early works, rule-based approaches, n-gram probabilistic methods, and lexical lookup methods are commonly employed to predict the type of the words surrounding a period, thus determining whether the period marked the end of the sentence.

Grefenstette and Tapanainen (1994) were the first to propose a rule-based algorithm to resolve the ambiguities caused by the usage of periods. Specifically, a set of rules are manually created using regular expressions to represent possible patterns where periods that do not occur as full-stops. The system then matches the surrounding context of every period in texts against the regular expressions to predict whether it is a sentence boundary. Their system obtains reasonably high accuracy with great computational efficiency. To enable automated extraction of rules, Stamatatos et al. (1999) applied transformation-based learning to SBD. Compared to the original transformation-based algorithm (Brill, 1995), which is discussed in detail in the POS tagging section, their method limits the number of possible transformations. This is achieved by maintaining two sets of rules for each punctuation mark. Initially, all punctuation marks are considered to be full-stops. Then, for each of them, the system automatically learns a set for rules that triggers the removal of a sentence boundary, and a set of rules that triggers the insertion of a sentence boundary. This mechanism ensures that the maximum number of transitions for each punctuation mark is two.

Sadvilkar and Neumann (2020) developed a rule-based SBD system called PySBD, where the reasoning mechanism of their system are explainable. The performance is comparable to statistical models. Unlike the other supervised SBD methods, PySBD is trained and evaluated with the Golden Rule Set (GRS). GRS is a language specific corpus designed for SBD. It contains sets of hand-crafted rules over a variety of domains that are carried out in a pipeline fashion. Despite the decent accuracy, there are some drawbacks to rule-based methods. Firstly, periods exhibit absorption properties, meaning when multiple periods occurs they are often marked as one. Therefore, it is challenging to build a comprehensive set of rules where they do not contradict each other. Furthermore, such systems are often developed using a specific corpus, thus further application to a corpus in another language or domain is difficult. Therefore, many machine learning methods are proposed for the SBD tasks to address these shortcomings.

To reduce the labor efforts in developing hand-crafted rules and features, Reynar and Ratnaparkhi (1997) proposed a SBD system based on MaxEnt (Ratnaparkhi, 1996). It requires only simple information about the candidate punctuation marks. For each candidate punctuation mark, the system utilizes morpholexical features of its trigram tokens, estimating their joint probability distribution as a sentence boundary. The performance of their system is comparable to those which require vastly more resources.

Schmid (2000) proposed an unsupervised learning method by manually listing all the possible scenarios where a period may denote a decimal point or an abbreviation. For each of the listed scenarios, a probability model is created accordingly to predict whether the period is a full-stop. Unlike previous token-based methods, which rely on the focal token itself and its local context to determine whether it is an abbreviation, their model computes the distribution of possible abbreviations by scanning through the whole corpus. This method can be referred to as a type-based approach (Kiss and Strunk, 2002).

Similarly, Kiss and Strunk (2002) introduced an innovative unsupervised type-based system that requires less human effort. Their approach views the identification of abbreviations as a collocation detection problem, which can be solved using log likelihood ratio. Specifically, an abbreviation can be seen as a collocation of the abbreviated word and the following period. Kiss and Strunk (2006) further expanded this system to create the Punkt system, which has a more complex structure. First, the input text is processed by the previous type-based algorithm to obtain the initial annotation, which marks a period as either part of an abbreviation, part of an ellipsis, or sentence boundary. Subsequently, a token-based classifier is applied to revise the initial annotation as well as to determine whether a period is both part of an abbreviation or ellipsis and the end of sentence, thus producing the final annotation. In the token-based classifier, log likelihood ratio is used for two heuristic. One is the collocation heuristic, which takes a pair of word surrounding a period and tests whether a collocation tie exists between them. The other is the frequent sentence starter heuristic, which searches for word types that form a collocation with a preceding sentence boundary based on the results from the type-based classifier. The frequent sentence starter heuristic helps counterbalance the collocation heuristic, which in some scenarios falsely identifies collocation across a sentence boundary.

Gillick (2009) proposed a SBD system based on SVM. The system takes the trigram contexts of periods as input. Specifically, for a trigram of the form "$L.R$", the problem is defined as the conditional probability of the binary sentence boundary class s, $P(s \mid "L.R")$. The system utilizes morpholexical features, e.g., capitalization and word length to train the SVM. To reduce the error rate caused by words such as $U.S$ and $N.Y$, the system adopts the type-based approach, taking into account the log count of lowercased appearances of L and R in the input texts. Rudrapal et al. (2015) focused on the problem of SBD in social media texts and explored three machine learning algorithms for the task, namely CRF, Naïve Bayes, and sequential minimal optimization (SMO) (Platt, 1999). Since social media texts likely use informal languages, identifying sentence boundaries in such texts is a even greater challenge. To mitigate the ambiguities, the corpus is first tokenized, using the tokenizer in CMU Twitter POS tagger (O'Connor et al., 2010; Gimpel et al., 2010). Subsequently, the machine learning algorithms are trained on the basic textual features of the tokenized corpus. The CRF method is able to extract simultaneously correlated lexical features, thus easier to incorporate different knowledge sources. The SMO method, on the other hand, is an iterative optimization approach to train SVM. Results show that SMO yields the best performance on social media texts, whereas Naïve Bayes comes close in the second place. On the Brown corpus, both Naïve Bayes and SMO obtain good performance, but the former is more accurate. CRF, however, performs marginally worse than the other two algorithms. Riley (1989) first introduced a tree-based algorithm to SBD, where a regression tree is generated using a set of hand-crafted morpholexical features from the trigram context of words with periods in them to identify decimal points and abbreviations. Although their model achieves accurate results, the model requires a vast amount of training data, making it too costly for a preprocessing task and unrealistic for languages and domains with limited resources.

Wong and Chao (2010); Wong et al. (2014) applied an incremental algorithm on SBD. Their system is based on i^+Learning principle, which is an incremental decision tree learning algorithm, making it flexible to changes and suited for online learning. The algorithm can be divided into two phases. First, the system constructs a top-down binary decision tree offline using the initial training data. The resulting tree acts as an optimal base for the second phase, which is a online procedure that adopts the tree transposition mechanism of Incremental Tree Induction (Utgoff et al., 1997) as a bias to grow and revise the base tree. This method dynamically revises the tree according to the new incoming data while preserving the essential statistical information. Thus it is able to adapt to texts in a different language or domain without the need to retrain from scratch. Their system is trained on morpholexical features extracted from trigram contexts.

All of the aforementioned SBD systems, with the exception of Kiss and Strunk (2002, 2006), use n-gram based technique to extract textual information, which leads to sparse vector space problems. To address this, Treviso et al. (2017) suggested word embedding as an alternative and verified which embedding induction method works best for SBD. They investigated word2vec (Mikolov et al., 2013a), Wang2Vec (Ling et al., 2015a), and FastText (Bojanowski et al., 2016). They used Continuous Bag-Of-Words (CBOW) and Skip-gram to train the vectors, respectively. They then tested the embeddings on a Recurrent CNN (RCNN) (Treviso et al., 2016) for sentence segmentation. Experiments show that word2vec consistently performs better than the other two methods. Additionally, the Skip-gram strategy generally yields better results than the CBOW for the Wang2Vec and the FastText, whereas for word2vec the better strategy depends on the corpus. Furthermore, they also compared the model performance between using only the extracted morpholexical features and adding morphosyntactic features for SBD. Interestingly, the results show that the explicitly added features do not make a difference in terms of accuracy. They theorized that the word embedding alone carries sufficient morphosyntactic information for SBD.

Knoll et al. (2019) utilized both word embeddings and character embeddings to further capture morpholexical features. Based on the observation that previous SBD systems perform poorly on a domain-specific corpus such as clinical texts, they proposed a deep learning algorithm to address this problem. First, the input text is tokenized and transformed into word embeddings using FastText. The text is also fed into a CNN layer (Collobert et al., 2011) to obtain character embeddings, which is summed with the word embeddings. The final word representation is passed through a BiLSTM layer (Graves et al., 2013) and a sigmoid-activated dense layer to output the log-probability of a word being the start of a sentence. They tested their algorithm on the Medical Information Mart for Intensive Care (MIMIC) corpus (Johnson et al., 2016) and a dataset drawn from the Fairview Health Services (FV). Results suggest that the deep learning approach indeed outperforms the traditional ones, especially on corpora from different domains and corpora where sentences are often not terminated by punctuation marks.

2.3.2 Syntax-based Approach

It is difficult for word-based approaches to robustly capture contexts larger than the focal token. Syntax-based SBD systems can solve this problem by utilizing POS tags. Intuitively, we know that the number of possible POS patterns for a bigram is much less than the number of possible word patterns, thus lessening sparsity problems. Additionally, the syntactic function of a word makes a big difference when identifying abbreviations. For instance, when predicting whether a capitalized word following a period is a proper name or a common word, taking into account the POS tags of its trigram context is more effective than relying only on their morpholexical properties.

Palmer and Hearst (1994) were the first to incorporate POS tagging in the task of SDB. They proposed an efficient and portable system based on feed-forward neural network. The core idea is to use POS probabilities of the tokens surrounding a punctuation mark as input to the feed-forward network, which outputs an activation value to determine what label to assign to the punctuation mark. First, the system uses a slightly modified version of the PARTS POS tagger (Church, 1989), which also produces the frequency counts of the POS tags associated with each token. The system then maps them into a more generalized set of 18 POS categories. Each input token in the n-gram context around a punctuation is then represented by a descriptor array indicating their probability distributions for these 18 categories, in addition of two flags that mark whether the word is capitalized and if it follows a punctuation mark. Subsequently, these descriptor arrays are fed into a fully-connected hidden layer with a sigmoid activation function, and then into an output unit to decide whether the punctuation marks are full-stops. The system also introduces two adjustable thresholds to leave room for difficult ambiguities. When the output score falls under the first threshold, the punctuation mark is not a sentence boundary. When the score is higher than the second threshold, it is a sentence boundary. If the score is in between these two thresholds, that means the system is under-informed to make a confident prediction and it'll be marked accordingly for later use.

Based on this algorithm, Palmer and Hearst (1997) further developed a SBD system called Satz, where the aforementioned n-gram descriptor arrays containing POS information can be fed into either a neural network as in their previous work, or a decision tree. The learning algorithm chosen for the system is a c4.5 (Salzberg, 1994) decision tree induction program, which iteratively constructs the tree using the descriptor arrays as input attributes. Each leaf node of the decision tree represents the value of the goal attribute, which in this case indicates whether a punctuation mark is a sentence boundary. After the decision tree is built, the algorithm prunes it by recursively examining each sub-tree to reduce errors and overfitting. Experiments show that the tree-based learning method achieves comparable accuracy as the feed-forward neural network. However, there are problems with the n-grams of generalized POS categories. First, the generalized categories are far sparser than the traditional Penn Treebank POS tags, thus requiring more training data. Furthermore, since words outside of the n-gram have no influence on the prediction, the n-gram must be of sufficient length to capture syntactic information. In the Satz System, n is set to be 6, which is also much sparser than the commonly adopted bigrams and trigrams.

The reason behind these weakness is that, this method is built on the premise that sentence boundaries must be obtained before POS tagging. Thus, to utilize syntactic information, the original POS tags have to replaced by the generalized POS categories. This dilemma is solved by Mikheev (2000), who suggested that POS taggers do not necessarily require predetermined sentence boundaries to operate. The simple solution is to break the input text into short text-spans so that it is easier for POS taggers to handle. Based on this notion, he proposed a SBD system using POS tagging framework. To achieve this, he made some minor adjustments to the Brown Corpus and the WSJ corpus. First, the period in abbreviation is tokenized separately from the rest of the abbreviations. Second, all the periods are marked accordingly to three types of tags, namely full-stop, abbreviation, or both. With this setting, SBD can be performed using a POS tagger, which is able to fully make use of the local syntactic information. The POS tagger chosen for this system is based on HMM and MaxEnt, which will be discussed in detail in the POS tagging section. He further introduced a document-centered approach (Mikheev, 2002), which can also be regarded as type-based. When using this approach, the system scans the entire document for contexts where the focal token is marked unambiguously. This approach is proven to be effective in distinguishing whether a capitalized word is a proper name or a common word, which works in complement with the POS tagger to determine sentence boundaries.

Stevenson and Gaizauskas (2000) applied memory-based learning to identify sentence boundaries in transcripts produced by an ASR system. It is more challenging than SBD with standard texts. For instance, the text generated by an ASR system is generally unpunctuated, in single case, and likely to contain transcription errors. Their algorithm is based on TiMBL (Daelemans et al., 2003), which memorizes a set of training examples. It classifies new instances by assigning them the class of the most similar learned instances. The system adopts both POS tags and morpholexical features such as capitalization and stopword flag. Results show that the proposed algorithm cannot effectively address the difficulties in ASR transcripts. Following the work of Reynar and Ratnaparkhi (1997), Agarwal et al. (2005) proposed a MaxEnt classifier that incorporates both lexical and syntactic information. They assigned each token with a binary End-of-Sentence tag and a POS tag, formulating SBD as a sequential labeling problem. Similar to the previous MaxEnt model, their classifier is also trained on the features of trigram contexts. Their work also concluded with the best tested feature set for the MaxEnt classifier.

2.3.3 Prosody-based Approach

Prosody-based approaches are proposed specifically for SBD in ASR. Previous methods (Stevenson and Gaizauskas, 2000; Treviso et al., 2017) took a word-based approach, using only textual features from speech transcripts. However, transcripts tend to be erroneous, and thus are unable to provide reliable cues to segment sentences.

Prosody-based approaches, on the other hand, incorporate prosodic information, e.g., pitch, pause length, pre-pausal lengthening, and energy patterns. Such prosodic features are assumed to be able to compensate the lack of punctuation and capitalization in speech. Furthermore, sentence structures are often more elusive in spoken language. A sentence-like unit (SU) in speech can be a sentence, part of a sentence, or a semantically complete component in a sentence (Strassel, 2003). Therefore, SBD in ASR faces the additional challenge of determining whether an SU is an actual sentence. Gotoh and Renals (2000) presented two finite state models to statistically extract sentence boundary information from program scripts, as well as transcriptions of broadcast speech produced by a vocabulary speech recognition system. The former is an n-gram type language model trained by the program scripts.The latter is a pause duration model, which is a statistical prosody model estimated from the outputs of an ASR system aligned with their corresponding program transcripts. The joint probability of a sequence of prosodic features are combined the corresponding sequences of word- and class-tokens to enhance the accuracy.

Liu et al. (2004) first built an HMM-based SBD system for continuous speech. Then, they proposed a MaxEnt posterior probability model, which aims to address two main shortcomings of the HMM system. Firstly, the training objective of HMM is to maximize the joint probability of observed and hidden events, which is not the best criterion to reduce classification error. Secondly, the n-gram language model underlying the HMM transition model requires a significant increase in the number of model parameters to incorporate highly correlated features. The former is solved by replacing the generative model with a posterior probability model. The latter is addressed by MaxEnt framework, which allows for a large number of overlapping features. Aside from word n-gram and its syntactic information, they also utilized prosodic features around each word boundary, e.g., duration, pitch, and energy patterns. The system is evaluated using the NIST[3] SU error rate. Results suggest that the proposed model and HMM have complementary strengths and weaknesses. The MaxEnt model is able to achieve better accuracy, whereas the HMM makes more effective use of prosodic features, and degrades less with word recognition errors.

Taking it one step further, Liu et al. (2005) applied CRF to SBD in speech, comparing its performance with the HMM model and the MaxEnt model. CRF has the advantages of being discriminatively trained and able to model the entire sequence. As in their previous work, textual and prosodic information is used to train the models. Results show that CRF indeed performs marginally better than the HMM and MaxEnt. The only shortcoming of CRF model is that, compared to other two models, it takes longer to train as the number of features increases. Liu et al. (2006) found that a SU in speech is less frequently a sentence than a non-sentence unit. Thus, a prosody model must be able to deal with the imbalanced dataset distribution. To address this problem, they modeled prosodic information with a decision tree classifier in their HMM-based system (Liu et al., 2004). They also investigated different types of sampling approaches, concluding that random downsampling is the most advantageous.

---

[3] https://nist.gov/speech/tests/rt/rt2003/fall

Table 2.5: Comparison between different SBD methods. Token stands for token-based SBD. Type stands for type-based SBD.

Method		Morpholexical	Syntactic	Prosodic	Token	Type	Neural	ASR
	Grefenstette and Tapanainen (1994)	✓			✓			
	Stamatatos et al. (1999)	✓			✓			
	Sadvilkar and Neumann (2020)	✓			✓			
	Reynar and Ratnaparkhi (1997)	✓			✓			
	Schmid (2000)	✓				✓		
	Kiss and Strunk (2002)	✓				✓		
Word	Kiss and Strunk (2006)	✓				✓		
	Gillick (2009)	✓				✓		
	Rudrapal et al. (2015)	✓			✓			
	Riley (1989)	✓			✓			
	Wong and Chao (2010)	✓			✓			
	Wong et al. (2014)	✓			✓			
	Treviso (2017)	✓					✓	✓
	Knoll et al. (2019)	✓					✓	
	Palmer and Hearst (1994)	✓	✓		✓			
	Palmer and Hearst (1997)	✓	✓		✓			
Syntax	Mikheev (2000)		✓		✓			
	Mikheev (2002)	✓		✓		✓		
	Stevenson and Gaizauskas (2000)	✓	✓			✓		✓
	Agarwal et al. (2005)	✓	✓		✓			
	Gotoh and Renals (2000)	✓		✓	✓			✓
Prosody	Liu et al. (2004)		✓	✓	✓			✓
	Liu et al. (2005)		✓	✓	✓			✓
	Liu et al. (2006)		✓	✓	✓			✓

2.3.4 Summary

In conclusion, existing SBD systems can be categorized into word-based, syntax-based, and prosody-based approaches. Table 2.5 presents a summary of the introduced systems, and the features and methods they used. In this chapter, for convenience, if a SBD system only use the word itself as lexical features, it is not counted as a method using morpholexical features. A comparison of all the system performance can be found in Table 2.6. The datasets used are listed in Table 2.4. On the Brown Corpus, word-based method proposed by Wong and Chao (2010) achieved the highest accuracy, exceeding the best syntax-based method (Mikheev, 2002) by 0.26%. On the WSJ dataset, the best word-based method (Gillick, 2009) obtained 99.75% accuracy, whereas the best syntax-based method (Mikheev, 2002) obtained 99.55%. For prosody-based approaches, Liu et al. (2005) obtained the lowest NIST SU error rate.

Word-based systems utilize morpholexical features, e.g., capitalization, suffix, word length, etc, which can be further divided into token-based, type-based, and neural network based. Given a focal word, token-based methods use rules or machine learning models to determine sentence boundary based on the features of the local context. Among rule-based systems, the work of Stamatatos et al. (1999) could automatically generate rules, and Sadvilkar and Neumann (2020) applied sets of accurate rules in pipeline.

Since the rule-based methods are rigid and labor intensive, Reynar and Ratna-parkhi (1997); Rudrapal et al. (2015) alleviated this by using statistical models that require simple features. Riley (1989) proposed the first tree-based SBD system, which is more comprehensible to human. To reduce the computation cost, Wong and Chao (2010); Wong et al. (2014) applied an incremental tree-based algorithm, which is also more flexible and suited for online learning. Type-based methods, compared to token-based ones, not only consider the morpholexical features of the token, but also compute its likelihood as a non-sentence boundary over the entire corpus. Hence, type-based approaches have the advantage of having the capacity to manage a large amount of unannotated data. Among type-based methods, Schmid (2000) implemented separate probability models for different scenarios. However, it requires the manual labor of listing such scenarios. Kiss and Strunk (2002, 2006) adverted this by framing SBD as a collocation detection problem of words and following periods. Gillick (2009) classified sentence boundary based on trigram context, only incorporating a type-based method for difficult instances, which is less computationally expensive. Treviso et al. (2017) suggested word embedding as a more effective alternative for n-gram and type-based approaches, bypassing feature engineering. Further on this direction, Knoll et al. (2019) utilized word-level and character-level neural networks to automatically extract morpholexical features.

Syntax-based systems are proposed based on the assumption that POS tags can provide context information where word-based approaches are lacking. The majority of syntax-based systems utilize syntactic features in conjunction with morpholexi-cal ones to enhance the performance. Palmer and Hearst (1994, 1997) are the first to incorporate POS information, but wrongly assume that sentence boundary is a prerequisite for POS tagging. Notably, although these systems consist of a feedfor-ward neural network, it is solely used for classification, not for feature extraction. Thus, they are not included in the neural category in Table 2.6. Mikheev (2000, 2002) solved the POS tagging dilemma, and developed a type-based method. Fol-lowing this setup, Agarwal et al. (2005) used a token-based probabilistic model, whereas Stevenson and Gaizauskas (2000) tackled SBD in ASR with memory-based learning, which can be seen as type-based. Lastly, since speech lacks the textual cues crucial to sentence boundary, prosody-based systems incorporate prosodic fea-tures for SBD in ASR. Gotoh and Renals (2000); Liu et al. (2004, 2005) integrated prosodic features using various machine learning models. However, their systems are not adept in distinguishing SUs and sentences. Liu et al. (2006) confronted this weakness by tackling the imbalanced data distribution.

SBD for standard texts has already been well-studied. However, domain-specific SBD still remains a challenge, as some domains tend to use punctuations differently from general formal texts. Griffis et al. (2016) reviewed several off-the-shelf models on biomedical and clinical corpora. Their error analysis showed that the semicolons, colons, and newlines heavily used in clinical text are extremely error-prone. Addition-ally, periods used in unknown abbreviations, names, and numbers are a significant cause of error as well. Fatima and Mueller (2019) attempted to solve the task of SBD in financial domain via a machine learning approach and an unsupervised rule-based approach.

Table 2.6: Performance of the introduced SBD methods. W stands for Word. S stands for Syntax. P stands for Prosody. NIST is for NIST SU error rate.

	Method	Experiment 1			Experiment 2		
		Dataset	Result	Measure	Dataset	Result	Measure
W	Stamatatos et al. (1999)	MG	99.4%	Acc.			
	Sadvilkar and Neumann (2020)	GRS	97.92%	Acc.			
	Treviso et al. (2017)	T-CTL	79%	F_1	T-MCI	74%	F_1
	Knoll et al. (2019)	MIMIC	98.6%	F_1	FV	99.2%	F_1
	Rudrapal et al. (2015)	BC	99.6%	Acc.	SMC	87.0%	Acc.
	Grefenstette and Tapanainen (1994)	BC	93.78%	Acc.			
	Reynar and Ratnaparkhi (1997)	BC	97.9%	Acc.	WSJ	98.8%	Acc.
	Schmid (2000)	BC	99.70%	Acc.	WSJ	99.62%	Acc.
	Kiss and Strunk (2002)				WSJ	99.05%	F_1
	Kiss and Strunk (2006)	BC	98.89%	F_1	WSJ	98.35%	F_1
	Gillick (2009)	BC	99.64%	Acc.	WSJ	99.75%	Acc.
	Riley (1989)	BC	99.8%	Acc.			
	Wong and Chao (2010)	BC	99.98%	Acc.			
	Wong et al. (2014)	BC	99.81%	Acc.	WSJ	99.80%	Acc.
S	Palmer and Hearst (1994)				WSJ	98.5%	Acc.
	Palmer and Hearst (1997)				WSJ	98.9%	Acc.
	Mikheev (2000)	BC	98.8%	Acc.	WSJ	99.2%	Acc.
	Mikheev (2002)	BC	99.72%	Acc.	WSJ	99.55%	Acc.
	Agarwal et al. (2005)	BC	97.7%	F_1	WSJ	97.8%	F_1
	Stevenson and Gaizauskas (2000)	T-News	76%	F_1			
P	Gotoh and Renals (2000)	T-R	70%	F_1			
	Liu et al. (2004)	BN	48.61%	NIST	CTS	30.66%	NIST
	Liu et al. (2005)	BN	46.28%	NIST	CTS	29.30%	NIST
	Liu et al. (2006)	BN	49.57%	NIST	CTS	32.40%	NIST

Unfortunately, the former fails to produce acceptable results, whereas the performance of the latter is acceptable but still leaves a lot to be desired. Sanchez (2019) examined several off-the-shelf algorithms for SBD on legal texts. Similarly, the results are not ideal, indicating that there is still a lot of room for improvement for these existing approaches. Therefore, a robust SBD system that is capable of handling domain-specific corpora is called for. Another challenging aspect is SBD in speech, especially SU boundary detection. Considering the lack of annotated resources in this domain, one possible direction for future work is semi-supervised learning (SSL). For instance, the machine can learn from annotated training samples that are manually labeled to precisely indicate SU boundaries, infer labels on the unannotated data, and fine-tune itself.

2.4 POS Tagging

POS tagging is a fundamental task in NLU, which aims to label each word in a given text with its POS tag, e.g., noun, verb, adjective, etc. It is an upstream task that preprocesses the input texts to assist more complex NLU applications. Since the POS of a word can affect its meaning and polarity, POS tagging is important for downstream tasks such as WSD (Taghipour and Ng, 2015a; Alva and Hegde, 2016), information retrieval (Mahmood et al., 2017), sentiment analysis (Asghar et al., 2014; Mubarok et al., 2017), metaphor detection (Mao et al., 2021a; Ge et al., 2022) and interpretation (Mao et al., 2018, 2022a). Numerous studies have been done on POS tagging for a variety of languages (Shao et al., 2017; Nguyen et al., 2017; Kanakaraddi and Nandyal, 2018; Darwish et al., 2018). Considering that the fundamental methodologies are similar across different languages, here we focus on introducing POS taggers in English.

English POS tagging is a well-studied problem. Following Church (1989), most early works utilized hand-crafted features, derived from the local contexts, e.g., rule-based learning (Brill, 1995), memory-based learning (Daelemans et al., 1999b), and other statistical approaches, among which the MaxEnt framework (Ratnaparkhi, 1996; Toutanvoa and Manning, 2000; Curran and Clark, 2003) and directed graphical models (Kupiec, 1992; Brants, 2000; McCallum et al., 2000) received the most attention. The performance of feature engineering approaches took a big leap with the novel CRF (Lafferty et al., 2001) and the perceptron algorithm (Collins, 2002), which addresses the parameter estimation problem of MaxEnt models. Attempts at applying bidirectional inference to the task also achieved remarkable improvements, for instance the cyclic dependency network (Toutanova et al., 2003) and guided learning framework (Shen et al., 2007). With the recent surge of interest in deep learning, neural networks such as CNN (LeCun et al., 1998; Collobert et al., 2011), recurrent neural network (RNN) and LSTM (Hochreiter and Schmidhuber, 1997; Graves et al., 2013) have been applied to the field of POS tagging, freeing the task of POS tagging from hand-crafted feature set and yielding even more promising results. Subsequently, many studies extended the previous methods to SSL to further boost the tagging accuracy (Clark et al., 2003; Suzuki and Isozaki, 2008; Zhou et al., 2018). The performance of POS taggers are measured by accuracy, or in its other form, e.g., error rate.

In this chapter, we first review the tagging schemas of POS tagging. Then, we divide existing POS taggers into the following categories: feature engineering approaches, deep learning approaches, and SSL approaches.

2.4.1 Tagging Schemas

In this section, we introduce the two most prevalent tagsets used in the field of POS tagging, namely: Penn Treebank POS tags and Universal POS tags.

Table 2.7: Widely used corpora for POS tagging.

Category	Dataset	Source	Reference
Formal	BC	the Brown Corpus	Francis and Kucera (1979)
	WSJ	Wall Street Journal	Marcus et al. (1993)
	Genia	Genia Biomedical corpus	Kim et al. (2003)
Social media	T-POS	X text corpus	Ritter et al. (2011)
	ARK	X text corpus	Owoputi et al. (2013)
	NPS	X text corpus	Forsyth (2007)

The Penn Treebank POS tagset (Marcus et al., 1993) is the most widely used tagging paradigm. Derived from the Brown corpus, the Penn Treebank has since replaced it as the standard for POS tagging in English. Compared to its predecessor, the Penn Treebank offers a more fine-grained syntactic distinction, containing 36 POS categories in total. For instance, a base form of a verb is always tagged as VB in the Brown corpus, whereas the Penn Treebank differentiates it as VB (imperative or infinitive) or VBP (non-third person singular present tense) depending on the context. However, as the need for multilingual and cross-lingual POS induction arises, the granular differentiation of the Penn Treebank becomes a drawback, since many languages do not follow the grammatical structure of English.

To address the weakness of the Penn Treebank and facilitate future research, Petrov et al. (2011) proposed a tagset with coarser syntactic POS categories based on the observed commonalities across most languages. The Universal tagset contains 12 universal POS categories. In addition, they also created a mapping from 25 Penn Treebank POS categories to their tagset. When used in conjunction with the Penn Treebank, the Universal tagset and mappings are able to account for 22 different languages. The commonly used corpora for POS tagging (in English) are listed in Table 2.7.

2.4.2 Feature Engineering Approach

Feature engineering approaches, as the name suggests, rely on hand-crafted feature templates to determine the POS tag for a word. Some commonly used features are lexical, contextual, and morphological. To avoid confusion, here we define lexical features as the frequencies of the observed POS tags in the training data, contextual features as the surrounding words and their predicted POS tags if available, morphological features as affixes, and existence of numerals, hyphens, capitalization, etc. The taggers utilize such features via rule-based or machine learning algorithms. Notably, graphical models such as HMM and CRF gain most attention because of their proficiency at capturing sequential dependency.

Church (1989) was the first to determine that features from two or less of the nearby tokens are significantly informative to predicting the POS tag of a given token. For example, the word *bear* can be a verb or a noun, but if it is observed to follow a determiner such as *the*, then the tagger can label *bear* in the word sequence *the bear* as noun. The tagger takes a lexicon-based approach – for each word in the corpus, their most frequent POS tags and the corresponding lexical probabilities are stored in a lookup table. The tagger also computes the contextual probability, which is the probability of observing a POS tag given the two POS tags following it. Then, given a sentence, the goal is to search for the tag sequence that optimizes the lexical and contextual probabilities.

Brill (1995) proposed a POS tagger with transformation-based learning. Initially, the tagger assigns each word the most likely POS tag according to the training corpus. When the prediction is incorrect, the tagger tries another transformation from the transformational rule templates, which is derived from the three preceding words, the three following words, and their POS tags. The learning procedure stops when no more transformation can be found to reduces the errors. This method is simple and effective, but unfortunately it requires a long training time. To address this problem, Ngai and Florian (2001) optimized the model by incorporating good and bad counts for each transformational rule, which avert repetition in the learning procedure. This method successfully reduces the training time while maintaining the accuracy.

Daelemans et al. (1999b) introduced a memory-based approach to POS tagging. Examples in training set are represented as a feature vector with an associated tag category. For each test data point, its similarity to all examples in the memory is computed. The category of the most similar instances is chosen as the predicted category for the test data points.

Abney et al. (1999) applied boosting to POS tagging. Boosting algorithm is similar to transformation-based learning discussed above, where the model combines template rules to produce the most accurate classification rule. They proposed two methods to deal with the multi-class problem. First they applied the AdaBoost.MH algorithm (Schapire and Singer, 1999), where each possible class is paired with the given word and assigned a binary label as a derived problem, which is then solved by binary AdaBoost. AdaBoost.MH is memory-consuming, therefore they proposed a novel AdaBoost.MI algorithm, which uses binary AdaBoost to train separate binary classifiers for each class, and combines their output by choosing the class with most confidence. Unlike AdaBoost.MH, here the predictions are selected independently for each class. The boosting approach is shown to be more accurate than transformation-based learning.

Nakagawa et al. (2001) employed SVM to specifically solve the unknown word problem in POS tagging. In this method, binary SVM classifiers are created for each POS tag based on the training corpus, which are then used to predict the POS tags of unknown words. Subsequently, Giménez and Marquez (2004); Giménez and Màrquez (2004) proposed SVMTool, which extends the binary SVM to cover multi-class classification. SVM classifiers are created for each POS tag that contains ambiguous lexical items. When labeling a word, the most confident prediction among all the SVM classifiers is selected.

Ratnaparkhi (1996) first successfully applied MaxEnt framework to POS tagging. A set of binary features is paired with every possible output label. Each feature-label pair is assigned a weight which is trained to maximize the entropy. When predicting the POS tag of a word, these weights are multiplied by the corresponding feature to estimate the probability of a label. The label with the highest probability is selected. Toutanvoa and Manning (2000) improved this method by modifying the set of features. They removed the lexical features derived from the preceding words, adding other hand-crafted features derived from a larger context window, which alleviates the proper noun problem and the ambiguous word form distinction problem. Curran and Clark (2003) proposed a re-implementation of the MaxEnt tagger (Ratnaparkhi, 1996) using the Generalized Iterative Scaling estimation algorithm and model smoothing technique. Their method proves to be much faster in training and predicting than the original MaxEnt tagger.

Directed graphical model is another well-studied statistical approach to POS tagging. Kupiec (1992) proposed a POS tagger based on a HMM. For POS tagging, the observations are that the words in a given input sentence x, and the states are the corresponding POS tags y. The objective is to estimate the joint distribution $P(x, y)$ by computing the transition probability and the emission probability. The former links the current state with the previous one, whereas the latter represents how likely a word is observed under a given label. The transition matrix and emission matrix are estimated from frequency counts using a tagged training dataset. Brants (2000) proposed a Trigram'n'Tags POS tagger based on second order Markov model combined with a smoothing technique. It is a generative model, similar to HMM. The states in the model represent tags. The outputs represent words. The transition probabilities depend on the states, whereas the output probabilities only depend on the most recent class. The trigram probability is computed using a smoothing paradigm that is a linear interpolation of unigrams, bigrams and trigrams. This POS tagger requires less time to train while achieving similar accuracy as a MaxEnt tagger (Ratnaparkhi, 1996).

McCallum et al. (2000) presented Maximum Entropy Markov model (MEMM), a novel Markovian sequence model combining HMM and MaxEnt framework. HMM tagger needs to enumerate all possible observation sequences to define the joint probability $P(x, y)$. Additionally, the inference is intractable. Therefore, it is not cost-effective for HMM to represent multiple interacting features or long-range dependencies of the observations. Conditional probabilistic sequence model such as MEMM is a good alternative to alleviate this difficulty, since it defines the probability of possible label sequence given an observation sequence, i.e., $P(y \mid x)$. It also allows observations to be represented as arbitrary overlapping features. The MEMM uses the MaxEnt framework to fit a set of exponential models, where the probability of a state depends on the observations and the previous state. Although MaxEnt models are effective and widely used for sequence labeling tasks, their parameter estimation methods have limitations. In MEMM, the transitions leaving a given state compete locally instead of globally, which causes a bias towards states with fewer outgoing transitions. Thus, it cannot accurately represent dependencies between consecutive states. Different methods were proposed to solve this label bias problem.

Collins (2002) proposed a perceptron-based (Rosenblatt, 1958) parameter estimation algorithm for global learning. The tagger predicts a sequence of tags from left
to right, where each tag is determined based on the input sequence and the previous
tags, also known as history. The tagging task is then represented as a feature-vector
representation of history-tag pairs. The history-tag pairs are seen as local representations, which sums up to the global representation of the input sequence. A score
function is defined based on the global representation, which is used to estimate the
parameters corresponding to each history. The highest scoring tag sequence for the
given sentence is found via Viterbi algorithm. In the training phase, if the highest
scoring sequence is incorrect, the parameters will be updated using simple addition
and subtraction.

Another more widely-used alternative to MaxEnt is CRF, which is a novel method
as opposed to the sequential classification approaches mentioned above. Lafferty
et al. (2001) is the first to introduce CRF to sequence labeling. CRF is a discriminative linear-chain graphical model. It uses a single exponential model for the joint
probability of the entire state sequence given an observation sequence. Thus, the
weights of different features at different states can be traded off against each other.
The graph model can be expressed as $G = (V, E)$, where V denotes the nodes, i.e.,
states and observations, and E denotes the undirected edges that represent the dependencies between variables. The features are defined over each edge in the graph.
$x = (x_1, x_2, \ldots, x_n)$ denotes the input sequence. $y = (y_1, y_2, \ldots, y_n)$ denotes the
output sequence. Then, each y_i is dependent on x and y_{i-1}, represented by a set of
binary feature functions F. The joint probability is thus computed as

$$P(y \mid x) = \frac{1}{Z(x)} \exp(\lambda \cdot F(y, x)),$$

where Z is the normalization factor, and λ denotes weights for each feature function.

Rush et al. (2012) proposed a POS tagger using Markov random fields (MRFs) to
model global constraints between sentences, which alleviates the accuracy problem
when labeled dataset is limited or out-of-domain. MRF is an undirected graph
model $G = (V, E)$, where V are nodes for every word in the given sentence. An
index set is constructed to include all valid label assignments in the corpus, which
acts as inter-sentence constrains. The objective is to find the best sentence-level label
assignments with regards to V and E that are consistent with elements in the index
set, incorporating corpus-level information to augment sentence-level labeling.

Although undirected graphical models such as CRF avert the label bias problem,
they have a few disadvantages against the sequential classification method. On one
hand, CRF is less efficient to train because they need to perform Viterbi algorithm
over the entire sentence in each iteration. On the other hand, sequential classification
method leaves room to employ a variety of machine learning algorithms as local
classifiers. Therefore, many works focus on improving the performance of sequential
classification methods by enriching the information for the local classifiers. Since
the previous directed graphical taggers take an unidirectional approach, one popular
way of enhancing features is utilizing future tags via bidirectional networks.

Toutanova et al. (2003) was the first to introduce bidirectional inference for POS tagging. They proposed a cyclic dependency network with a series of local conditional log-linear models to exploit information from both directions explicitly. Each node in the network represents a random variable with a corresponding local conditional probability model that considers the source variables from all incoming arcs. The tagger finds the sequence that maximizes the score via Viterbi algorithm, similar to previous MaxEnt and HMM models. The only difference between bidirectional inference and an unidirectional graphical model such as HMM is that, when the Markov window is at the time step i, the score it receives is $P(t_{i-1} \mid t_i, t_{i-2}, w_{i-1})$ instead of $P(t_i \mid t_{i-1}, t_{i-2}, w_i)$, t and w being output tag and input word respectively. Their model, however, suffers from collusion problem where the model lock onto conditionally consistent but jointly unlikely sequences. This is because the local classifiers encounter double-counting problem when using the information from future tags.

In order to avert this problem, Tsuruoka and Tsujii (2005) proposed an alternative bidirectional inference algorithm with an easiest-first strategy. The label sequence of a given sentence is the product of local probabilities. The proposed inference method is to consider all of the possible decomposition structures and choose the optimal structure to predict label sequence. The paper also proposed a more efficient alternative to bidirectional decoding algorithm, which adopts the easiest-first strategy. Instead of enumerating all the possible decompositions, the tagger tags the easiest word at each step, and repeating the procedure until all the words are tagged. To pick the easiest word, the appropriate local MaxEnt classifier is selected according to the availability of the neighboring labels, and used to output the probabilities. The word with the highest probability is deemed as the easiest word for the current step. Their bidirectional inference method is proven to be able to find the highest probability sequence with similar performance but lower computational complexity.

Shen et al. (2007) proposed a novel guided learning framework for bidirectional inference. Unlike the easiest-first strategy which only uses heuristic rule to determine the order of inference, their approach incorporates the selection of inference order into the training of the MaxEnt classifier for individual token labeling, combining the two into a single learning task. Specifically, a sub-sequence of the input sentence is called a span. Each span is associated with one or more hypotheses, which are started and grown via labeling actions. The tagger initializes and maintains a set P of accepted spans and a set Q of candidate spans. It repeatedly selects a candidate span from Q whose action score of its top hypothesis is the highest, moving it to P, until a span covering the whole input sentence is added to P from Q. For training, the tagger uses guided learning to learn the weight of action score. If the top hypothesis of a selected span is compatible with the gold standard, then the candidate span is accepted. Otherwise, similar to perceptron algorithm (Collins, 2002), the weight is updated by rewarding the feature weights of the gold standard action and punishing the feature weights of the action of the top hypothesis. Then, all the existing spans in Q are removed and replaced with new hypotheses for all the possible spans generated based on the context spans in P. This allows the tagger to simultaneously learn the individual classification and the inference order selection.

Ma et al. (2013) proposed an easy-first POS tagger with beam search. The easy-first POS tagger enumerates all possible word-tag pairs, choosing the most confident one to label according to the score function, marking the word as processed. Then, the tagger re-computes the scores of the unprocessed words based on the local context, repeating the selection procedure until all the words are marked as processed. For the easy-first tagging with beam search, a set of labeling action sequences is maintained and grows via beam search. At each step, the sequences in the beam β are expanded in all possible ways, and the top expanded sequences within the beam width are selected into β. The trainable weight vector in a score function is learned through perceptron-based global learning similar to the previously mentioned guided learning framework (Shen et al., 2007), however its performance on the WSJ dataset is not as good as the latter.

2.4.3 Deep Learning Approach

All the above-mentioned methods are dependent on hand-crafted features sets. With the development of deep learning, the application of neural networks to POS tagging makes it possible to avoid feature engineering and further improve the tagging performance. Notably, neural network models are widely-employed to automatically capture character-level patterns, which have to be modeled with morphological features in previous taggers, e.g., suffix, capitalization, presence of numerals, etc. Collobert et al. (2011) proposed a window approach for sequence labeling tasks, which assumes the tag of a word is mainly dependent on its neighboring words. Hence, it considers a fixed size window of words around the current word as local features. Given an input sentence, the tagger passes it through a lookup table layer. The resulting sequence of representations is fed into the convolutional layer, which can be regarded as a feed-forward neural network with L layers, to extract local features. That is, a concatenation of the word vectors in the focal window is inputted to L linear layers to perform affine transformations over their inputs. Finally, a softmax layer computes the probabilities for each labels given the output of the L-th layer. The POS tagger is trained with word-level log-likelihood, also more commonly known as cross-entropy.

Dos Santos and Zadrozny (2014) used deep nets to learn the character-level representations and combined them with the corresponding word representations to perform POS tagging. Prior to their work, the morphological or other intra-word information is given to the tagger via hand-crafted features. To reduce human effort, CharWNN is proposed as an extension of the previously introduced convolutional architecture (Collobert et al., 2011). CharWNN uses the convolutional layer to extract features from the input words and generates their character-level embedding at tagging time. The character-level embeddings are concatenated with word-level embeddings as word representations. Taking the window approach, a fixed window size of word representations in the input sentence are concatenated into a vector, which are then fed into two linear neural network layers to compute the scores.

Wang et al. (2015a) proposed a BLSTM-RNN model for POS tagging. The model first implements a linear layer as a lookup table to produce word embeddings, which are fed into a bidirectional LSTM (BiLSTM) layer and then a softmax layer to output the scores of tags. They also introduced a novel method to train word embedding on unlabeled data, where BLSTM-RNN takes a sentence with some words replaced by randomly chosen words as input, tagging the words in the sentence as correct or incorrect. Thus the lookup table in BLSTM-RNN is trained to minimize the binary classification error. Ma and Hovy (2016) introduced a BiLSTM-CNN-CRF model, which utilizes both word-level and character-level representations automatically, requiring no task-specific resources, feature engineering or data preprocessing. First, a CNN layer (Chiu and Nichols, 2016) is applied to encode character-level information of a word into its character-level representation. A dropout layer is applied before character embeddings are feed into CNN. Then, the model concatenates the character-level representations and word embeddings, feeding them into BiLSTM to model context information of each word. A dropout layer is also applied to the output vectors. Lastly, the output vector of BiLSTM is fed to a CRF layer to decode labels for the whole sentence.

Akbik et al. (2018) proposed Flair, a contextual string embedding for sequence labeling tasks. The Flair embedding is learned using a LSTM-based language model over sequences of characters instead of words. For optimal performance, the Flair embedding is stacked together with pretrained static word embedding GloVe (Pennington et al., 2014) and a task-trained character embedding learned by LSTM. The final embedding is passed onto a standard BiLSTM-CRF architecture to acquire the output label sequence.

Zhao et al. (2019) proposed a deep CNN architecture called Deep Gated Dual Path CNN (GatedDualCNN) for sequence labeling. The model first uses a CharCNN to extract character-level representations, which are then concatenated with the word embeddings and fed into a 1-D convolution (Conv1D) layer followed by rectified linear unit (ReLU) and batch normalization (BN) to get the inital hidden states. Thereafter, in order to stack up more convolutional layers while averting the vanishing gradient problem, the paper incorporated gate blocks, residual connection, and dense connection. The core component of a gate block is the gated linear unite (Dauphin et al., 2017), whose output is processed by a Conv1D layer with ReLU and BN to produce the successive hidden state. To encourage feature re-usage between gate blocks, residual connection is introduced to bypass the non-linear transformation in the gate block. On the other hand, dense connection serves the purpose of new feature exploration in a dense path.To combine these two connections, the model uses a dual path, where hidden state produced by each block is split row-wise, then fed into the residual path and the dense path respectively. The outputs are concatenated as the input of the next block. The final hidden state is then passed on to a CRF layer to decode the best sequence of tags. Yang et al. (2017) introduced transfer learning for deep hierarchical RNN POS tagger to alleviate the out-of-domain problem. The base model uses a character-level GRU (Cho et al., 2014) to obtain character embeddings, which are concatenated with word embeddings and passed on to a word-level GRU and a CRF layer to predict the tag sequence.

They described three different transfer learning architectures for this base model. Transfer model T-A, which shares all the model parameters and feature representations between domains, is used for cross-domain transfer where label mapping is possible. It only performs a label mapping step on top of the base's CRF layer. If the two domains have disparate label sets, then transfer model T-B learns a separate CRF layer for each tasks while sharing parameters in other layers. For cross-lingual transfer, model T-C only shares the parameters and representations in the character-level GRU, keeping two separate word-level GRU and CRF layers for the source task and the target task. The paper experiments with transferring from chunking and Name Entity Recognition (NER) to standard POS tagging on WSJ, and also tests transfer learning from WSJ to Genia Biomedical corpus (Kim et al., 2003) and X corpus T-POS.

Similarly, Meftah and Semmar (2018) presented a transfer-learning-based end-to-end neural model. Their base model uses a CNN layer to extract character embedding, a GRU layer to compute hidden states, and a fully-connected layer and softmax layer to output the scores for tags. Two transfer learning architectures are proposed based on this neural network. For cross-domain transfer, they used a parent network for source data and a child network for target data. The parent network is trained on annotated out-of-domain data, namely WSJ, whose parameters are transferred to the child network. Then, the child network is fine-tuned through training on labeled X datasets. For cross-task transfer, the parent network and the child network shares a set of parameters, jointly optimizing the two tasks, while maintaining separate task-specific parameters that are trained on the corresponding task. The task selected for the parent network is NER.

2.4.4 Semi-Supervised Approach

As researchers have easy access to large amounts of unlabeled texts online, it is a natural progression for POS taggers to utilize unlabeled data. In addition, POS taggers trained from annotated dataset generally achieve high accuracy on in-domain data, whereas their performance drops on out-of-domain data such as social media posts and conversational texts. Consequently, there have been many attempts to use unsupervised or SSL to alleviate the OOV problem that is largely responsible for the fall in tagging accuracy. Unsupervised learning POS taggers mostly take the word clustering approach. For example, Clark (2003) described an unsupervised POS tagger that incorporates morphological information into a distributional clustering algorithm to sort unlabeled text into lexical classes. Biemann (2006) introduced a graph-based word clustering method based on context similarity of high frequency words and log-likelihood of lower frequency words. Owoputi et al. (2013) proposed an unsupervised MEMM tagger for word clustering on X conversations. The SSL approach, on the other hand, utilizes both small amounts of labeled data and relatively large amounts of unlabeled data for training.

SSL is more scalable than unsupervised learning. It can be applied to POS tagging to solve the out-of-domain problem and improve accuracy. Clark et al. (2003) proposed bootstrapping POS tagger using co-training with unlabeled data. The main idea is to co-train a Trigram'n'Tags tagger (Brants, 2000) and a MaxEnt tagger (Curran and Clark, 2003), using the output from one as additional labeled data for the other. The model first learns two separate classifiers for each view of the task via a small amount of labeled seed data. Then, each classifier incrementally labels a subset of unlabeled data for the other to use as the new training dataset. Results show that when using little labeled training data and a much larger amount of unlabeled data, the accuracy of POS tagger improves. However, the performance drops when the amount of labeled training data increases.

Ando and Zhang (2005) proposed a semi-supervised multi-task learning (MTL) framework. The objective is to learn from unsupervised tasks, and transfer to the target supervised task. The framework consists of multiple task classifiers, each made of two alternatively optimized feature functions. One is task-specific and the other is trained to map into a low dimensional subspace common across all tasks. For POS tagging, two unsupervised auxiliary tasks are used. The first task is word prediction, which predicts the word at the current position. The second task is Top-2, which predicts the top-2 choices of the POS tagger trained with labeled data.

Toutanova and Johnson (2007) presented a latent Dirichlet allocation (LDA) model for semi-supervised POS tagging. Compared to previous semi-supervised approaches, the tagger is not given any labeled data, but a dictionary that constrains the possible tags of some words. The LDA-based model is a generative model that uses only observed context features to predict the tags of words. They incorporated a sparse prior on the distribution over tags for each word, and employed a Bayesian approach that maintains a distribution over parameters. The semi-supervised LDA is enabled by explicitly modeling ambiguity classes obtained from the dictionary constrain.

Spoustová et al. (2009) extended the average perceptron algorithm (Collins, 2002) for SSL. Specifically, they used an ensemble of taggers (Brill and Wu, 1998; van Halteren et al., 2001) to pre-tag a large unannotated corpus, which is then combined with labeled dataset and fed into an average perceptron POS tagger. Furthermore, they observed that it is beneficial to feed the tagger with different chunks of unsupervised data at each training iteration. Hence, they explored three types of selection mechanisms, namely, sequential chopping, random selection without replacements, and random selection with replacements. Based on the properties of the average perceptron algorithm, the selected chunk of unannotated data is inputted after the labeled dataset. Results show that the semi-supervised method indeed outperforms supervised ones, and the three mechanisms contribute identically for English POS tagging. Suzuki and Isozaki (2008) introduced a semi-supervised extension of CRF (Lafferty et al., 2001) that combines supervised and unsupervised probability models via a parameter estimation called Maximum Discriminant Functions sum. For a given input sequence x and its corresponding output y, C denotes the set of cliques in an undirected graphical model. y^c is the feature function and the output from the associated clique c.

Then, similar to standard CRF, the conditional probability for y is

$$P(y \mid x) = \frac{1}{Z(x)} \prod_{c \in C} \exp(\lambda \cdot \mathbf{F}^c(y^c, x)),$$

where $Z(x)$ is the normalization factor, $\mathbf{F}^c$ is the set of feature function for the corresponding clique c, and λ is the function weights. In this SSL-based architecture, the feature functions for clique c are the concatenation of $\mathbf{F}^c$ and the log likelihood of all the joint probability models. A set of model parameters θ is introduced to weight the joint models, which is estimated using unlabeled data via Maximum Discriminant Functions.

Subramanya et al. (2010) described another CRF-based algorithm for semi-supervised POS tagger. For graph construction, they used local sequence contexts as graph vertices V, which consists of a set V_l of n-grams that occur in the labeled data and a set V_u in the unlabeled data. The graph is built over types rather than tokens via a symmetric similarity function, thus named similarity graph. This similarity graph is used as a smoothness regularizer to train CRF in a semi-supervised manner. Specifically, given a set of CRF parameters, the tagger first computes marginals over tokens in the unlabeled data, which are then aggregated to marginals over types and used to initialize the graph label distributions. After running label propagation, the posteriors from the graph are used to smooth the state posteriors. Subsequently, the unlabeled data is decoded using Viterbi algorithm to produce a set of automatic annotations, which are combined with the labeled data to retrain the CRF through supervised learning. The procedure is repeated until convergence. They used WSJ as labeled source domain training data, and the QuestionBank (Judge et al., 2006) as test data. The unlabeled data are collected from Internet search queries in similar forms to the QuestionBank. Experiments show that their proposed algorithm indeed outperforms supervised CRF in other domains.

Søgaard (2010) introduced stacked learning as a way to reduce POS tagging to a classification task, thus simplifying semi-supervised training. The stacking approach here is to combine SVMTool (Giménez and Màrquez, 2004) and an unsupervised tagger (Biemann, 2006) into a single-end classifier, where the former predicts the POS tag of a given word and the latter sorts the word into a word cluster. SSL is achieved by tri-training with disagreement. Firstly, three classifiers of the same learning algorithm mentioned above are trained on three bootstrap samples of the labeled dataset, which ensures that the classifiers are diverse. Then, a data point in the unlabeled dataset is labeled for classifier c_1, if and only if the other two agree on its label assignment but c_1 disagrees, which strengthens the weakness of the classifier without skewing the labeled data by easy data points. This labeling process is repeated until the classifiers no longer change. Subsequently, the three classifiers are integrated by majority voting. Zhou et al. (2018) proposed a weakly supervised sequence tagging model with ECOC (Error-Correcting Output Codes) that can learn to predict the POS tag for a given word in a context, given a dictionary of words with their possible tags.

Most approaches prior to this paper are based on disambiguation, such as CRF and HMM, which suffers from the negative effects of false positive tags as the size of possible tags increases. The POS tagger is trained and tested based on constrained ECOC (Dietterich and Bakiri, 1994). First, a unique L-bit vector is assigned to each tag. The set of bit-vectors is regarded as a coding matrix, where each row represents a codeword, e.g., class, and each column specifies a dichotomy over the tag space to learn a binary classifier. In the encoding stage, for each column of the coding matrix, a binary classifier is built based on binary training examples derived from the dictionary of the words with their possible tags. In the decoding stage, the codeword of an unlabeled test instance is generated by concatenating the predictive output of the L binary classifiers. The predicted instance is the class with the closest codeword according to hamming distance or Euclidean distance. Thus, the proposed model which not only treats the set of possible tags as an entirety without resorting to disambiguation procedure, but also needs no manual intervention for feature engineering.

Gui et al. (2017) proposed a Target Preserved Adversarial Neural Network (TPANN) for POS tagging on $\mathbb{X}$ (formerly known as Twitter). WSJ is used as the labeled out-of-domain data, T-POS, ARK, and NPS as labeled in-domain data, and tweets collected via $\mathbb{X}$ API as unlabeled in-domain data. The objective is to learn common features between resource-rich domain and target domain while preserving some domain-specific features of the target domain. TPANN first extracts character embedding features via CNN and concatenates them to word embedding as input. The hidden states are produced by a BiLSTM layer. Subsequently, the hidden states are transferred to a POS tagging classifier and a domain discriminator, which are both standard feed-forward networks with a softmax layer. The POS tagging classifier maps the hidden states to their labels, whereas the domain discriminator maps the same hidden states to the domain labels so as to make the input features domain-invariant. By training this adversarial network, common features can be obtained, but some domain-specific features are weakened. Thus, the paper introduced a domain-specific auto-encoder to reconstruct target domain data. Specifically, at the $\mathbb{X}$ decoder side, the hidden state h_t is computed with $h_t = LSTM([h_0 \oplus z_{t-1}], h_{t-1})$, where h_0 is the last hidden state of the BiLSTM layer, $\oplus$ denotes the concatenation operation, and z_{t-1} is computed from h_{t-1} using a multiple perceptron function. In this way, the auto-encoder counteracts the adversarial network's tendency to erase target domain features by optimizing the common representation to be informative on the target domain data.

2.4.5 Summary

POS tagging is a well-researched problem in the field of NLU. Existing POS tagging models can be divided into feature engineering approaches and deep learning approaches. Nonetheless, to address the growing need for domain adaptable POS taggers, we introduce another category to introduce semi-supervised methods.

Table 2.8: Comparison between different POS tagging methods. FE stands for feature engineering. DL stands for deep learning. SS stands for semi-supervised. RW stands for rare word.

	Method	Lexical	Contextual	Morphological	RW	Character	Global	Neural
FE	Church (1989)	✓	✓					
	Brill (1995)		✓					
	Ngai and Florian (2001)		✓					
	Daelemans et al. (1999b)	✓	✓					
	Abney et al. (1999)	✓	✓	✓				
	Nakagawa et al. (2001)		✓	✓				
	Giménez and Màrquez (2004)		✓	✓				
	Ratnaparkhi (1996)		✓	✓	✓			
	Toutanvoa and Manning (2000)	✓	✓	✓	✓			
	Curran and Clark (2003)		✓	✓		✓		
	McCallum et al. (2000)		✓	✓				
	Kupiec (1992)	✓	✓	✓		✓	✓	
	Brants (2000)		✓	✓	✓		✓	
	Lafferty et al. (2001)		✓	✓			✓	
	Collins (2002)		✓				✓	
	Rush et al. (2012)		✓	✓			✓	
	Toutanova et al. (2003)		✓	✓			✓	
	Tsuruoka and Tsujii (2005)		✓	✓			✓	
	Shen et al. (2007)		✓	✓			✓	
	Ma et al. (2013)		✓	✓			✓	
DL	Collobert et al. (2011)					✓	✓	
	Dos Santos and Zadrozny (2014)					✓	✓	✓
	Wang et al. (2015a)						✓	✓
	Ma and Hovy (2016)					✓	✓	✓
	Akbik et al. (2018)					✓	✓	✓
	Zhao et al. (2019)					✓	✓	✓
	Yang et al. (2017)					✓	✓	✓
	Meftah and Semmar (2018)					✓	✓	✓
SS	Clark et al. (2003)		✓	✓	✓			
	Ando and Zhang (2005)	✓	✓					
	Toutanova and Johnson (2007)		✓	✓				
	Spoustová et al. (2009)		✓	✓			✓	
	Suzuki and Isozaki (2008)		✓	✓			✓	
	Subramanya et al. (2010)		✓	✓			✓	
	Søgaard (2010)	✓	✓					
	Zhou et al. (2018)		✓				✓	✓
	Gui et al. (2017)					✓	✓	✓

Early feature engineering methods predict the POS tag of a word based on its local n-gram context with various machine learning techniques, following Church (1989). Graphical models are commonly used for sequence labeling. Generative graphical models (Kupiec, 1992; Brants, 2000) estimate the joint distribution based on the explicit dependency between the states and the observations, and thus difficult to accommodate context features. Directed discriminative model (McCallum et al., 2000) addresses this weakness by modeling the conditional probability based on the dependencies between adjacent states and the observation sequence, while it suffers from the label bias problem. It is an undirected discriminative graphical model that is able to leverage decisions globally. However, it is disadvantageous in computational efficiency.

Collins (2002) proposed an alternative that enables global learning by modifying the parameter estimation process of directed graphical model. Others alleviate the local limitation by introducing bidirectional sequence classification (Toutanova et al., 2003; Tsuruoka and Tsujii, 2005; Shen et al., 2007). These methods fallen out of favor in recent years, because it is not as adaptable to deep learning as CRF. MRF (Rush et al., 2012) is an undirected skip-chain graphical model that makes global decision not only on sentence-level but also corpus-level. Computationally it is as costly as CRF. For deep learning, early methods take the window approach (Collobert et al., 2011; Dos Santos and Zadrozny, 2014) to extract contextual information. With the advancement of neural networks in NLP, it is replaced by RNN and RNN variants (Wang et al., 2015a), which is better suited for processing text. The most common approach is the BiLSTM-CRF architecture, with improvement achieved through exploration of embeddings (Ma and Hovy, 2016; Akbik et al., 2018). Zhao et al. (2019) proposed a CNN-based model that addresses the vanishing gradient problem of deep CNN, which outperforms BiLSTM-CRF when using the same embeddings. For transfer learning on cross-domain POS tagging, Meftah and Semmar (2018) relied on hard parameter sharing, whereas Yang et al. (2017) utilized hard and soft parameter sharing on different scenarios.

For SSL, Clark et al. (2003); Søgaard (2010); Spoustová et al. (2009) take the ensemble approaches. Clark et al. (2003); Søgaard (2010) co-trained multiple POS taggers, whereas Søgaard (2010) introduced a tri-training strategy to reduce errors. Spoustová et al. (2009) used ensemble to label unannotated data for a specific tagger, which risks error propagation. Ando and Zhang (2005) jointly trained two unsupervised auxiliary tasks with supervised POS tagging. Toutanova and Johnson (2007); Zhou et al. (2018) used dictionary constrains instead of labeled data. The former method was based on clustering, thus difficult to evaluate. The latter used constrained ECOC to solve the false positive tag problem of disambiguation-based methods. Suzuki and Isozaki (2008); Subramanya et al. (2010) extended CRF to SSL by introducing cliques of states and similarity graph respectively. The former focused more on incorporating unlabeled data, whereas the latter targeted cross-domain scenarios. Compared to other cross-domain taggers, TPANN (Gui et al., 2017) is advantageous in utilizing a large amount of unlabeled data through a domain-specific auto-encoder.

Table 2.8 illustrates a summary of the features and properties of all the discussed models. For feature engineering methods, there are mainly three types of features. First, lexical features are the observed POS tags of the focal word in the training corpus and their corresponding frequencies, or less commonly, other frequency counts relating to the word, e.g., capitalized form frequency (Toutanvoa and Manning, 2000). Second, contextual features are the words surrounding the focal word and their predicted POS tags if available. Third, morphological features are the focal word's suffix, prefix, and existence of numerals and special symbols. For some taggers less adept at handling overlapping features, only rare words in the corpus are given such morphological features. Thus we mark out the methods using rare word distinction. For methods using neural networks, on the another hand, such features can be automatically extracted via character-level encoding.

Table 2.9: Performance of the introduced POS methods.

| Method | Experiment | | |
	Dataset	Result	Measure
Feature Engineering			
Brill (1995)	WSJ	96.6%	Acc.
Ngai and Florian (2001)	WSJ	96.61%	Acc.
Daelemans et al. (1999b)	WSJ	96.6%	Acc.
Abney et al. (1999)	WSJ	96.68%	Acc.
Nakagawa et al. (2001)	WSJ	97.1%	Acc.
Giménez and Màrquez (2004)	WSJ	97.05%	Acc.
Ratnaparkhi (1996)	WSJ	96.63%	Acc.
Toutanvoa and Manning (2000)	WSJ	96.86%	Acc.
Curran and Clark (2003)	WSJ	97.27%	Acc.
Kupiec (1992)	WSJ	95.7%	Acc.
Brants (2000)	WSJ	96.7%	Acc.
Collins (2002)	WSJ	97.11%	Acc.
Lafferty et al. (2001)	WSJ	95.73%	Acc.
Rush et al. (2012)	WSJ	91.98%	Acc.
Toutanova et al. (2003)	WSJ	97.24%	Acc.
Tsuruoka and Tsujii (2005)	WSJ	97.24%	Acc
Shen et al. (2007)	WSJ	97.33%	Acc.
Ma et al. (2013)	WSJ	97.28%	Acc.
Deep Learning			
Collobert et al. (2011)	WSJ	97.37%	Acc.
Dos Santos and Zadrozny (2014)	WSJ	97.47%	Acc.
Wang et al. (2015a)	WSJ	97.40%	Acc.
Ma and Hovy (2016)	WSJ	97.55%	Acc.
Akbik et al. (2018)	WSJ	97.85%	Acc.
Zhao et al. (2019)	WSJ	97.59%	Acc.
Yang et al. (2017)	WSJ	97.55%	Acc.
Yang et al. (2017)	Genia	92.62%	Acc.
Yang et al. (2017)	T-POS	83.65%	Acc.
Meftah and Semmar (2018)	T-POS	90.90%	Acc.
Meftah and Semmar (2018)	ARK	92.01%	Acc.
Meftah and Semmar (2018)	NPS	93.20%	Acc.
Semi-supervised			
Clark et al. (2003)	WSJ	409	Perplexity
Ando and Zhang (2005)	BC	93.1%	Acc.
Toutanova and Johnson (2007)	WSJ	93.4%	Acc.
Spoustová et al. (2009)	WSJ	97.44%	Acc.
Suzuki and Isozaki (2008)	WSJ	97.35%	Acc.
Subramanya et al. (2010)	QuestionBank	86.8%	Acc.
Søgaard (2010)	WSJ	97.27%	Acc.
Zhou et al. (2018)	WSJ	92.91%	Acc.
Gui et al. (2017)	T-POS	90.92%	Acc.
Gui et al. (2017)	ARK	92.80%	Acc.
Gui et al. (2017)	NPS	94.10%	Acc.

Additionally, some methods determine the optimal output sequence by making the best local decision, whereas others are able to leverage predictions at different positions (see the global column in Table 2.8). Table 2.9 shows a comparison of the accuracy of the introduced POS taggers. Currently, most POS tagger yield stable and reliable performance on the WSJ dataset from the Penn Treebank, with accuracy slightly above 97%. Akbik et al. (2018) obtained the state-of-the-art result of 97.85%. On X text corpus, Gui et al. (2017) consistently achieved the best performance. Manning (2011) performed an error analysis and categorized the common errors from POS taggers into 7 classes. Results indicate that, a proportion of errors due to lexicon gap and unknown words can be addressed by semi-supervised methods. The errors caused by inconsistent or faulty gold standards can be fixed by correcting the WSJ dataset, for which he proposed a solution using deterministic rules. The other errors are mostly high frequency words that have odd properties, which is an inherently difficult problem most POS taggers are attempting to solve.

Another challenge for English POS tagging is informal or out-of-domain texts, which is the focus of the semi-supervised approach section. With the boom of deep learning in NLP, novel neural networks and pretrained language models (PLMs) can be expended to unsupervised or SSL to further ameliorate the efficiency and tagging accuracy for user generated texts. Noticeably, as a language preprocessing technique, POS tagging is ultimately intended for improving the performance of complex downstream tasks. Therefore, when pushing for higher accuracy for POS taggers, it is better to prioritize how we can eliminate the errors that have a strong influence on the downstream tasks. In the future, we hope to see POS taggers that are not only accurate and reliable, but also serve their purpose of enhancing more complicated NLU tasks.

2.5 Text Chunking

Text chunking is an NLU task that splits sentences into non-overlapping segments, such as Noun Phrase (NP) and Verb Phrase (VP). Chunking, also called shallow phrasing, can be applied as a preprocessing step before complete parsing. It helps the machine to learn the sentence structure and relation between words, e.g., recognizing names and syntactic components. Thus, it provides a useful foundation for downstream NLU tasks that require a general understanding of sentence components, e.g., NER (Collobert et al., 2011; Yang et al., 2017), text summarization (Gupta et al., 2016a), and sentiment analysis (Syed et al., 2014).

In this section, we first review different tagging schemas in text chunking, then investigating previous methods in tow categories: feature engineering approaches and deep learning approaches.

2.5.1 Tagging Schemas

2.5.1.1 IOB tagging schema

Ramshaw and Marcus (1999) proposed noun phrase chunking as a machine learning problem. Prior to their work, chunk structure was mostly encoded with brackets between words, which is often met with the problem of unbalanced brackets. To solve this problem, they introduced the IOB1 (also known as IOB) tagging schema to represent chunk structures, where "B" stands for the beginning of a chunk that immediately follows another chunk, "I" means the word is inside a chunk, and "O" stands for outside of any chunk. Thus, chunking is considered as a sequence labeling problem. The dataset they derived from the Penn Treebank is later referred to as baseNP and is used in some very early works. Thereafter, Sang and Buchholz (2000) introduced the widely-used dataset CoNLL-2000, extending the task of chunking from noun phrase to other types of chunks, such as verb phrases, prepositional phrases and adverb phrases. The dataset modifies the above-mentioned IOB1 encoding schema to IOB2 (also known as BIO), where "B" is simply used in the beginning of every chunk. It also contains the corresponding POS tag of every token assigned by a standard POS tagger from Brill and Wu (1998). The CoNLL-2000 dataset, along with F_1 score as metric, has become a standard for evaluating chunkers. Although the CoNLL-2000 comes with IOB2 encoding, it is not difficult to convert it into other schemas. A variety of encoding schemas are explored to study their effects on chunking performance.

2.5.1.2 IOE tagging schema

An alternative to IOB is IOE (Sang and Veenstra, 1999), where "E" represents the final word of a chunk immediately preceding another chunk in IOE1, or the final word on every chunk in IOE2. Sang and Veenstra (1999) split the baseNP dataset into two group and investigated the effectiveness of IOB and IOE. Results are inconclusive to determine which one can best improve the performance. Considering that IOB and IOE follow the same core concept to segment chunks, it is reasonable that they do not vary much in performance.

2.5.1.3 BIOES tagging schema

Another popular encoding schema is BIOES (BILOU) (Ratinov and Roth, 2009), where "E" stands for the ending token of a chunk, and "S" denotes a single element. Research shows that chunkers using BIOES outperform those using IOB significantly (Yang et al., 2018; Ratinov and Roth, 2009; Dai et al., 2015). This is likely because BIOES is more fine-grain then IOB, allowing the machine to learn a more expressive model with only a small amount of extra parameters.

2.5.2 Feature Engineering Approach

The feature engineering approaches rely on hand-crafted feature sets from the surrounding contexts, e.g., local lexical information, POS tags, and chunk tags of previous words. In this chapter, we further categorize the feature engineering approach into two groups: local classification approaches and global classification approaches. The local classification approaches view the chunking task as a sequence of classification problems, one for each of the word in the sequence, where the predicted tag at each position may depend on the features of the whole input sentence and the predicted tags of previous words. The global classification approaches, on the other hand, are able to trade off decisions at different positions to obtain a globally optimal label sequence. Chunkers in this category are mostly graphical models, such as HMM (Freitag and McCallum, 2000) and CRF.

2.5.2.1 Local Classification Approach

The local classification approaches predict the label of one word in a sequence at a time, utilizing different lexical and syntactic information as features, e.g., the word itself, its POS tag, its surrounding words and their POS tags, to make the best local decision. Ramshaw and Marcus (1999) proposed a chunker with transformation-based learning. The chunk structure is represented by IOB1 tag schema in non-recursive base NP distinction, and by BN, N, BV, V, P in noun/verb phrase separation. First, a baseline heuristic is learned using POS tags. It is then used to produce initial hypotheses for each site in training corpus. When the baseline prediction is incorrect, the rule templates generate candidate rules for different locations based on the identities of words within a neighborhood, their POS tags and the current chunk tags. The candidate rules are tested against the rest of the corpus and sorted based on their positive scores. This will eventually create an ordered sequence of rules that predict the features of words. In order to speed up the learning process, an index is constructed to link each candidate rule to its static locations in the corpus, and the rules are disabled and re-enabled based on their scores and changes. (Ngai and Florian, 2001) also applied transformation-based learning to chunking, whose method is previously introduced in the POS tagging section.

Daelemans et al. (1999a) proposed a memory-based learning method where POS tagging, chunking, and identification of syntactic relations are formulated as memory-based modules. The proposed model is a lazy learner, keeping all training data available for extrapolation. Thus, it is more accurate than greedy learners for NLU tasks. Memory-based learning constructs a classifier for a task by storing a set of examples. Each example associates a feature vector with one of a finite number of classes. The classifier extrapolates the classes of feature vectors from those of the most similar feature vectors in the memory. The syntactic analysis process is split into a number of classification tasks where input vectors represent a focused item and a dynamically selected surrounding context. Outputs of some memory-based modules are used as input by other memory-based modules.

Based on their work, Sang (2000) proposed a system-internal combination of memory-based learning classifiers to find base chunks. The main idea is to generate five different chunking models by using different chunking representations, namely IOB1, IOB2, IOE1, IOE2, and the bracket structures. Each classifier uses the memory-based learning algorithm IB1-IG (Daelemans et al., 1999a) for determining the most probable tag for each word. The training data is stored and a new item is classified by the most frequent classification among training items closest to the new item. Outputs of the five classifiers are combined using either voting methods, classifier stacking method or combination method. The paper also explored three processing strategies: single-pass, double-pass, and n-pass, where the data are processed once, twice or many passes to identify the correct chunk tags. According to the experiment, the best combination is the majority voting method and double-pass.

Similarly, Van Halteren (2000) proposed a chunking method using Weighted Probability Distribution Voting (WPDV) model (van Halteren, 2000). The proposed model has a three-stage architecture. In the first stage, five different base chunkers are trained, including a stacked TiMBL model (Daelemans et al., 2003), a WPDV model, a reverse WPDV model, a R&M WPDV model, and a LOB WPDV model. Subsequently, another WPDV model is used to combine the outputs of the five base chunkers. Lastly, corrective measures are applied to the systematic errors, which are mostly due to the determination of the start position of NPs. Koeling (2000) introduced a MaxEnt model for chunking, which is an exponential model that chooses the probability distribution with the highest entropy. MaxEnt operates on the intuition that if there is no evidence to favor one solution over the other, then both solutions are equally likely. Therefore, the probability distribution with the highest entropy should be chosen. Given a word, the probability of a candidate label is dependent on its history, containing the word itself and its tag context. The label with the highest probability is selected.

Kudo and Matsumoto (2000) employed SVM for text chunking. The paper applied a weighted voting method to 8 pairwise SVM classifiers, using contextual information as features for training. Training data is split into 4 different types of representations, namely IOB1, IOB2, IOE1, IOE2. For each representation, forward parsing and backward parsing are employed, thus creating a weight voting of 8 SVM systems. Four methods are proposed to determine the weight given to each system: uniform weights, cross validation, VC-bound, and Leave-One-Out. Zhang et al. (2001) applied a Winnow algorithm (Littlestone, 1988; Grove and Roth, 2001) to text chunking. Winnow is suitable for problems with a high dimensional feature space. The Winnow multiplicative update algorithm (Littlestone, 1988) updates the trainable parameter repeatedly when the algorithm cannot correctly classify an example. However, it may not converge when the data is not linearly separable. The paper proposes regularized Winnow, which converts the original Winnow into a numerical optimization problem that converges in both linear separable case and linear non-separable case, thus making it suitable for NLU tasks such as text chunking. Besides contextual information, the model also uses English Slot Grammar (McCord, 1990) as addition features. Dynamic programming (Punyakanok and Roth, 2000) is used to determine the best sequence of chunk tags.

Based on their work, Lee and Wu (2007) proposed a mask method based on SVM classifier that does not depend on external knowledge and multiple learners. The purpose of the mask method is to collect unknown word examples from original training data, so that the chunker can handle unknown words in testing data. First, a tokenizer and a POS-tagger is applied to produce POS tag for each token. Then, the feature selection component encodes the important features of context words. One-Against-All SVM classification method is employed to classify the IOB1 tag of the words. After lexicon-related features are derived, the training set is divided into 2 or more parts. New training examples can be generated by mapping the new feature dictionary set from each training part. This method emulates training examples that do not contain lexical information, which helps the model considers the effect of unknown words, and adjusts the weights on lexical related features.

Johnson and Zhang (2005) proposed a semi-supervised method for text chunking, which is based on the idea that good classifiers should have similar predictive structure, and thus learns good structure from an auxiliary classification problem can help improve performance on the target problem. The paper presents a linear prediction model for structural learning. Supposedly, there is a low-dimensional predictive structure shared by multiple prediction problems, which can be discovered through joint empirical risk minimization (ERM). The goal of this model is to discover the common low-dimensional predictive structure parameterized by the projection matrix in the predictor, i.e., to find the optimal projection matrix that minimizes the empirical risk summed over all the problems. This optimization problem is solved by alternating structure optimization (ASO) (Ando and Zhang, 2005). For SSL, the auxiliary prediction problems are generated automatically from unlabeled data. A classifier is trained with a feature map and labeled data, whose behavior is then predicted on the unlabeled data using another distinct feature map. After the training data for each auxiliary problem is created, the optimal projection matrix is computed from the training data via ASO, and the empirical risk on the labeled data is minimized.

2.5.2.2 Global Classification Approach

The objective of chunkers using the local classification approaches is to minimize functions related to labeling errors, where they make the best local decisions. As a result, they cannot trade off decisions at different positions against each other to obtain a globally optimal labeling. To solve this problem, many works attempt to establish global classifiers for text chunking. The earliest global classification approach relies on generative graphical models, such as HMM (Kupiec, 1992). Zhou and Su (2000) proposed an error-driven HMM-based text chunking tagger with context-dependent lexicon. The input token sequence is the product of the word sequence and the POS tag sequence. The chunking structure is represented by structural tags, which consists of structural relation, phrase category, and POS tags. The model uses the Viterbi algorithm to find a stochastic optimal tag sequence for the given token sequence.

The baseline system only uses the current POS as lexical entry to determine the current structural chunk tag. Then, the paper attempts to add more contextual information by adding lexical entries into the lexicon, such as the current and the previous words, and their POS tags. Adding more contextual information significantly improves the accuracy, however, it is difficult to merge all the above context-dependent lexicons in a single lexicon due to memory limitation. Thus, an error-driven learning approach is adopted to examine the effectiveness of lexical entries and reduce the size of lexicon.

Molina and Pla (2002) proposed an HMM-based tagging method where the model finds the sequence of states of maximum probability given the input sequence. This method can be used in many different shallow parsing tasks including text chunking, given the appropriate input information. When implemented for chunking, the model considers words and POS tags as the input. In addition, the paper suggests that the output tag set could be too generic to produce accurate models. Thus, for a chunking task, the model can be enriched by adding POS information and certain selected words into the chunk tags. These are achieved by applying a specialization function on the original training set. From this new training set, the Specialized HMM can be learned by maximum likelihood. The tagging process is carried out using the Viterbi algorithm.

Although the HMM-based algorithms are well-understood, they require strict conditional independence assumptions to work effectively, which makes it difficult to represent non-independent features, such as surrounding words. Attempts have been made to enable chunkers to handle more statistically correlated features of input tokens while obtaining global optimal labeling, e.g., the perceptron algorithm (Collins, 2002) and bidirectional inference algorithm (Tsuruoka and Tsujii, 2005), both previously discussed in the POS tagging section.

Another popular algorithm to address this problem is based on CRF, which is the most widely-used alternative to generative graphical models. As mentioned in the POS tagging section, first proposed by Lafferty et al. (2001), CRF is an undirected linear-chain graphical model. It uses a single exponential model for the joint probability of the entire tag sequence given an observation sequence. CRF can not only take in many statistically correlated features from the input data and train them discriminatively, but also trade off decisions at different sequence positions to obtain a globally optimal labeling. Hence, it averts the limitations while maintaining the advantages of the local classification approach and the HMM-based approach.

Sha and Pereira (2003) introduced the application of CRF to text chunking, proposing a novel CRF training algorithm with better convergence properties. For chunking task, the CRF labels are pairs of consecutive chunk tags, which establishes a second-order Markov dependency between chunk tags. The local feature is based on a predicate on the input sequence and current position, and a predicate on pairs of label. Instead of using iterative scaling as training algorithm (Lafferty et al., 2001), the paper experiments with two training methods to maximize the log-likelihood of the training set: preconditioned conjugate gradient (Shewchuk et al., 1994), and limited-memory quasi-Newton (Nocedal and Wright, 2006). Both utilize approximate second-order information to achieve high convergence speed.

Following their work, many alternative CRF or other second-order random fields algorithms are proposed for chunkers to model more complex dependencies. For instance, as described in the POS tagging section, Suzuki and Isozaki (2008) employed a SSL CRF, which is also tested to be effective at text chunking. Sutton et al. (2007) proposed dynamic CRF (DCRF), which is a generalization of the original CRF that repeats structure and parameters over a sequence of state vectors. Compared to conventional CRF, DCRF is able to represent distributed hidden states and complex interactions among states, such as factorial, second-order and hierarchical structure. To achieve this, DCRF introduces clique index c, which represents any state in the unrolled graph through a time step offset and its index in the state sequence y. Then, the set of variables in the unrolled version of clique index c at time step t can be denoted as $y_{t,c}$. Let C be a set of clique indices. Similar to standard CRF, given an input sequence x, the conditional probability $P(y \mid x)$ is computed as

$$P(y \mid x) = \frac{1}{Z(x)} \prod_t \prod_{c \in C} \exp(\lambda \cdot \mathbf{F}(y_{t,c}, x, t)),$$

where Z is the normalization factor, F is a set of feature function, and λ denotes weights for each feature function. The generalization of DCRF allows for complicated structure, such as the proposed factorial CRF (FCRF), which incorporates edges between co-temporal labels to explicitly model dependencies between different chains. Assume an FCRF has L chains, where $y_{l,t}$ is the variable in chain l at time t. The distribution over output sequence is computed within-chain and between-chain:

$$P(y \mid x) = \frac{1}{Z(x)} \left(\prod_{t=1}^{T-1} \prod_{l=1}^{L} \exp(\lambda \cdot \mathbf{F}(y_{l,t}, y_{l,t+1}, x, t)) \right)$$
$$\left(\prod_{t=1}^{T} \prod_{l=1}^{L-1} \exp(\lambda \cdot \mathbf{F}(y_{l,t}, y_{l+1,t}, x, t)) \right).$$

This factorized structure can be used to jointly train several sequence labeling tasks, such as POS tagging and text chunking, with shared information. Based on this, the paper further describes a marginal DCRF for joint learning between POS tagging and chunking, which is inspired by the notion that the main purpose of POS tagging is to help the prediction of chunking. Therefore, training by maximizing the joint likelihood is not ideal, since the model might trade off accuracy among the chunk tag to obtain accuracy among the POS tag.

The proposed marginal training encourages the model to prioritize learning the main task whilst retaining useful information from the other task. That is, in the training set, the observations of POS tag sequence are ignored, thus the model is able to focus on the conditional probability over y. Experiments show that joint training using marginal FCRF improves chunking accuracy slightly in comparison to cascaded training where the two tasks are learnt in sequence.

Sun et al. (2008) proposed a chunker based on Latent-Dynamic CRFs (LD-CRF) (Morency et al., 2007) and a coinciding inference algorithm named Best Lable Path (BLP), which can learn latent-dynamics explicitly. More specifically, latent-dynamics is the underlying structure of syntactic contexts, which is often too complex for chunk labels to encapsulate. For example, it is difficult for the BIO tagging schema to differentiate the latent-structures between the sequences "He is her -" and "He gave her-", where the former is likely to be followed by a I-NP tag and the latter a B-NP tag. LDCRF is able to mitigate this problem by explicitly modeling hidden state variables. Given an input sentence $x = (x_1, \ldots, x_n)$, the task is to learn the mapping between x and a sequence of labels $y = (y_1, \ldots, y_n)$. The model also assumes a vector of hidden state variables $h = (h_1, h_2, \ldots, h_n)$ for the sequence. To make training and inference more efficient, the model is restricted to have a disjointed set of hidden states $\mathbf{H}_{y_i}$ associated with each class label y_i. The conditional probability can be written as

$$P(y \mid x, \Theta) = \sum_{h \in \mathbf{H}_{y_1} \times \cdots \times \mathbf{H}_{y_n}} P(h \mid x, \Theta),$$

where $P(h \mid x, \Theta)$ is calculated as conventional CRF:

$$P(h \mid x) = \frac{1}{Z(x)} \exp(\Theta \cdot \mathbf{F}(h, x)),$$

where Θ denotes the model parameters. Due to the inclusion of hidden states, the best label sequence $\hat{y}$ cannot be found directly via Viterbi algorithm. Therefore, NLU inference is introduced to search for $\hat{y}$, where the top-k hidden paths over hidden states are chosen using $A*$ search (Hart et al., 1968), and the corresponding probabilities of hidden paths are produced. Subsequently, the estimated probabilities of various label paths can be computed as the sum of the probabilities of hidden paths.

Muis and Lu (2016) proposed a weak semi-Markov CRF model for NP chunking on user generated text, namely NUS SMS Corpus (Chen and Kan, 2012). Semi-Markov CRF (semi-CRF) (Sarawagi and Cohen, 2004) can be defined as conventional CRF with additional edges from one node to all the nodes up to L words away, representing a segment within which the words will be labeled with a single label. Based on Semi-CRF, the paper introduces a weaker variant that restricts each node to connect either to only the nodes of the same label up to L words away, or to all the nodes only in the next word. With this restriction, weak semi-CRF can decide the next segment length and type separately, whereas Semi-CRF is encouraged to make the decisions simultaneously. This restriction is achieved by splitting every original node into a Begin node and an End node. The End node connects only to the every next Begin nodes of any label. On the other hand, the Begin node connects only to the End nodes with the same label up to L next words. Thus, given the input sequence $x = (x_1, \ldots, x_n)$, the conditional probability of label sequence $y = (y_1, \ldots, y_n)$ can be defined as

$$P(y \mid x) = \frac{1}{Z(x)} \exp \left(\sum_{i=1}^{n} (\lambda f(y_{i-1}, y_i, x, i) + \sum_{k=1}^{L} \lambda g(y_i, x, i - k, i)) \right),$$

where $Z(x)$ is the normalization factor, λ is function weights, f is the feature function as in standard CRF, and $g(y_i, x, i - k, i)$ represents the feature vector on the edge between the Begin node with the current state y_i at position $i - k$ and the End node with state y_i at position i.

Motivated by the idea that the choice of encoding schema can affect the performance of sequence labeling models, Lin et al. (2020) proposed two Latent Variable CRFs (LVCRF) that can automatically choose the best encoding schema for a given input sentence. They tested the models on the most popular schema, IOB2 and BIOES. The latent variable CRF is achieved by inserting a set of variables between the input and the output based on the chain rule of probability. This allows the model to capture the latent structure between observations and labels. The first model LVCRF-I labels the input sentence by labeling it with two encoding schemas simultaneously and optimizing the parameters to maximize the probability of both schemas. In the training phase, the model is not informed on which encoding schema is better, but trained to maximize the probability of both schemas. The best encoding schema is determined implicitly during the decoding stage when the best labeling path is found via Viterbi algorithm. The second model LVCRF-II, on the other hand, chooses the encoding schema on a word-level, combining the labeling path in two encoding schemas. Therefore, LVCRF-II allows the transformation between the two encoding schemas. Experiments show that LVCRF-II indeed yields higher accuracy. It is also proven that the LVCRF framework can be used in neural network based chunkers such as LSTM to achieve even better result.

2.5.3 Deep Learning Approach

Many deep neural networks, e.g., CNN, RNN and LSTM, can be applied to text chunking. Unlike the previous methods, deep learning approaches are able to automatically extract features from the input texts, making it possible to avoid handcrafted feature templates.

As previously mentioned in the POS tagging section, Collobert et al. (2011) described a window approach for sequence labeling problems. Aside from the proposed model, they also suggested a novel training algorithm for tasks such as text chunking, where tags are organized in chunks and some tags cannot follow other tags. Named sentence-level log-likelihood, it is proposed as an alternative to softmax and CRF, thus allowing the consideration of scores over all possible tag paths for a given sentence. To achieve this, a transition score $A_{i,j}$ is introduced, which is a trainable variable for jumping from tag i to tag j in successive words. The score of a tag path is computed by summing the transition scores and the scores outputted by the neural network. The score over all possible tag paths is normalized via softmax and interpreted as a conditional tag path probability.

The advantage of this method over CRF is that it uses a non-linear neural network instead of a linear model to maximize the likelihood, which encourages the model to learn useful features according to the task of interest. Additionally, they jointly train POS tagging, text chunking, and NER using the proposed window approach, where all models share the same lexicon lookup table and the parameters of the first linear layer, and the training objective is to minimize the loss averaged across all tasks. Similarly, Yang et al. (2017) also utilized the correlation between text chunking, POS tagging and NER. They showed that transfer learning from the latter two tasks to chunking yields competitive performance. Their method was introduced in detail in the POS tagging section. Huang et al. (2015) proposed a BiLSTM model with a CRF layer (BiLSTM-CRF). The model efficiently utilizes past and future input features via a BiLSTM layer and sentence level tag information via a CRF layer.

Yang et al. (2016a) described a deep hierarchical RNN, which can encode both character level and word level sequential information. Many previous works (Dos Santos and Zadrozny, 2014; Santos and Guimaraes, 2015; Kuru et al., 2016) show that character level features help alleviate the out of vocabulary (OOV) problem in sequence labeling tasks, most of which rely on convolutional layer to extract such features. The proposed model, however, employs bidirectional GRU (Cho et al., 2014) to achieve this. The model stack multiple recurrent layers together to build a deep GRU. Such deep GRU is used on both character level and word level, together forming a hierarchical GRU. The word representations produced by the hierarchical GRU are fed into another deep bidirectional GRU to extract the context information in the word sequence. The resulting sequence of hidden states is used as input features for the next layer, where a CRF models the dependencies between tags in the sequence and predicts a sequence of tags.

Zhai et al. (2017) proposed a BiLSTM-based sequence chunking model where each chunk is treated as a complete unit for labeling. They also explored the idea of using pointer networks (Vinyals et al., 2015a) instead of IOB labels. The paper divides sequence labeling into two subtasks: segmentation and labeling. The former is to identify scope of the chunks explicitly, whereas the latter is to label each chunk as a single unit based on the segmentation results. The model employs an encoder-decoder framework where the decoder is modified to take chunks as inputs. The BiLSTM encoder is used to create a sentence representation as well as segment the input sequence into chunks. It uses a CNNMax layer to extract important information from words in the chunk, and utilizes context word embeddings of the chunks to capture context information. The decoder is a LSTM that takes all the information above to generate hidden states to label the segmented chunks. To further improve the accuracy, the model uses pointer network instead of IOB2 tags to identify chunks. It identifies a chunk, labels it, and repeats the process until all words are processed. At the beginning of a possible chunk, the pointer network determines which word is the ending point. After a chunk is identified and labeled, it serves as the input of the decoder in the next time step. With this setup, the model is able to utilize chunk-level features for segmentation. Experiments show that pointer network yields better performance than the IOB2 encoding schema.

Rei (2017) proposed a semi-supervised MTL framework for sequence labeling tasks. They used BiLSTM-CRF as baseline model, integrating unsupervised language modeling as the supplementary task, whose objective is to predict the next word in the sequence based on only the hidden state produced by the forward LSTM, and the previous word based on the hidden state from the backward LSTM. This additional language model objective encourages the model to learn more general patterns of semantic and syntactic composition.

Sun et al. (2020) proposed a hybrid neural CRF for multi-view sequence labeling, termed MVCRF, and adopted diverse neural networks for feature extraction of multiple views. The model not only considers the correlation between neighborhood labels and jointly decoding the best label sequence, but also combines multi-view learning by utilizing consensus and complementary principles. The model takes the word view and the POS view of the sequential data as input. Then, it uses BiLSTM to extract features from the word view x_1 with pretrained SENNA word embedding, and uses a linear network for the POS view x_2. The model then regularizes the log-likelihood by the consistency among distinct views to minimize the Euclidean distance between the features across two views in a joint representation space. Then, features from both views are fed into a MVCRF, where the output y is determined by the conditional probability $P(y \mid x1, x2)$.

Liu et al. (2020b) proposed a semi-CRF chunker based on stacked BiLSTM. Firstly, the word embedding and character encoding vector are concatenated and fed into the lower BiLSTM to obtain the sub-word representation. Subsequently, the parameters of the lower BiLSTM are extracted and loaded into the upper BiLSTM, which takes in the subword representation, and outputs the word representation $w = (w_1, w_2, \ldots, w_n)$. Based on existing semi-CRF (Sarawagi and Cohen, 2004), the paper describes a new semi-CRF (NSCRF). Let $c = (c_1, c_2, \ldots, c_r)$ be a label path of the given sentence, where $c_i = (b_i, e_i, l_i)$ denotes the i-th chunk, b_i represents the beginning word index, e_i the end word index, and l_i the chunk-level tag. Then, the conditional probability of NSCRF can be expressed as

$$p(c \mid w) = \frac{1}{Z(c)} \prod_{i=1}^{r} \exp \boldsymbol{F}(l_{i-1}, l_i, w, b_i, e_i),$$

where $Z(c)$ is the normalization factor, $\boldsymbol{F}$ is the feature function.

Wei et al. (2021) proposed a Position-aware Self Attention (PSA) mechanism to address the RNN-based method's limitation of capturing discrete relations in a sentence. The core of this model is the BiLSTM-based context encoder, which employs self-attention to encode relative positional information. Specifically, the context encoder contains two self-attentional context fusion layers: one for assigning weights to the initial inputs, the other for re-weighting the output of the BiLSTM. The self-attentional context fusion layer uses a PSA mechanism to learn context-aware representations. Alignment scores between tokens are computed via a feed-forward neural network, incorporating three positional biases: self-disabled mask bias, distance-aware Gaussian bias, and token-specific position bias. This enables the model to learn non-continuous token relationships.

Table 2.10: Comparison between different text chunking methods. FE stands for feature engineering. LC stands for lexical context. SC stands for syntactic context. PC stands for preceding chunk tag prediction. CL stands for chunk-level. Morph is short for morphological. Joint stands for joint learning with other tasks.

	Method	POS tag	LC	SC	PC	2nd order	CL	Morph	Character	Joint
	Ramshaw and Marcus (1999)	✓	✓	✓						
	Ngai and Florian (2001)	✓	✓	✓	✓	✓				
	Daelemans et al. (1999a)	✓	✓	✓	✓					
	Sang (2000)	✓	✓	✓	✓					
FE - local	Van Halteren (2000)	✓	✓	✓	✓					
	Koeling (2000)	✓	✓	✓	✓	✓				
	Kudo and Matsumoto (2000)	✓	✓	✓	✓					
	Zhang et al. (2001)	✓	✓	✓	✓	✓				
	Lee and Wu (2007)	✓	✓	✓	✓			✓		
	Johnson and Zhang (2005)	✓	✓	✓	✓			✓		
	Zhou and Su (2000)	✓	✓	✓						
	Molina and Pla (2002)	✓	✓	✓						
	Sha and Pereira (2003)	✓	✓	✓	✓	✓				
	McDonald et al. (2005)	✓	✓		✓					
FE - global	Sutton et al. (2007)	✓	✓				✓			✓
	Sun et al. (2008)	✓	✓	✓		✓				
	Muis and Lu (2016)	✓	✓		✓					
	Lin et al. (2020)	✓	✓	✓		✓				
	Collobert et al. (2011)									✓
	Yang et al. (2017)	✓								✓
	Huang et al. (2015)	✓	✓	✓			✓			
	Yang et al. (2016a)								✓	✓
Deep Learning	Zhai et al. (2017)							✓		
	Rei (2017)	✓								✓
	Sun et al. (2020)	✓								✓
	Liu et al. (2020b)	✓							✓	
	Wei et al. (2021)								✓	

2.5.4 Summary

In conclusion, existing text chunking algorithms can be categorized as feature engineering and deep learning. Notably, feature engineering methods can be further divided into local and global approaches. The latter is able to leverage global information whereas the former can only make the locally optimal decision. The local approach includes machine learning methods such as transformation-based learning (Ramshaw and Marcus, 1999; Ngai and Florian, 2001), memory-based learning (Daelemans et al., 1999a; Sang and Buchholz, 2000; Sang, 2000; Van Halteren, 2000), MaxEnt (Koeling, 2000), SVM (Kudo and Matsumoto, 2000; Lee and Wu, 2007), and window algorithm (Zhang et al., 2001). For global approach, early works utilize HMM. A significant limitation of HMM is that its strict independence assumptions makes it inefficient to incorporate contextual information. To mitigate this problem, Zhou and Su (2000) took an error-driven approach, whereas Molina and Pla (2002) transformed the training set through specialization functions. CRF (Lafferty et al., 2001; Sha and Pereira, 2003), as previously mentioned in the POS tagging section, is more flexible than HMM in terms of feature design.

However, its linear-chain structure limits the ability to capture dependencies between non-adjacent states. Some CRF variants are proposed to address this limitation. DCRF (Sutton et al., 2007) is a skip-chain graphical model that enables complex interactions between labels. Similarly, semi-Markov CRF (Muis and Lu, 2016) is a skip-chain CRF targeting informal text, which accommodates restricted dependencies between nodes. LDCRF (Sun et al., 2008) captures latent dynamics by establishing a set of hidden states for every class label. LVCRF (Lin et al., 2020) is able to automatically choose the optimal encoding schema by inserting variables between states and observations. Compared to feature engineering approaches, deep learning approaches are able to automatically extract features, and better facilitate joint training with other sequence labeling tasks.

The most commonly used architecture is RNN variations combined with CRF. Huang et al. (2015) implemented the standard BiLSTM-CRF structure. Yang et al. (2016a) utilized hierarchical GRU with CRF for better feature extraction. Sun et al. (2020) replaced the standard CRF with MVCRF, where the POS sequence is also incorporated as input. Similarly, Liu et al. (2020b) utilized semi-CRF, which adds helpful labeling restrictions to standard CRF. Wei et al. (2021) applied a self-attention mechanism to the BiLSTM-CRF structure to model position-aware dependencies. In short, the RNN-CRF structure consistently yields good performance, and the neural network component leaves room for exploration and development. However, such structure is computationally costly, because of the limitation of CRF. Zhai et al. (2017) reframed the chunking task as segmentation and labeling tasks, where a pointer-network is used to identify chunk span, averting the use of CRF. The weakness of this method is that there is limited interaction between the two subtasks to minimize error propagation. Joint learning is another direction of interest. Collobert et al. (2011) and Yang et al. (2017) jointly learned text chunking with other sequence labeling tasks. The former proposed a neural alternative to CRF, which is more adept to learn task-specific features in joint learning. The latter presented a transfer learning framework. Rei (2017) incorporated unsupervised language modeling as an auxiliary task to text chunking.

Table 2.10 shows a comparison of the features used by the introduced chunkers. Given a focal word, the commonly used features are its POS tag, lexical context (n words to its left and to its right), syntactic context (n POS tags to its left and to its right). Some chunkers are able to utilize second-order features, e.g., concatenation of the focal word and at least two preceding or following POS tags, or the current POS tag and at least two preceding chunk tags. Such features provide more context but can lead to sparsity problems, and thus are not widely adopted. A few chunkers attempt to capture the structure of a chunk and use chunk-level features, such as the distance from the verb to the chunk head (Daelemans et al., 1999a), or employing a pointer network to learn chunk segmentation before labeling each word (Zhai et al., 2017). Morphological features e.g., affix and capitalization are less prevalent in text chunking, as most of the information they contain are expressed with POS tags. Nonetheless, morphological information can still be helpful in the case of OOV words. Deep learning methods adopt character-level embedding to acquire similar effects.

Table 2.11: Performance of the introduced text chunking methods.

Method		Experiment		
		Dataset	Result	Measure
Feature Engineering - local	Muis and Lu (2016)	NUS SMS	76.62%	F_1
	Ramshaw and Marcus (1999)	baseNP	92.3%	F_1
	Daelemans et al. (1999a)	baseNP	93.8%	F_1
	Ngai and Florian (2001)	CoNLL2000	92.30%	F_1
	Sang (2000)	CoNLL2000	92.50%	F_1
	Van Halteren (2000)	CoNLL2000	93.32%	F_1
	Koeling (2000)	CoNLL2000	91.97%	F_1
	Kudo and Matsumoto (2000)	CoNLL2000	93.95%	F_1
	Zhang et al. (2001)	CoNLL2000	93.56%	F_1
	Lee and Wu (2007)	CoNLL2000	94.22%	F_1
	Johnson and Zhang (2005)	CoNLL2000	94.39%	F_1
Feature Engineering - global	Zhou and Su (2000)	CoNLL2000	93.68%	F_1
	Molina and Pla (2002)	CoNLL2000	92.23%	F_1
	Collins (2002)	CoNLL2000	93.53%	F_1
	Tsuruoka and Tsujii (2005)	CoNLL2000	93.70%	F_1
	Sha and Pereira (2003)	CoNLL2000	94.38%	F_1
	Suzuki and Isozaki (2008)	CoNLL2000	95.15%	F_1
	McDonald et al. (2005)	CoNLL2000	93.90%	F_1
	Sutton et al. (2007)	CoNLL2000	93.87%	F_1
	Sun et al. (2008)	CoNLL2000	94.34%	F_1
	Lin et al. (2020)	CoNLL2000	92.44%	F_1
	Muis and Lu (2016)	NUS SMS	76.62%	F_1
Deep Learning	Collobert et al. (2011)	CoNLL2000	94.32%	F_1
	Yang et al. (2017)	CoNLL2000	95.41%	F_1
	Huang et al. (2015)	CoNLL2000	94.49%	F_1
	Yang et al. (2016a)	CoNLL2000	95.41%	F_1
	Zhai et al. (2017)	CoNLL2000	94.72%	F_1
	Rei (2017)	CoNLL2000	93.88%	F_1
	Sun et al. (2020)	CoNLL2000	95.44%	F_1
	Liu et al. (2020b)	CoNLL2000	91.80%	F_1
	Wei et al. (2021)	CoNLL2000	95.15%	F_1
	Zhao et al. (2019)	CoNLL2000	94.80%	F_1
	Akbik et al. (2018)	CoNLL2000	96.72%	F_1

Furthermore, since text chunking is a relatively coarse-grained syntactic task that can benefit from other tasks e.g., POS tagging, NER, or language model, we specially mark out the algorithms that employ joint learning. A summary of the performance of the introduced algorithms are listed in Table 2.11. The current state-of-the-art model (Akbik et al., 2018) obtained the F_1 score of 96.72%. Compared to other sequence labeling tasks in NLU such as POS tagging and NER, text chunking has received relatively less attention. This has become more apparent in recent years, since most neural network models meant for sequence labeling are largely universal and can be extended to chunking.

It is worth exploring whether a model designed specifically for chunking can improve the performance. The more significant challenge, however, is that text chunking as a sub-system in complex applications remains quite rare. Although pushing for accuracy is important, it is also good to keep in mind how chunking as a syntactic preprocessing step can benefit higher-level NLU tasks such as semantics or pragmatics. With the development of neural network, the performance of sequence labeling model is advancing rapidly. Currently, the most popular approach to text chunking is the combination of different deep learning algorithms and CRF. In the future, one possible way to further improve the performance is to apply graph neural network to chunking models. Another way is to find a neural network alternative to CRF. Lastly, we hope to see more integration between text chunking and other more complex downstream tasks, e.g., aspect extraction, sentiment analysis, etc.

2.6 Lemmatization

Lemmatization is an NLU task that reduces the inflected forms of given words into their morphologically correct root forms. It is an essential preprocessing technique that extracts concepts and keywords for downstream applications, e.g., search engine (Halácsy and Trón, 2006; Balakrishnan and Lloyd-Yemoh, 2014) and dialogue system (Zhao and Gao, 2017; Altinok, 2018; Liu et al., 2019a). Another commonly used method is stemming, which also converts words into their base form, but does so by cutting off the prefixes or suffixes. Lemmatization, on the other hand, performs a morphological analysis based on the context of given words, and thus is able to preserve more syntactic information. For example, given the word "*studied*", a stemmer simply removes the suffix and returns "*studi*", whereas a lemmatizer is able to extract the proper lemma "*study*".

Lemmatization has received growing attention in recent years, especially for highly inflected languages such as Dutch, Latin and Arabic. As lemmatizing English words are relatively easy, here we cover lemmatizers for inflection-rich languages to provide more perspectives to this task. Annotated training corpora for lemmatization mostly include lexicons for the target languages, CoNLL-2007 (Nivre et al., 2007), CoNLL-2009 (Hajič et al., 2009), CoNLL-2018 (Zeman et al., 2018) and the Universal Dependencies (UD) treebanks (Nivre et al., 2016). The standard evaluation metric is accuracy. Relevant information about the commonly used lemmatization datasets can be viewed in Table 2.12.

Existing lemmatizers regard lemmatization as either a suffix and prefix transformation problem, or a string-to-string transduction problem. The former focus on the starting and ending letters, identifying recurring affixes and transforming them. The latter, on the other hand, considers the whole word form and generates the operations to convert it into its lemma. In this chapter, we introduce the previous works in three section: the transformation approaches, the statistical transduction approaches, and the neural transduction approaches.

Table 2.12: Widely used corpora for lemmatization.

Dataset	Description	Reference
CoNLL-2007	Dependency parsing corpus	Nivre et al. (2007)
CoNLL-2009	Syntactic and semantic dependencies corpus	Hajič et al. (2009)
CoNLL-2018	Multilingual parsing corpus	Zeman et al. (2018)
UD	The Universal Dependencies treebanks	Nivre et al. (2016)
Multext	Multilingual text tools and corpora	Ide and Véronis (1994)
MULText-EAST	The MULText-EAST Slovene lexicon	Erjavec (1998)
UniMorph	The Universal Morphology project	Kirov et al. (2018)
PDT	The Prague Dependency Treebank	Böhmová et al. (2003)
IFD	The Icelandic Frequency Dictionary	Helgadóttir (2012)
CELEX	Lexical databases of English, Dutch, German	Jongejan and Dalianis (2009)
Cast3LB	Spanish treebank	Civit and Martí (2004)
CESS-ECE	Multilingual and multilevel annotated corpus	Martı et al. (2007)
1984	Manually annotated G. Orwell's *1984*	Gesmundo and Samardzic (2012)
GML	The Middle Low German dataset	Peters and Nagel (2014)
PATB	Penn Arabic Treebank	Maamouri et al. (2004)
ARZ	Egyptian Arabic Morphological Annotation	Maamouri et al. (2012)
FTB	French Treebank	Seddah et al. (2013)
L-Lem	Modern Greek & English lexical databases	Lyras et al. (2007)
D-Lem	Afrikaans lexicon	Daelemans et al. (2009)
Lem-list	Lexical databases of 25 languages	Akhmetov et al. (2020)
RELIG	Religious texts in Middle Dutch	Van Kerckvoorde (2019)
CG-LIT	Literary texts in Middel Dutch	Van Kerckvoorde (2019)
WSD	Hindi health and tourism corpora	Khapra et al. (2010)
LT4HALA	Latin texts	Celano (2020)
T'uBa-D/Z	The T'uBa-D/Z treebank for German	Telljohann et al. (2004)
NoSta-D	German texts with non-standard variations	Dipper et al. (2013)
UD-EWT	The English Web Treebank	Silveira et al. (2014)
SIGMORPHON	The SIGMORPHON 2019 shared task	McCarthy et al. (2019)
SEJFEK	Polish economic lexicon	Savary et al. (2012)

2.6.1 Transformation Approach

Early lemmatizers learnt a set of classification rules that detected and modified suffix and/or prefix of a given word form to transform it into the corresponding lemma. The transformation approaches viewed lemmatization as a rule-based classification problem, where the class label assigned to a word was defined via giving the transformation to be applied on the word in order to get the normalized form. For instance, class labels could take the form of *(x to y)*, where x is the suffix of the word form and y is that of its lemma. Mladenic (2002) proposed two methods for mapping from words to their lemmas. The first on is letter-based representation using transformation-based learning, where the machine learns a set of classification rules from feature set comprised of suffixes. The other method is context-based representation using Naïve Bayes Majority classifier, where the features are n-grams of the given words.

Plisson et al. (2004) proposed a lemmatizer based on Ripple Down Rules (RDRs) (Compton et al., 1992) induction algorithm. As opposed to the if-then classification rules, RDR creates exceptions to existing rules, so that the addition of new rules is confined to the context and will not cause inconsistency in the rule base. New RDRs are added by creating "except" or "else" branches to the existing "if-then" rules, creating a tree-like decision structure. The model is trained on a lexicon of lemmatized Slovene words to learn what suffixes should be removed and/or added to get the normalized form. Results show that the RDR approach achieves better accuracy than the standard classification rules. Juršič et al. (2007) further explored the application of RDR in the automatic construction of lemmatizers, presenting the LemmaGen system. They improved the original RDR method by instantiating general concepts in lemmatization into domain-specific terms, such as organizing the training samples into *(word-form, lemma)* pairs, and restricting the form of rule condition and consequent. The LemmaGen is trained and tested on lexicons from multiple languages. It is proven to consistently outperform the original RDR.

Erjavec and Džeroski (2004) presented a lemmatizer for unknown Slovene words. Their method consists of two steps. First, a trigram tagger (Brants, 2000) performs morphosyntactic description tagging on the input text. Subsequently, the resulting morphosyntactic tag and the word itself are passed onto a first-order decision list learning system (Manandhar et al., 1998), which learns ordered sets of classification rules. The lemmatizer is trained on a lexicon containing lemmas with their full inflectional paradigms. A limitation of the rule-based methods above is that classification rules are learned based on fixed-length suffixes. An alternative is to find the LCS between a word form and its lemma to identify the possible suffix. Kanis and Müller (2005) constructed a lemmatizer automatically from the Full Form Lemma training dictionary using the LCS approach. They focused on three types of OOV problems, namely, missing full forms, compound words, and unknown words. To address the former two, they introduced the Hierarchical Lemmatization without Rule Permission Check (HLWRPC), where a new lemmatization algorithm without the lemmatization rule permission check is used when encountering missing full forms or compound words. To address the unknown words problem, a set of word applicable rules is assigned to every word. When lemmatizing an unknown word, the word applicable rules create a set of prefix and suffix patterns. Then, these patterns are matched in the relevant table. The rules with the highest count of winnings are applied on the unknown word.

Jongejan and Dalianis (2009) described a method that automatically generates classification rules to handle not only suffix but also prefix and infix changes to transform word forms into lemmas. Such affixes are identified via finding the LCS and their positions in relation to the LCS. During training, the system selects rules to lemmatize words. If a rule is incorrect, a new one is added to generate the correct lemma. Training ends when all words are correctly lemmatized. Then, the system returns a data structure containing a set of classification rules that a lemmatizer must traverse to arrive at one rule that is selected to fire. They explored two different data structures to store rules, namely directed acyclic graph (DAG) and plain tree structure with depth first, left to right traversal.

Daelemans et al. (2009) applied prototype-based Active Learning (AL) to memory-based lemmatization. AL (Cohn et al., 1996) is a type of unsupervised approach where criteria are investigated to allow ordering the unannotated data in a way that the instances potentially contributing most to the speed of learning can be annotated first. The proposed lemmatizer is inspired by a novel AL algorithm called Prototype-Based Active Classification (PBAC) (Cebron and Berthold, 2009), where a new labeled prototype is added in each learning iteration to fine-tune the classification. Prototypical examples are selected first, whereas examples at the classification boundary are only selected automatically when it becomes necessary. To apply the PBAC approach to lemmatization, they used word frequency and word length as features, assuming that longer or low frequency words are less prototypical than shorter or high frequency words. They hypothesized that contrary to the standard PBAC, less prototypical linguistic examples should provide better results faster in AL. Therefore, less prototypical instances are added to the memory-based classifier (Daelemans et al., 2003) at the start of the learning process. The classes are automatically generated based on the LSC. Experiments show that their algorithm indeed outperforms random data selection and other AL methods in lemmatization.

Gesmundo and Samardzic (2012) formalized lemmatization as a category tagging task, where a word-to-lemma transformation rule is encoded as a single label. Specifically, given a word form, its prefix and/or suffix are found using the LCS method. Then, a transformation rule is defined by a tuple that contains the four types of generic transformation, namely, removing a suffix, adding a new lemma suffix, removing a prefix, and adding a new lemma prefix. Such transformation rules are used as labels. With this setup, any supervised tagging model can be used as a lemmatizer for any languages.

2.6.2 Statistical Transduction Approach

As opposed to transforming word forms into lemmas by modifying suffix and/or prefix, the transduction approach views lemmatization as a character-level word-to-lemma transduction process. The introduction of edit tree paves the foundation for this approach, which converts an input string into the output string. Alternatives include edit distance, word vectors, etc. Chrupała (2006) was the first to propose a data-driven, context-sensitive lemmatizer based on a class-inference mechanism called the shortest edit script (SES), which is able to detect recurring patterns in the mappings from words to their corresponding lemma, thus automatically deriving the lemmatization classes from training data. Specifically, an edit script of two sequences is a set of instructions that specifies the transformations from one sequence to the other. These instructions take the form of inserting or deleting a character at a designated position. Hence, finding the LCS is integral to finding the SES between two sequences. Though unlike the previously mentioned LCS-based transformation approach, SES does not identify suffix or prefix based on its position to map onto that of the lemma, but generating character-level modifications instead.

This approach is proven to perform well for different languages, effectively reducing the labor to manually create full-paradigm inflectional lexicon. Based on the SES mechanism, Chrupała et al. (2008) proposed Morfette, which is a modular, data-driven, probabilistic system that jointly learns lemmatization and morphological tagging from morphologically annotated corpora. Morfette consists of two MaxEnt classifiers (Ratnaparkhi, 1996) that are trained to predict lemmas and morphological tags respectively, and another module that dynamically combines their predictions and outputs the probability distribution over tag-lemma pair sequences. The lemma classifier uses the SES method to induce classes automatically.

Taking inspiration from the above-mentioned works, Müller et al. (2015) presented LEMMING, a modular model that jointly learns lemmatization and morphological tagging at the token level. The proposed lemmatizer maps an inflected form into its lemma given its morphological attributes using a log-linear model. Following the induction method in Morfette (Chrupała et al., 2008), the lemmatizer selects candidates through a deterministic pre-extraction of edit trees. This formalization allows the integration of arbitrary global features. It is vastly used in later transduction-based models. For joint learning, the lemmatizer is combined with a morphological tagger MARMOT (Müller et al., 2013) in a tree-structured CRF. Experiments show that LEMMING yields significant improvements in joint accuracy compared to Morfette and other lemmatizers.

Lyras et al. (2007) implemented the Levenshtein edit distance in a dictionary-based algorithm for automatic lemmatization. The system calculates the similarity between the input word and all dictionary lemmas, selecting those with the minimum edit distance. To improve accuracy, they also used common suffixes in the target language, removing possible suffixes and recalculating distances. The system then compares all stored lemmas and returns the n-best ones with the lowest edit distance. Experiments showed that the suffix-removal method, though more labor-intensive, achieved higher accuracy than the baseline in both modern Greek and English.

Dreyer et al. (2008) presented a conditional log-linear model, which employs overlapping features over latent alignment sequences, and learns latent classes and latent string pair regions from incomplete training data. Given an input, the candidate output is selected by a sliding window over the aligned *(input, output)* pair. At each window position, the log probabilities of all possible alignments are accumulated, evaluating each alignment separately. To further improve the performance, new latent dimensions can be added to the *(input, output)* tuple. Toutanova and Cherry (2009) presented a global joint model for lemmatization and POS tagging trained on morphological lexicons and unlabeled data. The model consists of two components – a semi-supervised POS tagger (Toutanova and Johnson, 2007), and a lemmatizing transducer that is optionally given the POS tags of input word to inform the lemmatization. Taking a pipeline approach, given a sentence, the POS tagger first predicts a set of tags for each word. Subsequently, the lemmatizer predicts one lemma for each of the possible tags. The k-best predictions of tag sets and lemmas are chosen to be the output. Additionally, based on the idea that words with the same lemma have dependent POS tag sets, dependencies among multiple words are dynamically determined.

Nicolai and Kondrak (2016) presented three lemmatization methods leveraging inflection tables. The first one is a stem-based lemmatizer, which is built upon a word-to-stem transduction model by adding a stem-to-lemma model, both of which are adapted from the DIRECTL+ transducer (Jiampojamarn et al., 2010). The word-to-stem model is trained on morphological segmentations annotated by leveraging paradigmatic regularity in inflection tables, whereas the stem-to-lemma model is trained on character-aligned pairs of stems and lemmas. The second model is a stemma-based lemmatizer, consisting of a word-to-stemma model and a stemma-to-lemma model. The word-to-stemma model is an improved word-to-stem model, where inflection tables are further utilized by replacing stems with the corresponding stemmas, and affixes with inflection tags. In this way, the boundaries between stems and affixes are identified by tags. The stemma-to-lemma model is the same as the stem-to-lemma model except it is trained stemma-lemma pairs. The final method is direct word-to-lemma transduction, which is trained on word-forms paired with their lemmas and inflectional tags obtained from the inflection tables. Furthermore, the paper employs a re-ranking algorithm (Joachims, 2002) after the n-best lists of possible lemmas are generated, which is able to leverage large, unannotated word lists to improve the accuracy. Experiments show that the re-ranking algorithm indeed boost the performance, and the direct lemmatization method outperforms the other two models in most languages.

Barteld et al. (2016) presented a novel candidate ranking method to tackle two challenges in data-driven lemmatization. First, highly inflected languages can feature complex morphological changes such as prefix and/or infix modification. Secondly, spelling variations can appear in non-standard texts. To bring more emphasis on word-internal modifications, Lexical Correspondence (LC) is used for lemma candidate generation instead of the commonly used edit trees. An LC is defined as a tuple (T, L) where T and L are two sequences of constants and variables with the requirement that the same variables appear in both sequences. Given a word form and an LC, the constants in sequence T are matched with characters in the word form and the variables are replaced with the remaining substrings. Then, the corresponding lemma can be read off the second sequence L. To restrict over-generalization, insertions are anchored by their character offset either from the beginning or the end of the input word form. To deal with spelling variation, given an OOV word form, the lemmatizer generates lemma candidates from similar in-vocabulary word form. The similarity is measured by he Levenshtein distance.

Gallay and Šimko (2016) proposed a method that automatically constructs a lemmatizer by using text corpora to build vector models by means of principal component analysis (PCA). The core concept is that vector space word representation encodes not only syntactic and semantic patterns, but also morphological ones. Given an input word, the relevant reference *(word, lemma)* pairs are chosen from the reference lexicon by applying vector shifts to the input word to retrieve all candidate lemmas. Then, each candidate is assigned a weight based on their similarity with the input word. The correct lemma candidate is expected to appear in connection with different reference pairs. Therefore, the weights for the same candidate are summed together and the candidate with the highest total weight is selected as final lemma.

Rosa and Žabokrtský (2019) presented an unsupervised approach by performing agglomerative clustering of word forms with a novel distance measure. With the assumption that the inflected forms of the same lemma tend to be similar in both spelling and meaning, the proposed distance measure is a combination of string similarity that is given by Jaro-Winkler edit distance (Winkler, 1990), and the cosine similarity of FastText word embeddings. Akhmetov et al. (2020) introduced a language-independent lemmatizer based on the Random Forest classification algorithm. Firstly, all words are converted into vectors by a term frequency-inverse document frequency (TF-IDF) vectorizer (Luhn, 1957; Jones, 1972) in the form of a sparse matrix. Then, the transpose of this matrix is multiplied by itself to produce a character co-occurrence matrix, every row or column of which serves as character embedding. For every word-lemma pair in the given dictionary, the word form is encoded by the character co-occurrence matrix, whereas the lemma is encoded by the character ordinal numbers. For lemmatization, a Random Forest classifier with bootstrapping is adopted. Experiments show that this simple lemmatizer performs well across a wide range of languages.

2.6.3 Neural Transduction Approach

The recent advancement of neural networks brings a fresh perspective to the transduction approach. The encoder-decoder architecture especially receives a lot of attention, as character-level Seq2Seq model is exceptionally adept at handling the string-to-string transduction task, able to capture more contextual information than the statistical approaches. Kestemont et al. (2017) described a deep learning approach to lemmatization for variation-rich languages. The proposed system consists of two basic components. One is temporal convolutions that model the orthography of input words at the character level. The other is distributional word embeddings from Skip-gram model, which represents the lexical context surrounding the input. Given a word, the system feeds the focus token into the convolutional component, and its left and right tokens into two embedding components, respectively. The outputs of these three sub-nets, along with the one-hot encoding of the input token, are concatenated to form one single hidden representation, which is then passed through a final linear layer to produce the lemma.

Chakrabarty et al. (2017) introduced a composite bidirectional gated RNN (BGRNN) architecture for language-independent, context-sensitive lemmatization. Following previous works, the task is defined as detecting the correct edit tree representing the transformation between a word form and its lemma. The proposed architecture consists of two stages. First, a BGRNN is used to extract the character level dependencies. The outputs are combined with the corresponding word embedding given by word2vec (Mikolov et al., 2013a) to form the final word representations. Then, the representations are fed sentence-wise into the second BGRNN to capture the contextual information of the input word, and to learn the mapping from word embeddings to word-lemma transformations.

To further improve the performance, a subset of unique edit trees involving irregular transformations are sorted out from the corpus before training. This prior knowledge is represented as applicable tree edit vectors, and combined with the hidden states form the second BGRNN to be passed on to the final classification layer. For the BGRNNs, the paper explores the combination of two BiLSTMs and two bidirectional GRUs. Experiments show that the former achieves higher accuracy.

Bergmanis and Goldwater (2018) proposed a context-sensitive lemmatizer called Lematus, which is based on the standard attentional encoder-decoder architecture, and incorporated character-level sentence context. The system is based on an NMT model (Sennrich et al., 2017), which consists of a 2-layer bidirectional GRU encoder, and a 2-layer decoder with a conditional GRU (Sennrich et al., 2017) in the first layer and a standard GRU in the second layer. The input sequence is a space-separated character representation of a given word in its n-character left and right sentence context. With this method, context is modeled using solely the character contexts of the focus word, as oppose to additional POS information or word embeddings trained on large corpus.

To address the problem of low-resource languages, Bergmanis and Goldwater (2019) proposed a semi-supervised method that combines type-based learning and context-sensitive approach. Using the above-mentioned lemmatizer Lematus as a baseline, they incorporated unambiguous word-lemma pairs in the inflection table from the Universal Morphology project (UniMorph) (Kirov et al., 2018) as additional training data. They also collected sentence context for them from Wikipedia. Experiment results indicate that this data augmentation indeed improves a baseline lemmatizer trained on low-resource language corpus. Celano (2020) introduced a joint lemmatizer and POS tagger for Latin. Given a sentence, gradient boosting machine called LightGBM (Ke et al., 2017) predicts the corresponding POS tag sequence. Subsequently, the POS labels are combined with word embedding pretrained from large Latin corpora, and fed into a standard Seq2Seq model to output the lemma sequence.

Arakelyan et al. (2018) introduced an LSTM-based neural network that jointly learns lemmatization, POS tagging, and morphological feature extraction. Each word in a sentence is represented by three vectors: a pretrained word vector from FastText, a one-hot casing feature, and a character-level representation from a BiLSTM. This concatenated representation is processed through three LSTM layers to obtain hidden states for lemma, POS tags, and morphological features. POS tagging and feature extraction use a linear output layer, while lemmatization uses a GRU-based decoder per word, sharing weights and working in parallel. The output of each decoder is fed to another linear layer to produce the predicted lemma.

Pütz et al. (2018) proposed a morphologically-informed Seq2Seq-based lemmatizer. The model extends the standard encoder-decoder architecture with Luong-style monotonic attention (Luong et al., 2015; Raffel et al., 2017). The hidden state outputted by the encoder is concatenated with the embedded morphological and POS tags, passing through a feed-forward layer with SELU activation function (Klambauer et al., 2017), and fed into the decoder to generate the predicted lemma.

The proposed model is compared with LEMMING (Müller et al., 2015) in German, achieving competitive accuracy. The results also indicate that a log-linear lemmatizer such as LEMMING is preferable when dealing with misspelled words, whereas Seq2Seq lemmatizer is able to generalize and handle OOV words better. Kondratyuk et al. (2018) proposed LemmaTag, which is a featureless bidirectional-RNN-based architecture that jointly learns lemmatization and POS tagging. Given a sentence, a GRU outputs the character-level embedding for every word, which is summed with the word embedding to form the final word embeddings. The resulting sequence is passed onto two layers of BiLSTM with residual connections, producing a sequence of word representations with sentence-level connections. The POS tagger, made up of a fully-connected layer, predicts the tag values, concatenating them into a flat vector to pass onto the lemmatizer. The lemmatizer consists of a LSTM layer with character-level attention mechanism that takes in the final word embedding, the character embedding, and the POS features of the focus word to generate the corresponding lemma.

Malaviya et al. (2019) introduced a simple LSTM-based joint learning model for lemmatization and morphological tagging. Given a sentence, the morphological tagger obtains the word representation for each word using a character-level BiLSTM, which is then fed into a word-level BiLSTM to predict the corresponding morphological tag. For lemmatization, a string-to-string transduction model (Wu and Cotterell, 2019) is adopted, which is a Seq2Seq model with hard attention mechanism (Xu et al., 2015c; Rastogi et al., 2016). The joint probability of the input sentence is define as the product of the probability outputted by the tagger and all the probabilities outputted by the transducer. Yildiz and Tantuğ (2019) proposed Morpheus, a joint contextual lemmatizer and morphological tagger. Similarly to the above-mentioned works, given a sentence, firstly a character-level LSTM generates word vectors, which are then fed into a word-level BiLSTM to produce context-aware word representations. Subsequently, two separate LSTM decoders are employed to predict morphological tags and edit operations from word forms to lemmas. More specifically, to find the minimum edit operations, a dynamic programming method based on Levenshtein distance is used.

Manjavacas et al. (2019) also built upon the classic encoder-decoder architecture to acknowledge the difficulty of spelling variations in lemmatization for non-standard language. Specifically, they presented a hierarchical sentence encoder that is jointly trained for lemmatization and language modeling, adopting the attention mechanism (Bahdanau et al., 2015) to extract additional context. The hierarchical encoder consists of three levels. A bidirectional RNN first computes the character-level representation, which is fed into another bidirectional RNN to extract word-level features. Lastly, the final bidirectional RNN outputs the sentence-level features based on the word-level hidden state. To extract higher quality sentence-level features, a word-level bidirectional language model is added to the lemmatizer. Two softmax classifiers predict left and right tokens using forward and backward sentence-level hidden states. Jointly minimizing lemmatization and language modeling losses allows the lemmatizer to represent global sentence context without needing POS or morphological tags.

Kondratyuk (2019) presented a cross-lingual joint learning model for lemmatization and morphological tagging using multilingual BERT (Devlin et al., 2018). Given an input sentence, a pretrained multilingual cased-BERT is used to encode the tokens in the sentence. Subsequently, the encoded tokens are passed through two separate layer attention (Kondratyuk and Straka, 2019) to produce embeddings that are specific to each task. Furthermore, the model also applies BiLSTM character embeddings (Ling et al., 2015b; Kim et al., 2016) to generate enhanced morphological representations that are summed with the two task-specific embeddings, respectively. Following his previous work (Kondratyuk et al., 2018), the decoder for each task consists of two successive layers of word-level bidirectional residual LSTMs. For lemmatization, the final hidden state is fed into a feed-forward layer to map the focus token to one of the pre-computed SES (Chrupała, 2006) that transforms it to the correct lemma. Similarly for morphological tagging, a feed-forward layer takes in the hidden state and predicts the morphosyntactic description class.

Taking joint learning a step further, Zalmout and Habash (2019) proposed a Seq2Seq-based model that handles both lexicalized and non-lexicalized features. It jointly learns lemmatization, diacritization, normalization, as well as fine-grained morphological tagging. This is achieved by means of applying parameter sharing strategies on top on the standard LSTM encoder-decoder architecture. On one hand, different non-lexicalized features are modeled on word-level by a BiLSTM tagger that shares parameters with the encoder. One the other hand, all lexicalized features are handled on character-level by the same encoder, and then sent to their corresponding decoders separately. The proposed model gains improvement across all of the joint tasks in two Arabic corpora.

Unlike previous contextual neural network approaches, which takes into account the whole sentence, Chakrabarty et al. (2019) hypothesized that a limited context is sufficient for lemmatization. Based on this notion, they introduced a BiLSTM-CNN lemmatizer. Given a sentence, a BiLSTM is applied to extract character-level embeddings, which are then arranged into a matrix and processed by a CNN to generate positional embeddings. The size of filters in the CNN is determined by the length of the context to include, which is set to be the trigram of focus tokens in the paper. Experiments show that BiLSTM-CNN indeed outperforms the BiLSTM-BiLSTM lemmatizer (Chakrabarty et al., 2017).

Schmitt and Constant (2019) focused on the lemmatization of multi-word expressions (MWEs), combinations of several words that display the linguistic properties of a lexical unit. The main challenge of lemmatizing MWE is that its unique properties require different classification rules on top of simple-word lemmatization knowledge. To acknowledge this problem, they propose a deep encoder-decoder lemmatizer solely based on the internal context of MWE. Given an MWE, a GRU encodes the character-level word embeddings, which are passed onto another GRU as a sequence to extract the internal context. The final word representation in an MWE is a concatenation of the character-level word embedding, its left internal context, its POS tag, its position and the length of the MWE. Then, a character-level conditional GRU with an attention mechanism decodes the final word representation to generate the corresponding lemma.

Table 2.13: Comparison between different lemmatization methods. Trans is short for transformation. Statistical is short for statistical transduction, whereas Neural is short for neural transduction. SX stands for Syntactic. MP stands for morphological. CX stands for Context. Char stands for character. CR is classification rule. ES is edit script. SS is string similarity. LG is lemma generation. Joint is short for joint learning.

	Method	Affix	SX	MP	CX	Char	CR	ES	SS	LG	Joint
Trans	Mladenic (2002)	✓		✓			✓				
	Plisson et al. (2004)	✓					✓				
	Juršič et al. (2007)	✓					✓				
	Erjavec and Džeroski (2004)	✓	✓				✓				
	Kanis and Müller (2005)	✓					✓				
	Jongejan and Dalianis (2009)	✓					✓				
	Daelemans et al. (2009)								✓		
	Gesmundo and Samardzic (2012)	✓	✓				✓				
Statistical	Chrupała (2006)				✓	✓	✓				
	Müller et al. (2015)		✓	✓			✓				
	Lyras et al. (2007)	✓						✓			
	Dreyer et al. (2008)					✓		✓			✓
	Toutanova and Cherry (2009)	✓	✓			✓	✓				
	Nicolai and Kondrak (2016)	✓			✓	✓		✓			
	Barteld et al. (2016)		✓	✓		✓		✓			
	Gallay and Šimko (2016)	✓						✓			
	Rosa and Žabokrtskỳ (2019)							✓			
	Akhmetov et al. (2020)					✓			✓		
Neural	Kestemont et al. (2017)				✓	✓				✓	
	Chakrabarty et al. (2017)				✓	✓	✓				
	Bergmanis and Goldwater (2018)				✓	✓				✓	
	Bergmanis and Goldwater (2019)			✓	✓	✓				✓	✓
	Celano (2020)		✓							✓	✓
	Arakelyan et al. (2018)	✓	✓			✓				✓	✓
	Pütz et al. (2018)	✓	✓	✓						✓	
	Kondratyuk et al. (2018)	✓			✓	✓				✓	✓
	Malaviya et al. (2019)			✓	✓	✓				✓	✓
	Yildiz and Tantuğ (2019)			✓	✓	✓				✓	✓
	Manjavacas et al. (2019)				✓	✓				✓	✓
	Kondratyuk (2019)		✓		✓	✓			✓		✓
	Zalmout and Habash (2019)		✓			✓				✓	✓
	Chakrabarty et al. (2019)				✓	✓				✓	
	Schmitt and Constant (2019)	✓			✓	✓				✓	
	Zalmout and Habash (2020)				✓	✓				✓	✓
	Milintsevich and Sirts (2021)	✓			✓	✓				✓	

Zalmout and Habash (2020) presented a character-level Seq2Seq lemmatization model, utilizing several sub-word features, namely FastText n-grams, greedy stem, orthographic roots, and orthographic patterns. They used Lematus (Bergmanis and Goldwater, 2018) as a baseline. Instead of fixed-size character window, they utilized the characters of the n words surrounding the focus word, employing LSTM instead of GRU. They explored two configurations to incorporate sub-word features. At the encoder side, they condition on different sub-words by concatenating their embeddings with the character embeddings before feeding them into the encoder. The embeddings for sub-words can be learned as a part of the system, or pretrained on an external corpus. Alternatively, greedy stem, roots, and patterns can also be learned jointly as auxiliary tasks at the decoder side. In this configurations, lemmatization and the auxiliary tasks share the same encoder, whilst having specific decoders for each task. Experiments indicate that though all the proposed methods outperform the standard Seq2Seq lemmatizer, incorporating pretrained FastText n-grams at the encoder side yields the best result.

Milintsevich and Sirts (2021) proposed a novel hybrid approach that enhances the Seq2Seq lemmatizer with external resources. Their model is based on the lemmatization module from the Stanford parsing system (Qi et al., 2018, 2020), which takes the character-level representation and the corresponding POS tag into a BiLSTM encoder and a BiLSTM decoder with attention, outputting the predicted lemma using greedy decoding. To improve the performance of this baseline, they incorporated additional lemmas extracted from an external lexicon or a rule-based system. The model employs another encoder, which receives and processes the candidate lemmas from an external lexicon, or generated by a rule-based system. The decoder then learns via two separate attention mechanism to predict lemma using the combination of the classic string-to-string transduction as well as copying lemma characters from external candidates. Results indicate that the enhanced lemmatizer performs considerably better than a simple lexicon extension method based on the Stanford system.

2.6.4 Summary

In conclusion, we divide existing lemmatizers into two categories – transformation approaches and transduction approaches. Transformation approaches regard affix as the base unit for rule-based classifications. In early works, the machine can only learn rules that transform suffixes of fixed length (Mladenic, 2002; Plisson et al., 2004; Juršič et al., 2007; Erjavec and Džeroski, 2004). Such limitation can be solved by using LCS, which enables the machine to identify prefixes and suffixes of any length. Kanis and Müller (2005) specifically targeted missing full forms, unknown words, and compound words. Jongejan and Dalianis (2009) focused on detecting prefix and infix as well as suffix. Daelemans et al. (2009) utilized PBAC to enable efficient unsupervised learning. Gesmundo and Samardzic (2012) reframed lemmatization as a category labeling task, so that any tagging model can be applied as a lemmatizer.

Table 2.14: Performance of the introduced lemmatization methods. Trans is short for transformation. Statistical is short for statistical transduction. Neural is short for neural transduction. MSA is Modern Standard Arabic. EGY is Egyptian Arabic.

	Method	Experiment 1				Experiment 2			
		Language	Dataset	Result	Measure	Language	Dataset	Result	Measure
Trans	Mladenic (2002)	Slovene	MULText-EAST	74.2%	Acc.				
	Plisson et al. (2004)	Slovene	MULText-EAST	77.0%	Acc.				
	Juršič et al. (2007)	Slovene	MULText-EAST	94.38%	Acc.	English	Multext	94.14%	Acc.
	Erjavec and Džeroski (2004)	Slovene	MULText-EAST	92.0%	Acc.				
	Kanis and Müller (2005)	Czech	PDT	95.89%	F_1				
	Jongejan and Dalianis (2009)	Icelandic	IFD	71.3%	Acc.	English	CELEX	89.0%	Acc.
	Daelemans et al. (2009)	Afrikaans	D-Lem	91.0%	Acc.				
	Gesmundo and Samardzic (2012)	English	1984	99.6%	Acc.				
Statistical	Chrupała (2006)	Spanish	Cast3LB	92.48%	F_1	Catalan	Cast3LB	94.64%	F_1
	Müller et al. (2015)	English	Pen Treebank	98.84%	Acc.	Spanish	CoNLL-2009	98.46%	Acc.
	Lyras et al. (2007)	Modern Greek	L-Lem	95.03%	Acc.	English	L-Lem	96.46%	Acc.
	Dreyer et al. (2008)	German	CELEX	94.0%	Acc.				
	Toutanova and Cherry (2009)	English	CELEX	99.0%	F_1	Slovene	MULText-EAST	91.2%	F_1
	Nicolai and Kondrak (2016)	English	CoNLL-2009	93.3%	Acc.	German	CELEX	90.0%	Acc.
	Barteld et al. (2016)	German	GML	97.76%	Acc.				
	Gallay and Šimko (2016)	MSA	PATB	95.4%	Acc.	EGY	ARZ	83.3%	Acc.
	Rosa and Žabokrtský (2019)	English	UD v2.3	97.78%	Acc.				
	Akhmetov et al. (2020)	25 languages	Lem-list	84.05%	avg Acc.				
Neural	Kestemont et al. (2017)	Middle Dutch	RELIG	90.97%	Acc.	Middle Dutch	CG-LIT	91.67%	Acc.
	Chakrabarty et al. (2017)	Hindi	WSD	94.90%	Acc.	Spanish	CoNLL-2009	98.11%	Acc.
	Bergmanis and Goldwater (2018)	20 languages	UD v2.0	95.0%	avg Acc.				
	Celano (2020)	Latin	LT4HALA	94.6%	Acc.				
	Arakelyan et al. (2018)	English	CoNLL-2018	95.77%	Acc.	French	CoNLL-2018	86.28%	Acc.
	Pütz et al. (2018)	German	T'uBa-D/Z	97.02%	Acc.	German	NoSta-D	83.96%	Acc.
	Kondratyuk et al. (2018)	Czech	PDT	98.37%	Acc.	English	UD-EWT	97.53%	Acc.
	Malaviya et al. (2019)	20 languages	UD v2.0	95.42%	avg Acc.				
	Yildiz and Tantuğ (2019)	97 languages	UniMorph	97.85%	avg Acc.				
	Manjavacas et al. (2019)	20 languages	UD v2.2	96.28%	avg Acc.				
	Kondratyuk (2019)	100 languages	SIGMORPHON	95.00%	avg Acc.				
	Zalmout and Habash (2019)	MSA	PATB	97.6%	Acc.	EGY	ARZ	88.5%	Acc.
	Chakrabarty et al. (2019)	French	UD	95.47%	Acc.	Hindi	UD	97.07%	Acc.
	Schmitt and Constant (2019)	French	FTB	95.6%	Acc.	Polish	SEJFEK	88.9%	Acc.
	Zalmout and Habash (2020)	MSA	PATB	95.4%	Acc.	EGY	ARZ	83.3%	Acc.
	Milintsevich and Sirts (2021)	23 languages	UD v2.5	97.09%	avg Acc.				

Transduction approaches, on the other hand, view either character as the base unit for lemma generation, or the entire word as the base unit to map onto its lemma based on string similarity measures. Transduction-based methods are more flexible than the transformation-based ones, as they do not rely on an affix lookup table, manually-built or automatically-generated. Thus, the transduction approach is better suited to handle irregular changes. In this chapter, we further split the transduction approaches into statistical and neural-network-based methods, since the development of deep learning brings a new perspective to lemmatization. Statistical transduction approaches mostly rely on edit scripts, for instance, the data-driven SES mechanism (Chrupała, 2006; Chrupała et al., 2008). Müller et al. (2015) improved SES by proposing edit tree, which does not encodes the LCSs and thus more capable of generalizing. Following their work, Lyras et al. (2007); Nicolai and Kondrak (2016) enhanced lemmatization with stemming. Toutanova and Cherry (2009) incorporated POS tagging as a sub-system. Barteld et al. (2016) further improved edit tree with LC, which is more adept at handling word-internal inflections. Dreyer et al. (2008) proposed a window approach as an alternative to edit tree, which can handle more complex inflections, while it is restricted by a window size. Another alternative is to infer lemma from word embeddings (Gallay and Šimko, 2016; Rosa and Žabokrtský, 2019; Akhmetov et al., 2020), paving the way for neural approaches.

Neural transduction approaches rely on neural networks to automatically learn the character-level correspondence between lemmas and words. Earlier methods use various types of neural networks to produce character embeddings, concatenate it with word embeddings, and use another neural network to predict the corresponding lemma (Kestemont et al., 2017; Chakrabarty et al., 2017, 2019). Recent works mostly adopt the character-level encoder-decoder architecture (Bergmanis and Goldwater, 2018; Pütz et al., 2018). Schmitt and Constant (2019) modified the encoder-decoder structure to specifically target MWEs. Bergmanis and Goldwater (2019) applied SSL to the architecture for low-resource domain. Milintsevich and Sirts (2021) enhanced Seq2Seq lemmatization with external resources. Joint learning is often applied to improve the performance of lemmatization. Commonly used auxiliary tasks include POS tagging (Celano, 2020; Arakelyan et al., 2018; Kondratyuk et al., 2018), morphological tagging (Arakelyan et al., 2018; Malaviya et al., 2019; Yildiz and Tantuğ, 2019; Kondratyuk, 2019; Zalmout and Habash, 2019), and language modeling (Manjavacas et al., 2019). With a more linguistically-driven approach, Zalmout and Habash (2020) used subword modeling as an auxiliary task, and expanded the character-level representations to include the surrounding words of the focus word.

Table 2.13 shows a summary of all the introduced lemmatizers. Here, we mark out the methods that utilize affix, syntactic, morphological, contextual, and character information as features, either handcrafted or automatically learned. Specifically, syntactic features refer to POS tags or coarse-grained POS categories. Contextual features are word n-grams or global information. Additionally, we indicate the mechanisms used by the lemmatizers to find the target lemma, namely, classification rule, edit script, i.e., SES and edit tree, string similarity matching such as Levenshtein distance and vector shift, and character-level (or occasionally word-level) lemma generation. Among all, character-level generation garners the most interest due to the advancement of generative Seq2Seq models. Lastly, we list the lemmatization algorithms that adopt joint learning with other tasks, e.g., POS tagging, morphological tagging, and language modeling. The commonly used corpora are listed in Table 2.12. A comparison of performance can be found in Table 2.14.

With the help of neural networks and joint learning, lemmatization has achieved remarkable accuracy, especially for less inflection-rich languages such as English. The next step could be exploring the application of more recent pretrain language models such as GPT (Radford et al., 2018), or further improving multi-task frameworks for joint learning. On the other hand, lemmatization for resource-scarce languages is still lacking. Although neural network approaches are proven to effective in ancient languages compared to a rule-based lemmatizer, for instance in Ancient Irish (Dereza, 2018), there is still a lot of room for improvement. More research on unsupervised learning, SSL, or transfer learning is called for resource-scarce language lemmatization.

2.7 Conclusion

In this chapter, we provided an extensive review of syntactic processing techniques, namely: microtext normalization, SBD, POS tagging, text chunking, and lemmatization. We believe that syntactics processing is a significant step in developing neurosymbolic AI. There have been some attempts to use simple syntactic processing techniques to improve more complex NLU tasks. For instance, microtext normalization is applied to assist sentiment analysis (Satapathy et al., 2017b, 2019b). SBD can be incorporated as a sub-system in information retrieval (Krallinger et al., 2017). In sentiment analysis, it can help identify negation scope to improve the performance (Councill et al., 2010). POS tagger is used as part of the feature extraction process for cyber-bullying detection (Nandhini and Sheeba, 2015) and query-based information retrieval (Mahmood et al., 2017). Asghar et al. (2014) suggested that POS tagging is also an important feature extraction step in sentiment analysis. Mubarok et al. (2017); Mao and Li (2021) applied it to ABSA and indeed obtained better results. Feng et al. (2019b) proposed NMT decoding models that jointly predict target words and corresponding POS tag sequence, acquiring significant improvement. POS tagger is also used to provide syntactic structure in dialogue system (Meena et al., 2014; Bang et al., 2015). As for text chunking, Gupta et al. (2016a) used both POS tagging and chunking as subtasks to tackle text summarization. Song et al. (2005) utilized chunking technique to improve biomedical information extraction. Syed et al. (2014) integrated text chunking with sentiment analysis in Urdu. Lemmatization has long been used for information retrieval (Halácsy and Trón, 2006; Kanis and Skorkovská, 2010; Balakrishnan and Lloyd-Yemoh, 2014) and other tasks, e.g., text categorization (Camastra and Razi, 2020), sentiment analysis (Mhatre et al., 2017), and Tweet stance classification (Priyanshu et al., 2020). More recently, MTL and curriculum learning techniques have been applied as an ensemble for syntactics processing Zhang et al. (2024a).

2.8 Learning Resources

A. Reading List

- Xulang Zhang, Rui Mao, and Erik Cambria. Granular Syntax Processing with Multi-Task and Curriculum Learning. Cognitive Computation 16, 2024 (Zhang et al., 2024a)

- Iti Chaturvedi, Ranjan Satapathy, Curtis Lynch, and Erik Cambria. Predicting Word Vectors for Microtext. Expert Systems, e13589, 2024 (Chaturvedi et al., 2024)

- Xulang Zhang, Rui Mao, and Erik Cambria. SenticVec: Toward Robust and Human-Centric Neurosymbolic Sentiment Analysis. In: Proceedings of ACL, 4851–4863, 2024 (Zhang et al., 2024c)

B. Relevant Videos

- Labcast about Granular Syntax Processing: youtu.be/Fr9lZ6u3Fxs

- Labcast about Intonation Disambiguation: youtu.be/qbba_CvEXcQ

- Labcast about SenticVec: youtu.be/jhCTzeC1ebc

C. Related Code

- Github repository about Syntactics Processing: github.com/SenticNet/Syntactics

- Github repo about Microtext: github.com/SenticNet/microtext-normalization

- Github repository about SenticVec: github.com/SenticNet/SenticVec

D. Exercises

- **Exercise 1.** Create a Python program that reads a list of microtexts and outputs their normalized versions. Use a predefined dictionary of common microtext abbreviations and their full forms. Implement a function to replace abbreviations with full forms. Test your program with a given set of microtexts.

- **Exercise 2.** Propose three sample paragraphs that contain ambiguities in sentence boundaries. Each example should include explanations of the ambiguities and how you would resolve them. After creating the examples, discuss your reasoning and approach.

- **Exercise 3.** Enhance your grasp of POS tagging by identifying the POSs in a given sentence. Choose a random sentence of at least 100 characters from your favorite book, write it down, and determine the part of speech for each word. Rewrite the sentence, placing the POS tags in parentheses next to each word.

- **Exercise 4.** Test your text chunking skills by identifying and labeling noun phrases (NP) and verb phrases (VP) in the previous sentence. Read the sentence again, determine NPs and VPs, and then label each chunk accordingly.

- **Exercise 5.** Lemmatize the previous sentence by replacing each word with its base form or lemma. Then, compare your lemmatized sentence with the output produced by NLTK's WordNetLemmatizer to see how they align and identify any differences.

Chapter 3
Semantics Processing

Abstract Semantics processing is a fundamental research domain in computational linguistics. In the era of LLMs, the advancement of research in this domain appears to be decelerating. However, the study of semantics is multi-dimensional in linguistics. The research depth and breadth of semantics processing can be largely improved with new technologies. In this chapter, we analyze five semantic processing tasks: WSD, NER, concept extraction, anaphora resolution, and subjectivity detection. We survey theoretical research in these fields, advanced methods, and downstream applications, with the hope of encouraging scholars to focus on these fine-grained NLU tasks. The review of theoretical research may also inspire new tasks and technologies in this domain. Finally, we compare different semantic processing techniques and summarize their technical trends, application trends, and future directions.

Key words: Word Sense Disambiguation, Named Entity Recognition, Concept Extraction, Anaphora Resolution, Subjectivity Detection

3.1 Introduction

Semantics is a linguistic term, generally referring to the meaning of language. Unlike syntax which studies the structure of sentences, the significance of semantics lies in its ability to aid our comprehension of how meaning is conveyed through words, phrases, and sentences, as well as how language is used to express various ideas, thoughts, and emotions. Language is one of the important carriers of meanings. However, the term "meaning" encompasses multiple aspects of language. Palmer and Frank Robert (1981) argued that there is a lack of consensus regarding the nature of "meaning", e.g., which components should be considered part of semantics, and how it should be characterized. Thus, the study of "semantics" is also multi-dimensional in academia.

The evolution of semantic research reflects the rich connotation of semantics in linguistics. At the early stage, much attention is given to the study of lexical semantics. The first English dictionary, *Robert Cawdrey's Table Alphabeticall*, dates back to 1604 (Noyes, 1943). The construction of dictionaries, e.g., *The Oxford English Dictionary* (Simpson and Weiner, 1989) became one of the most significant symbols of lexical semantic research achievements. The research of lexical semantics covers word senses, polysemy, word formation, contrastive lexical semantics, and more. Next, another important research dimension of semantics emerged, termed structural semantics. Structural semantics emphasizes the analysis of sentence structures, including the relationships between words and the ways in which words contribute to the meaning of a sentence. The study of structural semantics includes but is not limited to analyzing the meaning of words by syntax, grammar, and pragmatics. Structural semantics elevates the study of semantics from the word level to the sentence level.

The later cognitive semantics further enrich the connotation of semantics. The tenets of cognitive semantics posit that the faculty of language is intricately intertwined with the broader cognitive capacity of human beings (Croft and Cruse, 2004). In other words, semantics is a reflection of how humans understand and make sense of the world around them. Under cognitive semantics, researchers extend to frame semantics (semantics is the reflection of encyclopedic knowledge), situation semantics (semantics reflects the relationships between situations) (Barwise and Perry, 1981), conceptual semantics (semantics reflects the structural perception of concepts) (Jackendoff, 1976), and more. Fig. 3.1 summarizes partial semantic research domains in linguistics.

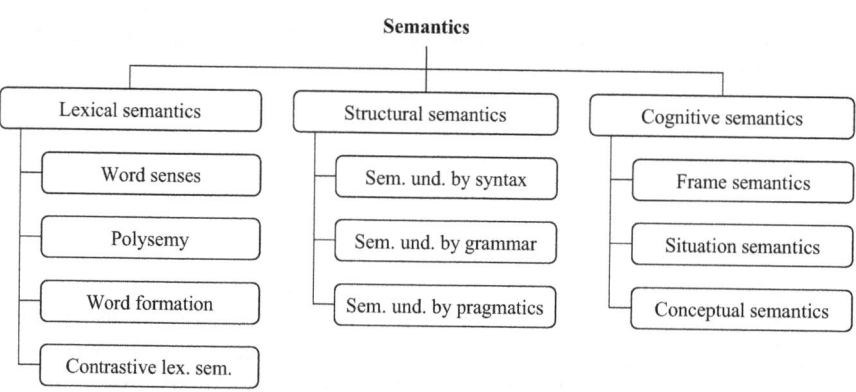

Fig. 3.1: Semantic research domains in linguistics. Lex. denotes lexical; sem. denotes semantics; und. denotes understanding.

The development of automatic semantic processing techniques has largely facilitated semantic research. Many useful tools and knowledge bases[1] were developed for WSD, NER, concept extraction, anaphora resolution, and subjectivity detection. These tools are the embodiment of many theoretical ideas in semantics. For example, WSD is an important task in lexical semantics. NER categorized named entities in texts by conceptually related classes, e.g., names, and locations. Similarly, concept extraction and subjectivity detection tasks also embody the cognitive properties of semantics. Anaphora resolution elucidates the relationship between the anaphor, which is the repetition of a reference, and its antecedent, which is the earlier mention of the entity.

In addition to improving semantic research, semantic processing techniques can also help other downstream NLU tasks with more complexity (see Table 4.1). For example, subjectivity detection can be an upstream task of sentiment analysis, because subjective expressions can be further categorized by positive, negative, and neutral expressions with different opinionated intensities. The semantic processing techniques that have been reviewed possess a range of potential applications, including the ability to generate features that are effective, as well as to be used as a parser in order to obtain desired categories of text. Additionally, these techniques have the potential to improve the explainability of downstream applications. The emergence of PLMs has greatly enhanced the semantic representation capabilities of deep learning models and the ability to fit downstream tasks (Devlin et al., 2018; Liu et al., 2019c; Lewis et al., 2020). Some LLMs, e.g., GPT-4[2] and Bard[3] even realize the functions of multiple complex NLP tasks by the means of dialogue, such as QA, translation, and text summarization. Many semantics processing studies have gradually faded out of the field of NLP. Then, in the era of PLMs and LLMs, an intuitive question is what is the motivation for studying semantic processing techniques?

As mentioned before, semantics reflects the multiple aspects of language. Besides understanding word senses, semantics is also the entrance to understanding the mechanism, and perception of language. Language intelligence encompasses more than just achieving a level of accuracy that is equivalent to or surpasses human accuracy for specific tasks. It also entails the capacity to unveil the nature of language and investigate the cognitive processes that underlie language. Much aforementioned semantic research in the context of linguistics has not been explored in computational linguistics to the best of our knowledge. Thus, we are motivated to propose a survey on semantic processing techniques to encourage future scholars that can expand the depth and breadth of semantic research, leading the public attention from the application value of NLP techniques to the research value of computational linguistics. Nevertheless, we also highlight the fusion of low-level semantic processing techniques and NLU techniques to demonstrate the application value of semantic processing techniques in different domains.

---

[1] A knowledge base normally refers to a collection of organized information that is machine-readable, and supportive for an intelligent system.

[2] https://openai.com/product/gpt-4

[3] https://bard.google.com

Table 3.1: Surveyed semantic processing tasks and their downstream applications. F denotes that the technique yielded features for a downstream task model; P denotes that the technique was used as a parser; E denotes that the technique improved the explainability for a downstream task. CE denotes concept extraction. AR denotes anaphora resolution. SD denotes subjectivity detection.

Downstream tasks	WSD	NER	CE	AR	SD
Polarity Detection	F, P, E		F, P, E	F	P
Information Retrieval	E		F, E		P
Machine Translation	F, P, E			F, E	
Summarization				F	
Textual Entailment				F	
Knowledge Graph Construction		P			
Recommendation Systems		F, P, E			
Dialogue Systems		P, E	F, P		
Commonsense Explanation Generation			F, E		
Hate Speech Detection					F, P
Question & Answering Systems					F, P

Given the broadness of semantics, this chapter's scope lies in semantic processing techniques for WSD, NER, concept extraction, anaphora resolution, and subjectivity detection. This is because these low-level semantic processing tasks reflect different aspects of semantics. In addition, there were many research works on these tasks in the field of computational linguistics. We focus on low-level semantic processing tasks, rather than high-level semantic processing tasks, e.g., sentiment analysis and natural language inference, because they provide fundamental building blocks for both high-level semantic processing tasks and higher-level NLU tasks.

Multiple semantic processing techniques were rarely surveyed in the same article. Salloum et al. (2020) surveyed several high-level semantic processing tasks, e.g., latent semantic analysis, explicit semantic analysis, and sentiment analysis. Unlike the work of Salloum et al. (2020), this chapter includes the latest research in low-level semantic processing techniques. Compared to the latest semantics processing surveys focusing on specific tasks (Ransing and Gulati, 2022; Poesio et al., 2023; Wang et al., 2022; Montoyo et al., 2012), we additionally reviewed important theoretical research and downstream task applications in these domains. These contents can help readers better understand the foundation of semantic research in linguistics, as well as potential application scenarios.

More importantly, theoretical research shows the big picture of a semantic processing task, which may inspire different research tasks in the computational linguistic community. The collection of multiple semantic processing techniques is helpful for readers to have a comprehensive understanding of a large field, inspiring more fusion research across different domains. Theoretical research of other tasks has the potential to inspire fresh perspectives among researchers who have been concentrating on a specific semantic research task.

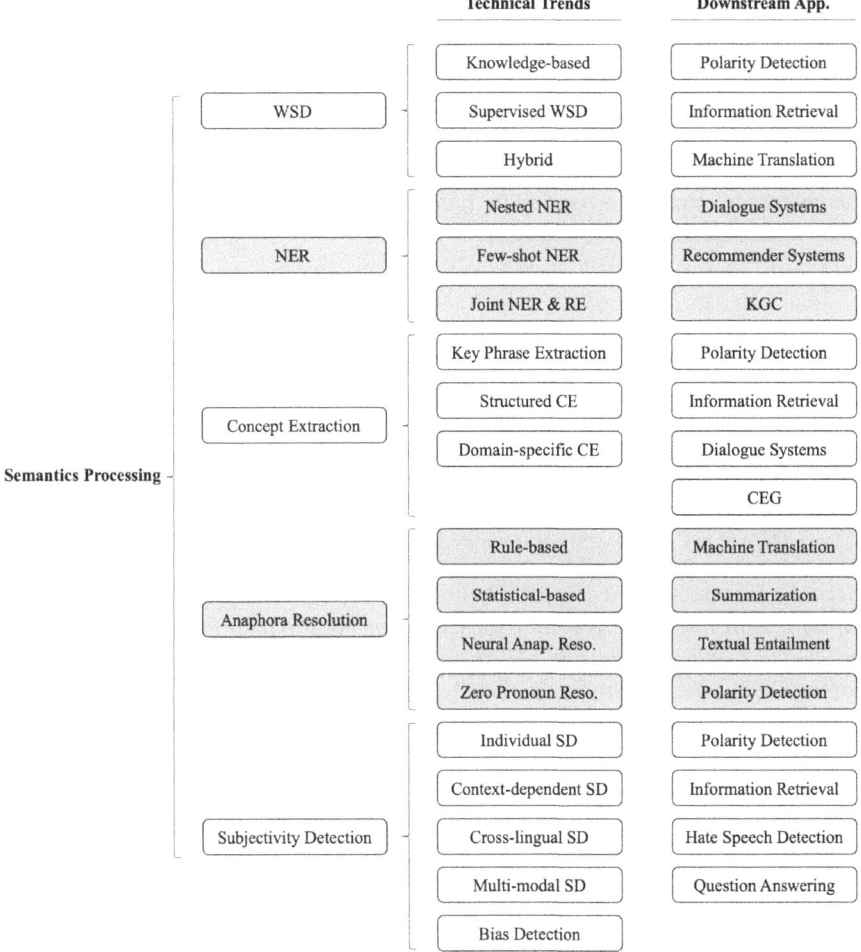

Fig. 3.2: Outline of this chapter. Each subtask is explained in terms of different technical trends and downstream applications.

The contribution of this chapter is threefold:

- We survey recent semantic processing techniques, annotation tools, datasets, and knowledge bases for five low-level semantic processing tasks.
- We highlight important theoretical research, and downstream applications to encourage deeper and wider research in the semantics processing domain upon the currently established task setups.
- We compare different semantic processing techniques, delineate their technical and application trends, and put forth potential avenues for future research in this domain.

In the following sections, we introduce five different semantic processing techniques, namely: WSD (Section 3.2), NER (Section 3.3), concept extraction (Section 3.4), anaphora resolution (Section 3.5), and subjectivity detection (Section 3.6). Fig. 3.2 presents an outline of the chapter: each task is organized in terms of theoretical research, annotation schemes, datasets, knowledge bases, evaluation metrics, methods, downstream applications, and a summary. Finally, Section 3.7 wraps up the chapter by proposing concluding remarks, a reading list, three relevant videos, some related code, and five exercises.

3.2 Word Sense Disambiguation

The complexity of human language is difficult for machines to understand it. One of the challenges is the ambiguity of word senses. In natural language, a word may have multiple senses, given different contexts. Consider the following example:

(1)   He got his shoes wet as he walked along the **bank**.

According to the Oxford English Dictionary, the major senses of "bank" include (a) *an organization that provides various financial services, for example keeping or lending money*; (b) *the side of a river, canal, etc. and the land near it*. With the context, humans can easily know that "bank" here refers to the sense (b). However, it is challenging for machines to do so because the interpretation made by humans is contingent upon their comprehension of the fact that the probability of getting one's shoes wet is higher when walking alongside a river bank as compared to a financial institution. Machines rarely take the commonsense into account when inferring the meaning of "bank"[4], because they do not have human-like cognition and reasoning abilities by nature.

There are two main technical trends in addressing the task of WSD, namely knowledge-based methods and supervised methods. Knowledge-based WSD utilizes the word relations from knowledge graphs, e.g., WordNet (Fellbaum, 1998) and BabelNet (Navigli and Ponzetto, 2012b) to achieve the disambiguation of word senses. In supervised methods, the WSD task is usually defined as a classification task by word senses. A WSD model is trained with annotated data. Two examples of supervised WSD are illustrated in Fig. 3.3. As shown in the figure, a naïve strategy of the knowledge-based WSD is that the sense that shares the most relations with the context words is selected as the best-matched one. For supervised WSD systems, the predictive model predicts the potential senses, given the target word and its context words as input. In recent times, the use of knowledge bases has proven advantageous for several modern supervised systems. As a result, there has been a growing trend in integrating knowledge-based and supervised methods to enhance their performance (Wang and Wang, 2020).

---

[4] Current methods likely disambiguate word senses by word co-occurrences. However, word co-occurrences are not commonsense.

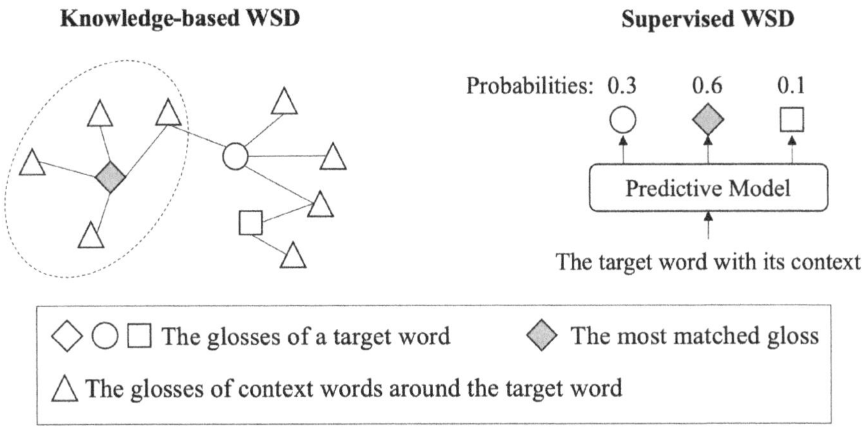

Knowledge-based WSD   **Supervised WSD**

Fig. 3.3: Simplified examples of the knowledge-based and supervised WSD.

WSD has been recognized as a crucial module in numerous NLU tasks that heavily rely on word senses, such as polarity detection, information retrieval, and machine translation. The application of WSD techniques has been demonstrated to be beneficial for these NLU tasks. While prior surveys (Bevilacqua et al., 2021; Navigli, 2009) have conducted extensive reviews for WSD, the works discussed in them are outdated.

3.2.1 Theoretical Research

3.2.1.1 Distributional Semantics

The hypothesis from distributional semantics (Firth, 1957) argued that word meanings can be inferred from word co-occurrences. Words that appear in similar contexts tend to have similar meanings. Such a hypothesis has been the most significant foundation of developing semantic representations in the computational linguistics community, e.g., vector space representations (Turney and Pantel, 2010; Mikolov et al., 2013b; Pennington et al., 2014) and PLMs (Devlin et al., 2018; Liu et al., 2019c). Based on such a hypothesis, dense semantic vectorial representation research commonly follows a similar training paradigm, e.g., using context words to predict a target word. Currently, ChatGPT further proves that learning to use words that have appeared before to predict the next possible word can achieve the skills of analogy and reasoning with the help of a very large Transformer (Vaswani et al., 2017)-based model.

3.2.1.2 Selectional Preference

Wilks (1973) proposed a concept of selectional preference. It is a procedure for representing the meaning structure of natural language. Compared to the "derivational paradigm" of transformational grammar and generative semantics, Wilks (1973) believed that selectional preference is a more efficient procedure in NLU. It focuses on determining preferences between various possible interpretations of a text, rather than identifying a solitary and unequivocally correct interpretation. Selection preference theory allows more flexibility and nuance in understanding word senses and language. Besides, the theory is computation-friendly. Wilks (1973) showed how the procedure could be computed and implemented. The work of Wilks (1973) supports that there are multiple possible meanings for a word. The meaning can be defined by the sectional preference of contexts.

3.2.1.3 Construction Semantics

Goldberg and Suttle (2010) argued that the meanings of words are frequently derived from larger language units, termed constructions. Constructions consist of a form and a meaning, ranging from single words to full sentences in size. The interpretation of a construction is reliant on both its structure and the situations in which it is employed. Goldberg and Suttle (2010) argued that semantic restrictions are better linked with the construction as an entirety rather than with the lexical semantic framework of the verbs. The work of Goldberg and Suttle (2010) highlights that the interpretation of meanings of language units can be extended from individual words to constructions. It shows the necessity of defining language units in WSD.

3.2.1.4 Frame Semantics

Fillmore et al. (2006) proposed frame semantics that provides a distinct viewpoint on the meanings of words and the principles behind language construction. Frame semantics emphasizes the significance of the surrounding context and encyclopedic knowledge in comprehending word meanings. Petruck (1996) explained that a "frame" refers to a collection of concepts interconnected in a manner that understanding any one concept depends on the understanding of the complete system. In frame semantics, the meaning of "cooking" is beyond its dictionary meaning. It also associates with the concept of "food", "cook", "container", and "heating instrument". Frame semantics motivates later ontology research, e.g., FrameNet (Ruppenhofer et al., 2016) and FrameNet-based WSD systems, significantly.

3.2.2 Annotation Schemes

For knowledge-based WSD, the data are normally presented as ontology, such as WordNet, FrameNet, and BabelNet, where words and concepts are connected by relations. The relations include hyponyms, hypernyms, holonyms, meronyms, attributes, entailment, etc. An explanation (gloss) and a few example sentences are given for each synset. Synsets of the same POS are connected under some relations independently. However, there exist relations when the basic concept of two words is the same but in a different POS (for example, "propose" and "proposal" were characterized as "derivationally related synsets" in WordNet). For supervised WSD, a particular word in a given sentence is annotated with a sense ID that corresponds to one of the potential senses in a knowledge base, such as WordNet. A sample of annotation is shown in the next section.

3.2.3 Datasets

Our surveyed datasets and their statistics can be viewed in Table 3.2. The biggest manually annotated English corpus currently accessible is SemCor[5] (Miller et al., 1993). It has 200K content terms tagged with their related definitions and around 40K phrases. Although SemCor serves as the principal training corpus for WSD, its limited coverage of the English vocabulary for both words and meanings is its most significant drawback. In essence, SemCor merely includes annotations for 22K distinct lexemes in WordNet, the most extensive and commonly employed computerized English dictionary, which corresponds to less than 15% of all words. To augment the coverage of words, Vial et al. (2019) incorporated the English Princeton WordNet Gloss Corpus (WNG)[6], which contains more than 59K WordNet senses, as a complemented data. The WNG is annotated manually or semi-automatically.

SemCor and its variations (Bentivogli and Pianta, 2005; Bond et al., 2012) lack an acceptable multilingual equivalent in the majority of global languages, which limits the scaling capabilities of WSD models beyond English. To address the aforementioned issues, numerous automatic methods for creating multilingual sense-annotated data have been developed (Pasini and Navigli, 2017; Pasini et al., 2018; Scarlini et al., 2019). In an English-Italian parallel corpus known as MultiSemCor (Pianta et al., 2002), senses from the English and Italian versions of WordNet are annotated. The Line-hard-serve corpus (Leacock et al., 1993) contains 4K samples of the nominal, adjective, and verbal words with sense tags. The data were sourced from Wall Street Journal (WSJ) corpus and the American Printing House for the Blind (APHB) corpus. The Interest corpus (Bruce and Wiebe, 1999) contains 2,369 occurrences of the term *interest* that have been sense-labeled. The data were sourced from the HECTOR word sense corpus (Atkins, 1992).

---

[5] https://web.eecs.umich.edu/~mihalcea/downloads.html
[6] https://wordnetcode.princeton.edu/glosstag.shtml

Table 3.2: WSD datasets and statistics. EMEA, KDEdoc, and EUB denote European Medicines Agency documents, KDE manual corpus, and the EU bookshop corpus, respectively.

Dataset	Source	# Samples	Reference
SemCor	WordNet	200,000	Miller et al. (1993)
MultiSemCor	WordNet, bilingual Collins	51,847	Pianta et al. (2002)
Line-hard-serve	WSJ, APHB	4,000	Leacock et al. (1993)
Interest	HECTOR	2,369	Bruce and Wiebe (1999)
DSO	Brown, WSJ	192,800	Ng and Lee (1996)
OMWE	Web	29,165	Chklovski and Pantel (2004)
OMSTI	UN documents	1,357,922	Taghipour and Ng (2015b)
SensEval-2	Unknown	2,282	Edmonds and Cotton (2001)
SensEval-3	Editorial, news story, & fiction	1,850	Snyder and Palmer (2004)
SemEval-2007	Brown, WSJ	455	Pradhan et al. (2007)
SemEval-2013	SMT workshop	1,644	Navigli et al. (2013)
SemEval-2015	EMEA, KDEdoc, EUB	1,022	Moro and Navigli (2015)

The Defence Science Organisation (DSO), based in Singapore, created the DSO corpus[7] (Ng and Lee, 1996), which contains 192,800 sense-tagged tokens from 191 words from the Brown and WSJ corpora. The Open Mind Word Expert (OMWE) dataset[8] (Chklovski and Pantel, 2004) is a corpus of sentences with 288 noun occurrences that were jointly annotated by Web users. One Million Sense-Tagged for WSD and Induction (OMSTI)[9] (Taghipour and Ng, 2015b) is a semi-automatically annotated WSD dataset with WordNet sense inventory. The data were sourced from MultiUN corpus, which is a collection of United Nation documents. The OMSTI includes 687,871 nouns, 412,482 verbs, and 251,362 adjectives and 6,207 adverbs after including selected samples from SemCor and DSO.

The SensEval and SemEval datasets are created from the SensEval/SemEval evaluation campaigns. Now, these datasets have been the most widely used benchmarking datasets in WSD. Raganato et al. (2017b) collected these datasets together[10] and developed a unified evaluation framework for empirical comparison. The statistics of the following datasets are from the collection of Raganato et al. (2017b). SensEval-2 (Edmonds and Cotton, 2001) used WordNet 1.7 sense inventory, including 2,282 sense annotations for nouns, verbs, adverbs and adjectives. SensEval-3 (Snyder and Palmer, 2004) employed WordNet 1.7.1 sense inventory, including 1,850 sense annotations. SemEval-2007 Task 17 (Pradhan et al., 2007) employed WordNet 2.1 sense inventory, including 455 nominal and verbal sense annotations.

---

[7] https://borealisdata.ca/dataset.xhtml?persistentId=doi:10.5683/SP2/QPOJSI

[8] https://web.eecs.umich.edu/~mihalcea/downloads/OMWE

[9] https://https://comp.nus.edu.sg/~nlp/corpora.html

[10] https://lcl.uniroma1.it/wsdeval/home

SemEval-2013 Task 12 (Navigli et al., 2013) used WordNet 3.0 sense inventory, including 1,644 nominal sense annotations. SemEval-2015 Task 13 (Moro and Navigli, 2015) utilized WordNet 3.0 sense inventory, including 1,022 sense annotations. It is worth noting that some of the SemEval tasks are multilingual, including SemEval-2013 and SemEval-2015, which facilitates multilingual WSD.

All of these corpora are annotated using various WordNet sense inventories, with the exception of the Interest corpus (tagged with LDOCE senses) and the SensEval-1 corpus. The Interest corpus and the SensEval-1 corpus were sense-labeled using the HECTOR sense inventories, a lexicon and corpus from a joint Oxford University Press/Digital project (Atkins, 1992) Generally, the data and labels in WSD datasets are organized in the following forms. Then, the task is to identify the sense classes, given contexts, and target words.

```
context: "You perform well in the exam, I will reward you.",
target word: "perform",
pos: "VB",
sense: "3"


context: "She worked in a renowned university for a long time.",
"target word": "university",
"pos": "NN",
"sense": "2"
```

3.2.4 Knowledge Bases

3.2.4.1 Machine-Readable Dictionaries

Machine-readable dictionaries (MRDs) have been a useful source for WSD due to their structured knowledge and easy access (Navigli, 2009). Dictionaries frequently contain extensive information about the various meanings of a word, as well as illustrative examples of their usage within context. Therefore, dictionaries can serve as valuable knowledge bases for the task of WSD. Additionally, MRDs may provide further information such as synonyms, antonyms, and related words, which can aid in facilitating a more comprehensive comprehension of a word's meaning. Through the analysis of this information, a system may make more precise determinations about which meaning is most fitting in a given context. There are many electronic dictionaries available for machines to refer to, such as the Longman Dictionary of Contemporary English (LDOCE) (Mayor, 2009), the Oxford Dictionary of English (Dictionary, 2010), Collins English Dictionary (Dictionary, 1982), and the Oxford Advanced Learner's Dictionary of Current English (OALD) (Hornby and Cowie, 1974).

Table 3.3: Useful knowledge bases for WSD. LDOCE means Longman Dictionary of Contemporary English. ODE means Oxford Dictionary of English. CED means Collins English Dictionary. OALD means Oxford Advanced Learner's Dictionary of Current English. Unstructured or structured means the knowledge base contains unstructured or structured lexical knowledge by concepts.

Name	Knowledge	# Entities	Structure
LDOCE 6th ed.	Lexical	230,000	Unstructured
ODE 2022	Lexical	600,000	Unstructured
CED 12th ed.	Lexical	722,000	Unstructured
OALD 8th ed.	Lexical	145,000	Unstructured
WordNet	Lexical	95,600	Graph
FrameNet	Lexical	13,687	Graph
BabelNet	Lexical & Multilingual	26,044,643	Graph
SyntagNet	Lexical	78,000	Graph

3.2.4.2 WordNet

WordNet (Miller, 1995) is a sizable, manually curated lexicographic database of English. It is arranged as a network with synsets, or collections of contextual synonyms, as nodes. A synset of synonyms each represents one of a word's senses. Through edges that express lexical-semantic links like meronymies (partof) and hypernymies (is-a), synsets and senses are connected to one another. WordNet additionally offers definitions (glosses) and uses examples for each synset as additional lexical information. The most current English WSD works use the 3.0 version, which was published in 2006 and has 117,659 synsets. Following the initial WordNet for English, many WordNets for other languages have been proposed, including languages such as Chinese (Wang and Bond, 2013), Arabic (Black et al., 2006), Dutch (Postma et al., 2016), and more[11].

3.2.4.3 FrameNet

FrameNet (Ruppenhofer et al., 2016) is an English lexical repository that is readable by both humans and machines, established by annotating real-life textual examples that depict the usage of words. It was developed based on the theory of frame semantics, containing 1,224 frames (a frame refers to a diagrammatic representation of a scenario encompassing diverse elements such as participants, props, and other conceptual roles), and 13,687 lexical units (lemmas and their POS) that evoke frames. In FrameNet, the lexical units of a sentence are associated with frame elements. Frame elements are the semantic role of lexical units. For example, given a sentence "I ate an apple this afternoon", "apple" would fill the role of "food" (a frame element).

---

[11] https://globalwordnet.org/resources/wordnets-in-the-world

3.2.4.4 BabelNet

BabelNet (Navigli and Ponzetto, 2012b) is a multilingual dictionary that covers both lexicographic and encyclopedic entries from 520 languages. These entries were created by semi-automatically mapping numerous sites, including WordNet, Multilingual WordNet, and Wikipedia. The topology of BabelNet is that of a semantic network, where the nodes are multilingual synsets (collections of synonyms that have been lexicalized in several languages), and the edges represent the semantic connections between them.

3.2.4.5 SyntagNet

SyntagNet (Maru et al., 2019) is a manually developed lexical resource that integrates semantically disambiguated lexical combinations, e.g., noun-verb and noun-noun pairs. The development of SyntagNet involved initially extracting lexical combinations from English Wikipedia and the British National Corpus, which were then subjected to a process of manual disambiguation, based on the WordNet. SyntagNet covers five major languages, e.g., English, German, French, Spanish, and Italian.

3.2.5 Evaluation Metrics

In the WSD task, given a sentence of n words $T = \{x_1, \ldots, x_n\}$, the model predicts a sense for each word given the dictionary. Normally, the F_1 score is adopted, which is a specialization of the F score when $\alpha = 1$:

$$F = \frac{1}{\alpha \frac{1}{P} + (1 - \alpha)\frac{1}{R}} \tag{3.1}$$

Where P denotes precision and R denotes recall:

$$P = \frac{\text{correct predictions}}{\text{total predictions}} \tag{3.2}$$

$$R = \frac{\text{correct predictions}}{n} \tag{3.3}$$

These metrics do not accurately represent how well systems can produce a level of confidence for a particular sensory choice. To this end, Resnik and Yarowsky (1999) developed an evaluation criterion that considers the discrepancies between the accurate and selected senses to weigh misclassification mistakes. Errors get penalized less severely than coarser sense distinctions if the chosen sense is a fine-grained distinction of the true sense. There have been evaluation metrics for even more precise measurements, including the receiver operation characteristic (ROC). These metrics, however, are not frequently utilized as precision, recall, and F_1.

3.2.6 Annotation Tools

LX-SenseAnnotator[12] (Neale et al., 2015) provides a user interface for manually annotating word senses. The software has the capability to process lexical data in any language, on the condition that the data is compliant with the format of Princeton WordNet. Human annotators can view the preprocessed text in three different modes, including the source text, sense-annotated text, and raw text, which can be switched between by using a tab widget. The source text mode displays the original text along with all tags, while the sense-annotated text mode displays the same text but with newly added sense tags. This allows the annotator to monitor the output file continually. Annotators can view the sense options in real time when annotating the sense for a word. LexTag[13] is another useful tool for WSD, it has been used to create a recent 10-language parallel dataset ELEXIS-WSD 1.0[14].

3.2.7 Methods

3.2.7.1 Knowledge-based WSD

Knowledge-based WSD utilizes knowledge bases to disambiguate word senses. Compared with supervised WSD, this class of WSD methods achieves lower performance but better data efficiency and transparency.

A. Semantic Space Matching

This matching algorithm looks for similarities between the context of a term whose sense needs to be disambiguated and its sense representation, such as the definition of a potential sense and its associated sense that was retrieved from a knowledge base. Lesk is a naïve knowledge-based WSD algorithm that looks for terms that are similar to the target word in the context of each sense (Lesk, 1986). It enumerates the intersections among lexicon definitions of the diverse connotations of every target word contained within a given sentence. Banerjee et al. (2003) proposed an advanced version of the Lesk where the standard TF-IDF method is employed for word weighting. Another improved version of Lesk (Basile et al., 2014) includes word embeddings to improve the accuracy of determining how close the definition and context of the target word are. Finally, $SREF_{KB}$ (Wang and Wang, 2020) is a vector-based technique that disambiguates word senses by using sense embeddings and contextualized word representations. It applies BERT to represent WordNet instances and definitions, as well as the automatically obtained contexts from the Web.

---

[12] https://nlx.di.fc.ul.pt/tools.html

[13] https://babelscape.com/lextag

[14] https://clarin.si/repository/xmlui/handle/11356/1674

A. Graph-based matching

This other matching algorithm creates a graph using the context and connections retrieved from knowledge bases. Synsets and the relationships between them become graph nodes and edges, respectively. Senses are then disambiguated based on the constructed graphs. A variety of graph-based techniques, such as LDA (Blei et al., 2003), PageRank (Brin and Page, 1998), Random Walks (Agirre et al., 2014), Clique Approximation (Moro et al., 2014b), and Game Theory (Tripodi and Navigli, 2019), are used to disambiguate the meaning of a given word using the created graph. Agirre and Soroa (2009) presented a graph-based unsupervised WSD system that employs random walk over a WordNet semantic network. They employed a customized version of the Page Rank algorithm (Haveliwala, 2002). The technique leverages the inherent structural properties of the graph that underlies a specific lexical knowledge base, and shows the capability of the algorithm to identify global optima for WSD, based on the relations among entities. Agirre et al. (2014) evaluated this algorithm with new datasets and variations of the algorithm to prove its effectiveness.

Navigli and Lapata (2007) also introduced a graph-based unsupervised model for WSD, which analyzed the connectivity of graph structures to identify the most pertinent word senses. A graph is constructed to represent all possible interpretations of the word sequence, where nodes represent word senses and edges represent sense dependencies. The model assessed the graph structure to determine the significance of each node, thus finding the most crucial node for each word. Babelfy (Moro et al., 2014b) is also a graph-based WSD method that uses random walk to identify relationships between synsets. It used BabelNet (Navigli and Ponzetto, 2012b) and performed random walks with Restart (Tong et al., 2006). In addition, it incorporated the entire document at the time of disambiguation. The candidate disambiguation is upon automatically developed semantic interpretation graph which used a graph structure to represent various possible interpretations of input text.

SyntagRank (Scozzafava et al., 2020) is a high-scoring knowledge-based WSD algorithm. It is an entirely graph-based algorithm that uses the Personalized PageRank algorithm to incorporate WordNet (for English), BabelNet (for non-English) and SyntagNet. SyntagRank is generally considered a stronger method than $SREF_{KB}$. BabelNet enabled SyntagRank to improve its ability to scale across a wide range of languages, whereas $SREF_{KB}$ has only been evaluated in English.

3.2.7.2 Supervised WSD

Currently, supervised approaches, especially deep learning-based supervised learning approaches, have become mainstream in the WSD community. Earlier deep learning-based approaches focused on architectures where WSD was defined as token classification over WordNet senses (Kågebäck and Salomonsson, 2016). Even though they performed well, these structures showed a lot of flaws, particularly when it came to predicting uncommon and invisible senses.

To address these issues, numerous works began to supplement the training data by utilizing various lexical knowledge, such as sense definitions (Kumar et al., 2019a; Blevins and Zettlemoyer, 2020), semantic relations (Bevilacqua and Navigli, 2020; Conia and Navigli, 2021), and data generated via novel generative methods (Barba et al., 2021b). In this section, we review representative works in supervised WSD.

A. Data-Driven Machine Learning Approaches

Data-driven machine learning approaches refer to methodologies and techniques in which the design, training, and optimization of machine learning models heavily rely on large amounts of data. In these approaches, the model's ability to generalize patterns and make predictions is learned directly from the provided data, rather than being explicitly programmed by humans. The data-driven nature of these approaches allows models to learn complex patterns, relationships, and representations present in the data itself. Singh et al. (2014) employed 5-gram and position features, and a decision tree algorithm to represent classification rules in a tree structure where the training dataset is recursively partitioned. Each leaf node indicates the meaning of a word. They developed a dataset, containing 672 Manipuri sentences to test their method. The sentences were sourced from a local newspaper, termed "The Sangai Express". O'Hara et al. (2004) proposed a class-based collocation method that integrates diverse linguistic features in a decision tree algorithm. For the collocation, three distinct word relatedness scores are used: the first is based on WordNet hypernym relations; the second is based on cluster-based word similarity classes; and the third is based on dictionary definition analysis. The authors also utilized POS and word form features. The It Makes Sense (IMS) WSD system (Zhong and Ng, 2010) used an SVM classifier. Different positional and linguistic features were considered, including nearby words, nearby words' POS tags, and nearby collocations. Later, word embeddings became important features in WSD. Taghipour and Ng (2015b); Rothe and Schütze (2015); Iacobacci et al. (2016) used IMS as the base model to examine word embeddings. Iacobacci et al. (2016) offered many approaches where different word embeddings were applied as features to test how many parameters impact the effectiveness of a WSD system. The authors found that word2vec which was trained with OMSTI can yield the strongest results on the three examined all-word WSD tasks.

B. Data-Driven Neural Approaches

Data-driven neural approaches refer to methodologies and techniques that utilize neural networks to learn patterns and representations directly from data. These approaches leverage the power of neural networks, which are computational models inspired by the structure and functioning of the human brain, to automatically learn complex relationships, features, and representations present in the provided data. Neural networks consist of interconnected nodes (neurons) organized in layers, allowing them to capture hierarchical and non-linear patterns.

Popov (2017) proposed to use BiLSTM (Graves and Schmidhuber, 2005), GloVe word embeddings, and word2vec lemma embeddings. Yuan et al. (2016) suggested another LSTM-based WSD approach that was trained in a semi-supervised fashion. SSL was achieved by employing label propagation (Talukdar and Crammer, 2009) to assign labels to unannotated sentences by assessing their similarity to labeled ones. The best performance on the SensEval-2 dataset can be observed from the model that was semi-supervision-trained with OMSTI and 1,000 additional unlabeled sentences. Additionally, Le et al. (2018) looked more closely at how many elements affect its performance, and several intriguing conclusions were drawn. The initial point to highlight is that achieving strong WSD performance does not necessitate an exceedingly large unannotated dataset. Furthermore, this method provides a more evenly-distributed sense assignment in comparison to prior approaches, as evidenced by its relatively strong performance on infrequent cases. Additionally, it is worth noting that the limited sense coverage of the annotated dataset may serve as an upper limit on overall performance.

With the development of self-attention-based neural architectures and their capacity to extract sophisticated language information (Vaswani et al., 2017), the use of transformer-based architectures in fully supervised WSD systems is becoming more and more popular. The WSD task is usually fine-tuned on a pretrained transformer model, which is a popular strategy. The task-specific inputs are given to the pretrained model, which is then further trained across a number of epochs with the task-specific objective. Likewise, in recent token classification models for WSD, the contextualized representations are usually generated by a pretrained model and then fed to either a feedforward network (Hadiwinoto et al., 2019) or a stack of Transformer layers (Bevilacqua and Navigli, 2019). These methods outperform earlier randomly initialized models (Raganato et al., 2017a). Hadiwinoto et al. (2019) tested different pooling strategies of BERT, e.g., last layer projection, weighted sum of hidden layers, and Gated Linear Unit (Dauphin et al., 2017).

The best performance on SensEval-2 is given by the strategy of the weighted sum of hidden layers, accounting for 76.4% F_1. Bevilacqua and Navigli (2019) proposed a bi-directional Transformer that explicitly attends to past and future information. This model achieved 75.7% F_1 on SensEval-2 by training with the combination of SemCor and WordNet's Tagged Glosses[15]. It is worth noting that, the categorical cross-entropy, which is frequently utilized for training, limits the performance. In reality, it has been demonstrated that the binary cross-entropy loss performs better (Conia and Navigli, 2021) because it enables the consideration of many annotations for a single instance in the training set as opposed to the use of a single ground-truth sense alone. In the above-mentioned approaches, each sense is assumed to be a unique class, and the classification architecture is limited to the information provided by the training corpus.

---

[15] https://wordnetcode.princeton.edu/glosstag.shtml

3.2.7.3 Knowledge-augmented Supervised WSD

The edges that connect the senses and synsets are a valuable source of knowledge that augments the annotated data. Traditionally, graph knowledge-based systems, such as those based on Personalized PageRank (Scozzafava et al., 2020), have taken advantage of this information. Moreover, utilizing WordNet as a graph has benefited many modern supervised systems. Thus, formally, knowledge-augmented supervised WSD is defined as a methodology that combines traditional supervised machine learning techniques with external knowledge resources to improve the accuracy and performance of WSD.

Wang and Wang (2020) used WordNet hypernymy and hyponymy relations to devise a try-again mechanism that refines the prediction of the WSD model. The SemCor corpus was utilized to acquire a supervised sense embedding for every annotated sense in their supervised method ($SREF_{Sup}$). Vial et al. (2019) reduced the number of output classes by mapping each sense to an ancestor in the WordNet taxonomy, then yielding a smaller but robust sense vocabulary. The authors used BERT contextualized embeddings. By training with SemCor and WordNet gloss corpora, the model achieved 79.7% F_1 on SensEval-2. Different variations also achieve outstanding performance on diverse WSD datasets.

Loureiro and Jorge (2019) created representations for those senses not appearing in SemCor by using the averaged neighbor embeddings in the WordNet. The token-tagger models EWISE (Kumar et al., 2019a) and EWISER (Bevilacqua and Navigli, 2020) both leveraged the WordNet graph structure to train the gloss embedding offline, where EWISER demonstrated how the WordNet entire graph feature can be directly extracted. EWISE used ConvE (Dettmers et al., 2018) to obtain graph embeddings. Conia and Navigli (2021) provided a new technique to use the same edge information by replacing the adjacency matrix multiplication with a binary cross-entropy loss where other senses connected to the gold sense are also taken into account. The edge information was obtained from WordNet. In general, edge information is increasingly used in supervised WSD, gradually blending with knowledge-based techniques. However, it can only be conveniently utilized by token classification procedures, whereas its incorporation into sequence classification techniques has not yet been researched.

It has also been extensively studied how to use sense definitions as an additional source for supervised WSD apart from the traditional data annotations. It considerably increased the scalability of a model on the senses that are underrepresented in the training corpus. Huang et al. (2019a) argued that WSD has traditionally been approached as a binary classification task, whereby a model must accurately decide if the sense of a given word in context aligns with one of its potential meanings in a sense inventory, based on the provided definition. define the WSD task as a sentence-pair classification task, where the WordNet gloss of a target word is concatenated after an input sentence. Blevins and Zettlemoyer (2020) used a bi-encoder to project both words in context and WordNet glosses in a common vector space. Disambiguation is then carried out by determining the gloss that is most similar to the target word.

Glosses are employed similarly by more advanced techniques like SensEm-BERT (Scarlini et al., 2020b), ARES (Scarlini et al., 2020a), and SREF (Wang and Wang, 2020). They used quite different approaches to find new contexts automatically in order to develop the supervised portion of the sense embedding. ARES achieved 78.0% F_1 on the SensEval-2 dataset by utilizing collocational relations between senses to get novel example sentences from websites. SensEmBERT leveraged BabelNet and Wikipedia explanations, achieving significant improvements on nominal WSD tasks over 5 major datasets. Barba et al. (2021a) proposed to solve WSD as a text extraction problem where, given a word in context and all of its potential glosses, models extract the definition that best matches the term under consideration. The authors demonstrated the advantages of their approach in that it does not require huge output vocabularies and enables models to take into account both the input context and all meanings of the target word simultaneously. By using sparse coding, Berend (2020) has demonstrated that it is also possible to make existing sense embeddings sparse. All of these methods handle each word independently of the others when disambiguating multiple words that co-occur in the same context. Thus, a word's explicit meaning is neither taken into account during word disambiguation nor does it have an impact on the disambiguation of surrounding words.

3.2.8 Downstream Applications

3.2.8.1 Polarity Detection

WSD has been applied in many Sentiment Analysis works to improve accuracy and explainability. Farooq et al. (2015) proposed a WSD framework to enhance the performance of sentiment analysis. To determine the orientation of opinions related to product attributes in a particular field, a lexical dictionary comprising various word senses is developed. The process involves extracting relevant features from product reviews and identifying opinion-bearing texts, followed by the extraction of words used to describe the features and their contexts to form seed words. These seed words, which consist of adjectives, nouns, verbs, and adverbs, are manually annotated with their respective polarities, and their coverage is extended by retrieving their synonyms and antonyms. WSD was utilized to identify the sentiment-orientated senses, such as the positive, negative, or neutral senses of a word in a sentence, because a word may have different sentiment polarities by taking different senses in different contexts.

Nassirtoussi et al. (2015) offered a novel approach to forecast intra-day directional movements of the EUR/USD exchange rates based on news headline text mining in an effort to address semantic and sentiment components of text-mining. They evaluated news headlines semantically and emotionally using the lexicons, e.g., WordNet and SentiWordNet (Baccianella et al., 2010). SentiWordNet is a publicly accessible lexical resource designed for sentiment analysis that allocates a positivity score, negativity score, and objectivity score to each synset within WordNet.

Nassirtoussi et al. (2015) found that both positive and negative emotions may influence the market in the same way. WSD worked as a technique to abstract semantic information in their framework. Thus, it enhances the feature representations and explainability in their downstream task modeling. SentiWordNet has served as a basis for various sentiment analysis models. In the work of Ohana and Tierney (2009), the feasibility of using the emotional scores of SentiWordNet to automatically classify the sentiment of movie reviews was examined. Other applications, e.g., business opinion mining (Saggionα and Funk, 2010), article emotion classification (Devitt and Ahmad, 2007), word-of-mouth sentiment classification (Hung and Lin, 2013; Hung and Chen, 2016) also showed that SentiWordNet as a semantic feature enhancement knowledge base can deliver accuracy gains in sentiment analysis tasks.

3.2.8.2 Information Retrieval

The impacts of using WSD for information retrieval have been examined in many works. Krovetz and Croft (1992) disambiguated word senses for terms in queries and documents to examine how ambiguous word senses impact information retrieval performance. The researchers arrived at the conclusion that the advantages of WSD in information retrieval are marginal. This is due to the fact that query words have uneven sense distributions. The impact of collocation from other query terms already plays a role in disambiguation. WSD was used as a parser to study this task. However, the findings from Gonzalo et al. (1998) are different. They examined the impact of improper disambiguation using SemCor. By accurately modeling documents and queries together with synsets, they achieved notable gains (synonym sets). Additionally, their study showed that WSD with a 40%-50% error rate could still improve information retrieval performance when using synset representation with synonyms.

Gonzalo et al. (1999); Stokoe et al. (2003) further confirmed the significance of WSD to information retrieval. Gonzalo et al. (1999) also found that POS information has a lower utility for information retrieval. Based on artificially creating word ambiguity, Sanderson (1994) employed pseudo words to explore the effects of sense ambiguity on information retrieval. They came to the conclusion that the high accuracy of WSD is a crucial condition to accomplish progress. Blloshmi et al. (2021) introduced an innovative approach to multilingual query expansion by integrating WSD, which augments the query with sense definitions as supplementary semantic information in multi-lingual neural ranking-based information retrieval. The results demonstrated the advantages of WSD in improving contextualized queries, resulting in a more accurate document-matching process and retrieving more relevant documents. Kim et al. (2004) labeled words with 25 root meanings of nouns rather than utilizing fine-grained sense inventories of WordNet. Their retrieval technique preserved the stem-based index and changed the word weight in a document in accordance with the degree to which it matched the query's sense. They credited their coarse-grained, reliable, and adaptable sense tagging system with the improvement on TREC collections. The detrimental effects of disambiguation mistakes are somewhat mitigated by the addition of senses to the conventional stem-based index.

3.2.8.3 Machine Translation

The challenge of ambiguous word senses poses a significant barrier to the development of an efficient machine translator. As a result, a number of researchers have turned their attention to exploring WSD for machine translation. Some works tried to establish datasets to quantify the WSD capacity of machine translation systems. Rios Gonzales et al. (2017) proposed a test set of 6,700 lexical ambiguities for German-French and 7,200 for German-English. They discovered that WSD remains a difficult challenge for NMT, especially for uncommon word senses, even with 70% of lexical ambiguities properly resolved. Campolungo et al. (2022) proposed a benchmark dataset that aims at measuring WSD biases in machine translation in five language combinations. They also agreed that state-of-the-art systems still exhibited notable constraints when confronted with less common word senses. Raganato et al. (2019) proposed MUCOW, a multilingual contrastive test set automatically created from word-aligned parallel corpora and the comprehensive multilingual sense inventory of BabelNet. MUCOW spans 16 language pairs and contains more than 200,000 contrastive sentence pairs. The researchers thoroughly evaluated the effectiveness of the ambiguous lexicons and the resulting test suite by utilizing pretrained NMT models and analyzing all submissions across nine language pairs from the WMT19 news shared translation task.

Some works analyzed the internal representations to understand the disambiguation process in machine translation systems. Marvin and Koehn (2018) examined the extent to which ambiguous word senses could be decoded through the use of word embeddings in relation to deeper layers of the NMT encoder, which were believed to represent words with contextual information. In line with prior research, they discovered that the NMT system frequently mistranslated ambiguous terms. Tang et al. (2019) trained a classifier to determine if a translation is accurate given the representation of an ambiguous noun. The fact that encoder hidden states performed much better than word embeddings suggests that encoders are able to appropriately encode important data for disambiguation into hidden states. Liu et al. (2018a) discovered that an increase in the number of senses associated with each word results in a decline in the performance of word-level translation. The root of the issue may be the mapping of each word to similar word vectors, regardless of its context. They proposed to integrate techniques from neural WSD systems into an NMT system to address this issue.

3.2.9 Summary

WSD as a computational linguistics task most closely related to lexical semantics research, has won extensive discussions among researchers from different fields. Linguists came up with important hypotheses to guide the modeling of word senses. We have observed that some hypotheses have been well grounded in NLU, e.g., learning and representing word meanings with their contexts and word co-occurrences.

However, we also observe some important linguistic arguments were rarely studied in the computational linguistic domain, e.g., defining the scope of linguistic units for WSD and integrating relevant concepts (frames) for word sense representations. The development of WSD datasets has greatly ignited the research enthusiasm of scholars in WSD. However, we also observed that the computational research on WSD is also limited by these well-defined datasets because WSD datasets generally follow a very similar labeling paradigm. Relevant linguistic studies have shown broader possibilities in WSD. Finally, we find that many of WSD modeling techniques do not link well with downstream applications. The research of WSD methods has intersections with downstream applications, whereas they cannot well cover the needs of downstream tasks. This also shows that the research opportunities in WSD can be largely extended besides word sense classification.

3.2.9.1 Technical Trends

Table 3.4 shows the technical trends of WSD methods. As seen in the table, earlier approaches likely used knowledge-based and supervised approaches. WordNet and BabelNet are useful knowledge bases that were frequently used by knowledge-based methods. Word embeddings, PLMs, and linguistic features, e.g., POS tags and semantic relatedness were frequently used by supervised methods. For old pure knowledge-based methods, the PageRank framework was likely used, because many knowledge bases are represented as graphs. PageRank is an algorithm used in graph computation to measure the importance of nodes in a graph. Classical machine learning techniques, e.g., decision trees and SVM, were commonly used by supervised WSD methods. Supervised learning algorithms demonstrate superior performance in comparison to knowledge-based approaches. Nevertheless, it is not always reasonable to assume the availability of substantial training datasets for different areas, languages, and activities. Ng (1997) predicted that a corpus of around 3.2 million sense-tagged words would be necessary in order to produce a high-accuracy, wide-coverage disambiguation system. The creation of such a training corpus requires an estimated 27 person-years of labor. The accuracy of supervised systems might be greatly improved above state-of-the-art methods with such a resource. However, the success of this hypothesis is at the cost of huge resource consumption.

We observe more hybrid approaches that leverage knowledge bases in a supervised learning fashion in recent years. This is because researchers have observed the limitations of typical supervised WSD in processing rare or unseen cases. Knowledge bases provide additional information to support the learning of unseen cases. Knowledge bases provide additional knowledge for the languages whose annotated data are scarce. In this case, multilingual knowledge bases can enhance the representations of word senses in a new domain. As a result, we can observe the accuracy of the hybrid approaches surpasses the pure knowledge-based or supervised approaches. Most existing WSD datasets define the task as a word sense classification task. Then, the following methodology research upon the datasets focused on improving the accuracy of mapping the sense of a word to its dictionary sense class.

Table 3.4: A summary of representative WSD techniques. Knwl denotes knowledge-based methods. Sup. denotes supervised methods. KB denotes knowledge bases. WN denotes WordNet. BN denotes BabelNet. DSM denotes Distributional Semantics Models. Prob. denotes probability. SE2013-EN denotes the SemEval-2013 English WSD task.

Task	Reference	Tech	Feature and KB.	Framework	Dataset	Score	Metric
Knwl	Lesk (1986)	Prob.	Statistics, OALD	Count def. overlaps	-	-	-
	Banerjee et al. (2003)	ML	Emb., WN	Score function	SensEval-2	34.6%	F_1
	Navigli and Lapata (2007)	Graph	Sense graph, WN	Connectivity measures	SemCor	31.8%	F_1
	Basile et al. (2014)	Prob.	Emb., BN	DSM	SE2013-EN	71.5%	F_1
	Wang and Wang (2020)$_{KB}$	DL	BERT, WN	Vector represent.	SensEval-2	72.7%	F_1
	Agirre and Soroa (2009)	Graph	WN	PageRank	SensEval-2	58.6%	Recall
	Moro et al. (2014b)	Graph	Sem. graph, BN	PageRank	SE2013-EN	69.2%	F_1
	Scozzafava et al. (2020)	Graph	WN, SN	PageRank	SensEval-2	71.6%	F_1
Sup.	Singh et al. (2014)	ML	5-gram, position	Decision Tree	Manipuri	71.7%	Acc
	O'Hara et al. (2004)	ML	Relatedness score	Decision Tree	SensEval-3	65.9%	F_1
	Zhong and Ng (2010)	ML	Position, POS	SVM	SensEval-2	68.2%	F_1
	Iacobacci et al. (2016)	ML	Emb., POS	SVM	SensEval-2	68.3%	F_1
	Popov (2017)	DL	Emb.	BiLSTM	SensEval-2	70.1%	Acc
	Yuan et al. (2016)	DL	Emb., label prop.	LSTM	SensEval-2	74.4%	F_1
	Le et al. (2018)	DL	Emb.	LSTM	SensEval-2	72.0%	F_1
	Hadiwinoto et al. (2019)	DL	BERT	Transformer	SensEval-2	76.4%	F_1
	Bevilacqua and Navigli (2019)	DL	Emb.	BiTransformer	SensEval-2	75.7%	F_1
Knwl + Sup.	Wang and Wang (2020)$_{Sup}$	DL	BERT, WN	Vector represent.	SensEval-2	78.6%	F_1
	Vial et al. (2019)	DL	BERT, WN	Transformer	SensEval-2	79.7%	F_1
	Loureiro and Jorge (2019)	DL	BERT, WN	Transformer	SensEval-2	76.3%	F_1
	Kumar et al. (2019a)	DL	Graph emb., emb., WN	BiLSTM, Att. ConvE	SensEval-2	73.8%	F_1
	Bevilacqua and Navigli (2020)	DL	BERT, WN	Trans., Struct. logit	5 datasets	80.8%	F_1
	Conia and Navigli (2021)	DL	BERT, WN	Transformer	SensEval-2	78.4%	F_1
	Huang et al. (2019a)	DL	BERT, WN	Transformer, sentence-pair classification	SensEval-2	77.7%	F_1
	Blevins and Zettlemoyer (2020)	DL	BERT, WN	Trasformer, Score Func.	SensEval-2	79.4%	F_1
	Scarlini et al. (2020b)	DL	BERT, BN, Wiki	Transformer, Context Retrieval	Nouns of 5 datasets	80.4%	F_1
	Scarlini et al. (2020a)	DL	BERT, WN, SN	Transformer, Context Retrieval	SensEval-2	78.0%	F_1
	Barba et al. (2021a)	DL	BERT, WN	Transformer, Extractive Sense Learning	SensEval-2	81.7%	F_1
	Berend (2020)	DL	BERT, WN	Transformer, sparse coding, PMI	SensEval-2	79.6%	F_1

However, should the research on WSD be limited to word sense classification? We have observed that many knowledge-based systems used existing knowledge bases to conduct word sense classification tasks. They have realized the importance of developing an effective knowledge base for WSD. However, it is rare to see that WSD research tries to improve the construction of knowledge bases according to the effectiveness of word sense classification. On the other hand, the meaning of WSD is much larger than detecting the definition of words in a dictionary. Mapping a word to a sense in a dictionary is just an aspect of WSD. Previous works rarely studied what is an appropriate linguistic unit for WSD; what concepts are associated with a word sense in a context. These are very interesting research topics from linguistic and cognitive aspects. However, these topics were not well studied in the computational WSD community.

Table 3.5: A summary of the representative applications of WSD in downstream tasks. ✓ denotes the role of WSD in a downstream task.

Reference	Downstream Task	Feature	Parser	Explainability
Farooq et al. (2015)	Polarity Detection	✓		
Nassirtoussi et al. (2015)	Polarity Detection	✓		✓
Ohana and Tierney (2009)	Polarity Detection	✓	✓	
Saggionα and Funk (2010)	Polarity Detection	✓	✓	✓
Devitt and Ahmad (2007)	Polarity Detection	✓		✓
Hung and Lin (2013)	Polarity Detection	✓	✓	
Hung and Chen (2016)	Polarity Detection	✓	✓	✓
Krovetz and Croft (1992)	Information Retrieval	✓		
Gonzalo et al. (1998)	Information Retrieval			✓
Gonzalo et al. (1999)	Information Retrieval			✓
Sanderson (1994)	Information Retrieval			✓
Stokoe et al. (2003)	Information Retrieval			✓
Kim et al. (2004)	Information Retrieval	✓	✓	
Blloshmi et al. (2021)	Information Retrieval	✓		✓
Rios Gonzales et al. (2017)	Machine Translation	✓		
Raganato et al. (2019)	Machine Translation	✓		
Marvin and Koehn (2018)	Machine Translation		✓	✓
Tang et al. (2019)	Machine Translation	✓		✓
Liu et al. (2018a)	Machine Translation	✓		

3.2.9.2 Application Trends

The WSD task was commonly defined as a word sense classification task. However, we observe that classifying words by sense classes is not the only need for downstream NLU tasks. There are three main tasks that are strongly related to WSD, e.g., polarity detection, information retrieval, and machine translation. One of the roles of WSD on the three tasks is to deliver or enhance features to gain improvements on the three tasks. On the other hand, we also observe many downstream works used WSD techniques as a parser to obtain words with different levels of word sense ambiguity or used WSD to gain insights into their model behaviors to improve the explainability of a study. In these cases, defining WSD as a sense classification task may be suboptimal for downstream applications.

WSD has a huge potential in NLU research. For example, disambiguating word senses in a large corpus can lead to a deeper understanding of language usage patterns and the semantic relationships between words. WSD is also a significant component in semantic explainable AI, because it helps researchers better understand the decision-making process of a model on the semantic level. Researchers can develop a more transparent and trustworthy model by explaining word senses in contexts. As a feature generator, a WSD may be more effective if it can generate contextualized word meanings in natural language, rather than predict a sense class that maps to a predefined gloss in a dictionary. However, research in these fields is rare in the WSD community.

Finally, according to Navigli (2009), the lack of end-to-end applications that utilize WSD can be attributed to the insufficient accuracy of current WSD systems. This suggests that in the future, more precise WSD systems may be developed, which could potentially enable the use of more semantics-dependent applications.

3.2.9.3 Future Works

The task of WSD can be broader than the current word sense classification task setup from either the theoretical research side or the downstream application side. Besides, the improvements in WSD accuracy can also attract more downstream applications. Thus, we come up with the following future work suggestions.

A. Extending the Form of WSD

WSD can have different learning forms, besides word sense classification, e.g., paraphrasing an ambiguous word into a less ambiguous one (Mao et al., 2018, 2022a), generating contextualized word senses in natural language. Such an extension may have significance in downstream applications. From the perspective of linguistic and cognitive research, studying how to define a language unit to better disambiguate word senses, or studying how to link a word to its associated concepts in a context can also improve the significance of WSD in the era of LLM-based NLP. Future works may study how to define the task of WSD to better support the research in different disciplines.

B. Rethinking Existing Knowledge Bases by WSD

Most of the existing knowledge bases were developed according to human-defined ontologies and word senses. These knowledge bases have been considered as an important resource for many knowledge-based systems. Although the knowledge bases have been used on different tasks, few works analyzed the weakness of the ontologies. Future WSD-related research may try to improve the knowledge bases by rethinking the sense definition, concept node connections, and coverage, rather than simply developing models to enhance the learning ability on a specific task.

C. Multilingual WSD

Most of the semantic representations are learned from monolingual corpora. As a result, the semantic representations are different between different languages. However, the disambiguation of meanings is not characterized by languages (Boroditsky, 2011). It will significantly improve multilingual semantic research if WSD research can break down language barriers from a cognitive perspective. As argued by frame semantics (Fillmore et al., 2006), the meaning of a word is beyond its dictionary definitions. It also associates with the concepts, interconnected with the word. Representing word senses by concepts may achieve a more robust multilingual WSD.

D. Learning WSD as a Pretraining Task

Recent years witness great success of PLMs in various domains. Existing PLMs followed the same hypothesis that the sense of a word can be learned from its associated context. However, there has not been a PLM that explicitly disambiguates word senses to enhance the learning of semantic representations. Naïvely learning the semantic representation of a target word by its associated context words cannot learn the conceptual association of the target word. For example, many words can associate with the word "apple". How can we know an apple as fruit is red or green, sweet, tree-growing, nutritious, etc? As an electronic device, Apple is associated with an operating system, a circuit board, a brand, etc. Disambiguating word senses before pretraining may build such connections between concepts.

E. Fusing WSD with other Tasks

As Bevilacqua et al. (2021) argued, WSD can also be integrated with an entity linking task (Moro et al., 2014b), where the model predicts associated entities to help WSD systems explore the related glosses and relations. Related fusion works also include fusing WSD for sentiment analysis (Farooq et al., 2015), information retrieval (Blloshmi et al., 2021) and machine translation (Campolungo et al., 2022). The future study of WSD can be grounded on an end task so that the end task can more effectively benefit from the fusion of a WSD model.

3.3 Named Entity Recognition

NER is a critical component of information extraction, which involves identifying entity mentions in text, defining their boundaries, and assigning them entity types. The most commonly recognized entity types by NER systems are Location, Person, and Organization, but there are many other types including time expressions (Zhong and Cambria, 2023). In the following example:

(2)   Steve Jobs is the founder of Apple.

An NER system would recognize the entities that "Steve Jobs" is Person; "Apple" is Organization. NER systems use pre-defined entity types, which may vary across different implementations. For example, Stanford's widely used NER software (Finkel et al., 2005) provides three versions that recognize three classes (Location, Person, Organization), four classes (Location, Person, Organization, Misc), and seven classes (Location, Person, Organization, Money, Percent, Date, Time), respectively. NER is a critical component in the field of NLU (Jinjie et al., 2022; Gao et al., 2022; He et al., 2021) and is often combined with other tasks, such as relation extraction, to serve as a foundation for various NLP applications.

Besides, NER is also used in various data mining tasks to recognize keywords, topics, and attributes (He et al., 2019a; Li et al., 2021d, 2019a). NER can be traced back to the third Message Understanding Conference (MUC) (Chinchor et al., 1993). The task for MUC-3 was designed to extract relevant information from the text and convert it into a structured format based on a predefined template, e.g., incident, the targets, perpetrators, date, location, and effects. Early NER systems that participated in MUC-3 primarily relied on rule-based approaches, which involved the manual creation of rules to identify named entities based on their linguistic and contextual features. However, with the dominance of deep learning in the NLP community, most NER tasks are now performed using neural networks.

One of the first neural networks for NER was proposed by Collobert and Weston (2008), which used a single CNN with manually constructed feature vectors. Later, this approach was replaced with high-dimensional continuous vectors, which were learned from large amounts of unlabeled data in an unsupervised manner (Collobert et al., 2011). With stronger models, now, the research in NER has been largely extended to nested NER (Su et al., 2022), few-shot NER (Huang et al., 2022b), joint entity and relation extraction (JERE) (Zhong and Chen, 2021). Compared to standard NER whose entity relationship is absent, entities in nested NER have a hierarchical or nested structure, where one entity is embedded within another entity. For example, given

(3)   The Ontario Supreme Court said ...

"Ontario" is a state entity that is embedded under the government entity of "Ontario Supreme Court" (Ringland et al., 2019b). Given the very expensive annotation costs, few-shot NER is also a very important research trend. It learns NER with a limited amount of labeled data. JERE tasks are established based on the needs of downstream applications. In many cases, people not only need to know what an entity is but also need to know the relationship between entities. Thus, JERE needs to identify named entities in text as well as extract the relationships that exist between them. In the following example

(4)   Greg Christie has been one of the greatest engineers at Apple.

For standard NER, "Greg Christie" should be identified as Person; "Apple" should be identified as Company. However, for JERE, besides the above entity recognition, an additional relationship label, "work_at" should also be predicted. Compared to identifying entities that are hierarchically structured within each other in nested NER tasks, the outcomes of JERE deliver another relationship dimension to connect entities. Both tasks are helpful in developing a comprehensive knowledge graph.

Due to the wide range of applications of NER, there have been several surveys conducted on this typical NLU task (Li et al., 2020b; Yadav and Bethard, 2018). Song et al. (2021) focused specifically on NER in the biomedical field, also known as Bio-NER. In this domain, the presence of meaningless characters in biomedical data presents a significant challenge, particularly with regards to inconsistent word distribution.

Similarly, Liu et al. (2022d) summarized and discussed the challenges specific to Chinese NER, rather than the more general English NER tasks. Meanwhile, Nasar et al. (2021) explored both NER and relation extraction tasks, as they are closely linked and are typically composed of pipeline tasks. The above mentioned surveys focus on the technical perspective of NER, based on deep learning technology, while this section takes a different approach. Specifically, we begin by examining the linguistic background of NER and its historical development, as well as current approaches and future directions. By taking a longer-term view and a more fundamental approach, we aim to provide a broader and more comprehensive perspective on the development of NER.

3.3.1 Theoretical Research

3.3.1.1 Prototype Theory

Rosch (1973) argued that our classification system, which includes the classification of named entities, is based on a central or prototype example. A prototype is a typical example of a category that represents the most common features or characteristics associated with the category. For example, the prototype of "bird" must associate the features, such as wings, feathers, and the ability to fly. Birds such as ostriches or penguins, which do not perfectly possess these characteristics, may be viewed as less typical examples. Rosch and Mervis (1975) discovered that individuals can identify typical category examples faster and with greater precision than atypical examples. Thus, learning from prototypes can help to quickly grasp the important features of a named entity with a few examples.

3.3.1.2 Graded Membership

Rosch et al. (1976) argued that the classification of categories is frequently determined not by strict boundaries, but by various degrees of membership. We can use this theory for NER because the NER task also categorizes entities by predefined classes. The idea of Graded Membership implies how humans perceive and categorize the world around us. Some categories, e.g., "vegetable", may be viewed as less distinct and vaguer. The theory suggests that the borders between categories may not be well-defined in some cases, leading to ambiguities when attempting to classify certain items, such as tomatoes or mushrooms. The ambiguity can be further compounded by cultural or regional differences in how categories are defined or classified.

3.3.1.3 Conceptual Blending

According to Fauconnier and Turner (2008), the act of blending different elements and their corresponding relationships is an unconscious process that is believed to be ubiquitous in everyday thought and language. This process involves the combination of various mental spaces or cognitive domains that are drawn from different scenarios and experiences. These scenarios may be derived from personal experiences, cultural practices, or societal norms, among others. Concept blending allows us to create a new concept by combining existing ones in novel ways. For example "SpaceX" may be mapped to mental spaces related to "aerospace" and "technology"; "Tesla" may be mapped to mental spaces related to "car" and "clean energy". Conceptual blending provides an explanation for the recognition and comprehension of newly named entities by mapping them onto existing mental spaces or concepts.

3.3.1.4 Grammatical Category

From the aspect of computational linguistics, the core issue of NER is how to define a named entity. Marrero et al. (2013) group the criteria of a named entity as grammatical category, rigid designation, unique identification, and the domain of applications. However, many of the entity definitions in the NER domain are imperfect. From the view of grammatical category, a named entity is traditionally defined as a proper noun or a common name for a proper noun. Previous work has described NER as the recognition of proper nouns in general. However, as pointed out by Borrega et al. (2007), the classic grammatical approach to proper noun analysis is insufficient to deal with the challenges posed by NER applications. For instance, in a toy QA task such as

(5)   Do crocodiles live in the sea or on land?

"crocodiles", "sea", and "land" are not proper nouns, while they are commonly recognized as the essential entities for a proper understanding of the question. Consequently, a proper noun is no longer considered a criterion for identifying named entities in current NER research.

3.3.1.5 Rigid Designation

The rigid designation is a concept in the philosophy of language which suggests that certain names or labels are inherently linked to the things they represent, e.g., "Barack Obama" rigidly designates the person who is the 44th President of the US, and it cannot be used to refer to any other person or entity. NER can be viewed as a form of rigid designation as it assigns labels to entities based on their intrinsic identity (Kripke, 1972), rather than on their usage in the text. However, LaPorte (2006) noted that not all expressions that appear to designate rigidly can be analyzed as directly referring to an object in every possible world.

This highlights the difficulty of defining entities with complex concepts in real-world applications. As a result, annotators likely make subjective judgments when labeling complex entities, which may be affected by entity descriptions and annotators' understanding.

3.3.1.6 Unique Identification

From the view of unique identification, MUCs require that NER tasks annotate the "unique identification" of entities for all expressions (Grishman and Sundheim, 1996). However, determining what is unique depends on contextual elements, and can be a subjective process. While this "unique identification" is typically considered to be the reference being referred to, the definition itself poses a challenge in terms of defining what is truly unique.

3.3.1.7 Domain of Applications

The definition of named entities was frequently grounded in the domain of applications. Entity definitions can be different between different NER tasks. For instance, in drug-drug interaction tasks (Deng et al., 2020), diseases may not be considered entities, whereas they are entities in adverse drug events (Demner-Fushman et al., 2019). Inconsistent entity definitions create challenges for machine learning. Because inconsistent entity definitions mean that for the same semantic unit, the machine has to summarize different entity representations to distinguish their labels under different tasks. This is also not conducive to training an all-around NER classifier on different application domains.

3.3.2 Annotation Schemes

NER is typically approached as a sequence labeling task, where each token in a sentence is assigned a label. Three common annotation schemes are shown in Table 3.6. The IO scheme is a classification task that distinguishes between two classes, namely "Inner" and "Other", to determine whether a token belongs to an entity or not. On the other hand, the BIO scheme employs three labels, namely "Beginning", "Inner", and "Other", to identify tokens that represent the start of an entity, tokens that belong to an entity, and tokens that do not belong to any entity. The BIOES scheme expands on the BIO scheme by incorporating two additional labels, namely "Single" and "End", to more precisely define the boundaries of entities. By employing the IO scheme, the binary classification of tokens is simplified, as each token is labeled as either belonging to an entity or not. This straightforward labeling system makes it easier to identify entities in a text, but it fails to specify the position of the entities within the text.

Table 3.6: The three common annotation schemes for NER.

Tokens:	West	African	Crocodile	are	semiaquatic	reptiles	that	live	in	Africa
IO	I	I	I	O	I	I	O	O	O	I
BIO	B	I	I	O	B	I	O	O	O	B
BIOES	B	I	E	O	B	E	O	O	O	S

In contrast, the BIO scheme provides more precise annotations by identifying the beginning and continuation of an entity in the text. This labeling system allows for more accurate recognition of entities in a text and better classification of individual tokens. The BIOES scheme further extends the BIO scheme by providing more precise boundaries for entities, thereby allowing for better recognition of entity boundaries in a text. The "Single" label is used to denote an entity that consists of a single token, whereas the "End" label is used to indicate the final token of an entity. By incorporating these additional labels, the BIOES scheme provides a more nuanced approach to entity recognition and annotation.

3.3.3 Datasets

The surveyed popular NER datasets and their statistics can be viewed in Table 3.7. The first NER-focused dataset was published in the 6th MUC (Grishman and Sund-heim, 1996). This task consists of three subtasks, including entity names, temporal expressions, and number expressions.

Table 3.7: NER datasets and statistics.

Dataset	Source	# Sample	Reference
MUC-6	Newswire	318 articles	Grishman and Sundheim (1996)
ACE-05	Social media	12,548 sentences	Walker et al. (2006)
TACRED	Newswire	106,264 instances	Zhang et al. (2017b)
CoNLL-2003	Reuters	1,499 articles	Sang and De Meulder (2003)
I2B2	ECI Corpus	1,600 patient records	Stubbs and Uzuner (2015)
ADE	MEDLINE	2,972 document	Gurulingappa et al. (2012)
DDI	DrugBank	1,025 document	Herrero-Zazo et al. (2013)
WNUT-17	Social media	2,295 documents	Derczynski et al. (2017)
OntoNote 5.0	Social media	-	Weischedel et al. (2013)
CPR	MEDLINE	-	Krallinger et al. (2017)
MultiNERD	Wikipedia	10 languages	Tedeschi and Navigli (2022)
HIPE-2020	Newspapers	17,553 mentions	Ehrmann et al. (2022)
NNE	Newswire	49,208 sentences	Ringland et al. (2019b)
GENIA	MEDLINE	18,546 sentences	Kim et al. (2003)

The defined entities include organizations, persons, and locations. The defined time expressions include dates and times. The defined quantities include monetary values and percentages. An example of this dataset is shown as follows.

```
text: "Taga Co.",
type: "ORGANIZATION".
```

MUC was replaced by Automatic Content Extraction (ACE) program after 1997. ACE05 (Walker et al., 2006) is another popular NER dataset published at ACE Conference. ACE05 is a multilingual dataset, which contains English, Arabic, and Chinese data. The corpus consists of data of various types annotated for entities, relations, and events. Its data source includes broadcast conversation, broadcast news, newsgroups, telephone conversations, and weblogs. An example of this dataset is shown below.

```
entity id: "NN_ENG_20030630_085848.18-E1",
type: "GPE",
subtype: "State-or-Province",
class: "SPC",
start: "82",
end: "91",
name: "california".
```

After MUC, the Text Analysis Conference (TAC) published the Knowledge Base Population challenge. In this challenge, the Stanford NLP Group developed TAC Relation Extraction Dataset (TACRED) (Zhang et al., 2017b), which contains 106,264 instances with annotated entities, relations and some other NLP tasks. An example of this dataset is shown as follows.

```
id: "e7798fb926b9403cfcd2",
docid: "APW_ENG_20101103.0539",
relation: "per:title",
token: "['At', 'the', 'same', 'time', ',', 'Chief', ...]",
subj_start: "8",
subj_end: "9",
obj_start: "12",
obj_end: "12",
subj_type: "PERSON",
obj_type: "TITLE",
stanford_pos: "['IN', 'DT', 'JJ', 'NN', ',', 'NNP', 'NNP', ...]",
stanford_ner: "['O', 'O', 'O', 'O', 'O', 'O', 'O', 'O', ...]"
stanford_head: "[4, 4, 4, 12, 12, 10, 10, 10, 10, 12, ...]",
stanford_deprel: "['case', 'det', 'amod', 'nmod', 'punct', ...]".
```

CoNLL-2003 (Sang and De Meulder, 2003) is another widely used NER dataset. This task concerned language-independent NER, which concentrates on four kinds of named entities: locations, persons, organizations, and names of miscellaneous entities that do not belong to the previous three kinds. The related data files are available in English and German. An example of this dataset is shown as follows.

```
text: "['U.N.', 'official', 'Ekeus', 'heads', ...], ",
pos: "['NNP', 'NN', 'NNP', 'VBZ', ...], ",
syntactic chunk: "['I-NP', 'I-NP', 'I-NP', 'I-VP', ...], ",
named entity tag: "['I-ORG', 'O', 'I-PER', 'O', ...]".
```

Besides the above famous datasets, MultiNERD (Tedeschi and Navigli, 2022), HIPE-2020 (Ehrmann et al., 2022), and NNE (Ringland et al., 2019a) are also popular NER datasets in general domain. NER tasks have garnered considerable attention in numerous specialized domains. Informatics for Integrating Biology and the Bedside (I2B2) (Stubbs and Uzuner, 2015) is a national biomedical computing project sponsored by the National Institutes of Health (NIH) from 2004 to 2014. I2B2 actively advocates mining medical value from clinical data and has organized a series of evaluation tasks and workshops for unstructured medical record data, and these evaluation tasks and open datasets have gained wide influence in the medical NLP community. I2B2 is maintained in the Department of Biomedical Information at Harvard Medical School and continues to conduct assessment tasks and workshops, and the project has been renamed National NLP Clinical Challenges (N2C2). Besides, there also exist many other biomedical datasets for specific medical NER tasks, including Adverse Drug Events (ADE) (Gurulingappa et al., 2012; Alvaro et al., 2017), Drug-Drug Interaction (Herrero-Zazo et al., 2013), and Chemical Protein Reaction (CPR) (Krallinger et al., 2017), and GENIA (Shibuya and Hovy, 2020).

3.3.4 Knowledge Bases

Table 3.8 illustrates useful knowledge bases for NER. The biggest ones are Wikidata[16] and Wikipedia[17], which are multilingual free online encyclopedias maintained by worldwide volunteers. There are also knowledge bases in a specific field. SNOMED CT (Systematized Nomenclature of Medicine – Clinical Terms) (Donnelly et al., 2006) is a systematically organized collection of medical terms that provides a standardized representation of clinical information, which is often used in NER tasks involving clinical data. MeSH (Medical Subject Headings) (Lipscomb, 2000) is another controlled vocabulary, developed by the U.S. National Library of Medicine. It is used for indexing and organizing biomedical literature.

---

[16] https://wikidata.org

[17] https://en.wikipedia.org

Table 3.8: Useful knowledge bases for NER.

Name	Knowledge	# Entities	structure
Wikipedia	World	13,489,694	unstructured
Wikidata	World	100,905,254	graph
DrugBank	Medical	over 500,000	structured
UMLS	Medical	16,857,345	structured
BioModels	Medical	unclear	structured
SNOMED CT	Medical	over 350,000	structured
ICD-10	Medical	unclear	structured
MIMIC-III	Medical	unclear	structured
MeSH	Medical	over 28,000	structured
GeoNames	Geographical	over 25,000,000	structured
EDGAR	Financial	unclear	structured
EduKG	Educational	5,452	structured

Other medical knowledge bases include UMLS (Unified Medical Language System) (Wheeler et al., 2007; Bodenreider, 2004), ICD-10 (Hirsch et al., 2016), MIMIC-III (Johnson et al., 2016), DrugBank (Wishart et al., 2018), and bioinformatics knowledge base BioModels (Li et al., 2010). GeoNames (Ahlers, 2013) is a comprehensive geographic knowledge repository that encompasses over 25 million geographical names and comprises over 11 million distinctive features, including cities, countries, and landmarks. EDGAR (Electronic Data Gathering, Analysis, and Retrieval) (Branahl, 1998) is a database maintained by the U.S. Securities and Exchange Commission (SEC), containing financial filings and reports from publicly traded companies. EduKG (Hu et al., 2016) is an educational knowledge base.

3.3.5 Evaluation Metrics

In the process of NER task evaluation, the main evaluation metrics are also Precision, Recall, and F-value.

3.3.5.1 Annotation Tools

One AI[18] is an online platform that offers NLP-as-a-service. The utilization of APIs enables developers to effectively analyze, manipulate, and transform natural language inputs within their programming code without requiring any specialized knowledge of NLP. One AI facilitates the interpretation of both the meaning and information conveyed in textual data, and can produce structured data in context via language processing.

---

[18] https://docs.oneai.com/docs

GATE Teamware[19] (Bontcheva et al., 2013a) is an integrated annotation tool for comprehensive language processing tasks, especially for Information Extraction systems. The University of Sheffield developed GATE Teamware that enables collaborative semantic annotation projects through a shared annotation environment. The software comprises several beneficial attributes such as the ability to load document collections, create project templates that can be used multiple times, initiate projects based on templates, assign project roles to individual users, monitor progress and obtain various project statistics in real-time, report project status, annotator activity, and statistics, and apply automatic annotations or post-annotation processing via GATE-based processing routines.

MAE[20] (Rim, 2016) (Multi-document Annotation Environment) is a general-purpose and lightweight natural language annotation tool. The tool enables users to specify and create their customized annotation tasks, annotate any text spans of their choice, utilize non-consuming tags, effortlessly establish links between annotations, and produce annotations in stand-off XML format. It also provides a simple adjudication process with a visualization feature that displays the extent tags, link tags, and non-consuming tags of any XML standoff annotated documents.

UIMA[21] (Ferrucci and Lally, 2004) (Unstructured Information Management Applications) is a framework that falls under the purview of the Apache Software Foundation. It serves as a comprehensive platform for managing language processing projects and is licensed under Apache's open-source license. With its versatile capabilities, UIMA can effectively handle a diverse array of language processing tasks and extract various types of information. The UIMA's Regular Expression Annotator is capable of identifying entities such as email addresses, phone numbers, URLs, zip codes, or any other entities based on the utilization of regular expressions and concepts. The tool can generate an annotation for each detected entity or update an existing annotation with relevant feature values.

Brat[22] (Browser-based Rapid Annotation Tool) is a free data labeling tool that offers a seamless browser-based interface for annotating text. It streamlines numerous annotation tasks related to NLU. With a thriving support community, Brat is a well-known and widely used tool in NER. It also offers the option of integrating with external resources, such as Wikipedia. Moreover, Brat enables organizations to establish servers that allow multiple users to collaborate on annotation tasks. However, implementing this feature does necessitate some technical proficiency and server management skills.

---

[19] https:///gate.ac.uk/teamware

[20] https://keighrim.github.io/mae-annotation

[21] https://uima.apache.org/sandbox.html

[22] https://brat.nlplab.org

3.3.6 Methods

3.3.6.1 Nested NER

A. Multi-label Method

Due to the fact that nested named entities can have multiple labels for a single token, traditional sequence labeling methods are not directly applicable to the recognition of nested named entities. To address this issue, researchers have attempted to convert the multi-label problem into a single-label problem or adjust the decoder to assign multiple labels to the same entity. Katiyar and Cardie (2018) proposed a method to address nested NER by modifying the label representation in the training set. Instead of using one-hot encoding, they used a uniform distribution over the specified classes as the label. During inference, a hard threshold is set and any class with probability above this threshold is predicted for the token. However, this approach has two limitations: it is difficult to determine the objective for model learning; the method is sensitive to the manually chosen threshold value.

Straková et al. (2019) changed nested NER from multi-label to single-label tasks by modifying the annotation schema. They combined any two categories that may co-occur to produce a new label (e.g., combine B-Location with B-Organization to construct a new label *B_Loc_Org*). One benefit of this approach is that the final classification task is still a single category because all possible classification targets had been covered in the schema. Nonetheless, this method brought about a proliferation of label categories in an exponential manner, leading to sparsely annotated labels that proved difficult to learn, particularly in the context of entities nested across multiple layers. In order to address the issue of label sparsity, Shibuya and Hovy (2020) proposed a hierarchical approach. If the classification of nested entities cannot be resolved in a single pass, the classification is continued iteratively until either the maximum number of iterations is reached or no new entities can be generated. Nevertheless, this approach is susceptible to error propagation, whereby an erroneous classification in a preceding iteration could impact subsequent iterations.

B. Generation-based Method

Li et al. (2020g) proposed a unified framework to accomplish flat and nested NER tasks by formulating NER as a machine reading comprehension (MRC) task (Liu et al., 2023a). In this approach, the extraction of each entity type corresponds to specific questions. For instance, when the model is given the question "which location is mentioned in the sentence?" along with the original sentences, it generates an answer such as "Washington". This approach is similar to Prompt Tuning (Liu et al., 2021c), which avoids the labor-intensive process of constructing manual questions. However, in this method, the generated tokens must be mapped to pre-defined named entity types.

Yan et al. (2021b) proposed a novel pointer generation network. Given an input sentence, the model generates the entity indexes in this sentence that belong to entities. In such a way, flat, nested, and discontinuous entities can be recognized in a unified framework. Skylaki et al. (2020); Fei et al. (2021a); Yang and Tu (2022); Su et al. (2022) are also following the idea of generating indexes of a sentence to recognize nested entities.

C. Hypergraph-based Method

A hypergraph is a generalized variant of a normal graph, which is characterized by an edge that can connect an arbitrary number of vertices (Feng et al., 2019c). It is widely used in the NLP community for the tasks of syntactic parsing, semantic parsing, and machine translation because it can accurately describe the relationship between objects with multiple associations. A set of objects with only binary relations can be described by a normal graph. However, when the objects are often related to each other in a more complex one-to-many or many-to-many, e.g., nested named entities, hypergraphs become a more appropriate data structure. A typical example of nested NER with a hypergraph solution is shown in Fig. 3.4.

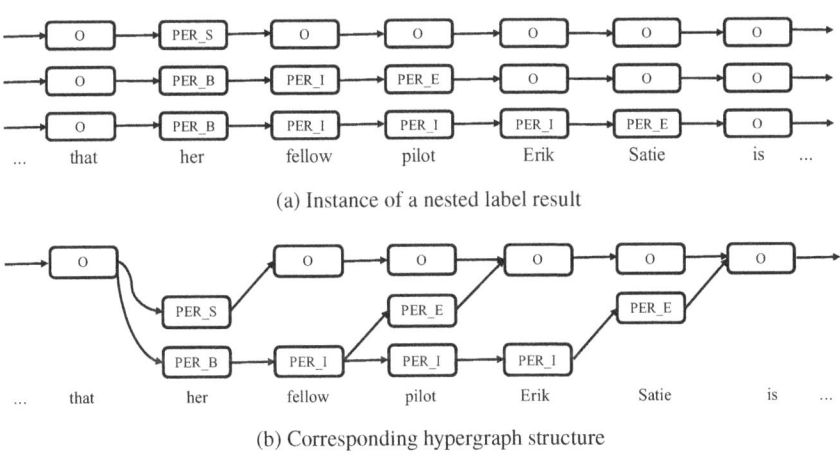

(a) Instance of a nested label result

(b) Corresponding hypergraph structure

Fig. 3.4: A typical example for nested NER with hypergraph solution

Finkel and Manning (2009) firstly introduced hypergraphs into nested NER tasks, named Mention Hypergraph. In their model, Mention Hypergraph utilized nodes and directed hyper-edges to jointly represent named entities and their combinations. To compute the training loss, the proportion of accurate structures was calculated and divided by a normalized term. This term was obtained using a dynamic programming algorithm that aggregated feasible nested sub-graphs for NER. However, the normalized terms obtained from this algorithm included fractions of pseudo-structures, which led to errors.

To deal with the problem of pseudo-structures, Muis and Lu (2017) proposed a gap-based marker model to identify nested entity structures by combining mention separators with features. In this method, the authors manually designed 8 types of mention separators for various scenarios. Based on the mention separators' states for any two consecutive tokens, they defined accurate and novel graph structures. However, since this approach only utilized local information to construct the graph structures, it may not be unambiguous for long-nested named entities. For instance, when presented with the nested entity "a West African Crocodile", which includes two separate entities, "West African" and "a West African Crocodile", their approach may also recognize "a West African" as a named entity.

This ambiguous problem was solved by Wang and Lu (2018), which proposes a segmental hypergraphs method. The method used an unambiguous ambiguity-free compact hypergraph representation to encode all possible combinations of nested named entities. Upon Mention Hypergraph (Finkel and Manning, 2009), segmental hypergraphs employed an inside-outside message-passing algorithm that can summarize the features of child nodes to the parent node and achieve efficient inter-ference. Besides the above work, Wan et al. (2021) introduced the concept of regional hypernodes and a combination method of graph convolutional network (GCN) and BiLSTM to generate hypernodes for each region. Yan and Song (2022) employed start token candidates and generated corresponding queries with related contexts, then used a query-based sequence labeling module to form a local hypergraph for each candidate.

3.3.6.2 Few-shot NER

A. Metric Learning

Metric Learning is a common technology in various few-shot tasks. Prototypical Networks (Snell et al., 2017) is a milestone in few-shot metric learning. Prototypical Networks compute the centroid of each category based on the support set. They de-termine the distance between the samples in the query set and the prototype center, followed by updating the model by optimizing this distance. Upon completion of the training phase, the embedding of each sample will be situated in closer proximity to the centroid of the corresponding category. Such an idea was largely inspired by Prototype Theory (see Section 3.3.1.1).

Fritzler et al. (2019) adopted the prototypical network into few-shot NER tasks. They argued that words in a sentence are interdependent and, therefore, the label-ing of adjacent words should be taken into account. To address this issue, they substituted the conventional token input of Prototypical Networks with complete sentences. However, this method ignores the problem of the Outside (O) class in NER tasks, which actually represent different semantic meanings. This problem would significantly affect the model's performance under few-shot settings.

To avoid the above issues, Yang and Katiyar (2020) followed the nearest neighbor inference (Wiseman and Stratos, 2019) to assign labels to tokens. In contrast to Prototypical Networks, which learn a prototype for each entity class, this study characterized each token by its labeled instances in the support set alongside its context. The approach determined the nearest labeled token in the support set, followed by assigning labels to the tokens in the query set that require prediction.

Das et al. (2022) proposed CONTaiNER, which optimized the inter-token distribution distance. CONTaiNER employed generalized objectives to different token categories based on their Gaussian-distributed feature vectors. Such a method has the potential to mitigate overfitting problems that arise from the training domains.

B. Prompt Tuning

Recently, prompt tuning has shown great potential on few-shot tasks by reformulating other tasks as mask language tasks (He et al., 2023a; Mao et al., 2023c; Schick and Schütze, 2021). Prompt tuning-based methods need construct prompts to obtain masked word predictions and then map predicted works into pre-defined labels, as shown in Fig. 3.5.

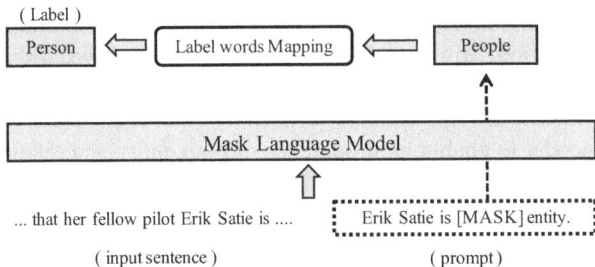

Fig. 3.5: A typical prompt tuning example for NER tasks.

Cui et al. (2021) proposed a template-based method for NER, which first applied the prompt tuning to NER tasks. However, their method had to enumerate all possible spans of sentences combined with all entity types to predict labels, which suffered serious redundancy when entity types or sentence lengths increased. Manually defined prompts were labor-intensive and made the algorithm sensitive to these prompts. To avoid the manual prompt constructions, Ma et al. (2022a) tried to explore a prompt-free method for few-shot NER. The present study introduced an entity-oriented language model that decodes input tokens into their corresponding label words if they belong to entities. In cases where the tokens are not entities, the entity-oriented language model decodes the original tokens. Nevertheless, this approach encounters difficulties in labeling word engineering. While this study proposed an automated label selection technique, the associated experiments revealed some degree of instability.

COPNER (Huang et al., 2022b) introduced class-specific words to construct prompt tuning. By comparing each token with manually selected class-specific words, this method needed neither manual prompts nor label words engineering. The selected class-specific words (a representative word corresponding to a class) were directly concatenated with original sentences as prompts. However, the manual selection of class-specific words is subjective, and a single word may not entirely capture the semantics of an entity category. To deal with such a problem, TFP (He et al., 2024a) encoded external description sentences to represent an entity class, thereby facilitating the integration of external knowledge into prompts and providing comprehensive representations of entity types. TFP also proposed a semantic-enhanced hybrid granularity contrastive loss to guide prompt tuning.

3.3.6.3 Joint NER and Relation Extraction

A. Parameter Sharing-based Multi-Task Learning

Considering that NER is usually combined with relation extraction tasks applied in various downstream tasks, jointly recognizing named entities and classifying relations is a hot topic in related fields. MTL is the most common solution in joint NER and relation extraction. Miwa and Bansal (2016) firstly employed a shared BiLSTM encoder to obtain token representations, and then fed encoded representations into NER and relation extraction classifiers, respectively. Sun et al. (2020) utilized a GCN as a shared encoder to enable joint inference of both entity and relation types. The core idea of the above study is that multi-task models can enhance the interactions between the learning of NER and relation extraction, and further alleviate the error propagation by sharing common parameters (He et al., 2021). However, this work cannot ensure that the sharing of information is useful and proper. NER and relation extraction might need different features to result in precise predictions. To this end, Yan et al. (2021c) proposed an information filtering mechanism to provide valid features for NER and relation extraction. Their method used an entity and relation gate to divide cell neurons into different parts and established a two-way interaction between NER and relation extraction. In the final network, each neuron contained a shared partition and two task-specific partitions.

B. Table Filling

While MTL can improve the interdependence between NER and relation extraction, the relation extraction process still requires the pairing of all entities from the NER tasks to classify relations, making it impossible to completely eliminate error propagation. To solve the problem, Miwa and Sasaki (2014) proposed a table-filling strategy to achieve joint NER and relation extraction by labeling input tokens in a table. The method utilized token lists of sentences to form rows and columns.

Sentence	The	United	States	president	Biden	will	visit	...
The	--	--	--	--	--	--	--	--
United		LOC_B	--	--	CP	--	--	--
States			LOC_E	--	CP	--	--	--
president				--	--	--	--	--
Biden					PER_S	--	--	--
will						--	--	--
visit							--	--
...								...

Fig. 3.6: The illustration of the table-filling strategy.

Then, they extracted entities using the diagonal elements and classified relations with a lower/upper triangular matrix of the table. This basic table-filling strategy can be seen in Fig. 3.6. Nonetheless, this approach involved the explicit integration of entity-relation label interdependence, which necessitated the use of intricate features and search heuristics. Gupta et al. (2016b) incorporated neural networks with a table-filling strategy via a unified multi-task RNN. This method detected both entity pairs and the related relations with an entity-relation table, which alleviated the need for search heuristics and explicit entity-relation label dependencies. Zhang et al. (2017a) further integrated global optimization and syntax information into the table-filling strategy to combine NER and relation extraction tasks. Ren et al. (2021a) argued that the above table-filling-based studies only focus on utilizing local features without the global associations between relations and pairs. Ren et al. (2021a) first produced a table feature for every relation, followed by extracting two types of global associations from the generated table features. Finally, the table feature for each relation was integrated with the global associations. Such a process is performed iteratively to enhance the final features for joint learning of NER and relation extraction tasks.

C. Tagging Scheme

The table-filling approach can mitigate issues related to error propagation. However, these techniques require the pairing of all sentence elements to assign labels, resulting in significant redundancy. To address the redundancy and avoid error propagation, Zheng et al. (2017) proposed a novel tagging scheme that converted joint NER and relation extraction into a united task. The idea was similar to the solution for nested entities Straková et al. (2019), which combined NER labels with relation extraction labels by modifying the annotation schema. For example, given the sentence "The United States president Biden will visit ...", by allocating the customized labels "Country-President_B_1", "Country-President_E_1" for tokens "United", "States", and "Country-President_E_2" for token "Biden", the proposed method can directly obtain the triplet (United State, Country-President, Biden).

Yu et al. (2020); Wei et al. (2019) proposed two similar methods. In contrast to conventional joint approaches for NER and relation extraction, which involve recognizing entities followed by relation classification, the two methods first identified all head entities. Next, for each identified head entity, they simultaneously predicted corresponding tail-entities and relations, achieving cascade frameworks combined with a customized tagging scheme. The typical joint NER and relation extraction tasks learn to model the conditional probability:

$$P(h, r, t) = P(s)P(t \mid h)P(r \mid h, t), \tag{3.4}$$

where h represent head entity; r represent relation; h represent tail entity. The above methods combined the last two parts in Eq. 3.4, yielding

$$P(h, r, t) = P(s)P(t, r \mid s). \tag{3.5}$$

3.3.7 Downstream Applications

3.3.7.1 Knowledge Graph Construction

Knowledge graphs are structured semantic knowledge bases for rapidly describing concepts and their interrelationships in the physical world, aggregating large amounts of knowledge by reducing the data granularity from the document level to the instance level (Yao et al., 2022). Thus, knowledge graphs enable rapid response and reasoning about knowledge. At present, the application of knowledge graphs has become prevalent in industrial domains, such as Google search. Generally, the construction of Knowledge Graphs consists of three main parts: information extraction, information fusion, and information processing. The task of information extraction involves the identification of nodes through NER and the establishment of edges via relation extraction. The task of information fusion is utilized for normalizing nodes and edges. The normalized nodes and edges need to go through a quality assessment with the task of information processing to be added to knowledge graphs.

He et al. (2021) proposed an MTL-based method for the construction of genealogical knowledge graphs. At first, He et al. (2019b) collected unstructured online obituary data. Then, they extracted named entities as nodes and classified family relationships for these recognized people as edges to construct genealogical knowledge graphs. Similarly, Jiang et al. (2020) utilized NER and relation extraction for obtaining the nodes and edge in biomedical knowledge graphs. They proposed a customized tagging schema to convert the construction of biomedical knowledge graphs into a sequence labeling task with multiple inputs and multiple outputs. Li et al. (2020c) proposed a systematic approach for constructing a medical knowledge graph, which involves extracting entities such as diseases and symptoms, as well as related relationships, from electronic medical records. Silvestri et al. (2022), Peng et al. (2019), and Shafqat et al. (2022) aimed to collect and utilize medical knowledge for NER.

Furthermore, constructing knowledge graphs requires the task of Entity Linking (Tedeschi et al., 2021a) to normalize entities with different names. Entity Linking and NER are typically performed as pipeline tasks to yield more nodes and edges for the constructed graphs. Additionally, Entity Linking can be seen as a downstream task for NER, as it further refines the identified entities by linking them to a specific reference entity in a knowledge graph.

3.3.7.2 Recommendation Systems

Recommendation systems can be classified into two primary categories based on their solutions, namely content-based recommenders and collaborative filtering-based recommenders (Batmaz et al., 2019). For both of these groups, gathering data on users and products is a crucial step in the entire process. In this regard, NER modules play a pivotal role. For example, Kim et al. (2012) introduced the 5W1H model, which utilizes NER to extract contextual information, specifically Who, Why, Where, What, When, and How, to generate contextual recommendations.

Zhou et al. (2020b) argued that recommendation systems currently in use suffer from a deficiency of contextual information in conversational data, as well as a semantic gap between natural language expressions and the preferences of individual users for specific items. To overcome these challenges, word- and entity-oriented knowledge graphs were incorporated to enhance the data representations. Mutual Information Maximization was adopted to align the word-level and entity-level semantic spaces. The aligned semantic representations were used to develop a knowledge graph-enhanced recommender component to make accurate recommendations, and a knowledge graph-enhanced dialogue component that can generate informative keywords or entities in the response text. An NER module is a crucial component in creating such a knowledge graph-enhanced system (Wu et al., 2023).

Iovine et al. (2020) proposed a domain-independent, configurable recommendation system framework, named ConveRSE (Conversational Recommender System framEwork). ConveRSE utilized various interaction mechanisms, including natural language, buttons, and a combination of the two. The framework comprised a dialogue manager, an intent recognizer, a sentiment analyzer, an entity recognizer, and a set of recommendation services. The entity recognizer component specifically focused on identifying relevant entities that were mentioned in the user's input, and linking them to an appropriate concept in the knowledge base. The ConveRSE framework's success is heavily reliant on the performance of the NER component, as it plays a crucial role in enhancing the system's overall performance.

Wang et al. (2019a) proposed RippleNet, an end-to-end framework that incorporates the knowledge graph into a recommender system. RippleNet overcame the limitations of previous embedding-based and path-based approaches to knowledge graph-aware recommendation by incorporating the knowledge graph as a form of supplementary information. RippleNet included both inward aggregation and outward propagation models. The inward aggregation version aggregated and incorporated neighborhood information when computing the representation of a given entity.

By extending the neighborhood to multiple hops away, it was possible to model high-order proximity, thereby capturing users' long-distance interests. On the other hand, the outward propagation model propagated users' potential preferences and explored their hierarchical interests in knowledge graph entities. Upadhyay et al. (2021) proposed an explainable job recommendation system by matching users with the most pertinent jobs, based on their profiles. The system also provided a human-readable explanation for each recommendation. The NER module was customized to extract pertinent details from both job postings and user profiles. These details were utilized to create comprehensible explanations for each recommendation. By identifying and categorizing entities, the NER module enhanced the accuracy and understandability of the textual explanations, providing a clear representation of the reasoning behind the recommendation system.

3.3.7.3 Dialogue Systems

Commonly, dialogue systems can be categorized into three main types, namely task-oriented, QA, and open-domain (Ni et al., 2022). NER plays a role in enhancing NLU for the three types of dialogue systems, organizing original user messages into semantic slots, and classifying data domain and user intention (Li et al., 2017b). Abro et al. (2022) proposed an argumentative dialogue system with NER and other NLU tasks. The approach can enhance comprehension of user intent by comprehending injected entities and relationships. For QA (Dimitrakis et al., 2020) and open-domain dialogue systems, NER also plays a crucial role in the part of intent recognition and knowledge retrieval. For example, Zhang et al. (2021a) developed a sequence of sub-goals with external knowledge to improve generation performance. External knowledge refers to a range of named entities and relationships that are associated with a conceptual entity. Leveraging external knowledge allows the dialogue system to deliver a more cohesive small talk from the open domain.

3.3.8 Summary

NER is a very important NLU subtask that enhances the accuracy of other subtasks, e.g., concept extraction. It is the manifest of cognitive semantics, because named entities are not simply categorized by their semantics. The classified named entities also reflect their inherent attributes in people's cognition. According to Prototype Theory (see Section 3.3.1.1), the inherent attributes of named entities can be represented by prototypes. It is gratifying to observe that a theory has had a significant influence on research related to few-shot NER. On the other hand, the ambiguity of named entity classification argued by Graded Membership (see Section 3.3.1.2) and Grammatical Category (see Section 3.3.1.4) was rarely analyzed from computational linguistic aspects.

We also do not see explainable NER studies that explain why an entity is classified into a particular category from the perspective of conceptual blending (see Section 3.3.1.3). NER research on these aspects is helpful for achieving human-like intelligence in categorizing named entities. The availability of numerous NER datasets, both in general and medical domains, has significantly enhanced computational research in this area. This may be attributed to the great application value of NER, as well as a wide range of data annotation tools. Encyclopedias knowledge and domain-specific knowledge also provide external information to help NER models better understand the context and commonsense. Now, NER has developed many practical task setups to the need of technical applications, e.g., nested NER, few-shot NER, joint NER and relation extraction, and downstream tasks, e.g., knowledge graph construction (KGC), recommendation systems, and dialogue systems.

3.3.8.1 Technical Trends

Due to the extensive research conducted on typical NER methods over the years, researchers are shifting their focus towards NER techniques that are more applicable to practical scenarios, for example, nested NER, few-shot NER, and joint NER and relation extraction. Recent technological trends for the aforementioned NER tasks are summarized in Table 4.21. Overall, nested NER can be addressed by multi-label, generation-based, and hypergraph-based methods. Among them, multi-label methods are straightforward and easy to implement. However, there are several limitations in the surveyed multi-label methods.

Table 3.9: A summary of representative NER techniques. The study with * means it cannot be compared with other studies since it did not report 5-shot results.

Task	Reference	Tech	Feature and KB.	Framework	Dataset	Score	Metric
Nested NER	Katiyar and Cardie (2018)	DL	Emb.	BiLSTM	ACE-05	70.2%	F_1
	Straková et al. (2019)	DL	Emb.	LSTM-CRF	ACE-05	84.3%	F_1
	Shibuya and Hovy (2020)	DL	Emb.	LSTM-CRF	ACE-05	84.3%	F_1
	Li et al. (2020g)	DL	BERT, Wikipedia	Unified Framework	ACE-05	86.9%	F_1
	Yan et al. (2021b)	DL	BERT	Pointer Networks	ACE-05	84.7%	F_1
	Finkel and Manning (2009)	Graph	Emb., Constituency Parsing	Hypergraph	GENIA	72.0%	F_1
	Muis and Lu (2017)	Graph	Emb., Multigraph Representation	Hypergraph	GENIA	70.8%	F_1
	Wang and Lu (2018)	Graph	Emb., Segmental Hypergraphs	Hypergraph	GENIA	75.1%	F_1
	Yang and Tu (2022)	DL	BERT	Pointer Networks	ACE-05	85.0%	F_1
	Su et al. (2022)	DL	BERT	Pointer Networks	CONLL04	88.6%	F_1
Few-shot NER (5 shot)	Fritzler et al. (2019)*	DL	Prototypical network	RNN+ CRF	Ontonotes	-	F_1
	Yang and Katiyar (2020)	DL	BERT	Nearest Neighbor	I2B2	22.1%	F_1
	Das et al. (2022)	DL	BERT	Contrastive Learning	I2B2	31.8%	F_1
	Cui et al. (2021)	DL	BERT	Prompt Tuning	I2B2	36.7%	F_1
	Huang et al. (2022b)	DL	BERT	Prompt Tuning	I2B2	43.7%	F_1
	He et al. (2024a)	DL	BERT, Wikipedia	Prompt Tuning	I2B2	52.7%	F_1
Joint NER and relation extraction	Miwa and Bansal (2016)	DL	Emb.	BiLSTM	ACE-05	55.6%	F_1
	Sun et al. (2020)	Graph	Emb.,	Bipartite Graph	ACE-05	59.1%	F_1
	Yan et al. (2021c)	DL	BERT	Partition Filter	ACE-05	66.8%	F_1
	Gupta et al. (2016b)	ML	Emb.,	Table filling	CoNLL04	72.1%	F_1
	Zhang et al. (2017a)	DL	Emb.,	Table filling	ACE-05	57.5%	F_1
	Zheng et al. (2017)	DL	Emb.,	Tagging scheme	NYT	49.5%	F_1
	Yu et al. (2020)	DL	Emb.,	Tagging scheme	NYT	59.0%	F_1
Task-driven NER	Shafqat et al. (2022)*	DL	Emb., ICD-10	BiLSTM	no public	-	F_1
	Hirsch et al. (2016)*	DL	Emb., UMLS	BiLSTM	no public	-	F_1
	Peng et al. (2019)*	DL	BERT, MIMIC-III	Fine Tuning	no public	-	F_1

For example, thresholds for multi-label selection are hard to decide empirically (Katiyar and Cardie, 2018); multiple labels are suffering sparsity (Straková et al., 2019) or error propagation (Shibuya and Hovy, 2020), which can lower model performance. Generation-based methods are flexible. By reformulating NER tasks as QA, they can generate any results which satisfied the pre-defined requirements (Shibuya and Hovy, 2020; Li et al., 2020g). These methods are used for handling Flat NER (Skylaki et al., 2020), nested NER (Yan et al., 2021b), and discontinuous NER (Fei et al., 2021a). However, a generation-based method is hard to control what is generated, even if some studies (Skylaki et al., 2020; Fei et al., 2021a; Yang and Tu, 2022; Su et al., 2022) have attempted to restrict the outputs of generation-based methods to a specific set of indexes (pointer network). The core point of the hypergraph-based method is about how to establish a hypergraph data structure to better represent interaction among all tokens in a sentence. These methods are good at modeling the interactions among all tokens in a sentence. It is important to note that the majority of hypergraph-based methods exhibit a task-specific nature, indicating a limited scope of applicability. These methods may not be universally applicable, and their effectiveness may be constrained by the specific task they are designed for.

Few-shot NER is usually achieved by metric learning and prompt tuning. Metric learning has demonstrated its effectiveness in various few-shot tasks (Kaya and Bilge, 2019; Fritzler et al., 2019). For few-shot NER tasks, some works predict the final labels by comparing token-to-token distance (Yang and Katiyar, 2020; Das et al., 2022) or token-to-prototype distance (Huang et al., 2022b). These methods have to decide different distance calculation functions according to different task (Kulis et al., 2013) and suffer instability introduced by insufficient data. By exploiting the full potential of language models, prompt tuning is proposed and demonstrated as a promising technology for few-shot tasks (Liu et al., 2021c; He et al., 2023a; Liu et al., 2022e). Prompt tuning reformulate NER as a mask language model task to reduce the gap between NER and employed pretraining LMs. The backward is that prompt tuning needs extra template construction and label word mappings and some studies have tried to deal with such problems (Huang et al., 2022b; He et al., 2024a).

For Joint NER and relation extraction tasks, we summarize related studies into three groups, including parameter sharing-based MTL, table-filling strategy, and customized tagging scheme. Parameter sharing is the basic idea in MTL, which can be used to enhance the interaction between NER and relation extraction (Li et al., 2017a; Bekoulis et al., 2018). This method can provide some relief from error propagation, but it cannot completely eliminate the issue. Also, this method has to pair every two entities for relation extraction, which introduces unnecessary redundancy. Table filling-based joint NER and relation extraction can completely eliminate error propagation by converting NER and relation extraction into a whole sequence-tagging task (Gupta et al., 2016b; Ren et al., 2021a; Ma et al., 2022b). However, these methods have to label every two token pairs in an input sentence in an enumerable fashion. If relation extraction is defined as an unidirectional task, the half of calculations are wasted.

Following the idea of the table filling strategy, tagging scheme-based methods also model the NER and relation extraction as an integrated task. The fundamental concept of the tagging scheme is to merge the labels assigned for NER with those assigned for relation extraction into a unified label (Zheng et al., 2017; Straková et al., 2019). Such a method has the potential to circumvent issues related to both error propagation and redundancy; however, it may also lead to a sparsity of labels.

3.3.8.2 Application Trends

We have discussed three main downstream applications of NER, including KGC, dialogue systems, and recommendation systems. Table 3.10 illustrate related studies. Usually, NER is the basic module for providing recognized entities for further utilization. In this case, an NER model works as a parser to mine knowledge from unstructured text. The recognized entities and relations can be used as nodes and edges for KGC. The entities can also serve as intent recognition methods in recommendation systems, and slot-filling methods in dialogue systems. For example, Wu et al. (2020a) proposed a pretrained task-oriented dialogue BERT, which significantly boosts the performance of a dialogue system by improving the intent detection subtask. Wang et al. (2020) proposed a method for recognizing related spans and value normalization with slot attention to improve the dialogue system. Besides, we also observe that using the identified named entities as features can also improve the performance of recommendation systems, because NER can help identify important entities that could be useful for making recommendations. The most common problem is error propagation between NER and other components in a downstream system. Kim et al. (2018) employed a two-step neural dialogue state tracker to alleviate the impact of the original error.

Table 3.10: A summary of the representative applications of NER in downstream tasks. ✓denotes the role of NER in a downstream task.

Reference	Downstream Task	Feature	Parser	Explainability
Yao et al. (2022)	Knowledge Graph Construction		✓	
He et al. (2021)	Knowledge Graph Construction		✓	
Jiang et al. (2020)	Knowledge Graph Construction		✓	
Li et al. (2020c)	Knowledge Graph Construction		✓	
Kim et al. (2012)	Recommendation Systems		✓	✓
Adomavicius et al. (2011)	Recommendation Systems	✓		
Zhou et al. (2020b)	Recommendation Systems	✓		
Iovine et al. (2020)	Recommendation Systems	✓		
Wang et al. (2019a)	Recommendation Systems		✓	
Li et al. (2017b)	Dialogue Systems		✓	
Abro et al. (2022)	Dialogue Systems		✓	
Dimitrakis et al. (2020)	Dialogue Systems		✓	
Zhang et al. (2021a)	Dialogue System		✓	✓

With the development of PLMs and LLMs, many downstream tasks are organized as end-to-end processing tasks to achieve higher accuracy and mitigate error propagation issues. However, we can still observe that NER can improve the explainability in recommendation and dialogue systems (Kim et al., 2012; Zhang et al., 2021a), which is also an important aspect of AI research. There is still a considerable untapped potential for integrating NER with other downstream tasks, e.g., explaining how concepts blend each other between different entities; what the inherent attribute of a group of entities the selected prototypes represent; how robust an identified named entity is.

3.3.8.3 Future Works

Open-domain NER

Compared with typical single-domain NER, open-domain NER has more categories. Besides, the entity classes are hardly defined in advance. For such reason, open-domain NER is more capable of handling rapidly expanding data, and mining more potential knowledge which is hidden in massive unstructured text data (Hohenecker et al., 2020; Kolluru et al., 2020). Open-domain NER is significant because it discovers and connects world knowledge via automatic text mining. Many manually developed lexical resources, e.g., WordNet can only cover limited concepts. When the concepts come to MWEs, manually mining, structuring and updating those concepts can result in the exponential growth of human efforts. Open-domain NER is helpful for mitigating human efforts and delivering a knowledge base that connects entities from different domains.

Multilingual NER

In light of the fact that a significant number of languages in existence lack sufficient annotated data, knowledge transfer from high-resource languages to low-resource languages can serve as a viable solution to compensate for the paucity of data (Rahimi et al., 2019; Tedeschi et al., 2021b; Zhou et al., 2022). Developing robust multilingual NER systems that can perform across multiple languages will achieve more comprehensive knowledge graphs, linking entities from different languages. It is valuable because it may lead to a united concept representation system covering different languages. On the other hand, the task of developing multilingual NER systems is fraught with difficulties, primarily due to the inherent dissimilarities in entity types and language structures across different languages. As a result, aligning entities and transferring knowledge learned from one language to another can present significant challenges for multilingual NER systems.

Unified Framework for NER

In real-world scenarios, there exist flat-named entities, nested entities, and discontinuous entities. Most NER-related studies only focus on the combination of flat with nested entities or flat discontinuous entities. Both of them cannot recognize all kinds of entities. Developing a unified framework to simultaneously handle such a problem becomes an urgent need for NER (Fei et al., 2021a). Hierarchical concept representation knowledge bases may provide a preliminary ontology that can be used for organizing entities and their relationships. However, most of the ontology systems were manually developed by experts. This manually constructed knowledge may be invalid in specific application scenarios. A potential avenue for future research in NER is the development of a unified and robust framework for organizing entities. Such a framework could create comprehensive knowledge graphs that capture entity relationships and support downstream tasks.

Continual-learning for NER

Humans exhibit a remarkable aptitude for transferring acquired knowledge from one task to another and retain their ability to perform the former task even after learning the latter. This ability is called continuous learning or life-long learning, which a regarded as an important characteristic of an intelligent system. Also, such ability can help us continue to use already deployed models when a new class of entity to be identified appears, rather than developed a new model from scratch (De Lange et al., 2021). There are some exploratory studies started to pay attention to such a problem. However, a satisfactory solution has not been found yet and existing methods still suffer the severe Catastrophic Forgetting (Monaikul et al., 2021; Xia et al., 2022; Vijay and Priyanshu, 2022). Continual learning is a critical skill for NER because NER is corpus-dependent. It is very important to update entity collections and the associated label sets, when a new corpus arrives (He et al., 2022b). Detecting new entities and labels with a previously trained NER model is a challenging yet promising research avenue.

3.4 Concept Extraction

Concept extraction is a process to extract concepts of interest from the text. To the best of our knowledge, the task of computational concept extraction was first proposed by Montgomery (1982). He argued that taxonomic hierarchies could be constructed to allow property inheritance of concepts, and therefore to perform rudimentary inference and analogical reasoning based on the taxonomies. He also highlighted two important subtasks of concept extraction for the next-generation knowledge-based systems from the perspective of 1982, namely lexicon development and conceptual structure construction.

Recent research on concept extraction has been conducted in various fields of AI research, including NLP and data mining (Miner et al., 2012). Keyphrase generation (Alami Merrouni et al., 2020) is one of the most common concept extraction tasks. It is a summarization task focusing on extracting keyphrases from a full passage to help readers quickly understand the passage, where keyphrases can be understood as the important concepts within a passage. Methods for keyphrase extraction can be both extractive (copying from existing words) and abstractive (not copying but summarizing and abstracting from existing texts).

The process of generating keyphrases facilitates the creation of a lexicon that corresponds to a specific set of concepts. Another stream of concept extraction aims at the development of ontological knowledge bases to represent, e.g., commonsense knowledge (Havasi and Speer, 2007), hypernym and synonym knowledge (Snow et al., 2006), and affective knowledge (Cambria et al., 2024). These tasks tried to extract concepts to fit into pre-defined knowledge structures. Then, the structured knowledge can be directly used in downstream tasks.

Current concept extraction research is also grounded on related application scenarios, such as clinical concept extraction (Fu et al., 2020b), course concept extraction (Pan et al., 2017), and patent concept extraction (Liu et al., 2020a). The main difference between concept extraction and NER tasks is that the extracted concepts or keyphrases are not identified by pre-defined entity classes. In contrast, they reflect the general idea of their contexts or target domain whose concepts are being discussed, while the goal of NER is to extract important factual information from the text. However, there are overlaps between NER and concept extraction when some concepts of interest, e.g., proper nouns can be also defined as named entities. Many domain-specific concept extraction tasks, e.g., clinical concept extraction, course concept extraction, and patent concept extraction can also be categorized as NER tasks because they aim at extracting concepts that are related to specific events. These events are also factual information. We review them in this section because they define themselves as concept extraction tasks in their original works. It also has become a trend of domain-specific concept extraction.

Another related field is relation extraction, which is a sub-field of information extraction. Relation extraction extracts information from raw text and represents it in the form of a semantic relation between entities (Kartik Detroja, 2023). The main difference is that, relation extraction targets at extracting relations between entities, while concept extraction targets at extracting noun entities. Concept extraction has also accelerated and contributed to multiple downstream applications, such as polarity detection (Cambria et al., 2024), information retrieval (Xiong et al., 2017), commonsense explanation generation (CEG) (Fang and Zhang, 2022). These applications mostly leverage explicitly extracted concepts. In this section, we provide a more comprehensive review on concept extraction.

3.4.1 Theoretical Research

3.4.1.1 Exemplar Theory

Medin and Schaffer (1978) argued that concepts are represented by a collection of particular exemplars or individual instances that are linked to the category. When we categorize an instance, we compare it with multiple specific exemplars of the category. This is different from Prototype Theory where a new instance is categorized by comparing the instance to the abstract prototype of the category (see Section 3.3.1.1). Medin and Schaffer (1978) formed the task of concept categorization as a classification task. Experiments showed that the classification judgments made by participants were impacted by various factors. These factors included the extent of resemblance between the probe item and exemplars previously acquired, the number of prior exemplars that shared resemblances with the probe item, and the similarity present both within and between the categories of the previously learned exemplars. For concept extraction, Exemplar Theory may suggest that models may take categorized instances into account when categorizing a new instance.

3.4.1.2 Semantic Primitives

Wierzbicka (1972) believed that it is possible to describe every human language by using a limited number of universal semantic primitives. These primitives are representative of fundamental concepts that form the basis of human communication and thinking. She established 64 universal semantic primes (basic words that cannot be defined in relation to more elementary concepts), which can be utilized to describe all other concepts present within a language. Semantic Primitives suggest that concepts should be organized as multiple layers from the concrete to abstract ones. Decision-making that runs on concrete concepts can be completed through the upper-level abstract concepts that contain those concrete concepts. Thus, it is critical to represent the hierarchical and linking relationships between concepts. There are other theories mentioned before, e.g., Frame Semantics (see Section 3.2.1.4), that may guide concept structure development. Frame Semantics highlights the connection of related concepts, while Semantic Primitives suggest the hierarchical relationships between concepts and the distinction between primitive concepts and others.

3.4.1.3 Conceptual Spaces

Gardenfors (2004) defined concept as the "theoretical entities that can be used to explain and predict various empirical phenomena concerning concept formation". The author believed that concept representations are multi-dimensional, where each dimension is indicative of a different characteristic or property associated with the concept. For example, one could represent the concept of a car within a conceptual space that includes dimensions such as size, speed, color, and shape.

This is very similar to current vectorial representations of words or entities in NLU, while the dimensionality of Conceptual Spaces is explainable by concept properties. Gardenfors (2004) also placed significant emphasis on the role of context in understanding and representing concepts. This is due to the fact that different contexts may emphasize different features or dimensions of concepts. Then, the connections between concepts are determined by the relationships between their property similarity in the conceptual space. For example, "dog" and "cat" are similar in the animal concept space, because their properties are similar; "mammals" can be separated from "reptiles" by a property difference boundary, although both are in the animal space. This may encourage concept extraction tasks to extract both concept entities and properties associated in contexts. This is because properties define how concepts are connected from the view of Gardenfors (2004).

3.4.2 Annotation Schemes

From the goal of the keyphrase annotation aspect, there are in general two types of annotation schemes for keyphrase extraction-like concept extraction. The first is to precisely select existing keyphrases from input text, but not to create semantically-equivalent phrases. The second is to both select existing keyphrases and create "absent keyphrases" that are necessary but do not exist in the input text (Hulth, 2003). From the format of assigned annotations aspect, there are in general two annotation schemes as well. The first scheme is to directly give the keyphrases existing in the source text. The second scheme treats the keyphrase extraction task as a sequence labeling task, and assigns a label to each of the tokens in source text (Hulth, 2003). The assigned labels in the current dataset follow a BIO scheme defined in Table 4.18. Specifically, three labels are used: B (Beginning), I (Inner), and O (Other).

```
abstract: "[ 'A', 'scalable', 'model', 'of', 'cerebellar', 'adaptive', 'timing',
'and', 'sequencing', ':', ...]"
doc bio tags: "[ 'O', 'B', 'I', 'O', 'B', 'I', 'I', 'O', 'O', 'O', ...]"
extractive keyphrases: "[ 'scalable model', 'cerebellar adaptive timing', ... ]"
abstractive keyphrase: "[ 'cerebellar sequencing', ...]"
```

3.4.3 Datasets

Table 3.11 lists some of the most popular concept extraction datasets, providing detailed statistics for each. It includes information such as task, source, and size of the dataset. Overall a main thread of dataset development is (1) larger scale of datasets; (2) attending to both extractive keyphrases and abstractive keyphrases; (3) more fine-grained annotations for tags; (4) more application domains.

Table 3.11: Concept extraction datasets and statistics. KE denotes Keyphrase Extraction. ClCE denotes Clinical Concept Extraction. CoCE denotes Course Concept Extraction. PCE denotes Patent Concept Extraction.

Dataset	Task	Source	# Samples	Reference
Inspec	KE	Inspec database	2,000	Hulth (2003)
NUS	KE	Google SOAP API	211	Nguyen and Kan (2007)
Krapivin	KE	ACM Digital Library	2,304	Krapivin et al. (2009)
SemEval-2010	KE	ACM Digital Library	244	Kim et al. (2010)
X	KE	X	1,000	Zhang et al. (2016)
KP-20K	KE	ACM Digital Library, ScienceDirect, and Web of Science	567,830	Meng et al. (2017)
CCF	KE	China Computer Federation	13,449	Wang et al. (2018c)
MLDBMD	KE	Academic Conferences	128.1k	Li et al. (2018a)
TempEval	ClCE	Mayo Clinic	600	Bethard et al. (2016)
i2b2-2010	ClCE	Clinical Records	826	Uzuner et al. (2011)
n2c2-2018	ClCE	Clinical Records	505	Henry et al. (2020)
MIMIC	ClCE	MIMIC-III Database	1,610	Gehrmann et al. (2018)
MOOCs	CoCE	Coursera and XuetangX	4375 videos	Pan et al. (2017)
EMRCM	CoCE	Chinese Textbooks	3,730 pages	Huang et al. (2019b)
USPTO	PCE	USPTO Database	94,000	Liu et al. (2020a)

Hulth (2003) proposed one of the first keyphrase extraction datasets, termed Inspec, which is based on scientific papers. Paper abstracts are used as the keyphrase extraction context and paper keywords are used as keyphrases. Each abstract has two sets of keywords: a set of controlled terms, i.e., terms restricted to the Inspec thesaurus; and a set of uncontrolled terms that can be any suitable terms that may or may not be present in the abstracts. They collected 1000 abstracts as a train set, 500 as a validation set, and 500 as a test set.

Nguyen and Kan (2007) proposed the NUS dataset with the motivation that keyphrase extraction requires multiple judgments and cannot rely merely on the single set of author-provided keyphrases. They first used Google Search API to retrieve scientific publications, and then recruited student volunteers to participate in manual keyphrase assignments. They finally collect 211 documents, each with two sets of keyphrases: one is given by the original authors of the paper, and the other is given by student volunteers. The data format of NUS is the same as Inspec.

Krapivin et al. (2009) proposed the Krapivin dataset, consisting of around 2,000 scientific papers as well as their keywords assigned by the original authors. The scientific papers were published by ACM in the period from 2003 to 2005, and were written in English. One of the novelties of this dataset is that the text data in the scientific papers were collected with three distinct categories: title, abstract, and main body. They finally collect 460 test data and 1.84k validation data. The data format is similar to Inspec but has a title and body in addition to the abstract.

SemEval-2010 Task 5 (Kim et al., 2010) is on automatic keyphrase extraction from scientific articles. Input for this task is a document from either of the four domains: distributed systems, information search, and retrieval, distributed AI, and social and behavioral sciences. Outputs are manually annotated keyphrases for the document. This dataset contains 144 documents as a train set, and 100 documents as a test set. It also selects 40 documents from the train set to compose a trial set. For each set, documents are evenly distributed from the four topics. The annotation follows the first scheme in Section 3.4.2. The data format is the same as Inspec.

Zhang et al. (2016) constructed a keyphrase extract dataset from $\mathbb{X}$ using an automatic text mining method. Their core assumption is that hashtags in a tweet can be used as keyphrases for the tweet. To construct the dataset, they first collected 41 million tweets, and then filtered them which contain non-Latin tokens. URL links, and reply tweets were removed. Thus, the remaining text only contains tweets and a hashtag. They finally kept 110K tweets. To evaluate the quality of the collected tweets, they sampled 1000 tweets and chose three volunteers to score them. As a result, 90.2% tweets are suitable, and 66.1% are perfectly suitable. The annotation follows the first scheme in Section 3.4.2.

```
tweet: "Hard to believe it but these are REAL state alternatives to taking
Obamacare funds from the gov't (via @Upworthy)"
keyphrase: "obamacare"
```

Pan et al. (2017) proposed a keyphrase extraction dataset, where data were sourced from online course captions. Labels are existing phrases in the captions. The courses are computer science and economics courses, selected from two famous MOOC platforms — Coursera and XuetangX. Labels were first filtered from captions using automatic methods and then annotated by two human annotators. A candidate concept was only labeled as a course concept if the two annotators were in agreement. As a result, they collected captions from 4375 videos, and 16720 labeled concepts.

```
course caption: "You might learn how to write a bubble
sort and learn why a bubble sort is not as good as a heapsort."
keyphrase: "[ 'bubble sort', 'heapsort' ]"
```

KP-20K (Meng et al., 2017) is a testing dataset, where the input texts are titles and abstracts of computer science research papers collected from ACM Digital Library. The labeled keyphrases are the keyphrases shown in the research papers. The annotation follows the second scheme in Section 3.4.2, since the keyphrases given by authors were not necessarily existing keyphrases in the papers. KP-20K has the same data format as Inspec. Huang et al. (2019b), instead, were motivated to automatically construct an educational concept map, showing concepts that will be learned in courses, as well as the temporal relation between the concepts (e.g., to learn concept A, it is a prerequisite to learn concept B; Concept A and concept B can help with the understanding of each other).

To construct the dataset written in Chinese, they first used OCR to obtain the text from textbooks, then manually labeled key concepts for each textbook (as "key concept" or "not key concept") and finally manually annotated the relationships among the labeled key concepts (as "w_i is w_j's prerequisite", "w_i and w_j has collaboration relationship", or "no relationship").

```
keyword: "[ 'average', 'weighted average', ... ]"
relation: "[ 'average : weighted average', ... ]"
```

As a result, they collected 3730 pages in textbooks, 1092 key concepts, 818 prerequisite relations, and 916 collaboration relations. However, in their GitHub repo, only keyphrases and relations between keyphrases can be found, while the text cannot be found. Other concept extraction datasets focused on specific domains such as clinical concepts, e.g., TempEval (Bethard et al., 2016), i2b2-2010 (Uzuner et al., 2011), n2c2-2018 (Henry et al., 2020), and MIMIC (Gehrmann et al., 2018), course concepts, e.g., MOOCs (Pan et al., 2017) and EMRCM (Huang et al., 2019b), and patent concepts, e.g., USPTO (Liu et al., 2020a). They also followed keyphrase extraction setups, whereas the targets are to extract concepts of interest.

3.4.4 Knowledge Bases

Besides classical lexicon resources such as WordNet, encyclopedias (including Baidu Encyclopedias and Wikipedia) can also be used to provide external knowledge for concepts (Pan et al., 2017). Methods for extracting concepts based on embedding techniques may encounter issues with low frequency, where some of the concepts have infrequent occurrences. Pan et al. (2017) utilized word embeddings (Mikolov et al., 2013b) trained on encyclopedias to obtain the semantic embedding for each concept. Inspec database is a scientific and technical database storing scientific papers. The papers of this database have been used to construct a keyphrase extraction dataset.

Table 3.12: Useful knowledge bases for concept extraction.

Name	Knowledge	#Entities	Structure
WordNet	Lexical	155,327	Tree
SenticNet	Sentiment	400,000	Graph & Dictionary
Baidu Encyclopedia	World	6,223,649	Unstructured
Wikipedia	World	9,834,664	Unstructured
Inspec	Science	20,000,000	Unstructured

3.4.5 Evaluation Metrics

The field of concept extraction also uses Precision, Recall, and F_1 score as evaluation metrics. Some keyphrase extraction research considered the task as an information retrieval task. Then, the information retrieval metric, e.g., mean average precision (MAP) was also used for keyphrase extraction as the main measure. It is calculated by taking the average of the average precision scores for each query in a dataset.

$$MAP = \frac{1}{n} \sum_{i=1}^{n} Avg\_Precision_i, \qquad (3.6)$$

where n is the total number of queries. $Avg\_Precision_i$ denotes the averaged precision of query i. In the context of keyphrase extraction, the MAP score is determined by comparing the generated list of keyphrases with a predefined gold standard set, and evaluating the average precision of the top n keyphrases, where n corresponds to the total number of keyphrases in the gold standard set. Each generated keyphrase is considered as a query. The gold standard set serves as the relevant document.

3.4.6 Annotation Tools

Numerous studies have investigated the utilization of pre-existing keywords in scientific publications (Hulth, 2003; Nguyen and Kan, 2007; Krapivin et al., 2009; Kim et al., 2010; Chen et al., 2018c) or hashtags in tweets (Zhang et al., 2016), whereby such in-context information can serve as labels without requiring additional annotation efforts, provided that the labels align with the research objectives. Given the similarities between the two tasks, NER annotation schemes can also be applied to annotating concept extraction data.

3.4.7 Methods

3.4.7.1 Keyphrase Extraction

The task of keyphrase extraction is to obtain keyphrases from a document to represent and summarize the document with such keyphrases. There are two main methods, namely: extractive and generative keyphrase extraction. Extractive methods appear first but have a systematic disadvantage in that they can only extract existing phrases in the documents. For example, Meng et al. (2017) argued that in addition to present keyphrases, there are also absent keyphrases, which can better summarize a document but are not explicitly present in it. Generative methods, instead, can generate every possible word but they require more sophisticated models to accurately catch the "semantic meaning" of a document to precisely generate relevant keyphrases.

A. Extractive Keyphrase Extraction

Zhang et al. (2016) focused on the task of keyphrase generation on $\mathbb{X}$ data, and framed this task as a sequence labeling task. They proposed a joint-layer RNN model. For each input token, the joint-layer RNN model outputs two indicators ($\hat{y}_1$ and $\hat{y}_2$), where $\hat{y}_1$ has two values *True* and *False*, indicating whether the current word is a keyword. $\hat{y}_2$ has 5 values *Single*, *Begin*, *Middle*, *End* and *Not* indicating the current word is a single keyword, the beginning of a keyphrase, the middle of a keyphrase, the ending of a keyphrase, or not a part of a keyphrase, respectively. Their experiments show that the joint-layer RNN model outperforms both the vanilla RNN model and the LSTM model. However, when $\hat{y}_1$ and $\hat{y}_2$ have contradictions, it might be hard to find an optimal strategy to determine which indicator to refer to. In addition, joint-layer RNN can only extract an existing sequence as a keyphrase, but cannot abstractively obtain a (non-existing but better) keyphrase.

Wang et al. (2018c) hypothesized that the performance of keyphrase extraction could be improved in the unlabeled or insufficiently labeled target domain by transferring knowledge from a resource-rich domain. They accordingly proposed a topic-based adversarial neural network (called TANN) that can learn transferable knowledge across domains efficiently by performing adversarial training. The experiment section shows that TANN largely outperforms joint-layer RNN (Zhang et al., 2016). Li et al. (2018a) proposed an unsupervised method for concept mining, which was motivated by the fact that supervised methods might be hard to generalize to unseen domains. They assumed that the quality of an extracted concept can be measured by its occurrence contexts and proposed a pipeline method for concept mining. The method first populates many raw concepts extracted from text, and then evaluates the concepts by comparing the embedding of concepts against the current local context.

Al-Zaidy et al. (2019) identified two limitations of previous supervised approaches: 1) They classify the labels of each candidate phrase independently without considering potential dependencies between candidate phrases; 2) They do not incorporate hidden semantics in the input text. Correspondingly, Al-Zaidy et al. (2019) addressed keyphrase extraction as a sequence labeling task, and proposed a model named BiLSTM-CRF that unite both the advantages of LSTM (capturing semantics) and CRF (capturing dependencies). Their results show that BiLSTM-CRF outperforms CopyRNN (Meng et al., 2017) by a large margin.

Fang et al. (2021a) hypothesized that previous extractive methods ignore structured information in the raw textual data (title, topic, and clue words), which might lead to worse performance. They accordingly proposed a model named GACEN that can utilize the title, topic, and clue words as additional supervision to provide guidance. GACEN also utilized CRF to model dependencies in the output. The experiment section shows that GACEN outperforms Joint-layer-RNN (Zhang et al., 2016) and CopyRNN (Meng et al., 2017).

B. Generative Keyphrase Extraction

Meng et al. (2017) were motivated that classic keyphrase generation methods can only extract the keyphrases that appear in the source text. Those methods are unable to reveal and leverage the full semantics for keyphrase ranking. Consequently, they proposed an RNN-based generative model incorporating a CopyRNN. Their method uses an encoder-decoder architecture to catch the semantics of the input text. Previous methods suffered from both coverage (not all keyphrases are extracted) and repetition (similar keyphrases are extracted) problems (Meng et al., 2017). For the coverage issue, Chen et al. (2018c) integrated a coverage mechanism (Tu et al., 2016) into their approach, which enhances the attention distributions of multiple keyphrases in order to cover a wider range of information within the source document and effectively summarize it into keyphrases. For the repetition issue, they constructed a target side review context set that contains contextual information of generated phrases.

Ye and Wang (2018) believed that although Seq2Seq models have achieved good performance, model training still relies on large amounts of labeled data. Correspondingly, they leveraged unsupervised learning methods such as TF-IDF and self-learning algorithms to create keyphrase labels for large amounts of unlabeled data. Then, they train their model with a mixture of self-labeled and labeled data together for training. They also used MTL to train their model. Experiments show that their performance outperforms previous works. Chen et al. (2019) argued that prior research on keyphrase generation has treated the document title and main body in the same manner, overlooking the significant role that the title plays in shaping the overall document. Hence, they proposed a Title-Guided Network (TG-Net) where the title is additionally employed as a query-like input to particularly assign attention to the title. The performance of TG-Net outperforms CopyRNN (Meng et al., 2017).

3.4.7.2 Structured Concept Extraction

Compared with keyphrase extraction-like concept extraction, structured concept extraction aims to develop an ontology where concepts are connected with each other by certain relationships. This approach allows for a more interconnected and hierarchical representation of knowledge. Here, we introduce three knowledge bases commonly used in concept extraction: WordNet, ConceptNet, and SenticNet.

WordNet (Fellbaum, 1998) is a manually developed knowledge base, where words and concepts are hierarchically organized. Snow et al. (2006) proposed a taxonomy induction method to expand WordNet 2.1 concepts by automatic noun hyponym acquisition. Compared to previous methods that relied on individual classifiers to uncover new relationships based on pre-designed or automatically extracted textual patterns, the proposed approach considers input from multiple classifiers to enhance the overall structure of the taxonomy and prioritizes the optimization of the entire taxonomy structure with a probabilistic architecture.

ConceptNet (Havasi and Speer, 2007) grew out of the Open Mind Common Sense (OMCS) project (Singh et al., 2002), which aimed at commonsense knowledge acquisition. Contributors delivered knowledge by fulfilling blanks within a sentence, For example, given "[ ] can be used to [ ]", the concepts, e.g., "ink" and "print" and the associated relationship "UsedFor" can be obtained. The nodes of ConceptNet are words and MWEs and its edges are one of the 21 basic relation types, such as "IsA", "PartOf", and "UsedFor". In the latest ConceptNet 5.5 (Speer et al., 2017), the relations are increased to 36. Concepts and predicates were obtained via pattern matching. Each collected sentence is compared with pre-defined regular expressions, e.g., "NP is used for VP"(UsedFor), "NP is a kind of NP"(IsA), "NP can VP" (CapableOf). NP (noun phrases) and VP (verb phrases) are concepts, while "Used-For", "IsA", and "CapableOf" are predicates. In the case of a complex sentence that contains several clauses, the patterns are employed to extract a simpler sentence from it, which can then be subjected to the concept and predicate extraction process. To evaluate ConceptNet, its assertions were compared with those in similar lexical resources to determine their alignment.

SenticNet (Cambria et al., 2024) is a commonsense knowledge base for concept-level sentiment analysis built by means of neurosymbolic AI. Concepts are extracted using a graph-based semantic parsing method termed Sentic Parser (Cambria et al., 2022), which analyzes text at both MWE-level and word-level (Fig. 3.7). The key novelties introduced by this parser with respect to previous work are that it leverages morphology for syntactic normalization and it uses primitives for semantic normalization. Sentic Parser is superior to standard concept parsers because it focuses on the extraction of root words from text and, hence, it normalizes complex natural language constructs to a sort of primitive-based protolanguage. This is done by leveraging a graph-based approach which enables nested affix handling, microtext normalization, and compound word processing. Preliminary experiments on other alphabetic languages showed that the proposed methodology could be language-agnostic.

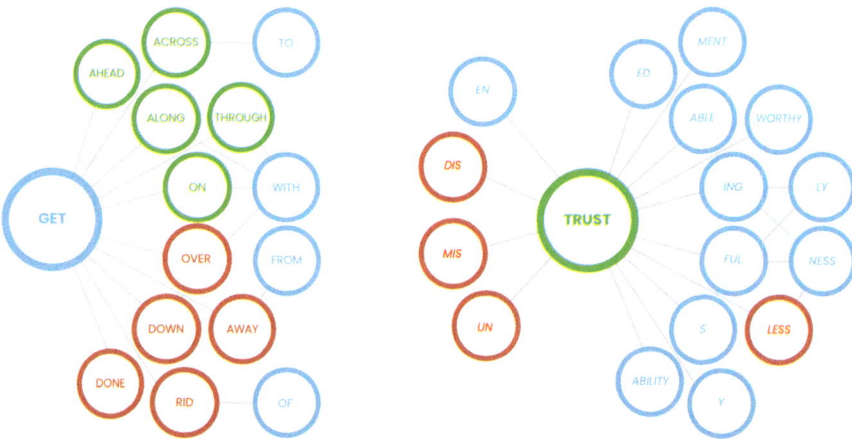

Fig. 3.7: Sentic Parser analyzes text at both MWE-level (left) and word-level (right).

An important task of concept extraction is to abstract concept representations from entities. Unlike SenticNet which obtains abstract concepts (primitives) by selecting the most typical entities from a group of extracted similar entities, Ge et al. (2022) proposed a conceptualization method that can directly abstract concepts from input text. The task is realized in the metaphor identification and interpretation domain. The authors aimed to generate concept mappings from metaphorical word pairs to explain the metaphoricity of the word pairs. For example, given "*blind* alley", "STREET is ADULT" can be automatically generated. This work is the realization of conceptual metaphor theory (CMT) (Lakoff and Johnson, 1980) (see Section 3.6.1.3) that the generated concept mapping explains the mapping of source (e.g., ADULT) and target (e.g., STREET) concepts of a metaphor. The conceptualization (e.g., from "alley" to "STREET") was achieved by selecting the most appropriate hypernym on the chain from the leaf node of "alley" to the root node "entity" in WordNet. The most appropriate hypernym is defined as the node that can cover the major senses of the leaf, meantime, keeping it as concrete as possible[23]. Subsequently, within MetaPro (Mao et al., 2023b), the conceptualization algorithm is synergistically integrated with sequential metaphor identification and interpretation techniques, culminating in the attainment of end-to-end concept mapping generation from full sentences.

3.4.7.3 Domain-specific Concept Extraction

A. Clinical Concept Extraction

The task of clinical concept extraction is to extract structural information from unstructured clinical narratives (Fu et al., 2020b). Li and Huang (2016) constructed a dataset for a seminal task called "UTA-DLNLP at SemEval-2016 Task 12" for clinical concept extraction. A system developed for this task should task raw clinical notes or pathology reports as input, and identify event expressions consisting of the "the spans of the expression in the raw text", "contextual modality", "degree", "polarity", and "type". As a baseline for this task, they propose a CNN to learn hidden feature representations for predictions, taking text and POS tags as input.

Liu et al. (2017c) adopted a BiLSTM to recognize the entity in clinical text. They found that BiLSTM outperforms the CRF baselines. Gehrmann et al. (2018) compared CNN with classic rule-based methods, bag of words, n-grams, and embedding-based logistic regression. They found that CNN is a valid alternative to rule-based and classic NLP methods, and should be further investigated. Yang et al. (2020b) comprehensively explored 4 widely used transformer-based architectures, including BERT (Devlin et al., 2018), RoBERTa (Liu et al., 2019c), ALBERT (Lan et al., 2020), and ELECTRA (Clark et al., 2020a).

---

[23] Intuitively, "entity" can cover all possible senses of the "alley" in WordNet, while it is not the ideal concept representation of "alley", because it is too abstract. Thus, the authors aimed at a concrete concept representation that can cover the majority senses of a word.

They compared the 4 models to LSTM-CRFs (Huang et al., 2015) baselines and found that transformer-based models are effective for clinical concept extraction tasks. Lange et al. (2020) proposed a joint model for both clinical concept extraction and de-identification tasks. De-identification is important since in some clinical concept extraction scenarios, the privacy of patients should be protected. They hypothesized that jointly modeling the two tasks can be beneficial, and proposed two end-to-end models. One is a multi-task model where the tasks share the input representation across tasks; the other is a stacked model, which used the privacy token predictions to mask the corresponding embeddings in the input layer and only use the masked embeddings for concept extraction. They found that the performance of the concept extraction model can be improved by training and evaluating it on anonymized data, thereby confirming their initial hypothesis.

B. Course Concept Extraction

In tasks involving the extraction of course concepts, the concepts are typically defined as the knowledge concepts that are taught in the course videos, as well as the related topics that aid in the students' comprehension of the course videos. Identifying course concepts at a fine level is very important, as students with different backgrounds need different concepts to quickly understand the main content of a course.

Pan et al. (2017) contributed the first attempt to systematically investigate the problem of course concept extraction in MOOCs. In the past, course concepts were presented by instructors at a general level, with only a few concepts being covered in an entire course video. However, they emphasized the significance of identifying course concepts at a granular level, i.e., automatically identifying all course concepts from each video clip, to facilitate easier comprehension. They identified a challenge for the task that the course concept appears at a low frequency mainly because the different courses have different concepts. They accordingly proposed to utilize word embedding to catch the semantic relations between words and incorporate online encyclopedias to learn the latent representations for candidate course concepts. They also proposed a graph-based propagation algorithm to rank the candidates based on learned representations.

Later, Wang et al. (2018b) argued that external knowledge must be involved to solve the concept extraction problem and proposed to utilize both the structured and unstructured data in Wikipedia to provide external knowledge to concept extraction. Their results show that their method outperforms the work of Pan et al. (2017).

C. Patent Concept Extraction

Liu et al. (2020a) developed a framework to extract technical concepts from patents. Patent documents have different structures than other documents. For instance, they have "title", "abstract", and "claim", which exhibit a multi-level of information. Motivated by this, the authors proposed a framework named UMTPE, which can effectively leverage multi-level information to extract concepts.

3.4.8 Downstream Applications

3.4.8.1 Polarity Detection

As discussed earlier, Sentic Parser (Cambria et al., 2022) can be used for polarity detection by leveraging the SenticNet knowledge base. Unlike deep learning techniques, SenticNet performs polarity detection in a completely explainable and interpretable way, e.g., by providing a list of key polar concepts that most contributed to the final polarity value.

Li et al. (2023b) proposed a neurosymbolic system for conversational emotion recognition. ConceptNet was used as a knowledge base to acquire commonsense knowledge out of context. For example, if a person mentions that he will "chop all onions we have and cry", another conversation participant expresses "disgust" emotion. This is because "onion IsA lacrimator" is a commonsense in ConceptNet. Such a commonsense cannot be obtained from the dictionary meanings of "onion" and the context, while ConceptNet commonsense knowledge provides the evidence and explainability to infer such an emotional status from the context. The authors used an utterance dependency parser and a neural network to learn symbolic knowledge to enhance the explainability and accuracy of their method.

By using the concept mapping method from the work of Ge et al. (2022), Han et al. (2022a) used concept mappings to support depression detection and explanation. The hypothesis is that depression patients may have similar cognition patterns that are reflected in their metaphorical expressions. Thus, they used concept mappings as additional features besides tweets. The concept mappings were generated from tweets that contain metaphors. They also proposed an explainable encoder that can identify significant concept mappings that contribute to depression detection. The concept mappings also improve the accuracy of depression detection, besides explaining the common concept mapping patterns.

3.4.8.2 Information Retrieval

Xiong et al. (2017) manually analyzed the potential problems of a literature search website SemanticScholar.org, and found that the issue of "Concept Not Understood" represents one of the most significant challenges. The reason is that previous methods measure similarity based on text, but not on their semantic embeddings. As a result, they proposed an embedding-based similarity matching method, which extracts the concepts in both query and documents and measures the similarity between these concepts to obtain the similarity between a query and a document. Liu et al. (2018b) used extracted knowledge concepts as one of the inputs to obtain a unified semantic representation for educational excises. The representation is further used to retrieve similar excises based on similarity with other representations.

3.4.8.3 Dialogue Systems

Young et al. (2018) integrated commonsense knowledge from ConceptNet in their dialogue system. They believed that in human dialogues, individuals responding to each other is not dependent on the most recent utterance only, but also on recollecting pertinent information related to the concepts addressed within the dialogue, e.g., commonsense. Thus, in retrieval-based dialogue generation, the model considers both the message content and relevant commonsense knowledge to effectively choose a suitable response. Huang et al. (2020a) proposed a new dialogue coherence evaluation metric, termed Graph-enhanced Representations for Automatic Dialogue Evaluation (GRADE). Liu et al. (2016) argued that traditional BLEU-like statistic-based metrics are biased in response coherence. Thus, Huang et al. (2020a) were motivated to propose a metric that measures the coherence by the topics of utterances. They believe that a cohesive exchange of dialogues is characterized by a seamless transition between topics. Thus, they used a ConceptNet-based method to construct topic-level dialogue graphs. Topic-level dialogue graphs were constructed by connecting the concepts that are extracted from utterances. The edge was weighted and undirected, which was derived from the shortest path between two nodes in Concept-Net. Such an evaluation metric can better represent the coherence of topics between utterances as it measures the relatedness of concepts from different utterances.

3.4.8.4 Commonsense Explanation Generation

Fang and Zhang (2022) grounded concept extraction in the context of CEG. CEG aims to generate an explanation in natural language to explain the reason why a statement is anti-commonsense. For example, given "he took a nap in the sink", the model aimed to generate "a sink is too small and dirty to take a nap in". The concepts, "small" and "dirty" (bridge concepts), are obtained via a prompt-tuning method. The authors developed a masked word prediction template to query the bridge concepts that are most likely to appear in the "mask" position. Then, they use a generator to generate the explanation with the concatenation of the original statement and the discrete bridge concepts. This method improves the explainability in explaining why a statement is anti-commonsense.

3.4.9 Summary

A concept is an abstract idea that is reflected in the mind. Concept extraction is the foundation of detecting the main idea of a context and developing conceptual knowledge bases. Related theoretical research showed that concepts may be abstracted from multiple specific exemplars (Medin and Schaffer, 1978) or prototypes (Rosch, 1973). There are limited primitives that construct human cognition and reasoning, which are the foundation of complex concepts (Wierzbicka, 1972).

According to Gardenfors (2004), conceptual space is multi-dimensional. The similarity between concepts can be measured by the similarity between concept properties. These theoretical research works frame the tasks of concept extraction from the perspectives of lexicon development and conceptual structure construction. On the other hand, current computational concept extraction methods divide this task into three categories, namely keyphrase extraction, structured concept extraction, and domain-specific concept extraction. We found that existing computational approaches inadequately address the tasks that have been put forth by the academic community's theoretical research. Although current concept extraction methods are limited, this task has greatly improved the explainability of downstream tasks such as polarity detection, information extraction, and counter-commonsense recognition.

3.4.9.1 Technical Trends

Within the domain of keyphrase extraction, generative keyphrase extraction takes advantage of generating "absent keyphrases", compared to extractive keyphrase extraction. Both tasks followed the general development of the NLP fields. They likely considered the task as a sequence labeling task (extractive keyphrase extraction) or a generation task (generative keyphrase extraction), and used typical NLP frameworks, e.g., sequence labeling and Seq2Seq frameworks. However, it is unclear if these general NLP frameworks have really learned how to summarize the main idea of context or just have learned by label distributions. There were no task-specific mechanisms proposed to explicitly learn the keyphrase extraction task on the concept level, with an explainable decision-making process. On the other hand, keyphrase extraction-based concept extraction is helpful for obtaining concept lexicons. However, compared to structured concept extraction, keyphrase extraction cannot learn the relationships between concepts.

The theoretical research of Conceptual Spaces from Gardenfors (2004) suggested that the similarity between concepts can be measured by their properties. It suggests that keyphrase extraction-based concept extraction should consider extracting properties together with keyphrases. Thus, the later works can use keyphrases and the associated properties to structure concepts by similarities. In contrast, structured concept extraction research likely utilized statistical learning and syntactic parsing methods. This is because the aim of structured concept extraction is to develop a large knowledge base or detect structured relationships between concepts.

Labeled data are insufficient in these areas. Thus, unsupervised methods are preferred. However, the concept knowledge base development is task-specific. As a result, the concepts in different knowledge bases share different relationships. For example, ConceptNet parses concepts through 36 commonsense relationships, Stanford WordNet (Snow et al., 2006) was expanded in synonyms and hypernyms relationships, SenticNet grouped concepts and extract primitives for polarity detection, and Ge et al. (2022) abstracted concepts for concept mappings.

Table 3.13: A summary of representative concept extraction techniques. KE denotes keyphrase extraction. CE denotes concept extraction. SL denotes statistical learning. Knwl. eng. denotes knowledge engineering. We do not show the evaluation results for structured concept extraction methods, because they all used very task-specific evaluation methods and datasets, where the results are not comparable.

Task	Reference	Techniques	Feature and KB	Framework	Dataset	Score	Metric
Extractive KE	Zhang et al. (2016)	DL	word2vec	Joint-layer RNN	X	86.40%	F_1
	Wang et al. (2018c)	DL	word2vec	BiLSTM, adversarial loss	CCF	29.60%	F_1
	Li et al. (2018a)	Pipeline	word2vec	Similarity matching	MLDBMD	97.00%	MAP
	Al-Zaidy et al. (2019)	DL	word2vec	BiLSTM-CRF	KP-20K	35.63%	F_1
	Fang et al. (2021a)	DL	word2vec	Attention; CRF	KP-20K	45.69%	F_1
Generative KE	Meng et al. (2017)	DL	word2vec	RNN	KP-20K	32.80%	F_1@5
	Chen et al. (2018c)	DL	word2vec	Seq2Seq	Krapivin	31.80%	F_1@5
	Ye and Wang (2018)	DL	word2vec	Seq2Seq, semi-supervised	KP-20K	30.80%	F_1@5
	Chen et al. (2019)	DL	word2vec	Seq2Seq, additional Title input	KP-20K	37.20%	F_1@5
Structured CE	Havasi and Speer (2007)	Knwl. eng.	textual patterns	Pattern matching	-	-	-
	Snow et al. (2006)	SL	feature vectors, WordNet	Probabilistic	-	-	-
	Cambria et al. (2022)	chunking, sem. pars., POS tag.	syntactic patterns	Syntactic parsing	-	-	-
	Ge et al. (2022)	SL	statistics, WordNet	Elbow algorithm	-	-	-
Clinical CE	Li and Huang (2016)	DL	token mention, pos tag, word shape	CNN	TempEval	78.80%	F_1
	Liu et al. (2017c)	DL	word2vec, character2vec	BiLSTM	i2b2-2010	85.78%	F_1
	Gehrmann et al. (2018)	DL	word2vec	CNN	MIMIC	76.00%	F_1
	Yang et al. (2020b)	DL	word2vec	Transformer	n2c2-2018	88.36%	F_1
	Lange et al. (2020)	DL	word2vec	Multi-task-biLSTM	i2b2-2010	88.90%	F_1
Course CE	Pan et al. (2017)	Graph	word2vec, Encyclopedia	Graph-based propagation	MOOCs	41.60%	MAP
	Wang et al. (2018b)	Graph	word2vec, Wikipedia	Graph-based propagation	MOOCs	47.50%	MAP
Patent CE	Liu et al. (2020a)	ML	self pretrained word2vec, DBpedia	Clustering	USPTO	43.37%	F_1

The evaluation of different concept extraction methods is different. Most of the evaluation was implemented on different downstream tasks. It shows domain-specific concept extraction is very similar to NER tasks. They used graph, machine learning methods, and external knowledge, e.g., encyclopedias and Wikipedia to discover concepts in a domain, e.g., clinical, course, or patent concepts. Similar to keyphrase-based concept extraction, these domain-specific concept extraction methods did not try to structure concepts after extraction. This is important because it distinguishes concept extraction from current NER tasks in specific domains.

3.4.9.2 Application Trends

Concept extraction methods have been widely used in downstream tasks, e.g., polarity detection, information retrieval, dialogue systems, and CEG. Compared to other low-level semantic processing techniques, the roles of concept extraction are more diverse in downstream applications. For all the surveyed downstream tasks, the products of concept extraction can be used as additional features to improve the performance of downstream tasks. Moreover, concept extraction can add interpretability to such downstream tasks, e.g., explaining anti-commonsense (Fang and Zhang, 2022) and concept mapping patterns of depressed patients (Han et al., 2022a).

Table 3.14: A summary of the representative applications of concept extraction in downstream tasks.

Reference	Downstream Task	Feature	Parser	Explainability
Cambria et al. (2022)	Polarity Detection	✓	✓	✓
Li et al. (2023b)	Polarity Detection	✓		✓
Han et al. (2022a)	Polarity Detection	✓	✓	✓
Xiong et al. (2017)	Information Retrieval	✓		✓
Liu et al. (2018b)	Information Retrieval	✓		✓
Young et al. (2018)	Dialogue Systems	✓		
Huang et al. (2020a)	Dialogue Systems	✓	✓	
Fang and Zhang (2022)	Commonsense Explanation Generation	✓		✓

In the era of PLM and LLM, it seems many complex tasks can be achieved from end-to-end with deep neural networks. However, black-box-like neural networks prevent humans from understanding their decision-making mechanisms. This may be contrary to the original intention of human beings to build AI, e.g., giving machines the ability to think like humans. Neurosymbolic AI which combines the knowledge of symbolic representations with neural networks, seems to be able to compensate for the lack of model interpretability of pure neural networks because symbolic representations in natural language, e.g., words and concepts are human-readable. We can explain a prediction by viewing what symbolic knowledge is activated. Meantime, symbolic knowledge can represent commonsense knowledge, which is difficult for neural networks to learn from corpora. As the fundamental technique of knowledge base development, concept extraction has a huge potential in downstream applications.

3.4.9.3 Future Works

A. Open Domain Concept Extraction

Prior research on concept extraction has primarily concentrated on extracting concepts within a particular domain, while other concept extraction efforts aimed at developing knowledge bases have focused on extracting concepts with predefined relations. These approaches severely limit the application scope of knowledge bases. It would be more practical to extract concepts and relations in an open domain, where both the concepts and relations are not focused on specific types. This requires an ontology study to guide the concept extraction, e.g., what can be defined as concepts and relations. It is a more challenging task than the joint NER and relation extraction task, because relationships and concepts are self-aware within a learning model, rather than pre-defined by humans.

B. Multimodal Concept Extraction

"Concept" is also very relevant to human visual recognition. It is argued that for humans, the ability of visual classification is obtained from concept learning, which learned the generalized concept description from sample observations such that a given observation can be identified as a learned concept (Seel, 2011; Xiong et al., 2021). On the other hand, the abstractness of concepts is strongly related to imagery (Paivio, 1965), because abstract concepts are those that are not applicable to tangible, perceptible objects that can be observed through touch, sight, hearing, or other sensory experiences (Löhr, 2022). Thus, learning the relationships between concepts and imagery can help concept extraction research hierarchically organized concepts, e.g., primitives, concepts, and entities. However, till now, there is a lack of research papers working on multimodal concept extraction to the best of our knowledge. It could be also interesting to investigate possible synergies in concept extraction between different modalities.

C. Concept Extraction Evaluation

Current concept extraction methods were evaluated on an application task, e.g., sentiment analysis to SenticNet or testing specific relationships, e.g., hypernym and hyponym relationship to ConceptNet and WordNet extension. The issue with such an evaluation method is that it can only reflect the effectiveness of a developed knowledge base or concept extraction method on a specific domain. Since different knowledge bases have different application targets, it is hard to evaluate and compare them with unified criteria. It would be valuable to propose a framework for knowledge base evaluation that is independent of specific tasks. It would be helpful to understand the quality of included concepts, relationships, and their representations.

D. More Concept Extraction Applications

Despite the attention some scholars have given to neurosymbolic AI, the body of related works remains relatively scant in comparison to end-to-end neural network models. One possible explanation for this disparity is that, at present, there is greater emphasis placed on the accuracy of the model rather than the transparency of its decision-making process. Thus, there is a need for more concept extraction applications, which can aid in enhancing the explainability of neural network-based models. It offers insights for the development of knowledge bases, prompting researchers to reassess how they extract and organize concepts in order to more effectively support subsequent applications.

3.5 Anaphora Resolution

In computational linguistics, Ruslan Mitkov defined anaphora as a *phenomena of pointing back a previously mentioned item in the text* (Mitkov, 2022). The pointing back phrase is called an *anaphor* while the previously mentioned item is called an *antecedent*. The concept of anaphora should not be confused with co-reference. On the one hand, either anaphora or cataphora (e.g., the phenomena of pointing ahead to a subsequently mentioned item) could be a kind of co-reference. On the other hand, an anaphor and its antecedent are not always co-referential. By definition, the difference between anaphora and co-reference is that anaphora does not require *identify-of-reference* while co-reference requires. In other words, anaphora may describe a relation between expressions that do not have the same referent. For example, in sentence (6), the anaphor "one" has the same sense as its antecedent "a dog", but they do not refer to the same dog.

(6)   Jack has a dog and Mary also has **one**.

Building on this, in relation to anaphora, both anaphor and its antecedent are not necessarily referring expressions. For instance, an anaphor can be a verb (henceforth, verb anaphora). In the following example from Mitkov (2014),

(7)   When Manchester United swooped to lure Ron Atkinson away from the Albion, it was inevitable that his midfield prodigy would **follow**, and in 1981 he **did**.

the anaphor "did" is a verb, having an antecedent "follow". Another example is the *bound anaphora* where the antecedent is a quantified expression (Reinhart, 1983):

(8)   Each manager exploits the secretary who works for "him".

The anaphor "him" refers to the quantified expression "each manager". Since antecedents in both above two examples are not referring expressions, neither of them is a co-reference.

Given the definition of anaphora, the task of anaphora resolution is to identify the antecedent of an anaphor. In this chapter, we decided to merely focus on anaphora resolution (rather than co-reference resolution) because, on the one hand, most semantic processing tasks only require identifying antecedents. On the other hand, we are not only interested in referring to noun phrases but also other phrases that an anaphor can refer to (e.g., verb phrases and quantified expressions; see the discussion above). It is worth noting that there have been reviews in the past 20 years about anaphora resolution from both computer scientists (Sukthanker et al., 2020; Liu et al., 2023c) and linguists (Mitkov, 2022; Poesio et al., 2023).

3.5.1 Theoretical Research

3.5.1.1 Constraints

When human beings resolute co-reference, there are semantic and syntactic constraints. As for the semantic constraints, agreements such as gender and number agreements are the strongest type (Garnham, 2001). However, most recently, agreement mismatch problems (especially for gender agreements) have been becoming more frequent since more people have started to use plural pronouns to avoid gender bias. As for syntactic constraints, according to the binding theory (Büring, 2005), in the sentence (a) of the following example, "John" cannot co-refer with "him" while in the sentence (b) "John" can.

(9)   a.   John likes him.
        b.   John likes him in the mirror.

3.5.1.2 Centering Theory

Centering Theory (Joshi and Kuhn, 1979; Grosz et al., 1983, 1995) was introduced as a model of *local coherence*[24] based on the idea of *center of attention*. The theory assumes that, during the production or comprehension of a discourse, the discourse participant's attention is often centered on a set of entities (a subset of all entities in the discourse) and such an *attentional state* evolves dynamically. It models transitions of the attentional state and defines three types of transitions: CONTINUE, RETAIN, and SHIFT. For each utterance, the transition is decided by its backward-looking center (defined as the most salient entity in the previous utterance that is also realized in the current utterance and denoted as C_b) as well as forward-looking center (defined as the most salient entity in the current utterance and denoted as C_f). Consider the following discourse adopted from Kehler (1997):

(10)   a.   Terry really gets angry sometimes.
        b.   Yesterday was a beautiful day and he was excited about trying out his new sailboat. [C_b = Terry, C_f = Terry]
        c.   He wanted Tony to join him on a sailing expedition, and left him a message on his answering machine. [C_b = Terry, C_f = Terry]
        d.   Tony called him at 6AM the next morning. [C_b = Terry, C_f = Tony]
        e.   Tony was furious with him for being woken up so early. [C_b = Tony, C_f = Tony]

where we annotate each utterance with its backward-looking and forward-looking centers. The transition from utterance (10-a) to (10-b) is a CONTINUE as both backward-looking and forward-looking centers are unchanged.

---

[24] Instead of focusing on the whole discourse, centering theory focuses only on the *discourse segment*.

The next one is a RETAIN transition since although the most salient entity changes (i.e., C_f), the forward-looking center stays the same, whereas the transition from utterance (10-d) to (10-e) is a SHIFT transition because of the change of backward-looking transition. Intuitively, a discourse with more CONTINUE transitions is more coherent than the one with more SHIFT transitions. Though Centering Theory is not a theory of anaphora resolution, anaphora resolution can directly benefit from modeling transitions, which provides certain information about the preference for the referents of pronouns.

3.5.1.3 Discourse Salience

A prominent strand of work in psycholinguistics investigates how human beings use anaphora. A referent is more likely to be realized as a pronoun if it is salient in a given discourse (Givón, 1983) (aka. *discourse salience*). Discourse salience is thought to be influenced by various factors, including givenness (Chafe, 1976; Gundel et al., 1993), grammatical role (Brennan, 1995; Stevenson et al., 1994), recency (Givón, 1983; Arnold, 1998), syntactic parallelism (Chambers and Smyth, 1998; Arnold, 1998), and many other factors. Similar to Centering Theory, most research on discourse salience is about the production of anaphora (McCoy and Strube, 1999; Orita et al., 2014, 2015; Chen et al., 2018a), but it also provides insights about an antecedent's relative likelihood for a given anaphor in a given discourse. In this sense, it is plausible to use the aforementioned factors as features to rank candidate antecedents of an anaphor (Lappin and Leass, 1994; Bos, 2003).

3.5.1.4 Coolness

Huang (1984) classified human languages into cool languages and hot languages. If a language is "cooler" than another language, then understanding a sentence in that language relies more on context. Specifically, cool languages (e.g., Mandarin) make liberal use of zero pronouns. Consider the following conversation:

(11)   a.   你今天看见比尔了吗? (Did you see Bill today?)
        b.   *pro*看见*pro*了。(*I* saw *him*.)

where a *pro* represents a zero pronoun[25]. The first zero pronoun refers to one of the speakers while the second zero pronoun refers to Bill. Zero pronouns of this kind are called anaphoric zero pronouns (AZPs). In addition to Mandarin, a number of other languages (i.e., cool languages) also allow zero pronouns, including examples like Japanese, Arabic, and Korean. The current theory suggests that the anaphora resolution of cool languages should also take AZPs into consideration, namely AZP resolution (AZPR) (Chen and Ng, 2013).

---

[25] In linguistics, a zero pronoun is a pronoun that is implied but not explicitly expressed in a sentence.

3.5.2 Annotation Schemes

In this subsection, we introduce two commonly used annotation schemes for anaphora resolution: MUC and MATE. There are also other schemes, for example, the Lancaster scheme (Fligelstone, 1992) and the DRAMA scheme (Passonneau, 1997).

3.5.2.1 MUC

MUC (Hirschman et al., 1997; Hirschman and Chinchor, 1998) is one of the very first schemes, which is used for annotating the MUC (Chinchor and Sundheim, 1995) and the ACE (Doddington et al., 2004) corpora and is still widely used these years. It is primary goal is to annotate co-reference chains in discourse, in which MUC defines and proposes to annotate the IDENTITY (IDENT) relation. Relations as such are symmetrical (i.e., if A IDENT B, then B IDENT A) and transitive (i.e., if A IDENT B and B IDENT C, then A IDENT C). Annotation is done using SGML, for example:

(12)   ⟨COREF ID="100"⟩Lawson Mardon Group Ltd.⟨/COREF⟩ said ⟨COREF
        ID="101" TYPE="IDENT" REF="100"⟩it⟨/COREF⟩ ...

The annotation above construct a link between the pronoun "it" and the noun phrase "Lawson Mardon Group Ltd.". MUC proposes to annotate co-reference chains following a paradigm analogous to anaphora resolution. Annotators are first asked to annotate markable phrases (e.g., nouns, noun phrases, and pronouns) and partition the phrases into sets of co-referring elements. This helps the annotation task achieve good inter-annotator agreement (i.e., larger than 95%). Nevertheless, it has been pointed out by Deemter and Kibble (2000) that MUC has certain flaws: MUC does not guarantee that the annotated relations are all co-referential. It includes non-identity-of-reference relations or bound anaphora, resulting in a corpus that mixes co-reference and anaphora.

3.5.2.2 MATE

Instead of annotating a single device INDENT, MATE (Poesio et al., 1999a; Poesio, 2004) was proposed to do so-called "anaphoric annotation" which is explicitly based on the discourse model assumption (Heim, 1982; Gundel et al., 1993; Webber, 2016; Kamp and Reyle, 2013). The scheme was first proposed to annotate anaphora in dialogues but was then extended to relations in discourse (see (Pradhan et al., 2012) for more details). Such a good extensibility is a result of the fact that MATE is a *meta-scheme*: It consists of a core scheme and multiple extensions. The core scheme can be used to conduct the same annotation task as MUC and can be extended with respect to different tasks. The annotation normally uses XML, but many of its extensions use other their own formats.

3.5.2.3 Zero Pronoun, Bridging Reference, and Deictic Reference

In addition to the "co-referential" relation discussed above, many are also interested in "hard" cases, each kind of which is often annotated as following an extension of MATE. These include the following three: (1) zero pronoun: Pradhan et al. (2012) annotated (both anaphoric and non-anaphoric) zero pronouns in Chinese and Arabic (see Section 3.5.1.4); (2) bridging reference: bridging anaphora is a kind of indirect referent, where the antecedent of an anaphor is not explicitly mentioned but "associated" information is mentioned (Clark, 1975). Identifying such a relation needs commonsense inference. Consider the following example from Clark (1975):

(13)   I looked into the room. The ceiling was very high.

"the room" is an antecedent of "the ceiling" because the room has a ceiling; (3) deictic reference: deixis (Webber, 1988) is a phrase that refers to the "speaker's position" (e.g., time, place, situation), which is always abstracted. For example, in

(14)   I went to school yesterday.

the first person pronoun "I" and the word "yesterday" are deictic references, which refer to the speaker and the day before the date when (14) was uttered, respectively. Schemes like ARRAU (Poesio and Artstein, 2008) extended MATE and is able to annotate bridging and deictic references.

3.5.3 Datasets

As we discussed when we introduced annotation schemes in Section 3.5.2, there is no clear cut between co-reference and anaphora in computational linguistics research. We hereby review either mainstream corpora utilized in anaphora resolution or co-reference resolution, while being mindful of the scope of each of them. The datasets and their statistics are summarized in Table 3.15.

Table 3.15: Anaphora resolution datasets and statistics.

Dataset	Source	#Samples	Reference
MUC	WSJ	200	Chinchor and Sundheim (1995)
ACE	News	1,800	Doddington et al. (2004)
GNOME	Multi-domain	505	Poesio (2000)
OntoNotes	Multi-domain	4,560	Hovy et al. (2006)
WSC	Manually Written	285	Levesque et al. (2012)
DPR	Manually Written	1,880	Rahman and Ng (2012)
GAP	Wikipedia	4,454	Webster et al. (2018)
NP4E	Reuters	104	Hasler et al. (2006)
ECB+	News	982	Cybulska and Vossen (2014)
ARRAU	Multi-domain	552	Poesio and Artstein (2008)

The 6th version of MUC (Chinchor and Sundheim, 1995) is the first corpus that enables the co-reference resolution, where the task of co-reference resolution and the MUC annotation scheme was first defined. Its texts are inherited from the prevision MUCs and are English news. An example of MUC-6 is shown in Example (12). (Chinchor, 1998) updated MUC-6 in 2001 and construct the MUC-7/MET-2 corpus. MUC-7 was designed to be multilingual (NB: data in Chinese and Japanese are included in MET-2, which has been considered as a part of MUC-7) and to be more carefully annotated than MUC-6 by providing annotators with a clearer task definition and finer annotation guidelines.

ACE is a multilingual (i.e., English, Chinese, and Arabic) multi-domain co-reference resolution corpus (Doddington et al., 2004). In terms of co-reference resolution, it was built with the same purpose as MUC[26] and the same problems pointed out by Deemter and Kibble (2000) (see Section 3.5.2 for more discussion). In addition to MUC and AEC, there are works following the MUC scheme, while targeting domains other than news, which include GENIA (Kim et al., 2003), GUM (Zeldes, 2017), and PRECO (Chen et al., 2018b).

The GNOME corpus was first proposed to investigate the effect of salience on language production (see Section 3.5.1.3) and then be used to develop and evaluate anaphora resolution algorithms (Poesio, 2003; Poesio and Alexandrov-Kabadjov, 2004) targeting especially the bridging reference resolution, in the course of which the MATE scheme was introduced (see Section 3.5.2). GNOME is an English multi-domain corpus. The initial GNOME corpus (Poesio et al., 1999b) consists of data from the museum domain (building on the SOLE project by Hitzeman et al. (1998)) and patient information leaflets (building on the ICONOCLAST project), which is then expended to include tutorial dialogues (Poesio, 2000). GNOME followed the MATE scheme. Each noun phrase is marked by an ⟨ne⟩ and its anaphoric relations (marked by) are annotated separately, for example:

```
⟨ne ID="ne07" ... ⟩
Scottish-born, Canadian-based jeweller, Alison Bailey-Smith⟨/ne⟩
...
⟨ne ID="ne08"⟩ ⟨ne ID="ne09"⟩Her⟨/ne⟩ materials⟨/ne⟩

⟨ante current="ne09"⟩
⟨anchor ID="ne07" rel="ident" ... ⟩
⟨/ante⟩
```

OntoNotes (Hovy et al., 2006) is a multi-lingual (i.e., English, Chinese, and Arabic) multi-domain dataset. It is one of the most commonly used anaphora/co-reference resolution and was used in the CoNLL 2012 shared task (Pradhan et al., 2012). It was annotated following an adapted version of the MATE, named M/O scheme by Poesio et al. (2023).

---

[26] Though, in terms of entity recognition, they do not have the same purpose.

Though it has been widely used in co-reference resolution tasks, many of its relations are not co-reference. For example, bound anaphora frequently appear (see the start of this section for more discussion). Additionally, OntoNotes annotates zero pronouns in its Chinese and Arabic portions (see Section 3.5.1.4). There are other corpora following M/O, but targeting different domains, including biomedical, e.g., CRAFT (Cohen et al., 2017), Wikipedia, e.g., GAP (Webster et al., 2018) and WikiCoref (Ghaddar and Langlais, 2016), and literary text, e.g., LitBank (Bamman et al., 2020), and different anaphorical phenomena, including bridging anaphora, e.g., ISNOTE (Hou et al., 2018), style variation, e.g., WikiCoref (Ghaddar and Langlais, 2016), and ambiguity, e.g., GAP (Webster et al., 2018).

ARRAU is an English multi-domain (i.e., dialogue, narrative, and news) anaphora resolution dataset, annotated following the MATE scheme (Poesio and Artstein, 2008; Uryupina et al., 2020). However, different from other corpora that also follow MATE, ARRAU extended MATE to annotate anaphoric ambiguity explicitly (recall that MATE is a meta-scheme). Poesio and Artstein (2008) introduced the *Quasi-identity* relation, which is used for the situation when co-refer is possible but not certain by annotators and allowed each anaphor to have two distinct interpretations. In the example sample below, the footnote "1,2" of the anaphor "it" means ambiguity exists and it can either refer to 'engine E2' or "the boxcar at Elmira".

```
(u1) M: can we .. kindly hook up ... uh ... [engine E2]₁ to [the boxcar at Elmira]₂
(u2) M: +and+ send [it]₁,₂ to Corning as soon as possible please
```

The Winograd Scheme Challenge (Levesque et al., 2012) focuses on the "hard" cases of CR, which often require lexical and commonsense knowledge. It can be traced back to Terry Winograd's minimal pair (Winograd, 1972):

(15) a. **The city council** refused the demonstrators a permit because **they** feared violence.
      b. The city council refused **the demonstrators** a permit because **they** advocated violence.

The antecedent of "they" changes from "the city council" to "the demonstrators" from (15-a) to (15-b). Levesque et al. (2012) introduced the WSC benchmark consisting of hundreds of such minimal pairs. Since then, many larger-scale WSC-like corpora have been constructed. This includes the DPR corpus (Rahman and Ng, 2012), the PDP corpus (Davis et al., 2017), and the Winogrande corpus (Sakaguchi et al., 2021). Following a similar paradigm, GAP (Webster et al., 2018), Winogender (Rudinger et al., 2018) and Winobias (Zhao et al., 2018) were proposed for "hard" cases that link to gender bias.

NP4E (Hasler et al., 2006) and ECB+ (Cybulska and Vossen, 2014) are corpora for investigating cross-document co-reference. They annotated both entities and events co-reference and both within and cross-document co-reference. These corpora were built by starting from a set of clusters of documents, the documents of each of which describe the same fundamental events.

Table 3.16: Useful knowledge bases for anaphora resolution.

Name	Knowledge	#Entities	Structure
WordNet	Lexical	155,327	Graph
COW	Lexical	157,112	Graph
ODW	Lexical	92,295	Graph
AWN	Lexical	≈10,000	Graph
SenticNet	Sentiment	400,000	Graph
Wikipedia	World	13,489,694	Unstructured
Wikidata	World	100,905,254	Graph
DBpedia	World	≈4,580,000	Graph
Freebase	World	≈2.4 B	Graph
YAGO	World	4,595,906	Graph
WikiNet	World	3,347,712	Graph
OMCS	World	62,730	Graph
Medical-KG	World	22,234	Graph

The corpora mentioned above are all in English, some of which have Chinese and Arabic portions. There are anaphora/co-reference resolution corpora that focus on languages other than them. These include ANCOR in French (Muzerelle et al., 2013), ANCORA in Catalan and Spanish (Taulé et al., 2008), COREA in Dutch (Hendrickx et al., 2008), NAIST in Japanese (Iida et al., 2007b), PCC in Polish (Ogrodniczuk et al., 2013), PCEDT in Czech (Nedoluzhko et al., 2014), and TUBA-DZ in German (Telljohann et al., 2004).

3.5.4 Knowledge Bases

Both lexical and world knowledge are useful for anaphor interpretation. See the following examples from Martin (2015):

(16) a. There was a lot of **Tour de France riders** staying at our hotel. Several of **the athletes** even ate in the hotel restaurant.
      b. She was staying at **the Ritz**, but even that **hotel** didn't offer dog walking service.

We need the lexical knowledge that indicates "riders" are "athletes" while need the world knowledge of the fact that "Ritz" is a "hotel". As discussed earlier, WordNet is a lexical knowledge base that could aid this process. Wikipedia has also been an important world knowledge source for many anaphora resolution systems. These knowledge bases consist of documents from Wikipedia as well as related meta-data. Typical examples include bases from those directly dumped from raw Wikipedia documents[27] to better-structured ones, such as Wikidata (Vrandecic and Krötzsch, 2014), DBpedia (Auer et al., 2007), and Freebase (Bollacker et al., 2008).

---

[27] https://dumps.wikimedia.org

Knowledge graphs have become popular in anaphora/co-reference resolution tasks because bases that build on raw Wikipedia are needed to be further processed (e.g., entity and relation extraction) before use. Popular knowledge graphs include those that build on Wikipedia, e.g., YAGO (Suchanek et al., 2007), and WikiNet (Nastase et al., 2010)), that are about commonsense, e.g., OMCS (Singh et al., 2002), and that are about expert knowledge, e.g., Medical-KG (Uzuner et al., 2012). Search engines, e.g., Bing and Google, were also used by a few works (Emami et al., 2018) to "hunt" knowledge for the target entities in order to resolve hard anaphora like those in WSC (see Section 3.5.3), in addition to the above knowledge bases in the strict sense.

3.5.5 Evaluation Metrics

Vanilla Precision, Recall and F_1. A plausible way to assess anaphora resolution systems is by viewing both mention detection and mention linking tasks as simple classification tasks and measuring the performance using vanilla precision, recall, and F_1 scores. A good evaluation metric needs to be both interpretable and discriminative. However, unfortunately, these measures cannot meet any of these criteria (Moosavi and Strube, 2016), especially for the mention linking task as they overlook the structure of these relations (most of which are chain-structured).

MUC and Beyond. Along with MUC-6 (see Section 3.5.3), Vilain et al. (1995) proposed the MUC score. It computes the recall and precision of anaphora/co-reference resolution outputs by considering co-reference chains in a document as a graph. Vilain et al. (1995) first defined two sets: a set of key entities $\mathcal{K}$, in which there are gold standard reference chains (NB: a chain is sometimes named as a class or a cluster), and a set of response entities $\mathcal{R}$, in which there are system generated chained. MUC score computes the recall based on the number of missing links in $\mathcal{R}$ compared to $\mathcal{K}$, formally:

$$\text{Recall} = \frac{\sum_{k_i \in \mathcal{K}} (|k_i| - |p(k_i, \mathcal{R})|)}{\sum_{k_i \in \mathcal{K}} (|k_i| - 1)} \tag{3.7}$$

where $|k_i|$ is the number of mentions in the chain k_i and $p(k_i, \mathcal{R})$ is the set of partitions that is constructed by intersecting k_i with $\mathcal{R}$. The computation of MUC precision is done by switching $\mathcal{K}$ and $\mathcal{R}$. However, it has been pointed out that MUC has certain flaws: on the one hand, since MUC is merely building on mismatches of links between the two sets, it is not discriminative enough (Bagga and Baldwin, 1998; Luo, 2005). For example, it does not tell the difference between an extra link between two singletons or two prominent entities. On the other hand, Luo (2005); Kübler and Zhekova (2011) argued that MUC prefers singletons. For instance, if we merge all mentions in OntoNotes into singletons, the resulting MUC will be higher than that of the state of the art (Moosavi and Strube, 2016).

Many metrics beyond MUC have been proposed by measuring recall and precision using mentions instead of links. Bagga and Baldwin (1998) proposed B^3, which considers the fractions of the correctly identified mentions in $\mathcal{R}$:

$$\text{Recall} = \frac{\sum_{k_i \in \mathcal{K}} \sum_{r_j \in \mathcal{R}} \frac{|k_i \cap r_j|^2}{|k_i|}}{\sum_{k_i \in \mathcal{K}} |k_i|} \tag{3.8}$$

The precision is also computed by switching $\mathcal{K}$ and $\mathcal{R}$. As pointed out by Luo (2005) and Luo and Pradhan (2016), B^3 still cannot fully properly handle singletons and, additionally, repeated mentions. To solve this, Luo (2005) proposed CEAF to incorporate measures of similarities between entities:

$$\text{Recall} = \frac{\sum_{k_i \in \mathcal{K}^*} \phi(k_i, g(k_i))}{\sum_{k_i \in \mathcal{K}} \phi(k_i, k_i)} \tag{3.9}$$

where $\mathcal{K}^*$ is the set of key entities that have the optimal mapping with $\mathcal{R}$, which is found by the Kuhn-Munkres algorithm, and $\phi(\cdot)$ is a similarity measure. Nevertheless, CEAF has two shortcomings: it overlooks all unaligned response entities (Denis and Baldridge, 2009) and weights entities equally (Stoyanov et al., 2009). In addition to the above mentioned based metrics, to handle singletons, Recasens and Hovy (2011) proposed BLANC to also consider non-coreference/non-anaphoric links. It measures the fiction of both correctly identified co-reference links and non-coreference entities, and averages them to obtain the final score. Moosavi and Strube (2016) conducted controlled experiments and proved that all the aforementioned computations of precision and recall are neither interpretable nor reliable as they suffer from the so-called *mention identification effect*. They proposed the LEA metric, which was claimed to be able to solve the above issues from two perspectives: (1) it considers both links and mentions; (2) it weights entities with respect to their importance.

3.5.6 Annotation Tools

Text Editors. In the early years, anaphora/co-reference were annotated using text editors or manipulation tools. For example, MUC-6 and ACE were annotated using plain text editors while GNOME was annotated using the XML manipulation tool developed by the University of Edinburgh[28].

Co-reference Annotation Tools. Later, linguists and computer scientists developed software that enables multi-layer annotation. The software that is designed for annotating co-reference or allows the annotations of relations between phrases can be used for anaphora/co-reference annotation tasks.

---

[28] https://ltg.ed.ac.uk/software

For example, ARRAU and PCC used MMAX2, which is a free, extensible, general-purpose, and desktop-based annotation tool. It allows users to annotate relations using fields in a form, and the form is customizable. The NP4E project used PALinkA and ECB+ used CAT (Bartalesi Lenzi et al., 2012). Both of them were designed for the event and reference annotation. More recently, co-reference annotation tools that provide better visualization, allow drag-and-drop annotation, and offer post-annotation analysis have been built. Typical examples include CorefAnnotator (Reiter, 2018), which is open-sourced and desktop-based, SCAR (Oberle, 2018), which is open-sourced and web-based, and LightTag, which is not fully free but provides good online teamwork services.

Annotation Tools with Advanced Functionalities. Some annotation tools provide extra services that help to make sure the annotation procedure is fast and reliable. We classify these services into three categories: (1) External Knowledge: BRAT (Stenetorp et al., 2012) and INCEpTION (Klie et al., 2018) integrate external knowledge bases, e.g., Freebase and Wikidata (see Section 3.5.4). Once an annotator identifies an entity, these tools would search the linked base and return related entry; (2) Pretrained Models: Tools such as TagEditor, Togtag, INCEpTION, and MyMiner (Salgado et al., 2012) can call embedded pretrained entity recognition models so that they can suggest positions of possible name entities during annotation, in which MyMiner was designed specifically for the medical domain (Neves and Ševa, 2021). Additionally, beyond name entities, TagEditor and INCEpTION can also suggest potential reference chains based on their integrated pretrained co-reference resolvers, enabling active learning for anaphora/co-reference resolution; (3) Cross-document Annotation: using CROMER (Girardi et al., 2014) and CoRefi (Bornstein et al., 2020), annotators can tag, link, or update entities across multiple documents. This is done by allowing annotators to cluster documents based on topics and annotate documents in a cluster together.

3.5.7 Methods

3.5.7.1 Rule-based Methods

A. Linguistically-inspired Approaches

Like many other tasks in NLP, early works on anaphora resolution built on rules that are rooted cognitively and linguistically. Here, the term "early" represents the age when systematic evaluations of anaphora resolution, e.g., MUC, had not been introduced. The very first algorithm is the naïve algorithm proposed by Hobbs (1978). It first does a breadth-first search from the parse tree of the sentence to search for identifying mentions and links mentions based on constraints introduced in Section 3.5.1.1.

Later on, a series of anaphora resolution systems were proposed together with computational investigations of the effect of salience (see Section 3.5.1.3). Based on a set of factors that proved to influence salience, Sidner (1979) introduced rules that are used to compute the expected focus of discourse and rules that are used to interpret anaphora. As a matter of fact, this work was built on the "centering view" rooted from Grosz (1977), which suggests that, during anaphora resolution, the searching of antecedents should be restricted to the set of centered entities. It could be seen as a prototype of the idea of "center of salience" of the centering theory (see Section 3.5.1.2), but the rules proposed by Sidner (1979) are extremely complex.

Starting from Sidner (1979), Carter (1987) focused on the rules about salience and developed a system coined Shallow Processing Anaphor Resolver (SPAR). SPAR maintains linguistically-inspired rules as domain knowledge and does commonsense inference over them. As pointed out by Carter (1987), since maintaining domain knowledge and reasoning rules is expensive, SPAR made them as simple as possible. That is why it was called "shallow processing". Carter assessed SPAR on a set of 322 test samples and found that SPAR could successfully resolve 93% pronominal anaphors and 87% non-pronominal anaphora.

Hobbs et al. (1988) formalized commonsense inference in anaphora resolution as abduction and introduced TACITUS. To do abduction, in TACITUS, knowledge (i.e., rules) is maintained in formal logic (FOL in this case). Focusing on salience, Lappin and Leass (1994) proposed the Resolution of Anaphora Procedure (RAP) algorithm. After selecting a set of candidate antecedents based on semantic and syntactic constraints, RAP contains a rule-based procedure for assigning values to several salience parameters, which are then used for resolute anaphors. An assessment on 360 hand-crafted texts containing pronouns showed RAP defeated the naïve algorithm by 2%.

Also starting from Sidner (1979), there were subsequent works that extended the idea of "focus" on the basis of the introduction of the concept of "centering". Brennan et al. (1987) introduced the BFP algorithm for anaphora resolution, which roughly has three stages: (1) construct a set of candidate antecedents with accordance to the rules of the semantic constraint; (2) filter and classify the candidates based on which action a candidate belongs to in centering theory (see Section 3.5.1.2); and (3) select the best candidate in according to a pre-defined preference over the actions. One limitation of the BFP algorithm is that its final choice is merely based on a linear preference order.

To optimize this selection process, Beaver (2004) marries BFP with the optimality theory. Another limitation is that, by only considering the center theory, BFP overlooked a key pattern of how human resolute pronouns, namely, incremental resolution (Kehler, 1997). In response to this problem, Tetreault (2001) proposed the Left-to-Right Centering (LRC) algorithm, which is an incremental resolution algorithm that adheres to centering constraints. An evaluation on the New York Time corpus (Ge et al., 1998) suggests that LRC outperformed both BFP and the naïve algorithm.

B. Knowledge-poor Approaches

After the introduction of the MUC-6 shared task, anaphora resolution systems are able to be evaluated on a large scale. However, the trade-off is that the anaphora resolution systems can no longer access inputs that are annotated with gold-standard semantic and syntactic knowledge. Building on this setting, "knowledge-poor" approaches were proposed and most systems of this kind prefer rules that have high precision but do not rely on knowledge. The most influential work is CogNIAC (Baldwin, 1997), which is a heuristic precision-first anaphora resolver that relies on rules that are almost always true. For example, CogNIAC contains a rule saying *if there is just one possible antecedent in entire the prior discourse, then that entity is the antecedent*. Its rules were selected based on the precision tested on a set of test sentences. It is worth noting that rules in CogNIAC are still used in many state-of-the-art practical anaphora resolution systems, e.g., the Stanford Deterministic Coreference Resolver (Lee et al., 2013).

C. Approaches with Approximate Knowledge

As pointed out by Poesio et al. (2023), this encourages two major changes in anaphora resolution: one this that instead of relying on perfect knowledge and doing reasoning on it, anaphora resolution systems started to syntactic parsers and approximate knowledge like WordNet. The other is that the focus of anaphora resolution models moved from being aware of only pronouns to all kinds of nominal phrases (that function as referring).

Kameyama (1997) proposed to resolve anaphors that are proper names, descriptions, and pronouns. It relies on syntactic and semantic constraints, but the related information came from a syntactic parser and morphological filter based on person, number, and gender features. Later on, approaches that marry rules with WordNet were introduced (Harabagiu and Maiorano, 1999; Liang and Wu, 2003). They made use of heuristic rules (as in CogNIAC), some of which consider lexical information from WordNet.

The most famous rule-based anaphora resolution system is the one proposed by Haghighi and Klein (2009), which is still frequently used as a strong baseline in today's research on anaphora resolution. In addition to aforesaid syntactic and semantic constraints, Haghighi and Klein (2009) makes full use of the parse trees. For example, it contains rules that rely on the distance between mentions, which is obtained from computing the shortest path between two mentions in the parse tree. It also uses Wikipedia as a resource for acquiring semantic knowledge of each entity.

One limitation of heuristic-based systems is that lower precision features often overwhelm higher precision features. In response to this, more recently rule-based systems (Raghunathan et al., 2010; Lee et al., 2013) categorized rules into sieves and made decisions with an ordered set of rules. These works are often called multi-sieve approaches.

3.5.7.2 Statistical-based Methods

The introduction of large-scale benchmarks also encourages the trend of using machine learning techniques in anaphora resolution. Basically, these learning-based models treat anaphora resolution as a series of classification problems. We categorize them on the basis of how they define the classification task.

A. Mention-pair Models

Mention-pair models train a classifier to determine whether two mentions co-refer or not. It was first introduced by Aone and Bennett (1995) and then perfected by Soon et al. (2001). To build a mention-pair model, there are five steps:

1) Identifying Mentions: As a practical anaphora resolution model, the first step of this framework is to identify mentions. Soon et al. (2001) break down the mention identification into two stages: they first used three statistical sequence taggers to do POS tagging, noun phrase identification, and name entity recognition, respectively. The outputs of them are noun phrases as well as name entities. Then, they designed rules to recognize nested noun phrases based on the identified noun phrases. For each discourse, the resulting set of mentions is the union of noun phrases, name entities, and nested noun phrases.

2) Feature Engineering: Akin to many statistical models, feature engineering is always needed. Soon et al. (2001) made use of not only syntactic and semantic features as usual but also lexical features with the help of WordNet. Many other works also used knowledge bases for feature engineering (Vieira and Poesio, 2000; Ponzetto and Strube, 2006). Later, Bengtson and Roth (2008) found that a simple model with good feature engineering can perform well.

3) Generating Training Examples: They used a heuristic-based method to generate training pairs (i.e., a pair of positive and negative examples). More specifically, a positive instance consists of an anaphor A_1 and its closest preceding antecedent A_2 while a negative instance consists of the same anaphor A_1 and the mention that intervenes A_1 and A_2. There has been a number of modifications to this strategy. For example, Ng and Cardie (2002b) forced that A_1 can only be a non-pronominal once A_2 is also a non-pronominal. Harabagiu et al. (2001); Ng and Cardie (2002a); Strube et al. (2002); Yang et al. (2003) further enhanced this process by applying rule-based or learning-based filters.

4) Building a Classifier: In this step, statistical machine learning techniques have been used. These include decision trees (Soon et al., 2001; McCarthy and Lehnert, 1995), random forests (Lee et al., 2017a), MaxEnt classifier (Berger et al., 1996; Ge et al., 1998), and memory-based learning (Daelemans et al., 2003).

5) Generating Co-reference Chains: The last step is to partition these anaphora into co-reference chains. Normally, clustering techniques are used in this step. These include closest-first clustering (Soon et al., 2001), best-first clustering (Ng and Cardie, 2002b), correlational clustering (McCallum and Wellner, 2004), and graph partitioning algorithms (McCallum and Wellner, 2003; Nicolae and Nicolae, 2006).

B. Entity-Mention Models

As a matter of fact, the task mention-pair anaphora resolution is counter-intuitive from the perspective of linguists and cognitive scientists. Additionally, Poesio et al. (2023) pointed out that mention-pair models also overlook features of entities (Ng, 2010). In response to this, entity-mention models were proposed. They directly link mentions to entities by clustering. Specifically, Cardie and Wagstaff (1999) trained a model to classify whether a mention belongs to a partially constructed cluster. However, according to the evaluation by Luo (2005), the performance of the models of this kind is not comparable to mention-pair models.

C. Mention-Ranking Models

Another problem of mention-pair models is that they only do binary classification without comparing different potential antecedents. To remedy this, Denis and Baldridge (2008) proposed an entity-ranking model, replacing the binary classification loss with a ranking loss. Rahman and Ng (2011) combined entity-ranking strategy with the entity-mention model, yielding good performance.

3.5.7.3 Neural Anaphora Resolution

A. Conventional Deep Learning Models

Wiseman et al. (2015) was the first to use deep neural networks in anaphora resolution. It is a non-linear mention-ranking model. Instead of conjunction features (as in statistical models), the model of Wiseman et al. (2015) uses a neural network to learn feature representations as an extension to the mention-ranking model. They defined two feature vectors, each of which is obtained from pretraining the model on any of the subtasks of anaphora resolution, namely, mention identification and mention linking. The final decision is made through a non-linear classification, based on these features. Both Wiseman et al. (2016) and Clark and Manning (2016b) augmented the work of Wiseman et al. (2015) by inducing global features, but they followed different schemes. Wiseman et al. (2016) ran an RNN to encode the representation of each sequence of mentions corresponding to an entity (i.e., a cluster) in the history. Whereas, Clark and Manning (2016b) first used a feed-forward neural network to encode each mention-pair of an entity and computed the entity representation by pooling over all mention-pairs. Later on, Clark and Manning (2016a) extended their previous work (Clark and Manning, 2015), which built up co-reference chains with agglomerative clustering. Each mention starts in its own cluster and then pairs of clusters are merged using imitation learning (a type of reinforcement learning technique) by assuming merging clusters are actions. Clark and Manning (2016a) replaced imitation learning with deep reinforcement learning (DRL).

Liu et al. (2023b) proposed an MTL framework for mention detection and mention linking tasks, because they found that the learning of mention detection task can enhance the learning of dependent information of input tokens, which is complimentary for mention linking detection. Such an approach achieved comparable performance to (Kocijan et al., 2019) with only 0.05% WIKICREM training samples.

B. End-to-End Models

A significant benefit of employing deep learning models lies in their capacity to operate without the requirement of handcrafted features, thus enabling the creation of end-to-end (End2End) systems. Lee et al. (2017b) proposed the first End2End anaphora resolution system. It needs no human-craft feature or parser and, more importantly, it learns to process mention identification and linking tasks jointly. To this end, the fundamental idea is to first view all spans in the previous discourse as candidate antecedents and do mention ranking (also known as span ranking as the spans it sent for rank are not always mentions). The inputs pass through an RNN and each span is represented by the concatenation of the RNN hidden states of the first token and the last token as well as the weighted sum of all tokens in the span using the attention mechanism (Bahdanau et al., 2015). The final decision of each pair is made using a feed-forward neural network. One limitation of this method is that since it searches over all possible spans, the search space would be extremely large. To remedy this, candidate spans are pruned by limiting the maximum span width, the number of spans per word, the maximum number of antecedents, and the length of input documents. This End2End model was tested on the OntoNotes dataset and outperformed all previous works.

Akin to mention-pair anaphora resolution systems, End2End anaphora resolution is problematic because it ranks every span-anaphor pair separately. In response to this problem, Lee et al. (2018) introduced a higher-order coarse-to-fine inference strategy for End2End anaphora resolution models (henceforth, C2F-AR), which, in short, does cluster ranking. It infers in an iterative manner. The antecedent distributions are used to update the span representations before doing inference, enabling later decisions conditioned on previous decisions. C2F-AR uses a coarse factor that can further prune candidate span during this higher-order inference,

More recent works focused on either improving span representations or selecting candidate spans. For example, Luo and Glass (2018) used a two-layer bi-directional RNN and combined the representations of adjacent sentences in order to improve span representation with cross-sentence dependency information. Zhang et al. (2018) proposed to enrich the span representations by training a mention identification model jointly assigning each candidate span an antecedent score. For each pair of spans, Kirstain et al. (2021) replaced span representations with a combination of lightweight bilinear functions between pairs of endpoint token representations. Wu et al. (2020b) formalized the End2End anaphora resolution as a QA task. A query is produced for each entity and predicts the positions of all spans in the co-reference chain.

C. Knowledge-based Models

Analog to classical rule-based and statistical-based approaches, works on neural anaphora resolution models also seek to integrate knowledge. In terms of the use of open knowledge bases, Aralikatte et al. (2019) used world knowledge to compute rewards for reinforcement learning-based anaphora resolution models. More specifically, they submitted the predictions to an OpenIE system and compared the predicted anaphora with the knowledge to compute the reward. Zhang et al. (2019) extracted knowledge triples related to each entity from knowledge graphs and used them to enrich span representations using a knowledge attention module.

It has been pointed out that PLMs can be considered as knowledge bases (Petroni et al., 2019). Many recent anaphora resolution models have incorporated PLMs, including BERT (Devlin et al., 2018), SpanBERT (Joshi et al., 2020), and Coref-BERT (Ye et al., 2020). There has been a line of work focusing on addressing mention linking in WSC-like corpora (see Section 3.5.3). As aforementioned, resolving these "hard" cases needs reasoning with world knowledge. Works of this line incorporate either external knowledge bases (Emami et al., 2018) or PLMs (Kocijan et al., 2019; Attree, 2019).

3.5.7.4 Anaphoric Zero Pronoun Resolution

As mentioned in Section 3.5.1.2, "cool" languages (e.g., Chinese, Japanese, Korean, and Arabic) contain AZPs, and many works have focused on resolving AZPs. As with other anaphora resolution tasks, early works on AZPR used rule-based approaches and statistical approaches. Theoretically, these works are built on the fact that speakers process zero pronouns in the same way as pronouns (Yang et al., 1999). Early on, most of the works are for Japanese because of the NAIST corpus (Iida et al., 2007b), in which AZPs are annotated. Kameyama (1985); Okumura and Tamura (1996) used center theory-based approaches for AZPR in Japanese. Statistical-based approaches were proposed with a focus on exploring useful features, including syntactic pattern features (Iida et al., 2007a), heuristic rules (Isozaki and Hirao, 2003), and features that had been considered in anaphora resolution systems (Nakaiwa et al., 1995; Nakaiwa and Shirai, 1996; Seki et al., 2001, 2002; Sasano et al., 2008; Sasano and Kurohashi, 2011). Meanwhile, there were also a number of Korean AZPR systems building on the Korean portion of Penn Treebank (Byron et al., 2006; Han, 2006).

Later on, the development of systems for Chinese (Zhao and Ng, 2007; Kong and Zhou, 2010; Chen and Ng, 2013) and Arabic AZPs became active after the introduction of OntoNotes (Aloraini and Poesio, 2020). From Chen and Ng (2016), AZPR systems also went into the age of deep learning. Most of the works were for Chinese AZPR, including approaches that use deep feedforward neural networks (Chen and Ng, 2016), RNNs (Yin et al., 2017a, 2019), attention network (Yin et al., 2018b), memory network (Yin et al., 2017b), DRL (Yin et al., 2018a) and BERT (Song et al., 2020a).

The training of AZPR systems shares the problem of lacking annotated training data. For example, the AZPR largest corpus, i.e., the Chinese portion of OntoNotes, contains only 12,111 AZPs. To incorporate more data into training, there have been three paradigms: (1) Joint modeling: Chen et al. (2021b); Aloraini et al. (2022) proposed to train a model that resolves either AZPs and non-zero pronouns jointly; (2) Multi-linguality: Iida and Poesio (2011); Aloraini and Poesio (2020) trained multi-lingual AZPR systems which were trained on AZPR data in multiple languages; (3) Data augmentation: Liu et al. (2017b) made use of large-scale reading comprehension dataset in Chinese to generate pseudo training data for Chinese AZPR. Aloraini and Poesio (2021) augmented Arabic AZPR data by a number of augmentation strategies, e.g., back translation, masking candidate mentions, etc.

3.5.8 Downstream Applications

3.5.8.1 Machine Translation

Stojanovski and Fraser (2018) provided the following example to illustrate how oracle anaphora singles can help machine translation systems.

(17)   a.   Let me summarize the novel for you.
        b.   It presents a problem.
        c.   er!@#$XPRONOUN It presents a problem.
        d.   Er prasentiert ein Problem.

Given the context (a) and the course sentence (b), based on the oracle anaphora information, Stojanovski and Fraser (2018) pre-pend the input sentence of machine translation with pronoun translation as shown in (c) and ask the system to translation with a target (d) in German. In this case, the pronoun "it" which refers to "novel" (in German "Roman") is translated to "er" (the German masculine pronoun agreeing with "Roman"). Without this information, they argued that machine translation will be hard to produce "er". The experiment on a number of NMT models suggested that would improve the BLEU scores by 4-5 points.

This argumentation was strengthened by the experiments conducted by Saunders et al. (2020), who concluded that NMT does not translate gender co-reference. Despite these theoretical studies, Le Nagard and Koehn (2010); Hardmeier and Federico (2010); Guillou (2012) focused on improving machine translation with anaphora resolution outputs. The solution is often using anaphora resolution outcomes to obtain features of each pronoun (including, gender, number, and animacy) in order to enhance the pronoun translation performance. Beyond these works, Miculicich and Popescu-Belis (2017) proposed to use clustering scores which are used for generating co-reference chains in anaphora resolution (see Section 3.5.7.2) as features for re-ranking machine translation results.

There has been a long tradition of studying the impact of AZPs on machine translation systems, especially when translating from a pro-drop language to a non-pro-drop language. For example, the Japanese-English machine translation in the 1990s had already been deployed an AZPR systems (Nakaiwa and Ikehara, 1992). Later systems followed a slightly different strategy. Instead of doing a full anaphora resolution, these systems only detect AZPs in the source language and directly translate them into the target language without further resolute them (Tan et al., 2019; Wang et al., 2019b).

3.5.8.2 Summarization

There are two major uses of anaphora resolution in text summarization (Steinberger et al., 2007). One is to help with finding the important terms while the other is to help with evaluating the coherence of the summarization. Many works have demonstrated that incorporating the information of co-reference chains contributes to both the faithfulness and the coverage of summarization systems (Bergler et al., 2003; Witte and Bergler, 2003; Sonawane and Kulkarni, 2016; Liu et al., 2021d). Nevertheless, it is also worth noting that there are also some studies that showed that anaphora resolution had negative effects (Orasan, 2007; Mitkov et al., 2007). One possible explanation is that the effect highly depends on the task the summarization system is addressing and the performance of the anaphora resolution systems.

3.5.8.3 Textual Entailment

For textual entailment, to understand the impact of anaphora resolution, Mirkin et al. (2010) manually analyzed 120 samples in the RTE-5 development set (Bentivogli et al., 2009). They found that for 44% samples anaphora relations are mandatory for inference and for 28% sample anaphora optionally support the inference. Based on this fact, many systems that got involved in the RTE challenge made use of anaphora resolution. Nevertheless, since anaphora resolution systems at that moment were not strong enough, errors they made would propagate to downstream textual entailment systems (Adams et al., 2007; Agichtein et al., 2008). As a consequence, the contribution of anaphora resolution was negative or not significant (Bar-Haim et al., 2008; Chambers et al., 2007).

3.5.8.4 Polarity Detection

Sukthanker et al. (2020) identified two ways anaphora resolution aids polarity detection: It enhances sentiment analysis of online reviews by linking pronouns to global entities and improves fine-grained ABSA by clustering entities into distinct aspects for better sentiment extraction.

The contribution of anaphora resolution in polarity detection tasks can be summarized as follows: it enables discourse-level sentiment analysis by linking mentions from different sentences. Many efforts have been carried out to demonstrate such an ability for anaphora resolution. Nicolov et al. (2008) conducted systematic experiments to understand the impacts of anaphora resolution on sentiment analysis. Specifically, they tried to incorporate anaphora information into a number of sentiment analysis models and assessed them on varieties of datasets. They concluded that, on average, anaphora resolution can boost sentiment analysis performance by 10%. Based on this finding, sentiment analysis systems that are assembled with anaphora resolution have been proposed (Jakob and Gurevych, 2010a; Ding and Liu, 2010; Le et al., 2016).

3.5.9 Summary

Anaphora resolution has been explored extensively by theoretical linguists, psycholinguists as well as computational linguistics. It is the manifest of structural semantics because the meaning of an anaphor elucidates the syntactic relationship between the anaphor and its antecedent. Early anaphora resolution models were inspired by theories and findings in linguistics, such as the theory of syntactic and semantic constraints from theoretical linguistics and the findings about factors that influence the choice of referential form from psycholinguists. Later on, by marrying these theories with computational models, linguists also gained insights regarding the comprehension and production of anaphora from anaphora resolution systems. For instance, we could understand better how each salience factor contributes to the use of anaphora through the importance analysis of a computational model that considers the factor.

Most recently, though most computational works focus on building End2End anaphora resolution systems based on deep learning techniques, linguistic theories about anaphora are still proven to play vital roles (Chai and Strube, 2022). Dataset is core for either practical or theoretical anaphora resolution research. Though many annotation schemes and datasets have been introduced, we found that they share two limitations: one is that due to the fact that anaphora is a complex concept, annotations of anaphora resolution datasets are always imperfect (Deemter and Kibble, 2000). The other is the lack of wide-coverage datasets that covers all kinds of anaphora. Finally, we found that anaphora resolution is useful in many downstream tasks, including major tasks of both NLU and natural language generation. It is always utilized as a producer of additional features for downstream tasks. Unlike other tasks in this chapter, we rarely see how anaphora resolution techniques help boost the explainability of downstream models, apart from the work of Saunders et al. (2020). We also have not observed that anaphora resolution techniques are used for constructing datasets for downstream tasks.

3.5.9.1 Technical Trends

As shown in Table 3.17, there are two clear technical trends. One is that the research interest in the realm of anaphora resolution has shifted from machine learning-based or rule-based anaphora resolution to neural approaches, especially the End2End neural anaphora resolution, which does mention identification and linking simultaneously. Another one is that, as previously elucidated in Section 3.5.7, there exist distinct shortcomings associated with each of the task formulations such as mention pair, entity mention, and mention ranking. Consequently, another tendency is to employ higher-order inferences (Lee et al., 2018) to directly rank clusters or entities, which allows for the incorporation of benefits from all the formulations.

To sum up, state-of-the-art anaphora resolution models are often *End2End cluster ranking models*. Most recent advances tended to further improve this paradigm from two angles, namely reducing the search space as an End2End anaphora resolution searches across all possible spans in its inputs for antecedents (Wu et al., 2020b); and equipping anaphora resolution systems with knowledge (which, recently, often large-scale PLMs) to boost their ability of reasoning (Joshi et al., 2019, 2020).

Table 3.17: A summary of representative anaphora resolution techniques. Rahman and Ng (2011) reported that the performance of Denis and Baldridge (2008) was 57.7% CEAF-F and that CoNLL-F is the average of MUC, B3, and CEAF scores. Stat. denotes statistics. DL denotes deep learning. Cha. Emb. denotes character embedding.

Task	Reference	Feature	Framework	Dataset	Score	Metric
Rule-based	Carter (1987)	Salience	Logic rules	self-collected dataset	93.00%	Acc
	Lappin and Leass (1994)	Salience	Logic rules	self-collected dataset	85.00%	Acc
	Brennan et al. (1987)	Semantic constraints	Centering theory	New York Times	59.40%	Acc
	Tetreault (2001)	Semantic constraints	Centering theory	New York Times	80.40%	Acc
	Baldwin (1997)	Syntactic, Semantic, Discourse	Logic rules	self-collected dataset	77.90%	Acc
	Liang and Wu (2003)	WordNet	Logic rules	Brown Corpus	77.00%	Acc
	Haghighi and Klein (2009)	Syntactic, Semantic	Logic rules	ACE	79.60%	MUC-F
Stat.-based	Soon et al. (2001)	Syntactic, Semantic, WordNet	Mention-pair	MUC-6	62.60%	MUC-F
	Cardie and Wagstaff (1999)	Lexical, Syntactic, Semantic	Entity-Mention	MUC-6	64.90%	MUC-F
	Denis and Baldridge (2008)	Linguistic & Positional	Mention-ranking	ACE	67.00%	CEAF-F
	Rahman and Ng (2011)	Lexical, Syntactic, Semantic	Mention-ranking	ACE	60.80%	CEAF-F
DL-based	Wiseman et al. (2015)	Syntactic, Semantic	Mention-rank., DNN	OntoNotes	82.86%	Acc
	Wiseman et al. (2016)	Syntactic, Semantic, Global Feature	Mention-rank., RNN	OntoNotes	64.21%	CoNLL-F
	Clark and Manning (2016a)	Syntactic, Semantic	DRL	OntoNotes	65.73%	CoNLL-F
	Clark and Manning (2016b)	Syntactic, Semantic, Global Feature	Mention-ranking, DNN	OntoNotes	65.52%	CoNLL-F
	Lee et al. (2017b)	Word & Cha. Emb.	End2End, LSTM, DNN	OntoNotes	68.80%	CoNLL-F
	Lee et al. (2018)	ELMo	End2End, LSTM, DNN	OntoNotes	73.00%	CoNLL-F
	Zhang et al. (2018)	Glove & Cha. Emb.	BiLSTM, Joint Learning	OntoNotes	69.20%	CoNLL-F
	Joshi et al. (2019)	BERT	Lee et al. (2018)	OntoNotes	76.90%	CoNLL-F
	Joshi et al. (2020)	SpanBERT	Lee et al. (2018)	OntoNotes	79.60%	CoNLL-F
	Wu et al. (2020b)	SpanBERT	QA	OntoNotes	83.10%	CoNLL-F
	Kocijan et al. (2019)	BERT_WikiCREM	DNN	DPR	84.80%	Acc
	Liu et al. (2023b)	BERT	Transformer, MTL	DPR	84.58%	Acc
AZPR	Okumura and Tamura (1996)	Salience	Center Theory	self-collected dataset	78.30%	Acc
	Sasano et al. (2008)	Salience	Probalistic	self-collected dataset	39.10%	F_1
	Chen and Ng (2016)	Syntactic, Lexical	DNN	OntoNotes	52.20%	F_1
	Yin et al. (2017a)	word2vec, Global	RNN	OntoNotes	53.60%	F_1
	Yin et al. (2018a)	Word Embedding	DRL	OntoNotes	57.20%	F_1
	Song et al. (2020a)	BERT	DNN, MTL	OntoNotes	58.49%	F_1

Furthermore, recent investigations on anaphora resolution have also led to advancements in various deep learning paradigms. DRL and MTL were employed for obviating the need for language-orientated hyperparameter tuning (Clark and Manning, 2016a), investigating the enduring impact of pronoun-candidate antecedent pairs (Yin et al., 2018a), and enhancing the dependency learning of mention pairs (Liu et al., 2023b). Meanwhile, there were also certain efforts that concentrated on resolving "hard" cases and multilinguality in anaphora resolution. As for the former one, people were aware of the models' capacity to resolve ambiguous pronouns and biases (especially, gender bias) learned by anaphora resolution models (Levesque et al., 2012; Rudinger et al., 2018).

State-of-the-art models of this line of work are often assembled with knowledge bases (Emami et al., 2018) or PLMs (Kocijan et al., 2019). As for the latter one, multilingual anaphora resolution systems were developed in order to either, theoretically, unify the theory of reference for different languages (Nedoluzhko et al., 2022), or, practically, enrich the datasets for low-resource anaphora resolution languages or tasks, e.g., AZPR (Aloraini and Poesio, 2020). In addition to these two trends for developing practical anaphora resolution systems, there is also a long tradition of studying how human beings understand and use anaphors with the algorithms introduced in this section from the age of rule-based methods (Sidner, 1979; Carter, 1987) to the most recent deep learning based methods (Chai and Strube, 2022; Same et al., 2022).

3.5.9.2 Application Trends

Many demonstrations were carried out approximately 15 years ago to validate the necessity of anaphora resolution for both language generation and understanding downstream tasks (Steinberger et al., 2007; Mirkin et al., 2010; Nicolov et al., 2008; Li et al., 2021d; He et al., 2022a). Nevertheless, practically, at that moment, anaphora resolution often had negative effects (Bar-Haim et al., 2008; Chambers et al., 2007; Orasan, 2007; Mitkov et al., 2007). This is mainly because anaphora resolution systems were not powerful enough and errors they made may propagate to their downstream tasks.

Recently, with significant advancements in the capabilities of anaphora resolution systems, more and more anaphora resolution systems have been used for providing anaphora information for downstream tasks (see Table 3.18). In short, anaphora resolution helps its downstream applications mainly in two ways. It links noun phrases in different sentences. As a consequence, these applications have better performance in comprehending discourse-level information. On the other hand, linking noun phrases helps downstream applications to do higher-level reasoning, e.g., extracting global entities (Sukthanker et al., 2020) and recovering the ellipses (Aralikatte et al., 2021). Most downstream task models utilize anaphora resolution as an additional feature to improve task performance. However, we did not see how anaphora resolution techniques help to explain how and why anaphora is used in a certain context.

Table 3.18: A summary of the representative applications of anaphora resolution in downstream tasks. ✓denotes the role of anaphora resolution in a downstream task.

Reference	Downstream Task	Feature	Explainability
Le Nagard and Koehn (2010)	Machine Translation	✓	
Hardmeier and Federico (2010)	Machine Translation	✓	
Miculicich and Popescu-Belis (2017)	Machine Translation	✓	
Saunders et al. (2020)	Machine Translation	✓	✓
Steinberger et al. (2007)	Summarization Evaluation	✓	
Bergler et al. (2003)	Summarization	✓	
Liu et al. (2021d)	Summarization	✓	
Agichtein et al. (2008)	Textual Entailment	✓	
Jakob and Gurevych (2010a)	Polarity Detection	✓	
Ding and Liu (2010)	Polarity Detection	✓	

3.5.9.3 Future Works

Developing Robust Annotation Schemes

Current annotation schemes for anaphora work practically but face theoretical issues due to the lack of unified rules on what is remarkable and the unclear distinction between co-reference and anaphora, despite clear boundaries in linguistic theory. Annotation schemes so far are imperfect to improve the practicality so that large anaphora/co-reference resolution datasets (that can be used for training and assessing data-driven anaphora resolution systems) could be constructed. In exchange, the resulting corpora are imperfect in terms of both quality (i.e., some annotated relations might not be anaphoras) and coverage (i.e., some kinds of anaphora are not covered). On a different note, anaphora resolution, which can also be seen as a pragmatics task, disagreement on how an anaphora is interpreted happens across different readers (Uma et al., 2022). Nonetheless, many datasets resolve disagreements through majority voting, while only a few works explicitly annotated ambiguities, which are the causes of the disagreements (Poesio and Artstein, 2008). In aggregate, it is plausible to design a scheme (probably by extending MATE) that not only handles disagreements but also balances quality, practicality, and coverage. Furthermore, it is important to empirically investigate how the errors and limitations inherent in the annotation scheme can impact the performance of anaphora resolution systems.

Anaphora Resolution Evaluation

Analogue to the disagreements in the anaphora annotation, one can expect that, for a single mismatch between an output and a reference answer, it might be an error for some readers but not an error for the rest. For different mismatches, they might have different severity. The impact of severity of errors has been studied for the production of reference (van Miltenburg et al., 2020).

For example, saying "a woman is a man" is more serious than saying "a red coat is pink"), but it has never been explored in the realm of anaphora resolution. This said, roughly computing the overlaps between model outputs and reference outputs might be problematic. On the one hand, due to discrepancies and varying degrees of errors in anaphora resolution, human evaluation (Martschat and Strube, 2014) is necessary to improve the analysis and evaluation of anaphora resolution models, as well as to establish benchmarks for developing more accurate evaluation metrics. On the other hand, when designing new evaluation metrics, disagreements, and error severity should be considered by data-driven methods.

Model Development

Regarding future advancements in anaphora resolution models, a significant area of focus should be on computational studies of anaphora resolution tasks that are firmly grounded in theory but have yet to be extensively explored. Examples of such tasks include but are not restricted to (1) bridging, deictic, and plural references, which are crucial aspects of referential language, yet their computational treatment has been limited, possibly due to a shortage of relevant annotated datasets; and (2) disagreement resolution, which involves learning from discrepancies in human interpretations of anaphoric expressions to better capture the pragmatic nuances of such references, and should be incorporated into future models (Uma et al., 2021); and (3) cross-document anaphora resolution, which is critical for downstream applications such as KGC and cross-document information extraction, yet has received insufficient attention in terms of data, methods, and evaluation metrics, particularly in relation to event resolution.

3.6 Subjectivity Detection

Conventionally, subjectivity detection is defined as a task to determine whether a text is subjective or not, where a subjective text expresses personal feelings, evaluations, and speculations (Wiebe, 1994), whereas an objective one merely delivers factual information. Generally, subjectivity can manifest in different forms, e.g., opinions, allegations, desires, beliefs, and suspicions (Liu et al., 2010) to express private states. It is not an easy task to identify the use of subjective language, as a subjective sentence does not always contain an opinion (Liu et al., 2010). Therefore, it is important for the subjectivity detection task to find reliable clues.

Aside from opinion-bearing words, syntax also provides essential clues in reporting private states, because grammaticalization involves the recruitment of items to mark the speaker's point of view (Traugott, 2010). Early works often equated the presence of subjectivity to the presence of subjectivity-bearing words in a sentence (Riloff and Wiebe, 2003; Kim and Hovy, 2005; He et al., 2023b; Bao et al., 2021). However, subjectivity is context- and domain-dependent. Some words are only subjective in certain contexts or domains.

Therefore, many researchers incorporated syntactic dependencies (Wilson et al., 2004; Xuan et al., 2012), interactions between neighboring sentences (Wiebe, 1994; Pang and Lee, 2004) or in discourse (Biyani et al., 2014) to extract different levels of contextual information. An alternative to this subjective-lexicon-based approach is the word-frequency-based approach (Rustamov et al., 2013; Kamil et al., 2018), which is completely domain-independent by learning from document-level information. However, this approach has difficulties capturing syntactic dependencies. By now, subjectivity detection research has been divided into several distinct tasks, each with its unique objectives. One such task is individual subjectivity detection, which focuses on detecting subjectivity at the sentence level. In contrast, context-dependent subjectivity detection aims to incorporate discourse information and a broader context in detecting subjectivity. Cross-lingual subjectivity detection, on the other hand, strives to identify subjectivity in various languages. Moreover, multi-modal subjectivity detection is concerned with identifying subjective expressions in different modalities such as audio and video. Finally, the bias detection task is centered on identifying biased statements in ostensibly impartial articles.

Subjectivity detection is commonly considered as a subtask of sentiment analysis since it serves as a filtering step for polarity detection (Liu et al., 2010). It can also be helpful for downstream tasks that require a distinction between opinionated and non-opinionated sentences, such as opinion and information retrieval (Zhang et al., 2007; Wiebe and Riloff, 2011), analyses in financial and political domains (Wang et al., 2021; Tang et al., 2014b; Al Hamoud et al., 2022), QA systems (Yu and Hatzivassiloglou, 2003; Li et al., 2008a), etc.

3.6.1 Theoretical Research

The broad range of subjective expressions, such as personal feelings, evaluations, and speculations, makes the related theoretical research in this domain extensive.

3.6.1.1 Subjective Elements

Early linguistic works studied subjective language extensively in third-person narrative text. Banfield (2014) defined the *SELF* of a sentence as the speaker in conversation or the narrating character in third-person fictional text. She identified a variety of morphological, lexical, and syntactic elements, termed subjective elements, that always express the private states, i.e., emotions and opinions, of the sentence's *SELF*. However, many linguistic elements are subjective only in certain conditions. Thus, Wiebe (1990) further defined a category termed potential subjective elements, which expanded the subjective elements with some linguistic elements that can report the private state of a character. She applied these findings to identify the subjective language in the non-fictional text, suggesting that potential subjective elements are also valid subjectivity clues for texts other than third-person narrative fiction.

3.6.1.2 Speech Acts

Speech acts have a strong connection with subjective expressions because speech acts perform actions, such as making a promise, giving an order, or expressing a belief. Austin (1975) argued that language is not just a tool for describing the world but also a means of accomplishing things in the world. Through speech acts, individuals can influence the world around them and the actions of others. In this sense, many seemingly objective expressions with speech acts can become subjective. For example, if someone says,

(18)   I promise to do it.

The utterance is not just conveying information but also performing the act of making a promise. A more subjective case is

(19)   I believe that it will rain tomorrow.

When individuals express belief in such a manner, they are essentially asserting their mental disposition or perspective towards a specific statement. This entails making a claim about their inner state or outlook toward a proposition. Austin (1975) argues that a considerable number of utterances possess illocutionary force, which signifies that their purpose is not merely to communicate information but also to accomplish something beyond that.

3.6.1.3 Conceptual Metaphor

Lakoff and Johnson (1980) argued that metaphors are not solely a linguistic phenomenon, but also mirror human cognition via concept mappings. When an individual uses a metaphorical expression, they employ a source concept to represent a target concept in a particular context, thereby conveying their cognitive attitude toward the target concept. This process, known as concept mappings, facilitates such representation. In instances such as the statement

(20)   Our love is a journey.

The individual utilizes the concept of a "journey" as the source to represent the target concept of "love", expressing their subjective feeling that their love is characterized by both ups (joy) and downs (sadness). "Our love is a journey" cannot be an objective statement, because the two concepts are from different domains, i.e., literally, love is not a journal. Thus, there is a semantic contrast between the literal and contextual meanings of a metaphor (Mao et al., 2019). The semantic disparities inherent in metaphors suggest that relying on the literal meanings of a statement alone is insufficient in substantiating its subjectivity. Even though the statement of Example (20) does not use any obvious opinionated words, e.g., "happy" and "sad", it also expresses a personal feeling. Thus, the pragmatics of statements must also be taken into account in subjective detection.

3.6.2 Annotation Schemes

For general subjectivity detection, it is sufficient for a dataset to annotate a sentence, snippet, or document as subjective (positive/negative) or objective (neutral). Nevertheless, Wiebe et al. (2005) proposed the MPQA scheme, which annotates text at the word and phrase levels. The MPQA scheme is suitable for fine-grained subjectivity detection that aims to identify the source, target, and properties of each expression of the private state. Wilson (2008) proposed the AMIDA Scheme for annotating subjectivity in speech. This scheme marks word spans that are in the following three main categories: subjective utterances, objective polar utterances, and subjective questions. A subjective utterance is a word span that expresses a private state. An objective polar utterance delivers positive or negative factual information without expressing a private state. A subjective question is a question in which the speaker is eliciting the private state of someone else. Each category is divided into finer classes that indicate the polarity and certainty of an utterance.

3.6.3 Datasets

A summary of all the introduced datasets can be found in Table 3.19. Generally, subjectivity detection data are organized in the following forms. A text is typically labeled as either subjective or objective, with the former category often further classified as positive, negative, or neutral. The following examples are from SemEval-2013 Task 2B: Sentiment Analysis on $\mathbb{X}$ (Nakov et al., 2013).

```
id1: "263732569508552704"
id2: "369152026"
text: "Kick-off your weekend with service! EV!'s Get on the Bus trip to the
Boys & Girls Club is Friday from 3-6! Hope to see you there :)"
label: positive

id1: "213342054351257601"
id2: "189656827"
text: "Desperation Day (February 13th) the most well known day in all mens life."
label: negative

id1: "263803288074477568"
id2: "396953010"
text: "It seem like Austin Rivers is tryin to had to get a bucket. I feel em tho my
1st game in the league I was trying hard too"
label: neutral
```

Table 3.19: Subjectivity detection datasets and statistics. ISD denotes individual subjectivity detection. CDSD denotes context-dependent subjectivity detection. CLSD denotes cross-lingual subjectivity detection. MMSD denotes multimodal subjectivity detection. BD denotes bias detection.

Dataset	Task	Source	# Samples	Reference
MPQA	ISD, CDSD	English news articles	9,700	Wiebe et al. (2005)
MPQA Gold	ISD, CLSD	Spanish sentences	504	Mihalcea et al. (2007)
MultimPQA	ISD, CLSD	Machine-translated MPQA	9,700	Banea et al. (2010)
Movie	ISD, CDSD	Rotten Tomatoes, IMDB	10,000	Pang and Lee (2004)
WebDoc	CDSD	English web documents	1,076	Chesley et al. (2006)
TREC	CDSD	WSJ	2,000	Yu and Hatzivassiloglou (2003)
Debate	CDSD	Political and ideological dataset	53,453	Al Hamoud et al. (2022)
Twitter1	ISD	English tweets	200,000	Barbosa and Feng (2010)
Twitter2	ISD	English tweets	498	Serrano-Guerrero et al. (2015)
Forum	CDSD	online forums	700	Biyani et al. (2014)
SemEval-2013	ISD	English tweets	12,002	Nakov et al. (2013)
NET	ISD	English nuclear energy tweets	2,308	Khatua et al. (2020)
MLT	ISD, CLSD	Multilingual nuclear energy tweets	7,700	Satapathy et al. (2017a)
TASS	ISD, CLSD	Spanish tweets	10,000	Villena et al. (2015)
Email	CDSD	BC3 corpus	1,800	Murray and Carenini (2011)
AMIDA	MMSD	AMI Meeting Corpus	13	Wilson (2008)
ICT-MMMO	MMSD	YouTube review videos	370	Wöllmer et al. (2013)
MOUD	MMSD	YouTube review videos	498	Morency et al. (2011)
Conservapedia	BD	Conservapedia statements	1,000	Hube and Fetahu (2018)
WNC	BD	Wikipedia sentence pairs	180,000	Pryzant et al. (2020)

For fine-grained subjective annotation, the labels are annotated at the span level. The following examples are from SemEval-2013 Task 2A: Sentiment Analysis on $\mathbb{X}$ (Nakov et al., 2013).

```
id1: "255732290246815744"
id2: "315400337"
text: "Billy Cundiff may be leaving Washington. Hopefully he won't miss the door
on the way out."
start id: "7"
end id: "7"
label: "positive"


id1: "255732290246815744"
id2: "315400337"
text: "Billy Cundiff may be leaving Washington. Hopefully he won't miss the door
on the way out."
start id: "9"
end id: "10"
label: "positive"
```

MultiParty Question Answering (MPQA) (Wiebe et al., 2005) is derived from 535 English news articles from a wide variety of news sources, manually annotated for subjectivity. The corpus contains 9,700 sentences, 55% of which are labeled as subjective and 45% as objective. The MPQA Gold (Mihalcea et al., 2007) contains 504 Spanish sentences manually annotated for subjectivity, where 273 sentences are subjective and 231 are objective. The MultimPQA (Banea et al., 2010) contains parallel corpora to the MPQA dataset in five languages other than English, namely, Arabic, French, German, Romanian, and Spanish.

The Movie Review dataset (Pang and Lee, 2004) contains 5,000 movie review snippets collected from Rotten Tomatoes[29], considered as subjective. Furthermore, 5,000 sentences are collected from plot summaries from the Internet Movie Database (IMDB)[30], considered as objective. All reviews and plot summaries are sourced from movies released post-2001, preventing overlap with the polarity benchmark dataset (Pang and Lee, 2004). A data sample, either sentence or snippet, is at least 10 words long. The Debate dataset (Al Hamoud et al., 2022) is derived from the political and ideological dataset (Somasundaran and Wiebe, 2010), containing 53,453 sentences from political and ideological posts and comments. The instances are automatically labeled for subjectivity by using lexicon-based and syntactic-pattern-based classifiers (Riloff and Wiebe, 2003).

Numerous microtext corpora exist that can serve as benchmark datasets for subjectivity detection. Barbosa and Feng (2010) presented a dataset containing 200,000 English tweets, where roughly 100,000 are subjective and the rest are objective. Serrano-Guerrero et al. (2015) manually annotated 498 English tweets as positive, negative, or neural. SemEval-2013 (Nakov et al., 2013) is a collection of 12,002 English tweets labeled as objective, positive, negative, or neutral. Nuclear Energy Tweets (NET) (Khatua et al., 2020) contains 2,308 English tweets about nuclear energy, manually annotated for subjectivity. The Multilingual Tweets (MLT) dataset (Satapathy et al., 2017a) is a collection of 12,719 tweets about nuclear energy in English, French, Spanish German, Malay, and Indonesian, 7,700 out of which are manually labeled for subjectivity. The Taller de Analisis de Sentimientos en la SEPLN (TASS) corpus (Villena et al., 2015) contains 10,000 tweets in Spanish, collected from posts by 150 public figures in fields of sports, politics, and communication during the period from 2011 to 2012. Each tweet is labeled as positive, neutral, negative, or without opinion.

The Web Document dataset (Chesley et al., 2006) contains 1,076 English web documents, sourced from traditional news websites and blog posts on diverse topics. Each document is manually annotated as objective, positive, or negative. The Text REtrieval Conference (TREC) dataset (Yu and Hatzivassiloglou, 2003) is a collection of 8,000 WSJ articles evenly distributed in the categories of editorial, letter to editor, business, and news. The articles and sentences from the former two categories are mapped as opinions (subjective), while the ones from the latter two are facts (objective).

---

[29] https://rottentomatoes.com

[30] https://imdb.com

The Forum dataset (Biyani et al., 2014) contains 700 threads from online forums Trip Advisor–New York[31] and Ubuntu Forums[32], manually annotated for subjectivity. Email (Murray and Carenini, 2011) contains 1,800 sentences derived from BC3 corpus (Ulrich et al., 2008), 172 out of which are labeled as subjective. For multimodal subjectivity detection, the AMIDA dataset (Wilson, 2008) consists of 19,071 dialogue act segments from 20 conversations from the AMI Meeting Corpus (McCowan et al., 2005), manually annotated with the AMIDA scheme. 42% of the dialogue act segments are tagged with at least one subjective annotation. The Institute for Creative Technologies Multimodal Movie Opinion (ICT-MMMO) dataset (Wöllmer et al., 2013) contains 370 YouTube review videos labeled as strongly negative, weakly negative, neutral, weakly positive, and strongly positive. Multimodal Opinion Utterances Dataset (MOUD) (Morency et al., 2011) is a collection of 80 YouTube review videos annotated as positive, negative, and neutral.

For the bias detection task, which aims to identify subjective bias in Wikipedia, the following datasets are widely used. Conservapedia (Hube and Fetahu, 2018) is a collection of 1,000 single-sentence statements from Conservapedia[33], manually annotated as biased or unbiased. Wiki Neutrality Corpus (WNC) (Pryzant et al., 2020) contains 180,000 aligned Wikipedia sentence pairs. Each pair consists of a sentence before and after bias neutralization by English Wikipedia editors.

3.6.4 Knowledge Bases

Lexicons of subjectivity clues and patterns are commonly used for subjectivity detection, as summarized in Table 3.20. The General Inquirer (Stone et al., 1966) is a lexicon consisting of 10,000 words sorted into 180 categories for content analysis. The Subjectivity Clues lexicon (Riloff and Wiebe, 2003) is a list of words that are subjective in most cases (strongly subjective) and words that may have subjective use in certain contexts (weakly subjective). MPQA Subjectivity Lexicon (Wilson et al., 2005) expanded the Subjectivity Clues using additional dictionaries and lexicons, containing over 8,000 subjectivity clues.

Knowledge bases that provide sentiment information are also widely used for subjectivity detection. WordNet-Affect (Strapparava et al., 2004) is a set of synsets derived from WordNet that effectively represents affective concepts. SentiWordNet, as introduced in the previous section, is based on WordNet. Each word in SentiWordNet is given three scores indicating its positivity, negativity, and objectivity. As discussed earlier, SenticNet (Cambria et al., 2024) is also a popular knowledge base for subjectivity detection.

---

[31] https://tripadvisor.com/ShowForum-g60763-i5-New_York_City_New_York.html

[32] https://ubuntuforums.org

[33] https://conservapedia.com

Table 3.20: Useful knowledge bases for subjectivity detection.

Name	Knowledge	# Entities	Structure
The General Inquirer	Sentiment labels	4,000	List
MPQA Subjectivity Lexicon	Subjectivity clues	8,000	List
SentiWordNet	Structured lexical knowledge by concept	100,000	Graph
WordNet-Affect	lexical knowledge	4,787	Graph
SenticNet	Sentiment scores	400,000	Graph

3.6.5 Evaluation Metrics

The performance of subjectivity detection is commonly evaluated via accuracy and F-measure.

3.6.6 Annotation Tools

The aforementioned NER annotation tools (see Section 3.3.5.1) can be used for subjectivity detection because these tools can annotate labels for spans (fine-grained subjectivity detection) and sentences (coarse-grained subjectivity detection).

3.6.7 Methods

3.6.7.1 Individual Subjectivity Detection

In individual subjectivity detection, the subjectivity of a sentence is evaluated in isolation and irrespective of any contextual factors. The primary methods used for addressing this task include lexicon-based, word frequency, and deep learning approaches.

A. Lexicon-based

Drawing on the premise that sentences that contain commonly-subjective expressions are more likely to be subjective, lexicon-based methods utilize a manually-constructed lexicon of subjective words, clues, or patterns to determine the subjectivity of a given sentence. Riloff and Wiebe (2003) introduced an unsupervised rule-based classifier that leverages the identification of subjective clues and patterns to detect subjective sentences, while also employing bootstrapping to recognize objective sentences based on the absence of such indicators. The clues were manually collected and annotated. The patterns were generated by the AutoSlog-TS algorithm (Riloff, 1996), based on pre-defined syntactic templates.

Wiebe and Riloff (2005) further improved this bootstrapping system by using the labeled sentence produced by the rule-based method as initial training data for a Naïve Bayes classifier. The major weakness of these methods is the unreliable assumption that the absence of subjective clues and patterns indicates objectivity, resulting in false-positive errors. Kim and Hovy (2005) first compiled lists of words that convey opinions and those that do not, which were manually annotated with corresponding classes and levels of strength. They expanded the lists with a common English word list by measuring the WordNet distance between a common word and the compiled seed lists. They further identified additional opinion words and non-opinion words from editorial and non-editorial WSJ documents by computing their relative frequencies. By detecting the subjectivity of a given sentence based on the presence of a single strong valence word, their method achieved 65% accuracy on MPQA.

Benamara et al. (2011) argued that sentence-level subjectivity detection cannot fully leverage context, because a sentence may contain several opinion clauses, and opinion expressions may be discursively related. As such, they proposed a segment-level annotation based on the Segmented Discourse Representation Theory (Asher and Lascarides, 2003), where segments are labeled as explicitly subjective, implicitly subjective, subjective non-evaluative, and objective. This fine-grained annotation can better enhance polarity detection, as segments in the latter two categories do not covey positive, negative, or opinion. However, the limitation of this method is that the four label classes are unbalanced in the corpus. Additionally, implicitly subjective segments are often nuanced and hard to identify. Thus, it would be challenging to design an appropriate classifier. The paper circumvented this problem by reframing the task as two parallel binary classification tasks and obtained 82.31% accuracy with a manually compiled French lexicon and SVMs as classifiers.

Merely detecting the existence of subjective keywords is often an insufficient indication of a sentence's subjectivity. Other works attempted to enrich the feature set by incorporating more sentence-level information. Relying on expert knowledge of parse tree, Xuan et al. (2012) manually constructed a set of syntax-based patterns from unigrams and bigrams to extract features. A MaxEnt model was employed as the classifier, obtaining 92.1% accuracy on the Movie dataset. Remus (2011) hypothesized that the readability of a sentence was related to its subjectivity. Hence, readability formulae such as Devereux Readability Index (Smith, 1961) and Easy Listening (Fang, 1966) were incorporated as features in addition to the MPQA Subjectivity Lexicon, obtaining 84.5% F-measure on Moive.

Many works proposed subjectivity detection systems that specifically targeted X (formerly known as Twitter). Given the word constraint imposed by X, a tweet is generally regarded as a sentence. Barbosa and Feng (2010) believed that using subjectivity detection as an upstream task would improve the performance of polarity detection on X text. Aside from conventional features such as subjective clues and POS tags, they leveraged Tweet-specific syntax features, e.g., links and upper case. An SVM classifier was employed, which achieved 81.9% accuracy on the X dataset, and improved the accuracy of polarity detection by 5.6%.

Following their footsteps, Sixto et al. (2016) incorporated more tweet-specific features that leveraged the structure of $\mathbb{X}$, e.g., the relationship between tweets, users, hashtags, and links. Using the stacking classifier proposed by Cotelo et al. (2015), their method obtained 89.8% on TASS. To reduce human effort, Keshavarz and Saniee Abadeh (2018) created an $\mathbb{X}$ subjectivity lexicon automatically through a meta-heuristic approach, i.e., a genetic algorithm, which produced separate lists of subjective and objective words. A Bayse network was employed to classify a given tweet based on its subjective and objective word counts, achieving 60.9% on SemEval-2013. Alternatively, Khatua et al. (2020) leveraged the concept-level knowledge base SenticNet as their lexicon, which is able to provide implicit meaning associated with commonsense concepts. Their method obtained 80.7% accuracy on the NET dataset.

The methods introduced above have a common limitation, i.e., the lexicons are lists of keywords, instead of word meanings. Some subjective clues in fact have both subjective and objective word senses, which are not distinguishable in keyword lexicons, leading to false-positive errors. This problem can be mitigated by incorporating a Subjectivity WSD (SWSD) system to build a sense-aware lexicon. Akkaya et al. (2009) trained a supervised targeted SWSD system using SVM. The training data was compiled using words that are both in the MPQA Subjectivity Lexicon and the sense-tagged SensEval corpora (Kilgarriff and Palmer, 2000; Preiss and Yarowsky, 2001; Litkowski, 2004). Alternatively, Ortega et al. (2013) applied an unsupervised, clustering-based SWSD system (Anaya-Sánchez et al., 2006) on SentiWordNet to label each subjective word with fine-grained sense. Both SWSD systems were applied to a rule-based classifier similar to the one proposed by Riloff and Wiebe (2003). The supervised one improved accuracy by 1.3% on MPQA, while the unsupervised one improved F-measure by 6.5% on the Movie dataset. A prominent limitation of lexicon-based methods is that they require external resources such as sentiment lexicon and knowledge base.

B. Word Frequency

Word-frequency-based methods detect subjectivity by modeling word presence or occurrence within a corpus. Therefore, compared to lexicon-based methods, they are language-independent and require neither manual annotation nor linguistic knowledge. They are also less computationally expensive due to the reduction of feature sets. Rustamov et al. (2013); Kamil et al. (2018) proposed a language-independent feature extraction algorithm with a novel statistical measure of word occurrence called Pruned ICF (Inverse-Class Frequency), which is proven to be more effective than the standard IDF (Inverse-Document Frequency). Additionally, they applied two widely-used methods for pattern recognition to detect subjectivity, namely Fuzzy Control System (FCS) (Helmi and AlModarresi, 2009) and Adaptive Nero-Fuzzy Inference System (ANFIS) (Fullér, 1995), achieving the accuracy of 91.3% and 91.66% on the Movie dataset, respectively. The latter obtained better performance due to the addition of a neural network layer.

Inspired by empirical evidence that hybrid systems improve the performance of NLP classifiers, Rustamov (2018) further integrated FCS, ANFIS, and HMM into a sequential hybrid system, where input sentences that are wrongly labeled by the prior classifier are passed onto the subsequent one. Using the same feature extraction method as the previous paper, this system increased the accuracy to 92.24% on the Movie dataset. Wang and Manning (2013) proposed a novel dropout algorithm to optimize the feature learning process. Conventional dropout training in neural network (Hinton et al., 2012) prevents feature co-adaptation by randomly sampling neurons and input features and setting them to zeros, which leads to slow training. The authors suggested fast dropout training as a more efficient alternative, using a Gaussian approximation to draw samples. They applied this dropout method to Naïve Bayes SVM (NBSVM) (Wang and Manning, 2012), which extracts features based on word presence. Their method not only achieved the accuracy of 93.6% on the Movie dataset and 86.3% on MPQA, but also greatly decreased the training time. Experiments also showed that fast dropout training could be applied to other loss functions and neural networks.

LDA (Blei et al., 2003) is a weakly-supervised generative model that assumes every document is a distribution of latent topics, which is determined by word frequencies. He (2010); Maas et al. (2011) suggested that subjectivity detection can be solved by LDA, based on the intuition that subjective sentences likely contain opinionated words. Hence, the paper modified conventional LDA so that the latent topics are word-level sentiment labels. An additional layer is inserted between word and document levels to model sentence-level subjectivity labels. Sentiment lexicons are incorporated to establish an informed prior distribution for word-level sentiment labels, achieving 71.2% accuracy on MPQA. On the other hand, Lin et al. (2011) argued that LDA likely discovers topics based on semantic similarities, instead of sentiment. Therefore, they modified LDA, so that it directly models word probabilities conditioned on topic distributions to capture semantic information. To explicitly extract sentiment information, they incorporated supervised sentiment analysis as an auxiliary task. Their method achieved 88.58% on the Movie dataset. The drawback of the word frequency approach is that the order of the words is not considered. Thus, syntactic information cannot be effectively learned using this approach.

C. Deep Learning

The acquisition of precise sentence representations is crucial for subjectivity detection, and as such, numerous studies have examined neural sentence modeling as a language-independent alternative to parse trees. Kalchbrenner et al. (2014) presented a Dynamic CNN (DCNN) that is able to capture short- and long-range relations. The core component of DCNN is dynamic pooling, which outputs the sub-sequence of k maximum values in the input sequence, where k can be dynamically chosen. Hence, DCNN produces a hierarchical feature graph that contains syntactic, semantic, and structural patterns of the input sentence. However, their sentence representations do not retain any intermediate information, e.g., word-level and phrase-level features.

To address this, Zhao et al. (2015) described a self-adaptive hierarchical sentence model named AdaSent. Inspired by gated recursive CNN (Cho et al., 2014), AdaSent forms a pyramid-shape DAG, where the bottom level is word representations and the top level is sentence representations. In this process, the gating network receives information from each level and selects the most appropriate representations for the given task. Their method obtained an accuracy of 95.5% on the Movie dataset and 93.3% MPQA. With similar motivation for higher-order dependencies, Chaturvedi et al. (2018) proposed a Bayesian Network-based Randomized Network (BNRN) framework for subjectivity detection. Randomized networks excel at inductive learning. However, the excessive number of their hidden neurons often leads to overfitting and slow performance. To address these weaknesses, Bayesian networks were introduced to model connections among the hidden neurons of a randomized network, as they can prune redundant and irrelevant hidden neurons and capture high-dimensional features. Furthermore, randomized networks cannot handle non-linear data such as sequences of sentences. Thus, an RNN layer was used to extract temporal features. Upon it, a fuzzy classifier was applied to achieve stability in case of noisy data, producing the output labels. Additionally, a deep CNN was employed prior to BNRN to provide low-dimensional features. The framework achieved the accuracy of 75% on MPQA Gold and 89% on TASS, outperforming previous models.

Likewise, Satapathy et al. (2017a) employed CNN and RNN to extract spatial and temporal information respectively. To make the model more robust, they incorporated reinforcement learning, namely Point-wise Probability Reinforcement (PPR) (Frénay and Verleysen, 2015), to regularize the learning process of CNN and reduce the influence of outliers. Specifically, convolutional layers in the CNN component were added iteratively, where the weight of each neuron was fine-tuned by the reinforced maximum likelihood of PPR. Their method did not perform very well on MPQA, obtaining 50% F-measure. However, it achieved a good performance of 76% F-measure on the multilingual $\mathbb{X}$ dataset MLT.

In the same vein, PLMs can also provide beneficial universal representations learned from a plethora of unlabeled text. For instance, Al Hamoud et al. (2022) fed GloVe embeddings to different types of RNN variants, among which LSTM with attention mechanism achieved the best accuracy of 89.53% on MPQA and 83.83% on their proposed political and ideological dataset, whereas BiLSTM with attention achieved the best accuracy of 92.8% on the Movie dataset. Kim (2014) fine-tuned pretrained word2vec with a simple CNN, obtaining accuracy of 93.4% on the Movie dataset and 89.5% on MPQA.

Furthermore, many works observed that it is complementary to combine PLM and MTL for more effective learning of text representations (Liu et al., 2019b; Sun et al., 2019; Mao and Li, 2021). Motivated by this, Huo and Iwaihara (2020) fine-tuned BERT using MTL, where the BERT layers are shared among subjectivity detection and three other text classification tasks. Similarly, Satapathy et al. (2022) proposed an MTL framework for subjectivity and polarity detection. The framework leverages BERT as embedding, which is fed into two separate self-attention BiLSTM layers. A neural tensor network (NTN) (Socher et al., 2013a) was used as the information-sharing layer.

Both methods employed a simple softmax classifier for each task. The former achieved 95.23% accuracy on the Movie dataset, while the latter obtained 95.1%. However, a shared limitation is that, despite their overall good performance, some of the tasks did not exceed single-task learning baselines. This is likely because both methods adopted hard parameter sharing MTL (Crawshaw, 2020), which emphasizes more on generalization rather than optimization.

Sagnika et al. (2021) presented an attention-based CNN-LSTM model for subjectivity detection, which served as a preprocessing step for sentiment analysis. The combination of CNN and LSTM enabled the model to capture both spatial and temporal information. Additionally, it utilized word embeddings enhanced by sentiment-related information (Sagnika et al., 2020). Initially, the training of the model was carried out with the Movie dataset, after which it was utilized to analyze the sentiment of the IMDB dataset. The objective sentences were eliminated from the dataset to form a modified set of reviews. Various models were tested as sentiment classifiers. The subjectivity detection model not only obtained 97.1% accuracy on the Movie dataset, but also consistently improved the performance of sentiment analysis.

3.6.7.2 Context-Dependent Subjectivity Detection

The method of individual detection categorizes each sentence without considering its context. However, subjectivity detection and sentiment classification are contextual problems since lexical items can affect each other in a discourse setting (Aue and Gamon, 2005; Polanyi and Zaenen, 2006). Pang and Lee (2004) were the first to leverage inter-sentence context information to filter out objective sentences, in order to better serve document-level polarity detection. Based on the hypothesis that adjacent text spans might have the same subjectivity label (Wiebe, 1994), suggested an algorithm known as the "minimum cuts algorithm" that aims to optimize the subjectivity status score for every sentence separately, while also punishing the assignment of different labels to two closely related sentences. These two sub-objectives are independent of each other, making the model more flexible for the addition of features. Context-dependent methods can be divided into two categories, namely, the feature engineering approach and the statistical approach.

A. Feature Engineering

A common way to incorporate document-level information is by designing relevant features. Das and Bandyopadhyay (2009) proposed a domain-independent rule-based algorithm, named theme detection. The model utilized document-level features, e.g., positional aspects (document title, first paragraph, last two sentences), the positions of subjectivity clues, and the distance between any two thematic words. As with many techniques at the sentence level, this approach also integrated syntactic characteristics and resources such as SentiWordNet and MPQA Subjectivity Lexicon. It achieved precision and recall of 76.08% and 83.33% on MPQA.

To automatically select an appropriate feature set, Das and Bandyopadhyay (2010) employed a genetic algorithm (Holland, 1992; Sastry et al., 2005), which is a probabilistic search method, to find the optimal range of values of every feature. To capture context information, positional aspects, word distribution, and document theme (Wiebe, 2000) were incorporated as discourse-level features, aside from the commonly-used lexical and syntactic features. The genetic algorithm then identified the globally optimal feature set by natural selection and computed the corresponding accuracy of the classifier through the fitness function. An advantage of the proposed method over other statistical classifiers is that the entire input sentence is encoded by the genetic algorithm and used as features, instead of using n-gram. Their method obtained the F-measure of 93.02% on MPQA and 95.69% on the Movie dataset.

Biyani et al. (2014) noticed a gap in subjectivity detection targeting online forums. Moreover, they argued that lexical features are highly dimensional, leading to the risks of overfitting and slow training. Thus, they presented a Forum dataset, and designed a set of non-lexical thread-specific features. Specifically, they leveraged thread structure and dialogue acts and utilized lexicons and tools such as MPQA Subjectivity Lexicon and SentiStrength (Thelwall et al., 2012) to extract sentiment features. With the addition of conventional lexical features, the logistic regression classifier obtained 77.01% accuracy on Forum.

B. Statistical Approach

To minimize human effort in designing features, a statistical approach automatically learns features from a given corpus using statistical models. Yu and Hatzivassiloglou (2003) simply implemented a Naïve Bayes classifier for document-level subjectivity detection, which achieved the F-measure of 97% on the TREC dataset proposed by them. Motivated by the observation that language models are adept at representing knowledge of the text they were trained on, Karimi and Shakery (2017) proposed a language-model-based document-level subjectivity detection method. During training, a subjective reference language model and an objective were built using labeled documents. During inference, a language model was constructed for each input document, which was compared with the reference language models using KL-divergence (Lafferty and Zhai, 2001), producing two similarity scores. The difference between these two scores was regarded as the subjectivity score of the document. The final output of the model was a sorted list of input documents, based on their subjectivity scores. To achieve language non-specificity, the paper also proposed a semi-supervised method where the reference language models were built on a lexicon divided into subjective and objective parts, based on polarity scores. The supervised method obtained 94.63% MAP on the Movie dataset, whereas the unsupervised obtained 53.61% MAP.

Word embeddings can only provide limited syntactic and semantic information (Belinkov et al., 2017). Therefore, to better initialize their model, Chaturvedi et al. (2016a) employed a Gaussian Bayesian Network (GBN) (Friedman et al., 1998) layer to capture long-range features among successive sentences, which were used to pretrain the CNN classifier.

The GBN layer converted the sentence sequence from the MPQA dataset into a time series of word frequency, captured second-order word dependencies with a time lag of 2, and generated a subset of sentences that contained the most significant words from the MPQA Subjectivity Lexicon. The model adopted a CNN sentence model with convolution kernels of increasing size, which combined the local word dependencies within the kernel size to model long-range syntactic relations. It was pretrained with the sub-set of sentences produced by GBN before being trained on the full dataset, obtaining the accuracy of 93.2% on MPQA and 96.4% on the Movie dataset.

3.6.7.3 Cross-Lingual Subjectivity Detection

A. Language-Independent Approach

For feature-engineering-based subjectivity detection, lexical resources and tools are often not readily available for non-English languages. A common approach to circumvent this problem is to use non-language-specific features that are based on the presence or occurrence statistics of a corpus, e.g., word frequency (Rustamov et al., 2013; Wang and Manning, 2013; Kamil et al., 2018; Blei et al., 2003; Lin et al., 2011; Belinkov et al., 2017) and language modeling (Karimi and Shakery, 2017). Mogadala and Varma (2012) further introduced language-independent feature weighing, leveraging unigram and bigram frequencies, and unigram word length. Entropy-based category coverage difference (Largeron et al., 2011) was employed as the feature selection method.

B. Translation Approach

Another solution is the translation approach, where lexical resources for the target language are automatically generated by translating the resources and tools available for English, usually with the help of SMT (Kim and Hovy, 2006; Mihalcea et al., 2007; Wan, 2009; Amini et al., 2019). Banea et al. (2010) conducted a study on English and five other highly lexicalized languages, proving that a multilingual feature space constructed through SMT improved the accuracy of subjectivity detection on all languages involved. However, the sentence translation process can lead to the loss of essential lexical information such as inflection and formality, which often served as an indicator of subjectivity.

Chaturvedi et al. (2016b) mitigated this information loss during translation by using a neural network to transfer resources from English to Spanish. They first translated the MPQA Subjectivity Lexicon into Spanish using an SMT system (Lopez, 2008). A MaxEnt-based POS tagger (Toutanvoa and Manning, 2000) and a multilingual WSD system (Moro et al., 2014a) were incorporated in the preprocessing stage to minimize the loss of lexical information during translation.

Their proposed model, named Lyapunov Deep Neural Network (LDNN), extracted spatial features from the input Spanish sentence and its translated English form using CNN, which were then combined with an RNN to capture the bilingual temporal features. To mitigate the vanishing gradient problem with RNN, a Lyapunov function was used as the error function of RNN for stable convergence. Utilizing the high-level features produced by Lyapunov-guided RNN, a multiple kernel learning (Subrahmanya and Shin, 2009; Zhang et al., 2010) classifier yielded the prediction. Their model obtained 84.0% F-measure on MPQA Gold, and 88.4% accuracy on TASS.

3.6.7.4 Multimodal Subjectivity Detection

While most studies on detecting subjectivity have concentrated on text-based data, the identification of subjective expressions in other modalities, such as audio and video, presents an important area for research. For instance, Murray and Carenini (2009, 2011) proposed an automatic pattern extraction method for subjective expression in spoken conversation, which is able to extract Varying Instantiation N-Grams (VIN) from labeled and unlabeled data. Unlike convention n-gram, a VIN is a trigram where each unit can be either a word or a POS label, which is a more robust alternative to syntactic parsers for fragmented and disfluent text, such as meeting transcripts. Combined with a large raw feature set, a MaxEnt classifier scored the F-measure of 52% on the AMIDA dataset. This method, however, did not leverage any information from other modalities.

Raaijmakers et al. (2008), instead, explored the effectiveness of lexical and acoustic features in speech subjectivity detection. Specifically, they investigated word, character, prosody, and phoneme n-grams. Following Wrede and Shriberg (2003); Banse and Scherer (1996), the prosodic features were extracted based on pitch, energy, and the distribution of energy in the long-term averaged spectrum. Word-, character-, and phoneme-level features were extracted from manual speech transcripts. A separate BoosTexter classifier (Schapire and Singer, 2000) was employed for each feature set, whose predictions were combined using a simple linear interpolation strategy to obtain the final output. The combination of the four types of feature sets achieved 75.4% accuracy and 67.1% F-measure on AMIDA. Experiments showed that word- and character-level features contributed the most.

3.6.7.5 Bias Detection

Bias detection refers to the task of identifying biased statements from supposedly impartial articles. Specifically, in Wikipedia, the Neutral Point of View (NPOV) is a core principle that ensures neutrality for controversial topics. Thus, the goal of this task is to detect sentences that violate NPOV policy on a Wikipedia page. Bias detection is closely related to subjectivity detection. Its development mirrors the technical trends of the latter.

However, it is considered to be more complex, because the linguistic cues of biased language are often nuanced, and depend heavily on the context. For lexicon-based approaches, Recasens et al. (2013) manually compiled a biased word lexicon and feature set that covered framing bias (use of subjective words or phrases that links to a particular point of view), and epistemological bias (linguistic cues that modify the credibility of a statement). However, their method focused only on detecting a single bias-inducing word in a known biased statement. Furthering their work, Hube and Fetahu (2018) constructed a more comprehensive biased word lexicon for sentence-level bias detection. To minimize human efforts, they leveraged word2vec to expand a seed word list by measuring the distance between word vectors. Aside from the lexicon, other syntactic and semantic features were incorporated, e.g., tri-gram, POS tags, Linguistic Inquiry Word Count (LIWC) (Pennebaker et al., 2001), framing bias features, and epistemological bias features. By using a Random Forest classifier, their method obtained 74% precision on their proposed Conservapedia dataset.

Aleksandrova et al. (2019) proposed a semi-automatic method to construct a multilingual bias detection corpus, consisting of Bulgarian, French, and English sentences from Wikipedia. Their method was applicable for building a corpus from a Wikipedia archive in any language, as it does not rely on language-specific features. Additionally, they provided the performance of three baseline models, namely BoW, fastText (Joulin et al., 2017), and logistic regression (Hosmer and Lemeshow, 2000), among which BoW achieved the best overall average F-measure of 59.57% across the three languages. For neural network approaches, Hube and Fetahu (2019) employed RNN to capture the inter-dependency of words and their context. To address the weakness of RNN in modeling long-range information, a hierarchical attention mechanism (Yang et al., 2016b) was adopted, which applied word-level attention on each sentence to compute sentence representations, upon which sentence-level attention was applied to learn biased cues from different samples. Following previous feature-based works, they concatenated GloVe embedding, POS tags, and LIWC features as word representations.

PLMs were also widely used in bias detection. Pryzant et al. (2020) extended the work of Recasens et al. (2013) by using a pretrained BERT-based detector to identify bias-inducing words and neutralizing them via an LSTM-based editor. A join embedding mechanism was employed to allow the detector control over the editor. They also introduced the WNC dataset for detecting and editing biased language, on which their model obtained 93.52% BLEU and 45.80% accuracy for the produced edits. However, a limitation is that they primarily targeted single-biased words. To mitigate this, Pant et al. (2020) enabled multi-word detection by identifying bias at the sentence level. They employed the weighted-average ensemble method on several BERT-based models to detect biased language, which obtained 71.61% accuracy and 70.40% F-measure on WNC.

3.6.8 Downstream Applications

3.6.8.1 Polarity Detection

The presence of objective text can decrease the accuracy of polarity detection, as sentiment classifiers are usually optimized for the binary classification task of labelling text as either positive or negative. Therefore, having a subjectivity detection as upstream task can greatly enhance polarity detection (Satapathy et al., 2017a; Das and Sagnika, 2020). For document-level sentiment analysis, Bonzanini et al. (2012) showed that subjectivity detection reduced the amount of data to 60% while still producing the same polarity classification results as full-text classification. Pang and Lee (2004); Das and Sagnika (2020) applied subjectivity detection to filter out objective sentences in reviews prior to classifying their polarity. Similarly, Kamal (2013) first extracted subjective sentences from customer reviews and then employed a rule-based system to mine feature-opinion pairs from the subjective sentences. Barbosa and Feng (2010); Soong et al. (2019) used subjectivity detection in sentiment analysis for X microtext. These works proved that removing objective content from the dataset indeed makes the learning of sentiment more effective.

3.6.8.2 Information Retrieval

Subjectivity detection can serve as a module in information retrieval systems too to determine whether a document is subjective or objective (Soong et al., 2019), because such systems normally aim to retrieve either opinionated or factual topic-relevant text from web sources, e.g., tweets, blog posts, and reviews webpages (Paltoglou and Giachanou, 2014). For opinion retrieval, it helps to select candidate opinionated documents. For instance, Zhang et al. (2007) first employed an SVM classifier that used unigram and bigram features to identify subjective documents. Then, they separated relevant documents from irrelevant ones. For factual information retrieval, on the other hand, subjectivity detection helps to filter out opinionated text such as allegations and speculations to prevent false hits. Wiebe and Riloff (2011) implemented a Naïve Bayes subjectivity classifier and a domain-relevant indicator for selective subjective sentence filtering. If a sentence was classified as subjective, it would be discarded unless it was also labeled as relevant by the indicator.

3.6.8.3 Hate Speech Detection

Hate speech detection is a task that identifies abusive speech targeting a person or a group based on stereotypical group characteristics, e.g., ethnicity, religion, or gender, on social media (Warner and Hirschberg, 2012). Since hate speech is often marked by its content, tone, and target (Cohen-Almagor, 2011), its detection is similar to that of polarity. Additionally, subjectivity clues tend to be surrounding the polarizing and arguing topics, which aligns well with hate speech detection.

As such, subjectivity detection can be used as a filtering subsystem in hate speech detection. For instance, Gitari et al. (2015) employed a rule-based subjectivity classifier that leveraged lexicons including MPQA Subjectivity Lexicon and SentiWordNet to identify subjective sentences. From the extracted sentences, they built a hate speech lexicon using bootstrapping and WordNet. Experiments showed that the addition of subjectivity detection significantly improved the performance of the hate speech classifier.

3.6.8.4 Question Answering System

QA systems generally encounter two types of questions: the ones that expect truth as answers, and the ones that expect opinions. Therefore, it is crucial for a QA system to distinguish opinions from facts, and provide the appropriate type depending on the question (Yu and Hatzivassiloglou, 2003). To achieve this goal, a QA system should operate in two stages. First, it must determine whether a question calls for a subjective or objective answer, which is its subjectivity orientation (Li et al., 2008a,b; Aikawa et al., 2011). Then, the system needs to consider subjectivity as a relevant factor in the information retrieval process. Subjectivity detection can be incorporated as a filter or feature set in a QA system. For instance, Stoyanov et al. (2005) modified the conventional QA system by applying a subjectivity filter and an opinion source filter on the initial information retrieval results, which improved the system significantly. On the other hand, Wan and McAuley (2016) leveraged subjective features from reviews to provide users with a list of relevance-ranked reviews, which improved the performance of answering binary questions from categories with abundant data.

3.6.9 Summary

Subjectivity detection is a cognitive semantic processing task. It categorizes statements by subjective and objective classes. Theoretical research indicates that subjectivity can be detected by certain subjective elements, e.g., morphological, lexical, and syntactic elements (Banfield, 2014). Thus, computational subjectivity research has developed lexical resources, e.g., Subjectivity Clues (Riloff and Wiebe, 2003), and MPQA Subjectivity Lexicon (Wilson et al., 2005). On the other hand, subjectivity can be also explained from the perspectives of pragmatics, e.g., speech acts (Austin, 1975) and conceptual metaphors (Lakoff and Johnson, 1980). Related subjectivity detection works defined the task as classification tasks. Although those classification tasks can be further divided into course-grained and fine-grained classifications, e.g., document-level, sentence-level, and span-level subjectivity detection, there have not been studies aimed at explaining the subjectivity from pragmatic perspectives, e.g., speech acts and metaphors. Subjectivity detection is crucial for many downstream tasks, especially those that mine opinions from text, as it greatly increases their accuracy by filtering out objective content.

3.6.9.1 Technical Trends

Subjectivity detection is a well-studied NLU subtask. There are five technical trends in this area, namely individual, context-dependent, cross-lingual, multimodal subjectivity detection, and bias detection. A summary of the trends can be found in Tables 3.21 and 3.22. For individual subjectivity detection (Table 3.21), the subjectivity of each sentence or snippet is determined only by the lexical, syntactic, and semantic information of the sentence itself. There are mainly three types of methods for individual subjectivity detection. First, the lexicon-based approaches rely on external lexicons that contain subjective and sentiment clues to predict the subjectivity of a sentence. The weakness of such an approach is that subjective clues are often not extensive and reliable enough to determine the subjectivity of a sentence. Some works attempted to address this issue by utilizing sentence-level features to extract syntactic information (Wilson et al., 2004; Xuan et al., 2012; Barbosa and Feng, 2010), or incorporating WSD to identify subjective clues according to context (Akkaya et al., 2009; Ortega et al., 2013). Nonetheless, these methods cannot fully extract the underlying sentence structure and contextual information. Word-frequency-based approaches, on the other hand, predict sentence subjectivity according to the word presence or occurrence in a given corpus, thus being able to adapt to new domains and languages. Additionally, this approach requires little external resources or human effort.

Table 3.21: A summary of representative subjectivity detection techniques (Part 1). ISD denotes individual subjectivity detection. SCSL denotes self-collected subjectivity lexicon. SWSD denotes subjectivity WSD.

Task	Reference	Techniques	Feature and KB	Framework	Dataset	Score	Metric
	Riloff and Wiebe (2003)	Rule	SCSL	Logic rules	-	-	-
	Kim and Hovy (2005)	Statistics	MPQA, WN	WN distance	MPQA	65.00%	Acc
	Benamara et al. (2011)	Statistics	Lexical, stylistic, syntactic, discursive features	SVM	Self-collected	82.31%	Acc
	Xuan et al. (2012)	Statistics	MPQA, syntax-based patterns	MaxEnt	Movie dataset	92.10%	Acc
	Remus (2011)	Statistics	MPQA, readability	SVM	Movie dataset	84.50%	F_1
	Barbosa and Feng (2010)	Statistics	MPQA, POS,tweet-specific features	SVM	Twitter1	81.59%	Acc
	Sixto et al. (2016)	Statistics	MPQA, tweet-specific features	Stacking classifier	TASS	89.80%	Acc
	Keshavarz and Saniee Abadeh (2018)	Statistics	SCSL	Genetic algorithm	SemEval-2013	60.90%	Acc
	Khatua et al. (2020)	DL	SenticNet	CNN	NET	80.70%	Acc
	Akkaya et al. (2009)	Statistics	MPQA, SWSD	SVM	MPQA	81.30%	Acc
ISD	Ortega et al. (2013)	Rule	MPQA, SWSD	Clustering, rules	Movie dataset	55.68%	F_1
	Kamil et al. (2018)	Statistics	Pruned ICF	ANFIS	Movie dataset	91.66%	Acc
	Rustamov (2018)	Statistics	Pruned ICF	FCS, ANFIS, HMM	Movie dataset	92.24%	Acc
	Wang and Manning (2013)	Statistics	Word presence	NBSVM	Movie dataset	93.60%	Acc
	Maas et al. (2011)	Statistics	Semantic and sentiment embeddings	Probabilistic model, LDA	MPQA	71.20%	Acc
	Lin et al. (2011)	Statistics	Sentiment	LDA	Movie dataset	88.58%	Acc
	Zhao et al. (2015)	DL	word2vec	CNN	Movie dataset	95.50%	Acc
	Chaturvedi et al. (2018)	DL	MPQA, POS	ELM, RNN, CNN, fuzzy classifier	MPQA Gold	75.00%	Acc
	Satapathy et al. (2017a)	DL	GloVe, MPQA	CNN, PPR	MPQA	50.00%	F_1
	Al Hamoud et al. (2022)	DL	GloVe	RNN, Att	Movie dataset	92.80%	Acc
	Kim (2014)	DL	word2vec	CNN	Movie dataset	93.40%	Acc
	Huo and Iwaihara (2020)	DL	BERT	MTL	Movie dataset	95.23%	Acc
	Satapathy et al. (2022)	DL	BERT	MTL, RNN,NTN	Movie dataset	95.10%	Acc
	Sagnika et al. (2020)	DL	Sentiment-enhanced word embedding	CNN, LSTM	Movie dataset	97.10%	Acc

Table 3.22: A summary of representative subjectivity detection techniques (Part 2). CDSD denotes concept-dependent subjectivity detection. CLSD denotes cross-lingual subjectivity detection. MMSD denotes multimodal subjectivity detection. BD denotes bias detection. SWN denotes SentiWordNet.

Task	Reference	Techniques	Feature and KB	Framework	Dataset	Score	Metric
	Pang and Lee (2004)	Statistics	SCSL	Minimum cuts, Naïve Bayes	Movie dataset	86.40%	Acc
	Das and Bandyopadhyay (2009)	Rule	MPQA, doc-level features, SWN	Logic rules	MPQA	79.54%	F_1
CDSD	Das and Bandyopadhyay (2010)	Statistics	MPQA, POS, doc-level features	Genetic algorithm	Movie dataset	95.69%	F_1
	Biyani et al. (2014)	Statistics	MPQA, SentiStrength, thread-specific features	Logistic regression	Forum	77.01%	Acc
	Yu and Hatzivassiloglou (2003)	Statistics	MPQA, POS	Naïve Bayes	TREC	97.00%	F_1
	Karimi and Shakery (2017)	Statistics	Language model	Rank by similarity	Movie dataset	94.63%	MAP
	Chaturvedi et al. (2016a)	DL	MPQA	GBN, CNN	Movie dataset	96.40%	Acc
	Banea et al. (2010)	ML	MPQA	SMT, Naïve Bayes	MultimPQA EN	74.72%	Acc
CLSD	Mogadala and Varma (2012)	Statistics	Unigram and bigram freq., word length	Naïve Bayes	MultimPQA EN	92.50%	F_1
	Chaturvedi et al. (2016b)	DL	MPQA, WSD	SMT, CNN, RNN	MPQA Gold	84.00%	F_1
MMSD	Murray and Carenini (2011)	Statistics	VIN, raw features	MaxEnt	AMIDA	52.00%	F_1
	Raaijmakers et al. (2008)	Statistics	Lexical, prosodic, and phonemic features	BoosTexter	AMIDA	75.40%	Acc
	Recasens et al. (2013)	Statistics	Biased lexicon, POS	Logistic regression	Self-collected	34.35%	Acc
	Hube and Fetahu (2018)	Statistics	word2vec, POS, LIWC, biased lexicon	Random Forest	Conservapedia	74.00%	Prec.
BD	Aleksandrova et al. (2019)	Statistics	Word frequency	BoW	Self-collected	59.57%	F_1
	Hube and Fetahu (2019)	DL	GloVe, POS, LIWC	RNN	Self-collected	77.10%	F_1
	Pryzant et al. (2020)	DL	BERT	LSTM	WNC	45.80%	Acc
	Pant et al. (2020)	DL	BERT	Ensemble, BERT	WNC	71.61%	Acc

Like lexicon-based approach, however, word frequency methods lack the ability to capture syntactic information. To address this limitation, deep-learning-based methods utilize neural networks to learn spatial and temporal dependencies. Specifically, PLMs are widely used for their ability to provide universal representations (Kim, 2014; Liu et al., 2019b; Sun et al., 2019). For context-dependent subjectivity detection (Table 3.22), the subjectivity of a sentence is determined with regards to its surrounding context, e.g., inter-sentence-level (Pang and Lee, 2004; Belinkov et al., 2017), document-level (Yu and Hatzivassiloglou, 2003; Das and Bandyopadhyay, 2009; Karimi and Shakery, 2017), or discourse-level (Biyani et al., 2014) information. In existing works, such information is typically captured through feature engineering or statistical means.

As a large part of subjectivity detection works relies on external subjective clues, cross-lingual subjectivity detection aims specifically to solve the lack of lexical resources for non-English languages. There are mainly two branches of thought to address this problem (Table 3.22). One is to make use of language-independent methods such as word frequency (Rustamov et al., 2013; Lin et al., 2011; Kamil et al., 2018; Belinkov et al., 2017) and language modeling (Karimi and Shakery, 2017). The other is to generate resources for the target language from English lexicons with the help of SMT systems (Banea et al., 2010; Chaturvedi et al., 2016b).

Multimodal subjectivity detection is a rising field of interest in accordance with the rising need for sentiment analysis in various media (Table 3.22). Existing works utilized lexical, prosodic, and phonemic features for subjectivity detection in spoken conversations (Murray and Carenini, 2011; Raaijmakers et al., 2008). Subjectivity detection in other modalities such as video remains mostly unexplored.

Bias detection is a task that is closely related to subjectivity detection (Table 3.22). It aims to identify biased statements from supposedly impartial articles such as Wikipedia. Despite its greater complexity, the identification of bias exhibits technical patterns that are akin to those found in subjectivity detection, e.g., lexicon-based (Recasens et al., 2013; Hube and Fetahu, 2018), deep learning (Hube and Fetahu, 2019; Pryzant et al., 2020), and cross-lingual (Aleksandrova et al., 2019) methods.

3.6.9.2 Application Trends

Due to its filtering nature, subjectivity detection is widely used as a parser for many downstream tasks, e.g., sentiment analysis (Pang and Lee, 2004; Barbosa and Feng, 2010; Kamal, 2013; Soong et al., 2019; Das and Sagnika, 2020), information retrieval (Zhang et al., 2007; Wiebe and Riloff, 2011), hate speech detection (Gitari et al., 2015), and QA systems (Stoyanov et al., 2005; Wan and McAuley, 2016). Most existing works take the pipeline approach, using the filtered results from subjectivity detection as the input of the target application. On the other hand, we also observe that subjectivity lexicons can also be useful features to support hate speech detection and QA systems.

A survey of literature pertaining to subjectivity detection reveals that the progress made in this research area has not kept pace with the advancements made in its downstream polarity detection tasks, e.g., sentiment analysis (Gandhi et al., 2023). This is likely because sentiment analysis may deliver more fine-grained classification outputs, which helps to gain business insights, e.g., sentiment polarities on product or service reviews.

Table 3.23: A summary of the representative applications of subjectivity detection in downstream tasks.

Reference	Downstream Tasks	Feature	Parser
Bonzanini et al. (2012)	Polarity Detection		✓
Pang and Lee (2004)	Polarity Detection		✓
Das and Sagnika (2020)	Polarity Detection		✓
Kamal (2013)	Polarity Detection		✓
Barbosa and Feng (2010)	Polarity Detection		✓
Soong et al. (2019)	Polarity Detection		✓
Zhang et al. (2007)	Information Retrieval		✓
Wiebe and Riloff (2011)	Information Retrieval		✓
Cohen-Almagor (2011)	Hate Speech Detection		✓
Gitari et al. (2015)	Hate Speech Detection	✓	✓
Li et al. (2008a)	Question Answering	✓	✓
Li et al. (2008b)	Question Answering	✓	✓
Aikawa et al. (2011)	Question Answering	✓	✓
Stoyanov et al. (2005)	Question Answering		✓
Wan and McAuley (2016)	Question Answering	✓	

However, it should be noted that while positive, negative, and neutral sentiment polarities represent subsets of subjective texts, there exists a substantial portion of texts that are objective in nature, presenting factual information. Objective texts are likely to be infrequent in reviews of products or services, as customers often use such platforms to express their opinions. However, in the context of opinion mining on social media, it is crucial to differentiate between subjective and objective statements, given that even statements with neutral sentiment polarities can be indicative of an individual's opinion. Thus, it is still necessary to conduct subjectivity detection before sentiment analysis.

3.6.9.3 Future Works

Fine-grained Subjectivity Detection

A sentence may contain several clauses with differing subjectivity. For instance, a sentence may present two or more opinions, or contain both opinions and factual information. Therefore, to better assist downstream applications, fine-grained subjectivity detection that identifies the particular opinion-bearing clauses is worthy of investigation. However, there is limited research on this issue. Benamara et al. (2011) proposed segment-level subjectivity detection. Wilson et al. (2004) proposed a method specifically for classifying the subjectivity of deeply nested clauses. There is scope for additional research to exploit the full potential of the fine-grained subjectivity annotation offered by the MPQA scheme (Wiebe et al., 2005).

Multimodal Subjectivity Detection

Subjectivity detection using information from multiple modalities remains largely unexplored. There is related multimodal research that might provide inspiration for future works. Wrede and Shriberg (2003) aimed to identify hot spots, which are regions in a meeting where participants are highly involved in the discussion, using solely a set of prosodic features. Hillard et al. (2003); Galley et al. (2004) both targeted the detection of agreements and disagreements in meetings. The former explored the combination of lexical and prosodic features, whereas the latter incorporated pragmatic features that captured the interactions between speakers.

Neiberg et al. (2006) recognized positive, negative, and neutral emotions in meetings using lexical and acoustic-prosodic features. Somasundaran et al. (2007) detected sentences and turns in meetings that express sentiment and arguing opinions using lexical and discourse features. Morency et al. (2011); Wöllmer et al. (2013); Tsai et al. (2019) conducted sentiment analysis on review videos using linguistic features, acoustic features, and visual features (face tracking).

Explainable Subjectivity Detection

While much of the subjectivity detection research has utilized lexical resources such as subjectivity and affective lexicons to explain the subjective nature of text based on individual words, these resources do not capture the pragmatic nuances of words within their contextual environment. This is because the utilized lexical knowledge is context-independent. Theoretical research has explained subjectivity from the perspective of pragmatics (Austin, 1975; Lakoff and Johnson, 1980).

It would be useful to distinguish between different types of objectivity, e.g., a piece of text that is completely factual versus one that contains both positive and negative sentiments towards the same opinion target and, hence, results in overall neutrality or ambivalence (Valdivia et al., 2018). Explainable subjectivity detection could push the development of more linguistics-inspired models that can account for the complexities of subjectivity and its expression in natural language. Additionally, there is potential for cross-disciplinary collaboration between linguistics, cognitive science, and computer science to further advance our understanding of subjectivity and its detection in various domains.

3.7 Conclusion

In this chapter, we reviewed WSD, NER, concept extraction, anaphora resolution, and subjectivity detection. We summarized useful datasets, annotation tools and knowledge bases that can facilitate research in these domains. We also explored their technical trends, related theoretical research, and applications. We found that the breadth and depth of semantics processing can be greatly extended, both from the perspective of the needs of theoretical research and downstream applications. This is because current computational semantic processing techniques are limited in their reliance on specific task settings and available datasets. Reviewing semantic processing applications may encourage further research into fusion methodologies to improve downstream task performance. Semantics processing methods can not only deliver effective features for these tasks, but also gain insights into analyzing model behaviors and studying linguistic and cognitive patterns.

As we continue to advance in the field of NLU, using powerful PLMs and knowledge graphs has become increasingly common to tackle more complex reasoning tasks. However, it is important to note that there is still great academic value in studying the low-level semantic tasks that these models are built upon. These tasks help us understand how language is presented and received, how it relates to human cognition and, hence, get one step closer to NLU. We observe that numerous contemporary semantic processing tasks have been translated into machine learning problems, which have somehow diminished linguistic motivations and intuitions from these computational studies.

Shaping semantic processing tasks into tasks that are more conducive to machine learning can indeed improve the accuracy of specific tasks. However, improving accuracy in a single-task setting is not the only pursuit of semantics processing. We should pay more attention to how semantic processing techniques can better serve humans and machines to explain language phenomena. We hope that this chapter can stimulate more research directions in the field of semantics processing and inspire researchers to place greater emphasis on the nature and cognition of semantics. With the development of more powerful tools such as PLMs and LLMs, it is perhaps valuable for our research to use these tools to address those fundamental linguistic challenges that were previously considered daunting. Regardless of the sophistication of tasks that can be performed by LLMs, basic semantic processing tasks remain crucial for comprehending and utilizing language effectively. These tasks serve as the foundation upon which our understanding of language is built.

3.8 Learning Resources

A. Reading List

- Xulang Zhang, Rui Mao, Kai He, and Erik Cambria. Neurosymbolic Sentiment Analysis with Dynamic Word Sense Disambiguation. In: Proceedings of EMNLP, 8772–8783, 2023 (Zhang et al., 2023b)

- Ran Zhou, Xin Li, Ruidan He, Lidong Bing, Erik Cambria, Luo Si, and Chunyan Miao. MELM: Data Augmentation with Masked Entity Language Modeling for Low-Resource NER. In: Proceedings of ACL, 2251-2262, 2022 (Zhou et al., 2022)

- R Satapathy, SR Pardeshi, E Cambria. Polarity and Subjectivity Detection with Multitask Learning and BERT Embedding. Future Internet 14(7), 191, 2022 (Satapathy et al., 2022)

B. Relevant Videos

- Labcast about Dynamic WSD: youtu.be/dRrxVWPBfVo

- Labcast about Low-Resource NER: youtu.be/CpHZzn1Ar0Y

- Labcast about Subjectivity Detection: youtu.be/PKeciTINEIc

C. Related Code

- Github repository about Dynamic WSD: github.com/SenticNet/Dynamic-WSD

- Github repository about Low-Resource NER: github.com/SenticNet/MELM

- Github repo about Subjectivity Detection: github.com/SenticNet/Subjectivity

D. Exercises

- **Exercise 1.** Implement the Lesk algorithm on the following sentences to determine the correct sense of each target word in the given context (use WordNet or any other lexical database for word definitions and examples): "He went to the bank to deposit his paycheck"; "She played the bass in the school orchestra"; "The bark of the tree was rough to the touch".

- **Exercise 2.** Design a simple NER algorithm that detects named entities and their corresponding types in the following sentences: "Apple is looking at buying U.K. startup for $1 billion"; "San Francisco considers banning sidewalk delivery robots"; "London is a big city in the United Kingdom"; "Elon Musk founded SpaceX in 2002"; "Barack Obama was born on August 4, 1961, in Honolulu".

- **Exercise 3.** Identify and list polar concepts (lemmatized key words and MWEs that capture the essence of content polarity) from the following text and then compare them with the concepts extracted using the concept parsing API (sentic.net/api/#concept): "I love affogato because I am a coffee addict and I like cold desserts: hot espresso on ice cream is such a delightful contrast in temperatures and flavors that I find irresistible".

- **Exercise 4.** Write a simple program to detect and resolve anaphoras in the following paragraph: "Roberto decided to bake a cake. He found a recipe online and gathered all the ingredients. It took him an hour to mix everything and put it in the oven. While it was baking, he cleaned up the kitchen and prepared some tea. When the cake was ready, Roberto let it cool before decorating it with frosting. He was proud of his creation and couldn't wait to share it with his friends".

- **Exercise 5.** Develop a simple algorithm that detects subjectivity based on SenticNet. Clean the text data by removing special characters, converting to lowercase, etc. Implement a function to score the subjectivity of a sentence based on the presence of words from SenticNet. Define a threshold to classify a sentence as subjective or objective based on its subjectivity score. Test your program with various sentences to evaluate its performance.

Chapter 4
Pragmatics Processing

Abstract Pragmatics, situated in the domains of linguistics and computational linguistics, explores the influence of context on language interpretation, extending beyond the literal meaning of expressions. It constitutes a fundamental element for NLU in machine intelligence. In the current era dominated by LLMs, the research focus in NLU has predominantly shifted towards high-level task processing, inadvertently downplaying the importance of foundational pragmatic processing tasks. Nevertheless, pragmatics serves as a crucial medium for unraveling human language cognition. The exploration of Pragmatics Processing stands as a pivotal facet in realizing linguistic intelligence. This chapter encompasses important pragmatic processing techniques, namely: metaphor understanding, sarcasm detection, personality recognition, aspect extraction, and polarity detection. It spans theoretical research, the forefront of pragmatic processing techniques, and downstream applications, aiming to underscore the significance of these low-level tasks in advancing NLU and linguistic intelligence.

Key words: Metaphor Understanding, Sarcasm Detection, Personality Recognition, Aspect Extraction, Polarity Detection

4.1 Introduction

Pragmatics, situated in the domains of linguistics and computational linguistics, explores the influence of context on language interpretation, extending beyond the literal meaning of expressions. While both semantics and pragmatics study meanings, according to Kroeger (2023), semantics studies the intrinsic meaning of words and sentences as linguistic constructs, isolated from their context. In contrast, pragmatics focuses on the dimensions of meaning contingent upon, or arising from, the actual usage of words and sentences.

Within the field of computational linguistics, the investigation of Pragmatics Processing has expanded its purview to employ NLU techniques for the contextual comprehension of implicit intentions and meanings within linguistic expressions. Examples of applications include metaphor understanding, sarcasm detection, personality recognition, aspect extraction, and polarity detection. The research significance of Pragmatics Processing lies in its close connection to uncovering human cognition. This is due to the fact that the comprehension of meanings within the pragmatic context extends beyond the interpretation of literal word meanings, encompassing the discernment of underlying purposes and intentions in the use of specific expressions.

Despite the remarkable success of LLMs in NLP complex reasoning tasks and multiple domains (He et al., 2023c), the question arises: why is fundamental research in Pragmatics Processing still of value in the era of LLMs? LLMs appear to have rapidly bypassed foundational research in computational linguistics, swiftly executing diverse NLP tasks and intricate reasoning in dialogical contexts. This can be predominantly ascribed to the success achieved in language modeling, which involves employing large neural networks and pretraining corpora to predict missing or subsequent words based on contextual cues. While the pretraining paradigm undeniably yields fluent and ostensibly coherent language based on probabilities, it does not align with the way humans comprehend languages. Relying on word association probability for understanding and generating language introduces a constraint wherein current LLMs do not explicitly learn to interpret the intention and implicit meaning conveyed in spoken language. Consequently, the comprehension of language by LLMs remains confined to modeling the relationships between tokens within a vector space. The relevant evaluation also shows that current LLMs fall behind expert systems in processing conventional NLP tasks (Mao et al., 2024a).

In contrast, humans comprehend language through various pragmatic contexts (Kasper and Schmidt, 1996). This involves understanding the contextual meaning of an expression and inferring the aim of the expression, e.g., expressing a specific sentiment. Furthermore, the pursuit of AI extends beyond the mere attainment of human-level accuracy in computational tasks. It encompasses the aspiration to cultivate a form of intelligence akin to human cognition, wherein AI systems emulate human thought processes and behavior. Pragmatics Processing constitutes crucial facets that contribute to progress aligned with these seven pillars. Thus, we contend that Pragmatics Processing research remains valuable in the era of LLMs. Such a research domain not only illuminates the state of human language usage and perception but also showcases mechanisms that offer potentially more scientific methodologies than the language modeling paradigms for pragmatic understanding. On the other hand, pragmatic processing techniques are beneficial for various downstream tasks (see Table 4.1), yet this area of study remains relatively overlooked. Challenges stemming from the scarcity of annotated data and the complexity of pragmatic modeling have led previous research to simplify pragmatic processing tasks using elementary learning paradigms. This approach enables the examination of machine learning algorithms in performing such tasks. However, the simplified task formulation also restricts the functional scope of pragmatic processing techniques.

Table 4.1: The surveyed pragmatic processing tasks and their downstream applications. F denotes that the technique yielded features for a downstream task model; P denotes that the technique was used as a parser; E denotes that the technique improved the explainability for a downstream task. MU denotes metaphor understanding. SD denotes sarcasm detection. PR denotes personality recognition. AE denotes aspect extraction. PD denotes polarity detection.

Downstream tasks	MU	SD	PR	AE	PD
Sentiment Analysis	P	F, P		F, P	
Knowledge Graph Construction	F, E	P		P	
Psychology Analysis	F, E	P			
Machine Translation	P				
Cognition Analysis	P, E				
Political Analysis	E				
Advertising		E			
Dialogue Systems		F			
Recommendation Systems			F, P		E
Social Network Analysis			F, P, E		
Human-robot Interaction				F, P	
Financial Prediction					F, E

For instance, in the domain of computational metaphor processing, where a shortage of extensive annotated datasets for metaphor interpretation exists, substantial focus has been placed on the foundational task of metaphor identification (Ge et al., 2023). However, considering the support for downstream NLP tasks and the exploration of human cognition, a more beneficial avenue may involve the research of linguistic and conceptual metaphor understanding rather than confining the investigation to metaphor identification. Thus, the formulation of tasks in a simplified manner and the limited availability of benchmark datasets consequently impose constraints on the research scope within Pragmatics Processing, resulting in the fact that the value of Pragmatics Processing has been seriously underestimated in the era of LLMs.

The motivation of this chapter (and textbook) is to draw scholars' attention to Pragmatics Processing and inspire broader and deeper pragmatic research works. Our objective is to provide a comprehensive overview of recent research developments in key pragmatic processing tasks, namely: metaphor understanding, sarcasm detection, personality recognition, aspect extraction, and polarity detection. By doing so, we aim to showcase the current task forms, and the cutting-edge advancements within this research domain. Furthermore, we undertake a review of relevant theoretical research in pragmatic processing tasks, approaching them from linguistic and cognitive perspectives, with the intention of encouraging researchers to broaden the scope of pragmatic processing tasks. Emphasizing the practical implications, we introduce downstream applications to underscore the importance of pragmatic processing techniques within the NLP community. Finally, we deliver insights into current technical trends and challenges, providing a foundation for future research endeavors to bridge existing gaps in the field.

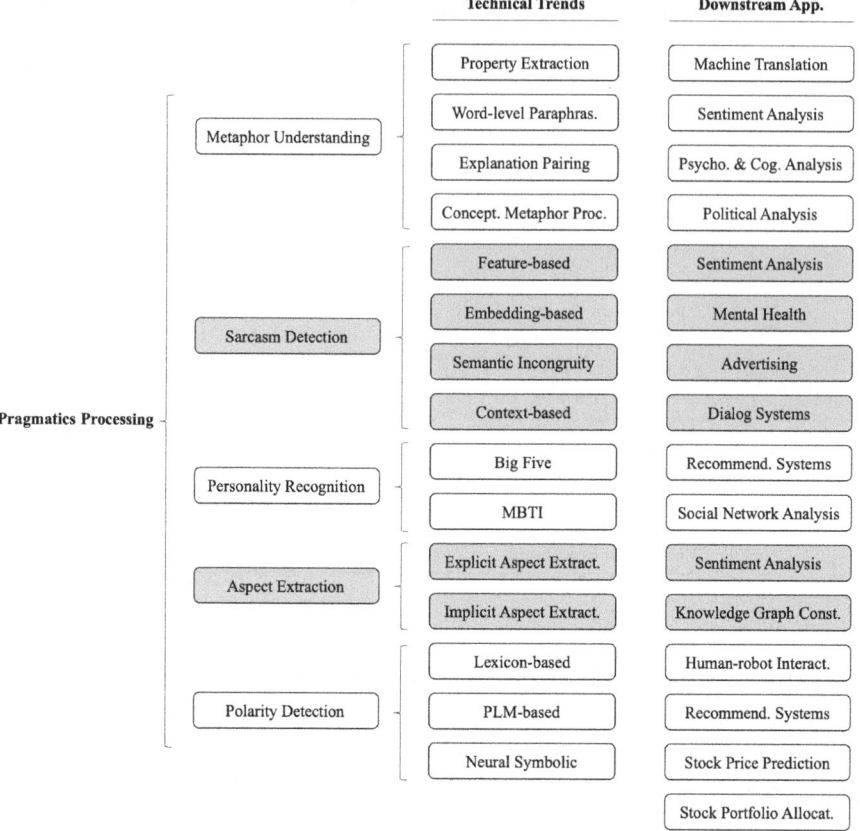

Fig. 4.1: Outline of this chapter. Each subtask is explained in terms of different technical trends and downstream applications.

The contribution of this chapter is threefold:

- It comprehensively examines contemporary Pragmatics Processing methodologies, datasets, and knowledge bases relevant to five specific pragmatic processing tasks.
- Emphasizing the significance of theoretical research and downstream applications, we aim to stimulate more extensive and profound investigations within the Pragmatics Processing domain, building upon the existing task setups.
- Through a comparative analysis of various pragmatic processing techniques, we elucidate their technical trajectories and applications, presenting insights into current trends. Additionally, we propose potential directions for future research in this domain.

In the following sections, we introduce five different pragmatic processing techniques, namely: metaphor understanding (Section 4.2), sarcasm detection (Section 4.3), personality recognition (Section 4.4), aspect extraction (Section 4.5), and polarity detection (Section 4.6). Fig. 4.1 presents an outline of the chapter: each task is organized in terms of theoretical research, annotation schemes, datasets, knowledge bases, evaluation metrics, methods, downstream applications, and a summary. Finally, Section 4.7 wraps up the chapter by proposing concluding remarks, a reading list, three relevant videos, some related code, and five exercises.

4.2 Metaphor Understanding

Metaphors constitute a form of figurative language wherein one or more words are employed to convey a meaning distinct from their literal interpretation. In pragmatics, metaphors are instrumental in conveying abstract concepts, expressing attitudes, and shaping the communicative intent of utterances. They serve as powerful tools for conveying meaning by drawing upon familiar, concrete imagery to illuminate more complex or abstract ideas. The interpretation of metaphors often involves an understanding of the contextual cues, shared knowledge between speakers, and the social and cultural background (Ritchie, 2004), making them a rich area of investigation in pragmatic analysis.

Additionally, metaphors are integral to various aspects of human communication, including humor, persuasion, and the conveyance of emotions (Charteris-Black, 2009), making their examination essential for a comprehensive understanding of pragmatic language use. On the other hand, metaphors pose a challenge for numerous NLP applications, such as machine translation and sentiment analysis (Mao et al., 2023b). The intricacy lies in the complexity faced by machine intelligence in deducing the real meanings embedded within metaphors (see errors in Fig. 4.2). All the aforementioned aspects highlight the significance of metaphor understanding in NLP and pragmatics.

Metaphor identification, positioned as an antecedent task for metaphor understanding, has garnered substantial attention within the research community (Mao and Li, 2021; Tian et al., 2023). This heightened interest is largely attributed to the creation of an extensive metaphor identification dataset, namely the VU Amsterdam (VUA) Metaphor Corpus (Steen et al., 2010). Conversely, the landscape of computational metaphor understanding research is comparatively less developed, primarily stemming from the scarcity of comprehensive annotated data encompassing diverse metaphor understanding tasks. In this endeavor, we survey important computational works in metaphor understanding with the aim of fostering increased engagement and contributions from researchers in this specialized research domain. This is because metaphor understanding is a direct gateway to pragmatic understanding, compared to metaphor identification.

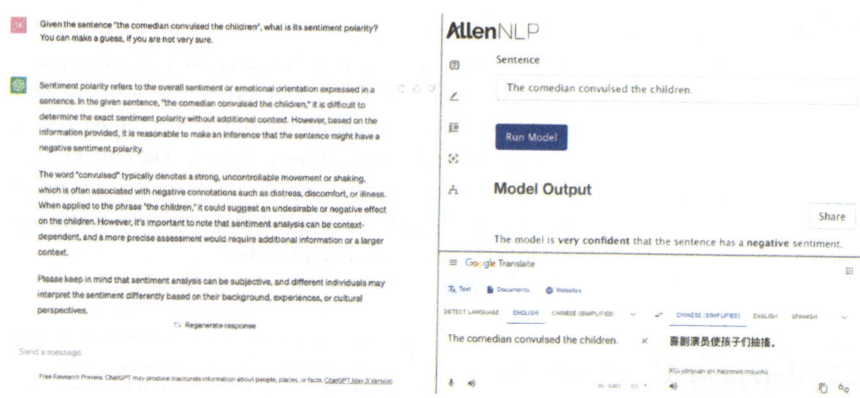

Fig. 4.2: Examples of errors caused by a metaphoric expression in ChatGPT (left), AllenNLP (top-right corner), and Google Translate (bottom-right corner).

Two distinct facets of metaphor understanding have been studied, namely linguistic metaphor understanding and conceptual metaphor processing. The former involves the generation of literal texts for the rephrasing or explanation of given metaphorical expressions from a linguistic standpoint. This category can be further delineated into three groups, namely property extraction, word-level paraphrasing, and explanation pairing. Conversely, conceptual metaphor processing endeavors to establish concept mappings that elucidate the target and source domains of a metaphorical expression from a conceptual standpoint. The source domain serves as the origin of metaphorical qualities, providing imagery, attributes, or characteristics that are subsequently applied to another domain. On the other hand, the target domain represents the subject of the metaphor, interpreting and incorporating the metaphorical qualities derived from the source domain. This dual framework illuminates the linguistic and conceptual dimensions integral to a comprehensive understanding of metaphorical expressions.

4.2.1 Theoretical Research

4.2.1.1 Conceptual Metaphor Theory

Lakoff and Johnson (1980) argued that metaphorical expressions reflect the human cognitive process in their CMT. They articulated the interpretation of metaphors by devising concept mappings, elucidating the cognitive mechanisms underlying metaphors.

(1)   I *spent* three days reading a book.

Given Example (1), the target concept TIME is metaphorically associated with the source concept MONEY through the verb "spent". Within the concept mapping of TIME IS MONEY, the attribute of MONEY, specifically its value and limited availability, is metaphorically transferred to the notion of TIME. This metaphor serves to underscore the conceptualization that TIME is regarded as a valuable and finite resource, akin to the attributes associated with MONEY.

(2)   She *attacked* his argument.

Metaphors also impact human behaviors. In Example (2), ARGUMENT is metaphorically projected into a WAR domain. When we view an "argument" through the lens of "war", it is easy to associate it with strategies, attacks, and defensive behaviors common in a war. This perception can lead people to forget the importance of collaboration and mutual benefit during an argument, resulting in an aggressive approach with raised voices to win the argument.

(3)   Encountering her *miraculously restored* my faith in love.

(4)   Our relationship has *reached a crossroads.*

As metaphors serve as reflections of cognitive mechanisms guiding human conceptualizations, variations in metaphoric expressions can signify distinct perceptions of a concept among individuals. For instance, a young couple may conceptualize LOVE as akin to MAGIC, influenced by the enchantment of their initial encounter and falling in love, e.g., Example (3). In contrast, an older couple might adopt the concept mapping of LOVE IS JOURNEY, embodying the cumulative experiences amassed over the years, e.g., Example (4). Thus, considering the frequent occurrence of metaphors in everyday communication (Shutova, 2015), metaphors can serve as an important entrance for studying human cognition.

4.2.1.2 Metaphor-Enriched Social Cognition

Landau et al. (2010) investigated the impact of metaphors on shaping social thought and attitudes, utilizing empirical evidence to illustrate that metaphors play a distinctive role in social information processing. They introduced the metaphoric transfer strategy and the alternate source strategy as empirical approaches to assess the influence of conceptual metaphor and embodied simulation on social information processing. Their findings reveal that manipulating psychological states associated with one concept induces metaphor-consistent alterations in how individuals process information related to seemingly dissimilar concepts. Furthermore, they observed that metaphors have bidirectional effects on social information processing. Manipulating abstract social concepts led to changes in perceptions of more concrete concepts. For example, changes in divinity-related images affected spatial memory biases, and different moods influenced attention shifts. Social exclusion also impacted perceptions, such as temperature sensations.

The study also discussed the differentiation between conceptual metaphor and embodied simulation. Landau et al. (2010) posit that while both are cognitive mechanisms involving bodily states in processing abstract concepts, they differ in their utilization of these bodily states. Conceptual metaphor operates as an inter-conceptual mechanism, mapping content and structure between apparently dissimilar concepts. It draws on representations of bodily states related to a concept in a manner distinct from how embodied simulations employ such representations. Conceptual metaphors can incorporate concepts representing common knowledge about bodily states, whereas embodied simulations exclusively involve specific bodily states occurring during experiences with abstract concepts. In contrast, embodied simulation functions as intra-conceptual mechanism, utilizing representations of bodily states associated with a given concept. It employs bodily states to simulate the experience of the concept, enabling individuals to understand it better.

4.2.1.3 Neural Theory of Language

Feldman (2008) undertook a neuroscientific exploration of language by delving into the neural mechanisms underpinning language processing and production. Within this context, he probed the intricate relationship between metaphor and meaning, asserting that the meanings of our cultural, abstract, and theoretical concepts are fundamentally rooted in metaphor. This assertion gains support from empirical evidence highlighting the robust connection between cultural knowledge and embodied experience. Exemplifying this linkage, the primary metaphor of affection, conceptualized as WARMTH, draws on the physical sensation of being held affectionately. This metaphorical correlation is evident in language, wherein affection is commonly articulated in terms of WARMTH, illustrating the formative role of embodied experiences in shaping our conceptual understanding of abstract concepts.

Moreover, Feldman investigated the metaphorical foundation of causation, specifically the metaphor positing causes as forces, grounded in our embodied encounters with FORCE in everyday life. This exploration showed the interplay between physical experiences and abstract notions such as causation. Expanding his inquiry, Feldman explored how simulation theory enables individuals to grasp discourse and infer meaning by leveraging their extensive experiences of the world and social systems. This exploration demonstrated the integral connection between embodied experiences and the comprehension of abstract and cultural concepts.

Feldman discussed complex systems of structural mappings connecting all knowledge domains to basic schemas. He argued that reasoning about complex and abstract topics stems from our embodied knowledge of actions, goals, and forces. The study examined how primary metaphors help children understand and express language about subjective experiences, providing a way to conceptualize cultural and abstract concepts. Embodied semantics plays a key role in interpreting narratives with new ideas and language, allowing people to use their experiential understanding of the world and social systems for inference.

4.2.2 Annotation Schemes

4.2.2.1 Metaphor Interpretation Annotation

Pragglejaz (2007) presented metaphor identification procedure (MIP) to annotate metaphors at the token level and eliminate the subjectivity issue during the annotation process. Steen et al. (2010) proposed an extension called MIP Vrije Universiteit (MIPVU) based on MIP and published currently the largest annotated token-level metaphor identification dataset, VUA. The main idea of MIP is to annotate metaphors by the semantic contrast between the contextual (dynamic) meaning and the basic (static) meaning of a target word. If there is a semantic contrast between the contextual and basic meanings, the target word is metaphoric, otherwise, it is literal. MIP provides a practical method for understanding the contextual and basic meanings of metaphors and annotating metaphoricity.

Inspired by MIP, Shutova (2010) asked annotators to provide meaningful paraphrases of metaphors or rate the produced paraphrases by the proposed model and the baseline system based on the context meanings. Nevertheless, achieving consensus on the basic meanings among annotators can prove challenging, even with the aid of dictionaries. The inherent subjectivity of annotators may introduce variability into the annotation process. Additionally, certain conventional metaphors are so pervasive that their meanings are incorporated into dictionaries. In such instances, the MIP faces limitations in accurately identifying or comprehending metaphors.

4.2.2.2 Concept Mapping Annotation

Shutova and Teufel (2010) presented a practical procedure for concept mapping annotations based on MIP. They built source and target concept domain lists containing a subset of domains from a master metaphor list (MML) (Lakoff, 1994) and novel domains annotated by themselves. They asked annotators to choose concept domains for the basic and contextual meanings of single-word verb metaphors.

(5)   If he asked her to post a letter or buy some razor blades from the chemist, she was *transported* with pleasure.

In Example (5). the basic meaning of the verb *transported* is "goods being transported/carried somewhere by a vehicle", while the context meaning of this sentence is "a person being transported by a feeling" (Shutova and Teufel, 2010). According to the different agents of the *transported* action, the source-target concept mapping is "EMOTION IS VEHICLE". However, this annotation scheme must provide the whole concept lists before the annotation, which is labor-costing and makes it difficult to update any novel metaphors. Mohler et al. (2016) used a similar annotation scheme to (Shutova and Teufel, 2010). In particular, they provided annotators with two source domains produced by their conceptual metaphor mapping system (Mohler et al., 2014).

4.2.3 Datasets

This section introduces existing datasets for metaphor understanding. Table 4.2 summarizes the basic information and statistics of the relevant datasets. Bizzoni and Lappin (2018) proposed a metaphor understanding dataset[1] in a paraphrasing task format. This dataset contains 200 sentence sets. Each set contains one metaphorical sentence and four literal sentences with different semantic similarities: one strong paraphrase, one loose paraphrase, and two non-paraphrases. The strict semantic similarity rule made this dataset manually constructed and a relatively small volume.

Table 4.2: Metaphor understanding datasets. EP denotes explanation pairing. CMP is conceptual metaphor processing. TP denotes token-level metaphor paraphrasing.

Reference	Task	Source	# Samples
Bizzoni and Lappin (2018)	EP	Manually created	200
Zayed et al. (2020)	EP	Dictionaries, X	1,500
Liu et al. (2022a)	EP	Manually created	10,256
Lakoff (1994)	CMP	Published books, papers, and research seminars	791
Shutova and Teufel (2010)	CMP	British National Corpus	761
Mohler et al. (2016)	CMP	ClueWeb09 corpus, Debate Politics Online Forum	80,100
Mao et al. (2024b)	TP	VUA	10,716

```
Metaphorical sentence: "The crowd was a river in the street"
Literal sentence 1: "The crowd was large and impetuous in the street.", #4
Literal sentence 2: "There were a lot of people in the street.", #3,
Literal sentence 3: "There were few people in the street.", #2,
Literal sentence 4: "We reached a river at the end of the street.", #1.
```

Zayed et al. (2020) developed a dataset for metaphor understanding through the generation of definitions, providing comprehensive interpretations of metaphorical expressions. The metaphorical expressions are in a verb-direct object form. The definitions are collected from multiple English dictionaries.

```
Metaphorical expressions: "release old emotional pain",
Definition: "to express feelings such as anger or worry in order to get rid of them",
Source: Oxford.
```

Liu et al. (2022a) created Fig-QA[2] metaphor interpretation dataset and proposed an exploration based on the Winograd schema (Levesque et al., 2012) to investigate the effectiveness of language models in metaphor interpretation. The aim is to prompt language models to select or generate implications for two metaphorical phrases with divergent meanings.

---

[1] https://github.com/yuri-bizzoni/Metaphor-Paraphrase
[2] https://github.com/nightingal3/Fig-QA

Fig-QA displays 10,256 distinct and creative metaphors, along with interpretations crafted by workers from Amazon Mechanical Turk (AMT). The dataset is structured as pairs of phrases embodying contrasting meanings, along with their potential interpretations.

```
Metaphorical sentence 1: "Her word had the strength of titanium.",
Interpretation 1: "Her promises can be believed.",
Metaphorical sentence 2: "Her word had the strength of a wine glass.",
Interpretation 2: "Her promises cannot be trusted."
```

Lakoff (1994) first attempted to compile an MML for source and target concept mappings. MML contains 791 nested concept mappings with corresponding metaphorical expressions. The data were sourced from published books and papers, student papers, and research seminars. However, MML has stayed unfinished up to now. The varying abstractness levels of concept mappings remain an issue (Shutova and Teufel, 2010).

```
Concept mappings: "CHANGE IS MOTION (Location)",
sub concept mappings: "STOPPING BEING IN A STATE IS LEAVING A LOCATION",
Metaphor expressions: "He came out of the coma with little long-term damage."
```

Shutova and Teufel (2010) extracted a subset of British National Corpus (BNC) (Consortium, 2007) and made annotators choose source and target concepts from provided lists. The three annotators are native English speakers with linguistic backgrounds. Mohler et al. (2016) made available a sizable collection of word pair metaphors in four languages (English, Spanish, Russian, and Farsi), termed Language Computer Corporation (LCC) metaphor datasets[3]. The annotation labels include metaphoricity rates on a 4-point scale (clear, conventionalized, weak metaphor, and literal), source and target concept domains, affect polarity, and intensity rates. The English data is collected from ClueWeb09 corpus[4] and Debate Politics Online Forum[5]. The annotators were asked to choose the most appropriate concepts for the given metaphorical expressions in a concept list generated by Mohler et al. (2014).

Mao et al. (2024b) proposed a human-annotated metaphor paraphrasing dataset (VMC-P) upon VUA. The dataset consists of 10,716 sentences, 11,880 metaphorical lexical units and the associated paraphrases. The lexical units include both metaphoric single- and MWEs. The dataset was developed for the purpose of end-to-end metaphor processing. To achieve this goal, the authors have made the following efforts: data were sourced from real-world text, metaphor identification and paraphrase labels were annotated on the token level, and annotations were delivered by a linguistic expert to improve the paraphrase consistency.

---

[3] https://languagecomputer.com/metaphor-data.html

[4] https://lemurproject.org/clueweb09.php

[5] https://debatepolitics.com

```
'ID': 'trn_1179',
'doc_ID': 'ahc-fragment60',
'sent_ID': '1221',
'sent': 'Relief was at hand .',
'metaphor_index_list': [[0], [2, 3]],
'pos_list': ['Assistance', 'nearby'],
'neg_list': ['Easement', 'Sculptural relief', 'Decrease', 'Ministration', ...],
'lemma': 'relief be at hand .',
'pos_tags': ['NN', 'VBD', 'IN', 'NN', '.'],
'open_class': ['NOUN', 'VERB', 'O', 'NOUN', 'O'],
'genre': 'news'
```

4.2.4 Knowledge Bases

Useful knowledge bases that can be used in metaphor understanding are WordNet and ConceptNet (introduced earlier in Section 3.4). Strzalkowski et al. (2013); Li et al. (2013); Dodge et al. (2015); Gagliano et al. (2016); Mao et al. (2018); Ge et al. (2022) utilized lexical relations in WordNet, such as synonyms, hyponyms, and hypernyms to obtain or expand candidate concept words. Mason (2004); Gandy et al. (2013); Strzalkowski et al. (2013) clustered concepts by WordNet functions, such as semantic categories and hypernyms. Shutova (2010) utilized WordNet to disambiguate the sense of concepts. Su et al. (2020) calculated relatedness between concepts by synonymous extension in WordNet.

FrameNet (Ruppenhofer et al., 2016) is another useful knowledge base for metaphor understanding, It serves as an English lexical repository designed for both human and machine readability. It was established through the annotation of real-life textual examples illustrating word usage and is grounded in the framework of frame semantics. The repository encompasses 1,224 frames, where each frame represents a diagrammatic depiction of a scenario, incorporating diverse elements such as participants, props, and other conceptual roles. Additionally, FrameNet comprises 13,687 lexical units, encompassing lemmas and their POSs that evoke frames.

Table 4.3: Useful knowledge bases for metaphor understanding. MWD denotes Merriam Webster's dictionary.

Name	Knowledge	# Entities	Domain	Structure
WordNet	Structured lexicon	155,327	Open	Graph
FrameNet	Structured lexicon	13,687	Open	Graph
ConceptNet	Structured lexicon & multilingual	8M	Open	Graph
SenticNet	Structured lexicon & multilingual	400,000	Sentiment	Graph & Dictionary
Wiktionary	Word senses	7,317,715	Open	Dictionary
MWD	Word senses	470,000	Open	Dictionary

Finally, dictionary resources are intuitive choices for metaphor understanding because language learners will open a dictionary when confused with unfamiliar materials (Zayed et al., 2020). Some studies utilized dictionaries for word senses or explanations. Gandy et al. (2013) extracted dictionary senses from Wiktionary[6] to find linguistic metaphors in the proposed model. Dodge et al. (2015) utilized Wiktionary to build the frame network pattern of metaphorical and literal words. Mao et al. (2022a) manually annotated an idiom dictionary to assist in metaphoric MWE interpretation. Other studies extracted lexical information, such as syntactic dependency relations from dictionaries. Li et al. (2013) collected syntactic information of word associations from Merriam-Webster's dictionary[7]. Table 4.3 shows relevant information and statistics of the above-mentioned knowledge bases.

4.2.5 Evaluation Metrics

Since multiple data structures in annotated datasets (see Section 4.2.3) and task forms (see Section 4.2.6) exist in the metaphor understanding domain, researchers developed a variety of evaluation metrics.

4.2.5.1 Automatic Evaluation

Accuracy (Shutova, 2010; Mao et al., 2018; Bizzoni and Lappin, 2018) and F_1 score (Gandy et al., 2013; Dodge et al., 2015; Ge et al., 2022) are frequently used automatic evaluation metrics for datasets with golden labels. When the proposed models can generate a list of possible interpretations, the higher ranks of correct interpretations show better performance of models. Mean reciprocal rank (MRR) is a metric traditionally used in QA systems and is appropriate for this task. Shutova (2010); Song et al. (2020b) applied MRR to evaluate the ranking quality of their models. Li et al. (2013); Song et al. (2020b) utilized the proportion of the correct interpretations among the top N predictions (Hits@N) to evaluate the performance comprehensively.

Each possible mapping in the work of Mason (2004) was evaluated regarding polarity, the number of word collocations instantiating the mapping, and the systematic co-occurrence. Gagliano et al. (2016) demonstrated the blending effect of generated concepts by cosine similarity. Bizzoni and Lappin (2018) used correlation to show the performance of the paraphrase ordering task. Other automatic evaluation methods involve grounding the evaluation tasks in downstream applications, such as metaphor identification (Ge et al., 2022), machine translation (Mao et al., 2018), and sentiment analysis (Mao et al., 2022a), to assess the efficacy of a metaphor understanding system.

---

[6] https://wiktionary.org

[7] https://merriam-webster.com

4.2.5.2 Human Evaluation

Researchers used human evaluation metrics for metaphor understanding tasks without golden labels to measure the performance of outputs from multiple perspectives. Li et al. (2013) defined "correct Top 3" to denote concepts within the top three positions of lists that, while not considered golden labels, are recognized as metaphors by human judges. Gandy et al. (2013) measured the meaningfulness of the generated concept mappings. Su et al. (2015, 2017, 2020) evaluated the acceptability of produced interpretations. Rai et al. (2019) chose appropriateness to show the performance of the proposed model. Sometimes the evaluation metrics could be too abstract to understand. Some researchers proposed detailed questions to help people for evaluation more understand the criteria. Strzalkowski et al. (2013) evaluated metaphoricity, affect, and force (inverse of commonness) by asking, "Q1: To what degree does the above passage use metaphor to describe the highlighted concept? Q2: To what degree does this passage convey an idea that is either positive or negative? Q3: To what degree is it a common way to express this idea?". Ge et al. (2022) evaluated the quality of the source and target domains with two questions: "Q1. Whether the noun is conceptually mapped to the source concept? Q2. Whether the basic meaning of the noun belong to the target concept?" Mao et al. (2022a) evaluated coherence, semantic completeness, and literality of generated interpretations with three questions: "Q1. Does the paraphrased word semantically and grammatically fit the context? Q2. To what extent does the paraphrased word represent the contextual meaning of the original target word? Q3. Is the paraphrased word a literal counterpart of the original target word?".

4.2.6 Methods

4.2.6.1 Property Extraction

Metaphor understanding systems based on property extraction seek to extract properties potentially connecting the source and target domains. For example, extracting an adjective attribute for two nouns from source and target domains represents the shared qualities of two words from source and target domains. Early works used semantic role labeling (SRL) for identifying the predicate-argument structures in a sentence and, hence, for extracting properties and roles of words in a metaphorical context. Su et al. (2015) proposed a Chinese metaphor understanding model by extracting an appropriate property of a source concept to connect with a target concept. The properties of source concepts were collected from Attribute Database[8] and Sardonicus[9].

---

[8] A database by NLP Lab of Xiamen University.

[9] An adjective taxonomy database from https://afflatus.ucd.ie

They extended properties with synonyms from a Chinese Thesaurus named Tongyi Cilin[10] and designed a relatedness score with cosine similarity between properties and target concepts, and between properties and contextual words. The property with the highest relatedness score was outputted as "target be property". This method deeply relied on external knowledge bases and ignored the direct interaction between contextual and target words. They evaluated the proposed method with the help of five Chinese volunteer native speakers on a self-collected Chinese dataset containing 80 instances of nominal metaphors from the web, blogs, and books. The annotators were asked to score the acceptability of predicted results, which reached the highest acceptability as the correct outputs. The proposed model achieved the highest accuracy, 0.84. Su et al. (2017) followed this idea but revised the relatedness score into cosine similarities between synonyms of properties and target words. Following the same evaluation guideline as Su et al. (2015), the test dataset extended to 100 Chinese nominal and 100 English nominal metaphors.

Rai et al. (2019) hypothesized that metaphors could be better understood with emotions and presented a metaphor understanding model regarding emotions as connections between source and target concepts. It started with searching properties of source concepts on the web. Then, it vectorized each property in six emotion dimensions by word2vec (Mikolov et al., 2013b), namely anger, fear, happiness, disgust, sadness, and surprise. Finally, it selected the most related properties in the most related emotion dimensions. The output was formulated as "target be <emotion: property>". The test data included 75 nominal metaphorical instances from Rai and Chakraverty (2017). The authors invited seven non-native English speakers to rate the appropriateness of the generated senses on a Likert scale (Likert, 1932) of 1 to 5 where 1 denoted highly inappropriate and 5 denoted highly appropriate. Ratings below an average of 3 were designated as incorrect.

Su et al. (2020) presented a hierarchical semantic model with the hypothesis that some metaphors are created with cultural implications. To broaden the property's coverage, they manually annotated a conceptual mapping knowledge base related to culture. The proposed method embedded word-level, perception-level, attribute-level, and context-level information. The perception layer utilized sentiment information from WordNet and Stanford CoreNLP (Manning et al., 2014). The attribute layer processed cultural information with three types: shared non-cultural attributes, non-shared non-cultural attributes, and cultural attributes. The global feature layer integrated three features to describe contextual semantics: cultural semantics, the theme of words, and the theme of discourse, which are from word2vec and LDA (Jo and Oh, 2011) models. They employed a random walk algorithm to find the properties with the highest relevance to target concepts. This paradigm concentrated on nominal metaphors in Chinese. The self-collected test dataset included 280 nominal metaphors, 190 of which are cultural metaphors and 90 of which are non-cultural metaphors.

---

[10] https://ir.hit.edu.cn

Song et al. (2020b) completed metaphor understanding in a knowledge graph completion pattern. The knowledge graph contains triples in the form of (source, attribute, target). The attribute is a shared property between source and target concepts. The metaphor understanding task is transferred to select an appropriate attribute based on the given source and target concepts. The final attributes were selected by knowledge graph embedding methods. The evaluation was conducted on a manually built metaphor dataset, compared with model variants based on different embedding methods and score functions. To demonstrate a complete model performance, the evaluation includes exact-match-based and synonymy-based metrics. Due to the uniqueness of the task form, no external baselines were included in the evaluation.

4.2.6.2 Word-level Paraphrasing

The objective of systems for word-level metaphor paraphrasing is to perform a word-by-word transformation of metaphors into their literal equivalents within the given context. These systems were initially designed to address primarily single-word metaphors. Shutova (2010) first defined metaphor understanding as a paraphrasing task, making outputs available for downstream applications. The proposed algorithm targeted word pairs: verb-subject and verb-direct object. The authors selected possible candidates from a large corpus by the co-occurrence of targets and syntactic dependency relations. Then, they filtered and ranked them by hyponyms in WordNet and selectional preference. The evaluation was on a subset of BNC with a brief version of the proposed model as a baseline. The evaluation settings are twofold. The first was to ask seven volunteer annotators to determine whether the generated paraphrases are literally used and deliver the same meanings as the metaphorical concepts. The second evaluation setting was to ask another five volunteers to write appropriate paraphrases as golden standards according to the given metaphorical expressions.

Mao et al. (2018) built an unsupervised metaphor identification and understanding method at the sentence level. Based on the hypothesis that the literal senses of words can occur more frequently than their metaphorical senses, they utilized the synonyms and hypernyms from WordNet as the candidate literal senses of the target words. They employed word2vec methods to capture the co-occurrences between contextual words and target words. This model was evaluated as a preprocessing step of the machine translation task, which largely enhanced the performance of Google and Bing Translator. Mao et al. (2022a) developed this model with PLMs, such as RoBERTa, ALBERT, and BERT. The human evaluation was managed on 100 sentences, randomly sampled from the VUA open-class testing set by three native English speakers from the UK and the US. The annotators were asked to measure the coherence, semantic completeness, and literality of produced paraphrases. The RoBERTa-based model achieved the highest scores in three dimensions.

To improve the learning of the end-to-end metaphor paraphrasing task, Mao et al. (2024b) proposed a new dataset (VMC-P) and a metaphor-tailored PLM. Compared to conventional language modeling methods, e.g., masked word prediction-based and next word prediction-based methods, Mao et al. (2024b) pretrained a model with an ALM paradigm. The authors randomly replaced some words, forming anomalous sentences, to simulate the selectional preference violation of metaphors. Then, they developed an MTL and contrastive learning framework to learn the anomalous detection and original word retrieval tasks, simultaneously, at the pretraining stage. Thus, the learned linguistic patterns can be inherited by the following metaphor identification and paraphrasing tasks. The authors found that such a pretraining paradigm exceeds parameter-size-comparable PLM baselines in the end-to-end metaphor interpretation task.

4.2.6.3 Explanation Pairing

Metaphor understanding systems based on explanation pairing seek to connect a metaphor with an explanation in the given context. Typically, dictionary definitions serve as the basis for explanations.

Martin (1990) introduced a framework known as Metaphor Interpretation, Denotation, and Acquisition System (MIDAS). In the initial stage, the interpretation system analyzed the input text to extract syntactic information. This syntactic data was then employed to identify potential candidates, which underwent scrutiny for coherence and abstractness through constraint testing. The interpretation process followed a recursive pattern, addressing metaphors based on language grammar principles and leveraging rich specialized information. KODIAK (Wilensky, 1986) was used by MIDAS, which is a knowledge representation language wherein facts are depicted as nodes interconnected by primitive links. The three categories of nodes encompass "absolutes", "relations", and "aspectuals". The associations of metaphors can be linked to other concepts and organized within abstraction hierarchies through the inheritance mechanisms facilitated by KODIAK.

Bizzoni and Lappin (2018) discussed the limitations of current metaphor understanding task forms and introduced two new task forms, namely binary classification and paraphrase ordering tasks. They argued that the sentiment might not be reserved after single-word replacement, so the new task forms with less constraint could be closer to human cognition of metaphors. The authors suggested a CNN-LSTM framework to capture the semantic representations of metaphorical and literal expressions at different levels. The inputs of the model were one metaphor sentence and one literal candidate. The outputs were binary labels as well as the sigmoid value. The binary labels (metaphorical or literal) were for the binary classification task. The sigmoid value could be used in the paraphrase ordering task. Experiments were conducted on a self-produced dataset containing 200 sets of sentences. However, these task forms needed rich language resources and considerable annotation labor, making them less liable for further research and application.

Mao et al. (2022a) presented a dictionary and rule-based method to identify and understand metaphorical MWEs. The method set up three rules with dependency pairing and lemma information to identify MWEs. The interpretation was completed with the help of semantic similarity and a manually annotated idioms dictionary. The selected explanation achieved the highest semantic similarity with the source sentence. The output added an explanation as a clause of the source sentence, which downstream tasks can directly use. They tested the MWE detection model on Formal Idioms Corpus[11]. The proposed method outperformed machine learning baselines on an unseen case evaluation task.

4.2.6.4 Conceptual Metaphor Processing

Current conceptual metaphor processing methods aim at generating source-target concept mappings of metaphorical expressions. Generally, concept mappings are formulated as "a target concept is a source concept", such as "ARGUMENT IS WAR.

A. Feature-based Methods

Mason (2004) proposed a system termed CorMet to generate concept mappings for conventional metaphors. It searched documents related to two specific domains, extracted representative verbs, and clustered selectional preferences by WordNet nodes. The step of searching documents online might introduce unnecessary noise for representative verbs. Those verbs could be unrelated to the specific domains. Due to the unavailability of a large corpus containing golden labels, the system was assessed on a subset of MML. Gandy et al. (2013) developed a rule-based heuristic algorithm to identify linguistic metaphors, discovered nominal analogies, and clustered them to find meaningful conceptual metaphors. The candidate concepts were extracted from WordNet and selected by linguistic features, e.g., point-wise mutual information and abstraction scores. They clustered candidates and self-defined a linguistic score to select a meaningful concept from WordNet hypernyms. The evaluation was delivered by 21 volunteers to validate whether each conceptual metaphor reflected a valid and meaningful interpretation of metaphorical expressions about GOD, GOVERNANCE, GOVERNMENT, MOTHER, and FATHER domains.

Strzalkowski et al. (2013) generated concept mappings from a passage in a specific domain – GOVERNANCE with topical structure, imageability scores, and a clustering framework. They hypothesized that metaphorical words are typically employed with strong imageability and beyond the topical framework of a sentence. They asked annotators from AMT to evaluate the concept mappings in the aspects of metaphoricity, affect, and commonness scores. The evaluation was conducted on English, Spanish, Russian, and Farsi datasets without baselines.

---

[11] https://github.com/prateeksaxena2809/EPIE_Corpus

Li et al. (2013) proposed the first unsupervised data-driven model for metaphor identification and concept generation. They prepared a word pair dataset by searching "like-a" and "is-a" syntactic patterns in a web corpus. They designed a context-related formula with selectional preference to obtain the implicit source or target concepts among the candidates from the metaphorical corpus. The testing dataset contained 83 sentences expressing metaphors by subject-verb or verb-object relations from Lakoff and Johnson (1980). The baseline was a variant of the proposed model with a replacement of a metaphor database. They applied two labeling criteria. The generated concept mappings are marked as "match" if they are the same as or subsumed by gold labels. If the generated concept mappings are not "match" but are considered metaphors by at least two annotators, they are marked as "correct".

Dodge et al. (2015) put up a framework termed MetaNet for bridging theory- and corpus-driven approaches to metaphor processing. Together with linguists, they created a hand-crafted repository. The repository of metaphor expressions and frame-works included several relations like "subcase of" and "incorporates as a role" as well as concept domains. Then, they looked for phrases using a list of grammatical constructs. Through the relational network of the repository and with the aid of WordNet, FrameNet, and Wiktionary[12], the source and target terms were matched with concept domains in the repository. By using pre-defined syntactic patterns from language characteristics and knowledge sources, this approach is restricted in its ability to identify metaphors.

Rosen (2018) proposed a source domain mapping model expressing the inter-action between a target word and its construction-based context. It extracted multiple self-defined features related to dependency relationships and fed the features into a deep neural network classifier. The features were interactions derived from dependency-parsed inputs, such as target domain x subject and target domain x object. The evaluation was conducted on LCC (Mohler et al., 2016). However, the source domains were represented as one-hot vectors in outputs, leading to data sparsity somewhat. The model could only output one of 77 known and limited source domains.

B. Embedding-based Methods

Gagliano et al. (2016) applied word2vec to identify the connection between two words in the semantic space to generate a figurative relationship. Based on word2vec vectors of a target concept and an attribute, the addition and intersection models independently derived candidates of a source concept. They designed quantitative and qualitative analyses to evaluate the generated terms. In the quantitative analysis, they used the cosine similarity of word2vec vectors to measure the semantic repre-sentations. In the qualitative analysis, AMT employees were asked to select the final source concepts in the provided list and generate a metaphorical expression. The result showed that candidate concepts with balanced cosine similarity between two words could boost the effect of integrating two semantic spaces.

---

[12] https://wiktionary.org

Fu et al. (2020a) designed an image metaphor understanding task containing three parts: literal concept detection, literal-implied concept mapping, and metaphor captioning. This study aimed to establish literal-implied concept mapping, wherein the identification of suitable target concepts was performed by analyzing source and contextual concepts present in the images. They constructed an undirected reference graph, in which the nodes represented candidates found by rule-based query searching, and edges reflected the compatibility between candidates and contextual concepts. The final implied concepts had the highest compatibility among all the concepts. The evaluation was conducted on a manually-built image metaphor dataset[13] with 54 images collected from stock image websites. They utilized Hit@N which denoted the proportion of correct concepts in the top N predicted concepts. Experiments compared multiple model variants, such as different ways for reference graph construction or external knowledge, with a baseline barely calculating the cosine similarity of word2vec representations between metaphorical terms and predicted concepts. The approach that yielded the most favorable outcomes involved the integration of word2vec-based co-occurrence and DeepWalk-based knowledge base.

Ge et al. (2022) proposed an MTL model for metaphor identification and concept mapping generation. They extracted hypernyms from WordNet and frequent associations of target words from Wikipedia and applied a knee algorithm (Satopaa et al., 2011) to select candidate concepts. Due to the absence of large annotated concept mapping datasets, they proposed novel a dynamic reward mechanism to push the learned concepts toward more accurate metaphor identification. The model could learn concept mappings with a broad conceptual spectrum from metaphor identification. Compared with previous works, this model can automatically identify metaphoricity and generate concept mappings in natural language instead of clustering concepts. The automatic evaluation was managed on an out-of-domain dataset[14], GUT, to examine whether the model can deliver accurate concept mappings for metaphor identification. The proposed model obtained higher scores in source, target, and mapping accuracy in human evaluation.

MetaPro (Mao et al., 2023b) integrated the works of Mao and Li (2021); Mao et al. (2022a) and Ge et al. (2022) to process concept mapping from end to end. The system first identified metaphors from a full sentence. Next, the identified metaphors are paraphrased into their literal counterparts. The original metaphors and their literal counterparts served as the basis for the source and target concepts. Finally, output is organized in the form of "the target concept is the source concept". To the best of our knowledge, MetaPro is the only system that can generate metaphor paraphrases and concept mappings from end to end.

---

[13] https://github.com/ADaM-BJTU/Metaphor-Image-Dataset
[14] https://bit.ly/1TQ5czN

4.2.7 Downstream Applications

Researchers have discovered that metaphorical expressions are difficult for the learning of downstream NLP tasks. Since the semantics of metaphorical expressions differ from those of literal ones, NLP systems likely make mistakes while comprehending them. This is why metaphor understanding is a key subtask of the NLU suitcase model (Fig. 1.12). In sentiment analysis, the literal meaning of the phrase "she devoured his novels" is not positive, while its metaphorical connotation is positive. A sentiment classifier may deliver an inaccurate answer for expressions without metaphor understanding (Mao et al., 2022a). Due to cultural variations, it might be difficult for language learners to understand metaphors. The literal translation of the expression "she devoured his novels" in the same case does not make sense in Chinese (Mao et al., 2018). Therefore, transforming metaphors into their literal counterparts can enhance the comprehension of metaphors for both humans and machines.

On the other hand, metaphors also provide an entry point for investigating human cognition. These psycho-linguistic expressions offer a distinctive window into the subconscious minds of humans (Goatly, 2007). By their very nature, metaphors extend beyond literal language, tapping into the rich reservoir of symbolism and imagery ingrained in our thoughts and emotions. When individuals employ metaphors, they tap into a collective repository of shared cultural, social, and personal experiences, unveiling deeper cognitive processes that might not be immediately apparent in explicit language. One structured way of representing metaphorical cognition processes is through concept mappings, where the projection from a target to various source domains highlights distinctions between different cognitive frameworks. Concept mappings have been utilized in various traditional diagnostic psychological assessments, such as the Word-association Test, Thematic Apperception Test, and Rorschach Test, each employing distinct forms of concept mappings (Rapaport et al., 1946). Consequently, the understanding of conceptual metaphors holds potential for downstream research in psychology and cognitive science.

4.2.7.1 Machine Translation

Mao et al. (2018) evaluated the proposed metaphor understanding model on a machine translation task. The testing dataset contained 50 metaphorical sentences and 50 literal sentences randomly sampled from Mohammad et al. (2016)[15]. The text identified as metaphors was paraphrased by the metaphor understanding model before going into the machine translation system. The results showed that paraphrasing metaphors into their literal equivalents boosted 26% accuracy in Google translation and 24% in Bing translation for an English-Chinese metaphor translation task.

---

[15] https://saifmohammad.com/WebPages/metaphor.html

4.2.7.2 Sentiment Analysis

Mao et al. (2022a) assessed MetaPro on a news headline sentiment analysis dataset (NHSA) from SemEval-2017 Task 5 (Cortis et al., 2017). The findings demonstrated that MetaPro could boost the performance of state-of-the-art sentiment analysis classifiers by an average of 4.0% F_1 scores. They also demonstrated that the sentiment classifier gained additional benefits in the non-metaphorical (paraphrased) dataset if the train and test sets were paraphrased first before training and evaluating.

4.2.7.3 Psychology and Cognition Analysis

Han et al. (2022a) believed that metaphorical concept mapping features could enhance the performance and the explainability of a depression detection task. They presented an explainable hierarchical attention network (HAN) that could extract depression-featured tweets and their corresponding concept mappings. Since the individual had implicitly delivered these concept mapping patterns in everyday social media contact, the generated concept mappings could provide insight into the inner reality of a depressed person. MetaPro was employed to generate the concept mapping features. The evaluation was conducted on a publicly available X dataset called MDL (Shen et al., 2017). Their experiments showed that introducing the concept mapping features could increase the accuracy of the depression detection task. The authors also demonstrated common concept mapping patterns among the depressed group.

Mao et al. (2023a) leveraged MetaPro to uncover the cognitive patterns of financial analysts under different market conditions. They examined MetaPro-generated concepts from 1.4 million financial reports spanning a decade, encompassing 1,000 analysts and 6,000 stocks in the US market. Analyzing these mappings revealed distinct patterns corresponding to various market conditions. In rising markets, reports often emphasized target concepts like POSSESSION, DIRECTION, MOTION, ACT, and ACTION through metaphorical expressions. However, these concepts were less prominent in turbulent markets. Conversely, during turbulent conditions, metaphors were frequently used to elucidate new target concepts such as COMMERCE, FORCE, SOCIAL_GROUP, BELIEF, ENVIRONMENT, CONSISTENCY, and ATTEMPT. Furthermore, there was a higher frequency of both optimistic and cautious concept mappings in comparison to down days. Conversely, on down days, report concept mappings tended to be more encouraging and persuasive.

With MetaPro, Mao et al. (2024d) also analyzed the public perception towards four types of weather disasters, e.g., floods, hurricanes, tornadoes, and wildfires. They parsed concept mappings from tweets. By statistical analysis, they found that disaster management is often likened to a strategic battle, focusing on mental preparedness, resource assessment, and adaptive planning. Disasters are viewed as dynamic and transformative, requiring urgent and intense responses. The public emphasizes proactive disaster response, including early warning systems and community resilience.

4.2.7.4 Political Analysis

Metaphor use in the text can reveal fine-grained sentiment and underlying cultural tendencies. Researchers in political analysis provided a perspective to deconstruct text by metaphors. Hu and Wang (2021) analyzed the differences and similarities in using conceptual metaphors in two government reports from China and the United States. The similarity between the two nations reflected their political systems' acceptance of commonsense. For instance, metaphors from the JOURNEY source domain conveyed that achieving the goals would take time. Diverse rooting cultures were mostly responsible for the variances. The USA political report, for instance, used metaphors from the THEATRE source domain to connect to American entertainment.

Mao et al. (2024e) also used MetaPro to analyze the relationship between metaphor cognition and voting behaviors at the United Nations. They first parsed metaphorical concept mappings from the debates at the United Nations Security Council from January 1995 to December 2020. The concept mappings were embedded by GloVe. The annual representations of the metaphorical cognitive pattern of a country were derived by averaging the concept mapping embeddings from the same country for a given year. Next, the pairwise consistency of voting between two countries at the United Nations General Assembly in a given year was measured using Cohen's Kappa (Cohen, 1960). The tests conducted on the United States, China, and Russia revealed moderate correlation coefficients between their metaphorical cognition and voting consistency. The authors concluded that metaphorical cognition can impact voting behaviors at the United Nations. They also provided a summary of the evolution of cognitive patterns over the 15-year analysis period.

4.2.8 Summary

Relevant theoretical research highlights the connection between metaphor understanding and human cognitive processes. CMT (Lakoff and Johnson, 1980) provided a practical framework for representing the concept mapping from a target domain to a source domain; Landau et al. (2010) further affirmed that metaphors play a role in shaping social cognition; Feldman (2008) suggested that embodied experiences influence the comprehension of abstract and cultural concepts, leading to varied usage of metaphorical language. Given the above theoretical findings, we argue that human perception of concepts is mutually affected by metaphors. External expressions of metaphor can influence how individuals structure the mapping of concepts and shape their social cognition. Concurrently, personal experiences are reflected in diverse metaphorical language usage. Thus, metaphors emerge as a crucial medium for studying both the personal cognition of individuals and the social cognition of broader populations.

4.2.8.1 Technical Trends

Table 4.4 summarizes the major technical trends by task setups that were introduced in Section 4.2.6. The property exaction systems normally use unsupervised and graph learning methods with linguistic features, such as POS tags, knowledge bases, and word embeddings. However, methods for property exaction are only effective when interpreting nominal metaphors. Generated properties can succinctly describe the link between source and target nouns, while property-based approaches barely stretch to other POS of metaphors, such as verbs, adjectives, and adverbs. Given the complexity of syntax in real-world text, the interpretation outputs defined as "target be property" are not valuable for assisting downstream operations.

Word-level paraphrasing-based approaches can be utilized as a text preprocessing tool to enhance the capacity to grasp semantic concepts for various downstream NLP activities such as machine translation and sentiment analysis. Given the absence of large annotated datasets, previous paraphrasing-based metaphor understanding systems commonly learn the task by word co-occurrence modeling. To improve the accuracy, they also leverage linguistic features, such as POS tags and WordNet knowledge base.

Table 4.4: Technical trends in metaphor understanding. PE denotes property extraction. WP denotes word-level paraphrasing. "Unsup." is for unsupervised learning. DL denotes deep learning. KB denotes knowledge base. GE denotes graph embedding. "co-occ" is for co-occurrence. MI denotes mutual information. Cos denotes cosine similarity. SPV denotes select preference violation (Wilks, 1975). KGE denotes knowledge graph embedding. MR denotes mean rank. LM denotes language modeling. TSG denotes the dataset is a collection of Tsvetkov et al. (2014); Shutova et al. (2016); Gutierrez et al. (2016). MT denotes the task was evaluated on English-Chinese machine translation, based on Bing Translator. "Acc." is for accuracy. PC denotes Pearson correlation. "$\mathcal{A}$" is for acceptability. "$\mathcal{P}$" denotes appropriateness. "C" is for coherence.

Task	Reference	Tech	Feature	Framework	Dataset	Score	Metric
	Su et al. (2015)	Unsup.	POS, KB, word2vec	Cos	Self-collected	0.84	Acc.-$\mathcal{A}$
	Su et al. (2017)	Unsup.	POS, KB, word2vec	Cos	Su et al. (2015)	0.86	Acc.-$\mathcal{A}$
PE	Rai et al. (2019)	Unsup.	POS, KB	Cos	Rai and Chakraverty (2017)	0.97	Acc.-$\mathcal{P}$
	Su et al. (2020)	Graph	POS,WordNet,word2vec,concept mapping	Random walk	Self-collected	0.77	Acc.-$\mathcal{A}$
	Song et al. (2020b)	Graph	POS, KB, GE	KGE	Self-collected	11.6	Hit@1
	Shutova (2010)	Unsup.	POS, Co-occ,WordNet	SPV	Sub-BNC	0.81	Acc.
WP	Mao et al. (2018)	Unsup.	POS, WordNet, word2vec	LM	Mohammad et al. (2016)	0.65	MT-$\mathcal{A}$
	Mao et al. (2022a)	DL	POS, WordNet, RoBERTa	LM	Sub-VUA	4.48	Acc.-C
	Mao et al. (2024b)	DL	POS, WordNet, ALM	LM	Mao et al. (2024b)	0.40	F_1
	Martin (1990)	Rule	KODIAK	Logic rules	Self-collected	-	-
EP	Bizzoni and Lappin (2018)	DL	word2vec	CNN-LSTM	Bizzoni and Lappin (2018)	0.67	Acc.
	Mao et al. (2022a)	Rule	POS, Dict.	Depd. Rules	Saxena and Paul (2020)	0.99	Acc.
	Mason (2004)	Unsup.	POS,WordNet	SPV, Cluster	Lakoff (1994)	-	-
	Gandy et al. (2013)	Rule	POS, Abst., Dict., MI, WordNet	Logic rules	Self-collected	0.65	Acc.-$\mathcal{A}$
	Strzalkowski et al. (2013)	Rule	POS, MI, Imag., Affect, Co-occ,WordNet	Cluster	Self-collected	0.71	Acc.
	Li et al. (2013)	Unsup.,Rule	POS, Lemma, Co-occ, WordNet, Dict.	SPV	Self-collected	0.43	MatchTop1
CMP	Dodge et al. (2015)	Rule	POS, WordNet, FrameNet, Lemma, Dict.	Logic rules	Self-collected	-	-
	Rosen (2018)	DL	POS	DNN	LCC	0.8	Acc.
	Gagliano et al. (2016)	Unsup.	POS, Lemma, WordNet, word2vec	Cos	Self-collected	-	-
	Fu et al. (2020a)	Graph	POS, CN, Co-occ, word2vec, GE	Cos,deep walk	Self-collected	0.7	Hit@1
	Ge et al. (2022)	DL	POS, Lemma, Co-occ, WordNet, RoBERTa	MTL	TSG	0.64	Acc.-$\mathcal{A}$

Table 4.5: Metaphor understanding downstream task applications.

Reference	Downstream Task	Feature Parser	Explainability	Employed technique
Mao et al. (2018)	Machine Translation	✓		Mao et al. (2018)
Mao et al. (2022a)	Sentiment Analysis	✓		Mao et al. (2022a)
Han et al. (2022a)	Psychology Analysis	✓	✓	Mao et al. (2023b)
Mao et al. (2023a)	Cognition Analysis	✓	✓	Mao et al. (2023b)
Mao et al. (2024d)	Cognition Analysis	✓	✓	Mao et al. (2023b)
Mao et al. (2024e)	Cognition Analysis	✓	✓	Mao et al. (2023b)
Hu and Wang (2021)	Political Analysis		✓	Statistics

Word-level paraphrasing techniques cannot capture the nuanced distinctions between metaphors and their literal counterparts. This deficiency leads to the loss of sentimental and emotional information after paraphrasing. With a newly developed metaphor paraphrasing dataset (vmc-p) and a metaphor-tailored ALM (Mao et al., 2024b), the limitations, e.g., end-to-end training and MWE paraphrasing have been largely improved. Explanation pairing generates a clause or full sentence to understand metaphors. Current methodologies exhibit fewer constraints compared to other task formats, rendering them effective for various types of metaphors, e.g., MWEs. For example, Mao et al. (2022a) developed a dictionary and rule-based methods for processing MWEs in diverse syntactic contexts. However, such an explanation pairing method needs a very large knowledge base, leading to difficulty in generalization.

Conceptual metaphor processing aims to generate concept mappings to comprehend metaphorical expressions from a cognitive perspective. Multiple metaphorical expressions with similar source and target domains can be connected with concept mappings. Concept mappings reveal how metaphors are created and what metaphors are compared to. Feature-based methods utilized hand-crafted linguistic features, such as syntactic information and WordNet hypernyms, to capture semantic representations in unsupervised patterns. Embeddings-based methods applied word embedding techniques including PLMs to generate concept mappings with the assistance of deep learning models. Concept mappings can be expressed as clauses or additional features and utilized for understanding human cognition for metaphors and other psycho-linguistic research. However, current approaches cannot process the concept mapping tasks for MWEs.

4.2.8.2 Application Trends

Since metaphors are frequently shown in human communications, metaphor understanding has close associations with downstream applications and other psycholinguistic research. The application trends are summarized in Table 4.5. Metaphor understanding can be a key upstream subtask for many NLU tasks. For example, Mao et al. (2018, 2022a) demonstrate the effectiveness of metaphor understanding for sentiment analysis and machine translation.

Metaphor understanding models can also be used as parsers for several NLU applications, including mental health (Han et al., 2022a), finance (Mao et al., 2023a), international relationship (Mao et al., 2024e), and weather disasters (Mao et al., 2024d). These studies not only highlight the relevance of metaphor research but also offer practical methodologies, empirically substantiating the utility of concept mapping in supporting research within psychology and cognition. Finally, Hu and Wang (2021) examined concept mappings of metaphorical expressions in government reports, revealing similarities and differences in political attitudes. This analysis underscores the critical role of metaphor comprehension in effective communication. In conclusion, examining cognition through metaphors provides a means to conduct extensive research with large samples, as this method does not require interviews to uncover cognitive patterns. This noninvasive approach, such as studying cognition using data mining techniques, can yield more reliable results by mitigating the risk of subjects not sharing their genuine thoughts during psychological tests.

4.2.8.3 Future Works

Metaphor understanding faces several challenges in becoming a more developed and standard NLU task. The main challenge is creating large annotated datasets for various metaphor understanding tasks, such as property extraction, paraphrasing, explanation pairing, and concept mapping. Besides the time and effort required for annotation, achieving consensus on consistent and comparable labels across different tasks is equally important. Current metaphor understanding studies exhibit a variety of task descriptions and dataset formats, yet the community has not yet established a unanimous agreement on task and dataset forms. This lack of consensus hinders the comparison of proposed methods across different contexts. It is envisioned that in the future, one or more standard task definitions will be established, incorporating theoretical insights to address these challenges.

Another challenge is how to appropriately measure the performance of proposed methods in different metaphor understanding setups. Whether the generated interpretations or concept mappings deliver the original meanings of metaphorical expressions is highly subjective. Researchers utilized multiple linguistic features, such as coherence and metaphoricity in human evaluation, to measure the outputs from different perspectives. However, these linguistic features have yet to be widely agreed upon in the community. Meanwhile, comparing with external baselines by human evaluation is not fair enough considering subjectivity.

Finally, the current integration of linguistic knowledge lags behind machine learning techniques. Since metaphors involve linguistic rules, finding a way to seamlessly combine linguistic knowledge with machine learning could be a future direction for the community. Furthermore, metaphor understanding can enhance downstream NLU tasks. While there has been thoughtful use of metaphors in academia, practical applications, like plug-in modules for preprocessing or feature extraction, are still lacking. We believe that this sort of solution has the potential to significantly impact the NLU community.

4.3 Sarcasm Detection

In psycholinguistics, sarcasm is defined as "a form of verbal irony used in a hurtful or critical way" (McDonald and Pearce, 1996; Colston, 2000). The affiliation of sarcastic language with pragmatics is evident as sarcasm inherently hinges on contextual factors, speaker intention, and mutual understanding between communicators. Functioning as a type of verbal irony, sarcasm entails expressing words with an intended meaning opposite to the literal interpretation, frequently leveraging contextual cues, intonation, and the collective background knowledge shared among communicators. In our daily lives, understanding sarcasm helps to avoid miscommunication. Sarcasm detection is crucial to the understanding of humans' sentiments and emotions (Eke et al., 2020; Maynard and Greenwood, 2014).

Sarcasm detection in computer science refers to the task of automatically determining whether or not text is bearing sarcasm using computational methods. It is typically formulated as a binary classification problem. Input can be uni- and multimodal. For uni-modal sarcasm detection, textual data is usually employed. For the multimodal task, existing works leveraged combinations of textual, visual and acoustic modalities. Given input data and its representations, a computational model discerns the label associated with each input, such as determining whether it is sarcastic or non-sarcastic.

4.3.1 Theoretical Research

4.3.1.1 Mental Clue

McDonald and Pearce (1996) explore psycholinguistics theories of sarcasm via experiments conducted with brain-damaged subjects. This empirical study reports that successful detection and interpretation of sarcasm is a multifactorial, interactive process. To understand sarcasm, the comprehension of the literal meaning of counterfactual remarks and rejecting it are necessary. However, such processes are not sufficient to fully understand sarcasm. Rather, other inferential reasoning processes, such as contextual inference of a speaker's attitude and the understanding of emotional information implied in the speaker's tone, are required to understand the conceptual relevance of counterfactual remarks in sarcastic utterances.

A more recent study (Gibbs Jr, 2002) lends a new perspective on the role of literal meanings in understanding indirect, figurative language such as sarcasm. This study argues that a complete comprehension of literal meanings and the rejection such meaning may not necessarily be performed either before or in parallel to the interpretation of figurative, implicit meanings of sarcasm given that there exist sufficient, appropriate pragmatic factors such as background knowledge and context. Various contextual factors, which may be useful for understanding a speaker's actual intention, include mutual beliefs, attitudes, and knowledge shared between the speaker and the listener.

4.3.1.2 Communicative Clue

Kreuz and Glucksberg (1989) focus on the communicative function of sarcasm. In particular, the study proposes the "echoic reminder theory" of sarcastic irony in which a speaker says counterfactual (i.e., not literally true) remarks (i.e., reminders) to convey disapproving attitude towards a target. In this theory, a listener recognizes *positive* remarks conveying sarcasm by perceiving implicit or explicit norms or expectations that are shared with a speaker. For example, a listener can interpret the sarcastic intention of a remark, "you're a nice friend", said when they failed to be helpful to the speaker. On the other hand, *negative* remarks require explicit context to be correctly interpreted as sarcasm. For example, a sentence, "you're an awful friend", cannot be used sarcastically when the listener was being helpful to the speaker. The same sentence can be interpreted as sarcasm or gentle mockery when the listener was being very helpful to the speaker in contrast to the speaker's low or negative expectation for the listener. In this case, the remark reminds the listener of the speaker's expectation.

Katz and Lee (1993) support the findings of Kreuz and Glucksberg (1989) in that sarcasm is perceived when an echoic reminder is used to ascertain a speaker's intention to reject the figurative truth-value of an utterance. Specifically, when the literal meaning of a remark is incompatible with the perceived context, the listener would infer that the speaker is rejecting a figurative resemblance between the utterance and the actual state of the ongoing event, and hence is speaking sarcastically. Moreover, the authors argue that contextual information (e.g., mutual knowledge) is required to interpret sarcasm.

The empirical findings of such psycholinguistics studies suggest that there are different types of sarcasm. In general, it is challenging to interpret sarcastic statements that convey a reminder of antecedent affairs or the speaker's expectations in comparison to those explicitly expressing an ironic situation. Those studies also argue that contextual information and the identification of incongruity between the meaning of a statement and observational evidence are crucial to the understanding and interpretation of sarcasm.

4.3.1.3 Neural Clue

Matsui et al. (2016) utilized functional magnetic resonance imaging (fMRI) to explore the neural mechanisms associated with sarcasm comprehension. Examining 21 participants, the researchers observed brain activity while subjects listened to both sarcastic and sincere comments within varied contexts. Analysis of the fMRI data focused on identifying regions exhibiting increased activity during the processing of sarcastic remarks compared to sincere ones. Specifically, attention was directed towards the left inferior frontal gyrus (IFG), a region recognized for its involvement in processing linguistic input incongruent with accompanying affective prosody. The hypothesis posited that the left IFG would exhibit heightened activity when affective prosody conflicted with a discourse context conducive to sarcasm perception.

The investigation pinpointed the left rostro-ventral portion of the IFG, specifically Brodmann's Area 47, as crucial in integrating discourse context, utterance, and affective prosody for sarcasm comprehension. This area demonstrated a significant interaction effect, underscoring its role in context-dependent perception of the utterance. Then, the researchers observed functional asymmetry in incongruity processing, with the left IFG implicated in integrating statement, context, and prosody, while the right IFG responded to negative prosody, indicating incongruity detection between the statement and prosody. Aligning with behavioral findings, the fMRI data affirmed that incongruity between discourse context and overall utterance meaning influenced sarcasm perception. Positive prosody with positive semantic content elevated the overall positive valence of utterance meaning, while negative prosody combined with positive semantic content diminished the positive valence, intensifying incongruity and enhancing sarcasm perception. These results elucidate the neural substrates involved in integrating affective prosody, semantic content, and discourse context during sarcasm comprehension, with a particular emphasis on the left IFG as a pivotal region in this cognitive process.

4.3.2 Annotation Schemes

There are two main approaches for annotating datasets for sarcasm detection: manual and weakly supervised annotation. In both approaches, the task is framed as a binary classification task, specifically, discerning whether a given target is sarcastic or not. However, sarcastic statements can be further annotated via subtypes, e.g., sarcasm, irony, satire, understatement, overstatement, and rhetorical question. Sarcasm is often hard to discern even by humans and is subjective. Due to such inherent characteristics, leveraging human supervision is a natural choice for high-quality sarcasm labeling. Normally, human annotators are not authors of utterances, which means that manually annotated datasets contain sarcasm perceived by annotators which are not necessarily the same as that intended by authors (Oprea and Magdy, 2020). Inter-annotator agreement (Artstein and Poesio, 2008) is often measured to ensure the consistency of data labeling.

Despite the higher quality of manually curated datasets, the manual labeling of large-scale, noisy data for sarcasm detection is highly labor-intensive and time-consuming in comparison to some annotation tasks which can be done without human supervision such as image annotation (Wallace et al., 2014). Furthermore, annotating sarcasm is challenging because it requires a thorough understanding of context and cultural background as discussed in Section 4.3.1. To overcome such limitations of manual annotation, weakly supervised annotation is used to increase efficiency and improve the usability of massive amounts of unlabelled data. Most corpora for sarcasm detection are generated using social media data based on the hypothesis that texts containing certain indicators express sarcasm.

X[16] and Reddit[17] are common social media platforms for studying sarcasm. For X, several hashtags, such as #sarcasm, #irony, #sarcastic, #cynicism and #not, are leveraged as markers of sarcasm (Kunneman et al., 2015). For Reddit, the marker "/s" at the end of posts is used to collect sarcastic statements (Khodak et al., 2018). However, gathering data by hashtags can inevitably introduce biases, such as those more subtle, sarcastic expressions without obvious hints will not be annotated due to the absence of hashtags.

4.3.3 Datasets

Commonly used sarcasm detection benchmarking datasets are shown in Table 4.6. Oprea and Magdy (2020) proposed a sarcasm detection dataset, iSarcasm[18], which consists of 777 sarcastic and 3,707 non-sarcastic English tweets. To collect intended sarcasm rather than perceived sarcasm, a subset of X users was asked to provide one sarcastic and three non-sarcastic tweets. For each sarcastic tweet, an explanation of why it is sarcastic and a non-sarcastic paraphrase were also collected. Moreover, a linguist was employed to further categorize each sarcastic tweet into one of the following seven categories: 1) sarcasm, 2) irony, 3) satire, 4) understatement, 5) overstatement, 6) rhetorical question, and 7) invalid/unclear.

Table 4.6: Sarcasm detection datasets and statistics.

Reference	Task	Source	# Samples
Oprea and Magdy (2020)	Multi-class	X	4,484
Khodak et al. (2018)	Binary	Reddit	533M
Van Hee et al. (2018)	Binary/Multi-class	X	3,000
Castro et al. (2019)	Binary/Multimodal	TV shows	690
Joshi et al. (2016)	Binary	TV show	17,338
Oraby et al. (2016)	Binary	Online forums	17,338
Cai et al. (2019)	Binary/Multimodal	X	24,635

```
Category: sarcasm

Text: Thank @user for being so entertaining at the Edinburgh signings! You did not
disappoint! I made my flight so will have plenty time to read @user

Explanation: I went to a book signing and the author berated me for saying I was
lying about heading to Singapore straight after the signing

Rephrased: I would have said 'here is the proof of my travel, I am mad you
embarrassed me in front of a large audience'!
```

---

[16] https://x.com

[17] https://reddit.com

[18] https://github.com/silviu-oprea/iSarcasm

```
Category: irony
Text: Staring at the contents of your fridge but never deciding what to eat is a
cool way to diet
Explanation: I wasn't actually talking about a real diet. I was making fun of how you
never eat anything just staring at the contents of your fridge full of indecision.
Rephrased: I'm always staring at the contents of my fridge and then walking away
with nothing cause I can never decide.
```

The Self-Annotated Reddit Corpus[19] (SARC) (Khodak et al., 2018) is a large-scale, self-annotated Reddit corpus comprising 533 million comments, 1.3 million of which are sarcastic. Initially, comments with the marker "/s" at the end are collected and several noise filtering approaches are applied to the initial collection. Despite the attempts to remove noise in the dataset, this corpus contains several false positive (i.e., non-sarcastic statements labeled as sarcastic due to the presence of the /s marker) and false negatives (i.e., sarcastic statements labeled as non-sarcastic due to the absence of the /s marker).

```
Label: non-sarcasm
Text: You do know west teams play against west teams more than east teams right?
Author: Shbshb906
Subreddit: nba
Comment score: -4


Label: sarcasm
Text: gotta love the teachers who give exams on the day after halloween
Author: DEP61
Subreddit: CFBOffTopic
Comment score: 3
```

SemEval-2018 Task 3 dataset (Van Hee et al., 2018) was constructed using both weakly supervised and manual annotation schemes. After collecting English tweets using several hashtags related to irony (i.e., #irony, #sarcasm, and #not), the corpus was further manually labeled by three human annotators. As a result, the dataset consists of 1,728 verbal irony with polarity contrast, 267 other types of verbal irony, 401 situational irony, and 604 non-ironic tweets.

```
Label: non-sarcasm
Text: Had no sleep and have got school now


Label: sarcasm
Text: I just love when you test my patience!!
```

---

[19] https://nlp.cs.princeton.edu/SARC/2.0

MUStARD[20] (Castro et al., 2019) was manually annotated, including three modal data (i.e., textual, audio, and visual features) extracted from TV shows. This dataset consists of 345 sarcastic and 345 non-sarcastic utterances. Each utterance consists of one or more sentences and its conversational context is also provided.

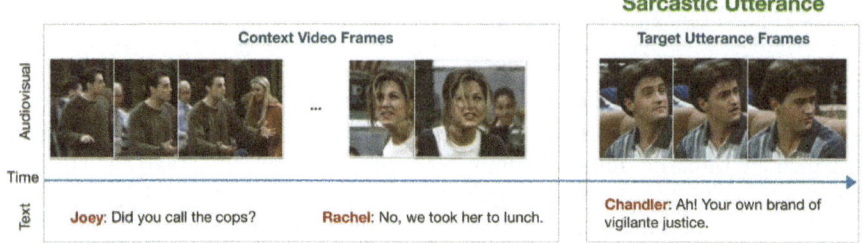

Fig. 4.3: A data example from the MUStARD multimodal sarcasm detection dataset, developed by Castro et al. (2019).

Friends dataset (Joshi et al., 2016) consists of transcripts of a TV show, *Friends*, manually annotated by two experienced linguists. Specifically, it consists of 913 sets multi-turn conversations. The average number of utterances per set is 18.6. 1,888 out of 17,338 utterances were labeled as sarcastic.

Sarcasm Corpus V2[21] (Oraby et al., 2016) was built on online debate forum conversations and consists of three types of sarcasm: 1) general sarcasm (GEN); 2) hyperbole (HYP); 3) rhetorical questions (RQ). Each type contains sarcastic and non-sarcastic posts. The GEN, HYP, RQ sets consist of 3,260, 582, and 851 posts per class. This dataset was annotated via AMT.

```
Label: non-sarcasm
Id: 11
Text: So will they be in front of or behind a leaf?


Label: sarcasm
Id: 2662
Text: That would be one hell of a kiss...you would probably drown first though :)
```

Multimodal sarcasm detection dataset[22] (Cai et al., 2019) includes sarcastic tweets that were collected using a set of hashtags. Non-sarcastic tweets in the dataset do not contain such hashtags. Each tweet has its associated image. After applying filtering rules, 10,560 sarcastic and 14,075 non-sarcastic tweets remain in the final dataset.

---

[20] https://github.com/soujanyaporia/MUStARD

[21] https://nlds.soe.ucsc.edu/sarcasm2

[22] https://github.com/headacheboy/data-of-multimodal-sarcasm-detection

Image:

Text: What a beautiful couple   What wonderful weather!
        #goldenglobes

Label: non-sarcasm   sarcasm

Fig. 4.4: A data example from the multimodal sarcasm detection dataset, developed by Cai et al. (2019).

4.3.4 Evaluation Metrics

As the task of sarcasm detection is normally formulated as a binary or multi-class classification problem, the most common evaluation metrics are accuracy, precision, recall, and F_1 score. Ghosh et al. (2017); Khodak et al. (2018) also exploit crowdsourcing to evaluate sarcasm detection results. Human evaluators are given a subset of prediction results along with context and evaluate them. Then, a majority vote is applied to obtain the final evaluation results. AMT is one of the most popular crowdsourcing methods used for the task.

4.3.5 Methods

Based on a thorough search of the literature on sarcasm detection, we introduce four strands of methods for the task: 1) feature-based methods use hand-crafted features; 2) embedding-based methods utilize (contextualized) embeddings; 3) semantic incongruity-based methods leverage contrast between the meaning of an utterance and contextual evidence or sentimental incongruity; 4) context-based methods employ conversational context.

4.3.5.1 Feature-based Methods

Conventional methods for sarcasm detection aim to distinguish sarcastic utterances from non-sarcastic ones based on manually curated features. Several studies emphasize the importance of extracting and utilizing linguistic features to accurately interpret sarcasm, recognizing that comprehending the meaning of remarks, whether literal or figurative, is pivotal for this purpose.

Potamias et al. (2019) proposed an ensemble model, termed Deep Ensemble Soft Classifier (DESC), which combined BiLSTM, LSTM with attention, and dense neural networks for tweet-level irony and sarcasm detection and sentiment analysis of figurative languages (i.e., irony, sarcasm, and metaphor). The authors argued that understanding syntactic, semantic, and emotional differences between figurative and literal languages can be helpful for identifying individuals' intentions. Hence, 44 hand-crafted features categorized into five major groups (i.e., syntactic, demonstrative, sentiment, mood, and readability) were integrated into DESC. Specifically, given an input tweet, these features were extracted and fed into dense neural networks. BiLSTM and LSTM, incorporating attention mechanisms, are employed in this study with the tweet's GloVe (Pennington et al., 2014) embedding serving as input. However, this work lacks an ablation study that specifically assesses the individual contributions of hand-crafted features and the attention mechanism. Consequently, the precise impact of each component on the classification performance remains unclear.

Pandey et al. (2021) proposed a hybrid attention-based LSTM (HA-LSTM) network. The model learned 16 hand-crafted linguistic features extracted from textual contents for sentence-level sarcasm detection, namely entropy, lex diversity, number of nouns, number of adjectives, number of verbs, sum of verbs available in the given sentence, the Flesch Reading Ease formula, Dale Chall readability, difficult words, length, set length, stopword count, wrong word, one-letter word, two-letter word, longer letter word, $< li >$ tags (hyperlinks in the text). The authors argued that combining the advantages of manually curated features with those of deep learning techniques will improve sentence-level sarcasm detection performance. Such an approach has been employed by several works in different social media research areas (Gao et al., 2020; Mou and Lee, 2020; Liu et al., 2022b). The model consists of two parts. One module processes a single sentence by utilizing GloVe embeddings as input for its analysis. These word vectors were fed into two LSTM layers followed by an attention layer. The output of the attention layer was concatenated with a hand-crafted feature matrix. The join representation of the input sentence was fed into a softmax layer which outputs a binary label (i.e., sarcastic or non-sarcastic). The authors reported that integrating hand-crafted features into a deep learning model improves the performance of the model in sarcasm detection. Similarly to (Potamias et al., 2019), this work also did not computationally evaluate how much the integration of hand-crafted features into a deep learning architecture can contribute to its performance.

Hand-crafted features have been widely used in various tasks in NLP and social media research. However, the effectiveness of solely relying on hand-crafted features is limited. Some studies in different computer science research areas (Gao et al., 2020; Wu et al., 2018; Bogacsovics et al., 2022) reported that integrating such features can make complementary contributions to the effectiveness of machine learning models when they are used with features automatically learned via deep learning techniques. There have been few attempts to apply such a finding to sarcasm detection, but most of them failed to conduct appropriate experiments to prove the hypothesis.

4.3.5.2 Embedding-based Methods

Approaches reliant on hand-crafted features often necessitate domain-specific knowledge, which may not always be accessible, and involve labor-intensive and time-consuming efforts in feature engineering. To overcome these limitations, several works leveraged recent advances in deep learning techniques such as embeddings obtained using PLMs. With advances in deep learning (e.g., Transformers) and language models, embedding-based methods have been widely used in recent research on sarcasm detection. Shrivastava and Kumar (2021) presented a brief overview of methods and datasets for sarcasm detection and proposed a BERT-based model for the task. A preprocessed input sentence was further processed by a BERT tokenizer. The tokenized vector was fed into a pretrained BERT followed by a feed-forward layer. The proposed model achieved an F_1 score of 0.705 on the SemEval-2018 Task 3 (Van Hee et al., 2018) dataset. However, the comparative assessment was conducted with a single benchmark dataset.

Potamias et al. (2020) discussed the limitations of previous studies on sarcasm detection. The authors emphasized the importance of using multiple benchmark datasets for fair comparative evaluation. To address such limitations, this work proposed a sentence-level, binary sarcasm detection model, based on BiLSTM and RoBERTa pretrained embeddings. Given an input sentence, RoBERTa learned its contextualized embedding and its last hidden state was fed into a BiLSTM layer. The output of LSTM and the learned embedding were concatenated. The final representation of the input was fed into a feed-forward network followed by a softmax layer, which output a binary label. The proposed model achieved F_1 scores of 0.80, 0.78, and 0.90 for the SARC (Khodak et al., 2018), SemEval-2018 Task 3 (Van Hee et al., 2018), and Riloff (Riloff et al., 2013) datasets, respectively, outperforming the state-of-the-art models for sentence-level sarcasm detection.

Like Potamias et al. (2020), Babanejad et al. (2020) argued that the use of hand-crafted features and traditional, non-contextualized word embeddings (e.g., word2vec) are sub-optimal. The authors hypothesized that understanding context and sentiments are useful for sarcasm detection. This work proposes two BERT-based models (i.e., ACE1 and ACE2), utilizing affective and contextual embeddings for sentence-level, binary sarcasm detection. A BiLSTM with multi-head attention and a BERT were used to obtain affective and contextual embeddings of an input sentence, respectively. In ACE1, another BERT took input as the concatenation of two features and output a task-specific, join representation of the input. In ACE2, two embeddings were simply concatenated. In both models, the final representation of the input was fed into a fully connected layer and a softmax layer. For evaluation, five benchmark datasets were used and several state-of-the-art models were compared with the proposed models. Experimental results showed that ACE1 outperforms and ACE2 and the baselines by achieving F_1 scores of 0.89 and 0.85 on the SARC (Khodak et al., 2018) and SemEval-2018 Task 3 (Van Hee et al., 2018) datasets, respectively. Such results indicate that fusing affective features with sentence embeddings is more effective that a simple concatenation for sarcasm detection.

Yao et al. (2021) aimed to simulate how humans process and interpret sarcasm given multimodal information. To this end, they proposed a multimodal, binary sarcasm detection model for X data. It incorporated multiple stacks of gate mechanisms, guide attention, attention pooling, and multi-hop processing of guide attention and attention pooling, in the hope that these components could effectively learn interactions among tweet text, an embedded image, text in the image, and image caption. Multimodal sarcasm detection dataset (Cai et al., 2019) was used for evaluation. The proposed model is compared with single- and multimodal models for tweet-level sarcasm detection. This work hypothesized that image captions help a sarcasm detection model learn context and incongruity. The image caption model (Xu et al., 2015c) utilized in the study was not deemed state of the art, and the rationale behind not opting for a more contemporary model was not substantiated in the research. In fact, the ablation study results showed that the effect of the image caption modality is marginal.

Liang et al. (2022a) argued that not every part of an image equally contributes to the performance of a multimodal, binary sarcasm detection model. The authors also believed that learning affective relationship between key visual and textual information is useful. Based on such hypotheses, a model based on GCNs and attention mechanism was proposed for multimodal sentence-level sarcasm detection. Given a sentence and an image paired with it, the textual and visual representations of the inputs were obtained using a pretrained BERT (Devlin et al., 2018) and Vision Transformer (ViT) (Dosovitskiy et al., 2021). A cross-modal graph in which nodes were the obtained multimodal representation. Edges were affective similarity weights calculated using SenticNet. Then, the adjacency matrix of the cross-modal graph was fed into multi-layer GCNs. To capture attention information, the simple concatenation of the input representations was passed through an attention layer. Computed attention weights were combined with the output of the final layer of the GCNs to form the final representation of the multimodal input, which was fed into a softmax layer.

Kamal and Abulaish (2022) focused on the binary classification of sarcasm associated with self-deprecation, based on a hypothesis that comprehending such sarcasm can be useful for improving marketing strategies. This work assumed that sarcastic tweets in which authors refer to themselves express self-deprecation. In addition to 6 benchmark datasets for tweet-level sarcasm detection, this work constructed its own corpus using hashtags. Specifically, sarcastic tweets were collected using "#sarcasm" and non-sarcastic tweets were collected using "#not", "#education", "#politics", "#love", and "#hate". To select tweets containing self-reference, filtering methods based on a pre-defined set of regular expressions and clustering were applied to the 7 datasets. In the proposed model, the GloVe embedding of an input tweet passed through a convolution layer, BiGRU and a sigmoid layer. Experimental results showed that their model outperforms several deep learning models and two models developed by existing works on sarcasm detection in terms of accuracy, precision, recall and F_1 score.

Cai et al. (2019) introduced a multimodal sarcasm detection model encompassing text, images, and image attributes, represented by 5 keywords. Raw and guidance embeddings were generated for each modality, and guided weights were computed using guidance embeddings. Modality-specific feature vectors were obtained through a weighted sum of raw vectors, and attention mechanisms facilitated the fusion of different modality representations. Textual information was processed using BiLSTM, while image attributes were predicted with ResNet-50 V2 (He et al., 2016), and a fine-tuned ResNet was employed for attributes. The proposed model demonstrated an F_1 score of 0.8018 on their proprietary dataset.

Finally, Liang et al. (2024) investigated multimodal sarcasm detection from a novel perspective, where a multimodal graph contrastive learning strategy is proposed to fuse and distinguish the sarcastic clues for textual modality and visual modality. Specifically, they first utilized object detection to derive the crucial visual regions accompanied by their captions of the images, which allows better learning of the key visual regions of visual modality. In addition, to make full use of the semantic information of the visual modality, they employed OCR to extract the textual content in the images. Then, based on image regions, the textual content of visual modality, and the context of the textual modality, they built a multimodal graph for each sample to model the intricate sarcastic relations between modalities.

4.3.5.3 Semantic Incongruity-based Methods

As discussed in Section 4.3.1, several theoretical studies have demonstrated that identifying contradictions between the meaning of a statement and observational evidence plays an important role in understanding and interpreting sarcasm. Some works have been motivated by such a finding and developed methods that identify sarcasm by modeling (semantic) incongruity. Ren et al. (2020) argued that sarcastic statements involve the incongruity of sentiment polarity within a sentence and contrast between the conveyed sentiment and the situation (context) of the statement. To integrate such observations, this work proposed a two-level memory network in which the first-level network obtained sentiment semantics of an input sentence using SenticNet and the second-level one learned incongruity between the sentiment semantics and the constituent words of the input by computing their intra-attention. The ablation study on the $\mathbb{X}$ corpus proved their hypothesis that sentiment semantics and memory networks are useful for sarcasm detection. Specifically, removing the first-level network and both memory networks reduced the model's performance by 1.05% and 1.84%, respectively.

Pan et al. (2020) argued that modeling incongruity between or within modalities helps sarcasm detection. The proposed multimodal architecture learned images, tweet textual content, and hashtags in tweets. Self-attention mechanism was employed to learn incongruity between images and tweet text. Incongruity between tweet text and hashtags was learned using affinity matrices which learn similarities between data points. The two incongruity representations were concatenated and fed into a softmax layer.

A multimodal sarcasm detection dataset (Cai et al., 2019) was used in the experiments and the proposed model achieves an F_1 score of 0.8292. An ablation study showed that incorporating inter- and intra-modality incongruity learning modules is useful for sarcasm detection. However, tweets expressing sarcasm may lack hashtags altogether, or the presence of hashtags does not guarantee a direct association with sarcasm. A more thorough analysis of the dataset is required to argue that the model can actually capture 'incongruity' between tweet textual content and hashtag.

Wu et al. (2021c) were motivated by a hypothesis that sarcasm involves incongruity between a positive word in text and either a negative facial expression or tone. The authors proposed a multimodal architecture that jointly learns such incongruities at word level and utterance-level features. It achieved F_1 scores of 0.745 and 0.7 on speaker-dependent and speaker-independent sets in the MuStARD (Castro et al., 2019) dataset, respectively. An ablation study showed the effectiveness of word-level incongruity learning in sarcasm detection. The authors believed that sarcasm does not involve incongruity between visual and acoustic modalities and that between a negative word and positive context by definition. Different researchers have given different definitions of sarcasm and its true meaning is in dispute. Additionally, sarcasm exists in a wide range of forms and structures (Eke et al., 2020).

Chen et al. (2022) argued that sarcasm involves exaggeration and incongruity. They propose a model consisting of two modules for learning these characteristics of sarcasm. For learning exaggeration, LSTM with a self-attention mechanism was applied to an input statement and output a sentiment-aware embedding. For incongruity, an input sentence was segmented into sequential context chunks, and then, the average of the pretrained embeddings of such chunks was fed into BiLSTM. Finally, the concatenation of forward and backward hidden states was obtained. The output vectors of the two modules were concatenated into a single vector, which was fed into a softmax layer predicting the label of the input. The ablation study shows that removing exaggeration and incongruity modules decreases F_1 scores by 2.31% and 2.71%, respectively. Although the authors hypothesized that their modules learn exaggeration and incongruity in sarcastic statements, it is not clear how each module is suitable for capturing a specific linguistic characteristic of sarcasm, and thus, their method and experimental results do not fully support their hypothesis.

Yue et al. (2023) proposed a cognition-inspired model, KnowleNet. They argued that achieving effective multimodal sarcasm detection relies on prior knowledge and recognizing the semantic contrast between modalities. The identification of semantic disconnection among modalities necessitates the incorporation of commonsense information, particularly given that implicit expressions of emotion often demand a nuanced conceptual understanding of words within diverse contexts. Leveraging conceptual knowledge from ConceptNet, the authors introduced a novel method for cross-modal semantic similarity detection, aiming to assess the interrelatedness between visual and textual information. To enhance feature differentiation between positive and negative samples, contrastive learning was employed, and the visualization results provided supporting evidence for the efficacy of this contrastive learning approach.

4.3.5.4 Context-based Methods

In the case of online sarcasm detection, contextual information typically refers to information obtained from conversations involving at least two interlocutors. Social media messages, particularly tweets, are short in length and contain limited context. Conversational threads can provide a computational model with contextual factors, such as a speaker's attitude and intention; emotional information implied in the speaker's tone; knowledge and beliefs shared between the speaker and a listener (Li et al., 2023b), which is also useful for identifying sarcasm. Consequently, exploiting conversational context can help sarcasm detection models better understand characteristics of sarcasm.

Kumar et al. (2021) argued that comprehending situational context and contextual incongruity in conversations is useful for identifying sarcasm in real-world scenarios. This work proposed an explainable model for conversational sarcasm detection, based on eXtreme Gradient Boosting (XGBoost) (Chen and Guestrin, 2016) and TF-IDF. For post-hoc explainability analysis, the architecture incorporated Model-agnostic Explanations (LIME) (Ribeiro et al., 2016) and Shapley Additive exPlanations (SHAP) (Messalas et al., 2019). Chauhan et al. (2022) argued that the presence and learning of emojis in text can facilitate the understanding of intended sarcasm. Their model combined textual, visual, acoustic, and emotional modalities. Textual input included a target utterance and its context (i.e., its preceding utterances in a dialogue). For each utterance, embeddings of its constituent words were obtained by using fastText (Bojanowski et al., 2016). Each utterance also had an emoji label, whose associated emoji's embedding was obtained via emoji2vec (Eisner et al., 2016). The four modalities were concatenated and fed into Gated Multimodal Units (GMUs) with attention. The authors extended MuStARD (Castro et al., 2019) dataset by adding emoji; sentiment (i.e., positive, negative, and neutral); and fine-grained emotion labels (e.g., happy, anger, etc.). The proposed model outperformed two state-of-the-art models with similar experimental set-ups by achieving a maximum F_1 score of 69.8. The ablation study results proved that leveraging emojis improves the performance of their model.

While most existing multimodal, context-based methods were proposed for English corpora, Bedi et al. (2021) proposed a code-mixed dataset, including Hindi and English dialogues, for multimodal sarcasm and humor detection tasks and a contextualized model for such tasks. The dataset consists of over 15K utterances from an Indian comedy TV show and was manually annotated. The proposed model employed two LSTMs and hierarchical attention mechanisms to learn textual and acoustic features of conversational utterances. Librosa (McFee et al., 2015) and fastText (Joulin et al., 2016) were used to extract textual and acoustic features, respectively. This work conducted extensive experiments with several variations of the proposed model and ablation studies. The proposed model outperformed existing models for utterance-level sarcasm detection and emotion recognition in conversations tasks by achieving F_1 score of 0.711 and 0.820 for sarcasm and humor classification tasks, respectively. This work is a good attempt to shed light on multi-lingual, multimodal, and multi-tasking sarcasm detection.

4.3.6 Downstream Applications

Sarcasm detection is an integral component of human communication. Despite its importance, practical applications of the task have not been widely explored. This section outlines a few attempts at applying sarcasm detection in downstream tasks in different domains.

4.3.6.1 Sentiment Analysis

Sarcasm detection plays a key role in sentiment analysis. Lunando and Purwarianti (2013) are one of the early works which considered sarcasm in the task of sentiment analysis. The proposed architecture first identified the sentiment label (i.e., positive, negative, and neutral) for an input text and then performed sarcasm detection for the texts labeled with the "positive" label. For the sarcasm detection component, the model employed two features (i.e., manually computed topic negativity score and the number of interjection words) on top of the three linguistic features, used for the first component.

Bouazizi and Ohtsuki (2015) attempted to examine whether sarcasm can improve sentiment classification on $\mathbb{X}$ corpora. To this end, they used hand-crafted features such as the number of laughters, that of interjections, and the existence of common sarcastic expressions. Experimental results showed that adding such features could improve the classification performance of machine learning models. Yunitasari et al. (2019) exploited the output of binary sarcasm detection to improve the performance of sentiment classification for Indonesian tweets. This work assumed that sarcasm is associated with negative sentiment. Therefore, when a sarcastic statement was classified as "positive", the model changed its label to "negative". For sarcasm detection, the architecture employed hand-crafted features associated with sentiment and punctuation as well as lexical and syntactic features. Experimental results failed to prove their hypothesis, which indicates that sarcasm is not necessarily associated with negative sentiment. More advanced contextual information is required to comprehend sarcasm.

El Mahdaouy et al. (2021) proposed an MTL model that performs sarcasm and sentiment classification for the Arabic language. The model was capable of learning the relationship between the two tasks by incorporating attention layers, while most existing MTL models for sentiment analysis performed different tasks independently. The ArSarcasm Shared Task dataset consisting of 15,548 tweets was used in the experiments. The results showed that their model outperformed single-tasking models in sarcasm detection in terms of macro-average F_1 score, while a single-tasking model without attention achieved the best performance for sentiment classification. As for class-specific F_1 score, the proposed MTL model performed best. Their findings indicated that utilizing relational information between sarcasm detection and sentiment classification could boost the classification performance of each task.

4.3.6.2 Mental Health

Larsen et al. (2016) investigated how Huntington's disease gene expansion carriers perform in sarcasm detection, emotion recognition, and theory of mind compared with healthy controls. This study used sarcasm detection as a measure to assess social-cognitive impairment. It was assessed using the Social Inference Minimal (SI-M) test of The Awareness of Social Inference Test (TASIT) (McDonald et al., 2003). The SI-M test evaluated the understanding of sincere and sarcastic interactions. The results indicated that the higher the severity of patient's illness is, the poorer their performance on sarcasm detection is. Their findings offered insights into how sarcasm detection could aid the early detection of mental health diseases and the tracking of their progression as well as support patients and clinical and health professionals.

Several studies and research articles have reported that exposure to cyberbullying is associated with several mental disorders such as depression, self-harm, and anxiety (Schodt et al., 2021; Skilbred-Fjeld et al., 2020; Maurya et al., 2022). Chia et al. (2021) applied a machine learning model trained on a sarcasm detection dataset to a cyberbullying detection dataset to examine the extent to which sarcasm detection could be useful for cyberbullying detection. Experimental results showed that a sarcasm detection model surpassed the performance of a model trained on a cyberbullying detection dataset, as evidenced by a higher F_1 score. This indicates that sarcastic expressions are commonly used in instances of cyberbullying. It highlights the importance of developing better tools for detecting sarcasm, as this could significantly improve the identification and understanding of online cyberbullying cases. Enhanced sarcasm detection could help in accurately identifying harmful content and protecting individuals from online abuse.

The COVID-19 pandemic has affected people's mental health globally. Rothermich et al. (2021) studied changes in the usage of humor and sarcasm among 661 adults (164 with mental disorders and 497 asymptomatic individuals) during the pandemic via self-report questionnaires. Experimental subjects were asked how their usage of sarcasm and humor has changed in the past month. Chi-square tests were used to evaluate relationships between the frequency of change in the usage of humor and sarcasm and the anxiety/depression severity of the subjects. The results indicated that the individuals with depression tend to use it more frequently while those with anxiety showed no significant change in the usage of humor.

As for sarcasm, individuals with mental disorders used more sarcasm than the asymptomatic ones did. The authors reported that it should be further investigated whether the motivation behind the increase in the usage of sarcasm is associated with self-enhancement or self-deprecation. The findings and future research directions described in this study offered useful insights into a better understanding of coping strategies of individuals with mental disorders during crises, which can potentially benefit patients and health professionals.

4.3.6.3 Other Applications

Some studies attempted to apply sarcasm detection in other domains apart from sentiment analysis and mental health. Although these areas have yet to be explored extensively, they have the potential to benefit businesses and human well-being. Danielyan (2022) studied the usage and impact of sarcasm in advertising campaigns from a pragmalinguistic perspective. This work reported that sarcastic expressions in advertisements drew people's attention and enhanced user engagement, thereby helping companies gain a positive brand reputation. The paper acknowledges potential drawbacks and adverse effects associated with the use of sarcasm in advertising. One concern is the risk of misinterpretation or misunderstanding by the target audience, potentially resulting in confusion or negative reactions. Furthermore, the caustic or bitter tone of sarcasm can be perceived negatively by the audience. The paper also highlights ongoing debates among scholars and linguists regarding the appropriateness of employing sarcastic language in social contexts, with some asserting that it may have toxic or detrimental effects on relationships.

Zhou et al. (2021b) proposed a dialogue system that generates sarcasm-aware responses. The authors argued that sarcasm-aware dialogue systems are essential to avoid misunderstanding and more human interactions. The proposed model first detected sarcasm from text and speech signals, using existing sentiment analysis tools. Next, the output of the sarcasm detection module was encoded as a two-dimension one-hot vector, in which the first and the second respectively represented whether an input utterance is sarcastic and whether each of the expected responses is sarcastic. The sarcasm-aware representation was concatenated with other types of input embeddings and the joint representation was fed into a deep learning model that generated a response for the input. For evaluation, the SARC (Khodak et al., 2018) dataset and BLEU (Papineni et al., 2002) which is originally designed to measure how well a machine-translated text correlates with human judgment, were used. BLEU was also used as an indicator of the adequacy and fluency of a natural language generation model. Some examples of sarcasm-aware responses generated by the proposed model were given in this work and the authors claimed that their model is capable of generating more interesting dialogues.

4.3.7 Summary

Sarcasm detection is an active research area and has great potential to improve the performance of computational methods for a wide range of NLU tasks. Findings of the theoretical research into sarcasm provide useful clues into the computational learning and modeling of sarcasm from the perspectives of cognition, communicative behavior, and neural mechanisms. For example, several studies emphasized that the understanding of contextual information and the identification of contradiction between the literal/figurative meaning of a statement and observational evidence are the keys to the successful interpretation of sarcasm.

The latest computational research actively takes such signals into the learning process of sarcasm, leading to the development of models equipped with diverse modality learning capabilities. We also reviewed two main approaches for annotating datasets for sarcasm detection. Manual annotation is a natural choice for constructing high-quality corpora, although it is difficult to generate large-scale datasets for labor- and time-costs. Some researchers have advocated for novel data labeling approaches that demand minimal human supervision, advocating for the adoption of weakly supervised annotation methods. These methods typically involve the generation of annotations through sets of pre-defined keywords, such as hashtags on $\mathbb{X}$, deemed relevant to sarcasm.

4.3.7.1 Technical Trends

Feature-based methods usually identify sarcastic utterances by inputting hand-crafted features into machine learning classifiers and paved the way for more advanced computational learning of sarcasm. The works under this technical trend believe that manually developed linguistic features can complement conventional embeddings, e.g., GloVe. Potamias et al. (2019) proposed 44 hand-crafted features encompassing syntactic, demonstrative, sentiment, mood, and readability attributes. Pandey et al. (2021) proposed 16 hand-crafted linguistic features, e.g., entropy, lex diversity, number of nouns, number of adjectives, number of verbs, sum of verbs available in the given sentence, the Flesch Reading Ease formula, Dale Chall readability, difficult words, length, set length, stopword count, wrong word, one-letter word, two-letter word, longer letter word, <li> tags. In the prevailing focus on deep learning within the academic landscape, the study highlights the enduring value of research dedicated to feature engineering. Despite the transformative impact of deep learning across multiple domains, the acknowledgment of traditional methods, particularly feature engineering, remains imperative. This recognition underscores the enduring significance of feature engineering in providing valuable insights and contributions. By fostering a nuanced understanding of language intricacies, feature engineering emerges as a fundamental tool for deciphering complex linguistic expressions. Moreover, its role extends to the broader domain of human-centric computing, emphasizing its continued relevance in shaping our comprehension of language phenomena.

Embedding-based methods exploit (contextualized) word embeddings, which are usually generated using PLMs, to represent input data. They have produced state-of-the-art results across a wide range of NLP tasks. Contextualized embeddings are advantageous to NLP tasks in that they can automatically encode word semantics in different contexts and discern polysemy and homonymy. Methods based on embeddings have demonstrated exceptional performance with the advancement of deep learning techniques, exemplified by models like BiLSTM, GCN, CNN, and Attention mechanisms. Leveraging pretrained embeddings, while proficient in capturing semantic information, falls short in directly acquiring pertinent knowledge beyond semantics for effective sarcasm detection.

Table 4.7: Technical trends in sarcasm detection. Feat., emb., SI, and cont. denote feature-, embedding-, semantic incongruity- and context-based trends, respectively.

Trend	Reference	Tech	Feature	Framework	Dataset	Score	Metric
Feat.	Potamias et al. (2019)		Hybrid hand-crafted features + GloVe	BiLSTM+Attention	Riloff et al. (2013)	0.87	F_1
	Pandey et al. (2021)		Hybrid hand-crafted features + GloVe	LSTM+Attention	Misra and Arora (2019)	0.88	F_1
Emb.	Shrivastava and Kumar (2021)	DL	BERT	BiLSTM+Attention	Van Hee et al. (2018)	0.85	F_1
	Yao et al. (2021)	DL	Multimodal embeddings	Gating+Attention	Cai et al. (2019)	0.80	F_1
	Liang et al. (2022a)	DL	BERT+ViT	GCNs	Cai et al. (2019)	0.87	F_1
	Kamal and Abulaish (2022)	DL	GloVe	CNN+BiGRU	Ptáček et al. (2014)	0.89	F_1
	Cai et al. (2019)	DL	RestNet-50 V2+GloVe	BiLSTM+Modality Fusion	Cai et al. (2019)	0.80	F_1
SI	Ren et al. (2020)	DL	SenticNet	Memory Network	Ptáček et al. (2014)	0.87	F_1
	Pan et al. (2020)	DL	RestNet+BERT	Attention networks	Cai et al. (2019)	0.83	F_1
	Wu et al. (2021c)	DL	RestNet+BERT+OpenSmile	Attention networks	Castro et al. (2019)	0.75	F_1
	Chen et al. (2022)	DL	RestNet+BERT+OpenSmile	MTL	Khodak et al. (2018)	0.74	F_1
	Yue et al. (2023)	DL	RestNet+BERT+ConceptNet	Matrix transf.+Similarity	Cai et al. (2019)	0.74	F_1
Cont.	Kumar et al. (2021)	ML	TF-IDF	XGBoost+LIME+SHAP	Castro et al. (2019)	0.92	F_1
	Chauhan et al. (2022)	DL	fastText+emoji2vec	GMUs+Attention	Castro et al. (2019)	0.70	F_1
	Bedi et al. (2021)	DL	fastText+Librosa	LSTM+Attention	Bedi et al. (2021)	0.71	F_1

Semantic incongruity-based methods are cognition-inspired. They leverage the contrast between the meaning of an utterance and contextual evidence or sentimental incongruity to detect sarcasm. Capturing incongruity is challenging. (Yue et al., 2023) tried to detect the clash between the factual and stated information from text and images. Interpreting facts from images However, since there exist different types of incongruities in sarcasm, and hence, effectively modeling the incongruity between facts and statements still remains several open research directions.

Finally, context-based methods learn contextual information obtained from conversations. The advantage of detecting sarcasm from conversions is that the multi-turn conversational interactions between a speaker and a listener provide more clues for detecting the clash between facts and statements. Using conversational threads can enhance the performance of machine learning models in diverse NLP tasks by providing additional context such as propagation patterns of a target event (Han et al., 2019; Gao et al., 2020; Sawhney et al., 2022; Vuong et al., 2021). While most sarcasm detection models in the other categories are trained and evaluated on social media datasets, these context-based methods usually employ transcripts of TV shows and movies, which are written by professional screenwriters.

4.3.7.2 Application Trends

Sarcasm is prevalent in human communication and correctly interpreting sarcastic utterances can unlock the potential of machine learning techniques for a wide range of NLU tasks. This chapter introduced a few attempts at sentiment analysis, mental health, advertising, and dialogue systems. Detecting sarcasm plays a crucial role in sentiment analysis, with existing studies often employing sarcasm detection techniques to enhance model performance through the integration of supportive features. Additionally, there have been works to utilize sarcasm detection as a parsing mechanism for sentiment analysis. These parsing-oriented sentiment analysis systems typically adhere to a simplistic rule, such as assigning the opposite sentiment polarity when confronted with a sarcastic input.

Table 4.8: Sarcasm detection downstream task applications.

Reference	Downstream Task	Feature	Parser	Explainability
Lunando and Purwarianti (2013)	Sentiment analysis		✓	
Bouazizi and Ohtsuki (2015)	Sentiment analysis	✓		
Yunitasari et al. (2019)	Sentiment analysis		✓	
El Mahdaouy et al. (2021)	Sentiment analysis	✓		
Larsen et al. (2016)	Mental health		✓	
Chia et al. (2021)	Mental health		✓	
Rothermich et al. (2021)	Mental health		✓	
Danielyan (2022)	Advertising campaigns			✓
Zhou et al. (2021b)	Dialogue system	✓		

Bouazizi and Ohtsuki (2015) claimed that some hand-crafted features are associated with sarcasm and adding such features improved the performance of the proposed model for sentiment analysis. Yunitasari et al. (2019) studied sentiment classification based on an assumption that sarcasm carries negative sentiment. However, their experimental results did not support such an assumption. El Mahdaouy et al. (2021) investigated how MTL on both sentiment and sarcasm could improve the performance of each task. They did not assume certain aspects of sarcasm and, hence, their method was more generalizable than the two studies introduced earlier.

Some studies explored the correlation between sarcasm and mental health conditions. In this context, the identification of sarcasm serves as a parsing mechanism, enabling subsequent mental analysis based on the recognition of patterns in sarcastic expressions, particularly concerning mental health issues. Larsen et al. (2016) investigated the cognitive functioning of patients with Huntington's disease via sarcasm detection and observed that patients lost their ability to understand and interpret sarcasm as their disease developed. This study did not exploit automatic, computational sarcasm detection. Chia et al. (2021) studied the usage of sarcasm in cyberbullying and reported that sarcastic expressions were commonly used in cyberharassment. Rothermich et al. (2021) researched the impact of the global COVID-19 pandemic on people's mental health by statistically analyzing changes in the usage of humor and sarcasm. This study reported that people with different mental disorders exhibited different patterns in the usage of humor and sarcasm.

Furthermore, the identification of sarcasm holds promise for enhancing explainability in advertising and improving efficacy in dialogue systems. Danielyan (2022) reported that sarcasm could be an effective tool in advertising campaigns as it helped to draw people's attention and increase customer engagement. In this context, the use of sarcastic language serves as a tool for explaining customer behaviors. Zhou et al. (2021b) proposed a sarcasm-aware, generative dialogue system. The dialogue system aims to employ the identified sarcastic label as a signal, prompting the dialogue system to generate appropriate responses to sarcastic expressions. Although this study shed light on the importance of sarcasm understanding in dialogue systems, the results generated using the proposed model showed that there was great room for improvement.

4.3.7.3 Future Works

Research into automatic sarcasm detection has gained popularity due to its importance in our daily communication. Despite several efforts to comprehend sarcasm and make computers understand and use sarcasm in different disciplines, several challenges have yet to be tackled. Firstly, understanding sarcasm requires diverse contextual factors. Several studies in psycholinguistics, pragmatics, and neuroscience have delved into how context could help humans perceive, process, and comprehend sarcasm. Ensuring the context awareness and generalizability of sarcasm detection methods is crucial. Computational research in automatic sarcasm detection usually employs sarcastic utterances obtained from social media (e.g., $\mathbb{X}$), which have limited context in nature due to short lengths. To overcome this limitation, it is recommended to investigate sarcasm detection using multiple modalities, broader contexts, and external knowledge. Employing multimodal data enables the detection of incongruities beyond text, while a broader context, such as shared long-term experiences and cultural background, offers clues for identifying inconsistent information in sarcastic expressions. Additionally, incorporating external knowledge, such as commonsense, provides information not directly obtainable from sarcasm datasets. Moreover, future research should further investigate how to generate and leverage effective representations of a wide range of contextual factors.

Secondly, there is a need to create multimodal sarcasm detection datasets with separate sarcasm annotations for distinct modalities. Current studies in multimodal sarcasm detection commonly utilize the dataset from the work of Cai et al. (2019), where a unified sarcasm label is assigned to a text-image pair for multimodal computing. While this label is reasonable for multimodal analysis, it poses challenges in ablation studies when text-only or image-only inputs share the same label derived from the multimodal context. This is problematic because some sarcastic expressions have labels annotated based on the combination of text and image, whereas the independent text or images may not be sarcastic. As illustrated in the example on the right in Fig. 4.4, a combination of a raining image and the statement "what wonderful weather" might be defined as sarcasm, but the sarcasm may not be evident in each modality. Therefore, it is unreliable to assert that a multimodal model outperforms a unimodal model if sarcasm labels annotated for multimodal data are employed for unimodal evaluation. To address this issue, a multimodal dataset with independent annotations for different modalities is essential in the sarcasm detection research community.

Thirdly, existing studies on sarcasm detection (Liu et al., 2014; Meriem et al., 2021; Mishra et al., 2016) mentioned that sarcasm identification could benefit several real-world applications such as marketing, advertising, opinion mining, dialogue systems, and review summarization. Although several researchers have acknowledged the importance of sarcasm detection in downstream tasks, practical applications of the task have not been widely studied. The complexity inherent in detecting sarcasm across diverse domains contributes to the challenges associated with this task. This heightened difficulty in sarcasm recognition poses additional obstacles to its effective application in downstream tasks.

Existing research in applications often incorporates sarcastic features as input or employs a dedicated sarcasm classifier to parse text with sarcastic elements. However, there remains a notable gap between comprehensive sarcastic analysis and research in psycholinguistics. Subsequent investigations could focus on refining the formulation of practical and impactful sarcasm detection tasks. For instance, incorporating explainability features could elucidate the cognitive mechanisms behind sarcastic expressions, thereby augmenting the practical utility of sarcasm detection techniques. An alternative avenue of exploration involves broadening the task formulation of sarcasm detection beyond a simplistic binary classification, encompassing a nuanced analysis across multiple dimensions of sarcastic language.

4.4 Personality Recognition

Personality encompasses an individual's distinctive patterns of thought, emotion, and behavior, inclusive of the underlying psychological mechanisms, whether overt or concealed, that give rise to these patterns (Funder, 1997). According to Ozer and Benet-Martinez (2006), personality is personality is linked to individual happiness, both physical and psychological well-being, spiritual beliefs, and self-perception. On an interpersonal level, it influences the nature of relationships with peers, family, and romantic partners. At a social institutional level, personality is tied to occupational decisions, satisfaction, and performance, along with community engagement, criminal behavior, and political ideology.

Automatic personality recognition, also known as personality recognition (Majumder et al., 2017) or personality identification (Kaushal and Patwardhan, 2018), has been one of the three emerging research topics of personality computing (Vinciarelli and Mohammadi, 2014). It aims to identify personality traits from machine-detectable distal cues, including textual, audio, and video sources (Mushtaq and Kumar, 2023) for individuals. It adopts approaches, based on social signal processing, affective computing, AI and methodologies in other domains. Personality recognition has been applied in many real-life scenarios (Mushtaq and Kumar, 2023; Remaida et al., 2020), such as recruitment, education, social network analysis, recommendation systems, and sentiment analysis, which attracts wide attention from academia and industry. Currently, there are two main personality recognition models, namely: Big Five (also known as OCEAN because the five factors or dimensions are Openness, Conscientiousness, Extraversion, Agreeableness, and Neuroticism) and Myers-Briggs Type Indicator (MBTI) personality recognition.

4.4.1 Theoretical Research

4.4.1.1 The Big Five and Its Instruments

One of the most famous personality factor models, the Big Five, is an inter-individual personality structure which originated from the five factor model (FFM) (Norman, 1963). Goldberg (1981) designated the FFM as the Big Five, based on the research of Wiggins (1968) who dubbed "Extraversion (E)" and "Neuroticism (N)" the Big Two. Costa Jr (1980) added a dimension "Openness to Experience (O)" and later in 1985 and 1989, they proposed measure scales of "Agreeableness (A)" and "Conscientiousness (C)". Finally, the Big Five dimensions, shown in Table 4.9, were compiled, showing the putative assignment of standard personality factors or scales. The descriptions of the binary values (Yes / No) utilize the Big Five trait definition from the works of McCrae and John (1992). Most personality traits are measured from two fundamental sources: *self-reports* and *observer ratings*. Using both instruments, i.e., questionnaires and trait adjectives (lexical approach), in self-reports or observer ratings originates in the 1980s, of which has been justified the feasibility by a series of research papers (Norman, 1969). Remarkably, all five factors were shown to have discriminant and convergent validity across personality instruments and observers (McCrae and Costa, 1985; Goldberg, 1989), enduring across decades in adults (McCrae and Costa, 1990). These studies justified empirically the correspondences between the named factors in the two traditional sources.

Personality assessment research has extensively utilized questionnaire instruments based on the five factors. These questionnaires involve individuals providing self-ratings or observers rating the target person on a Likert-type scale, ranging, for example, from -2 to 2 or from 1 to 5, to gauge the degree of agreement from *disagree strongly* to *agree strongly*. The top popular questionnaires include the Ten-Item Personality Inventory (TIPI, 10 items) (Gosling et al., 2003), the Big-Five Inventory (BFI, 44 items) (John et al., 1991), the NEO Five-Factor Inventory (NEO-FFI, 60 items) (McCrae and Costa Jr, 2004), and the NEO-Personality-Inventory Revised (NEO-PI-R, 240 items) (Costa Jr and McCrae, 1995). The first-person questionnaires (e.g., "I tend to find fault with others") lead to self-report, i.e., people's *identities*, while the third-person questionnaires (e.g., "this person tends to find fault with others") come from observer ratings, i.e., people's reputation. The quantitative measures of personality traits set the stage for representing personality in a numerical model.

Table 4.9: Big Five personality traits model (also known as OCEAN).

Factor	Description	
	Yes	No
Openness (O)	innovative and curious	dogmatic and cautious
Conscientiousness (C)	organized and professional	careless and sloppy
Extraversion (E)	active and sociable	reserved and solitary
Agreeableness (A)	trustworthy and modest	unreliable and boastful
Neuroticism (N)	sensitive and anxious	secure and confident

4.4.1.2 Myers-Briggs Type Indicator

Another popular personality typology was developed by Myers et al. (1998), termed MBTI. MBTI is rooted in personality typology (Jung and Beebe, 2016) and was formulated to assess individuals' psychological preferences in cognition and decision-making. It classifies people into 16 distinct personality types, determined by their preferences across four dichotomous dimensions: Extraversion-Introversion (E-I), Sensing-Intuition (S-N), Thinking-Feeling (T-F), and Judging-Perceiving (J-P) (see 4.10 for details). The primary objective of developing MBTI during World War II was to create a pragmatic tool facilitating self-awareness and interpersonal understanding, particularly in realms such as career planning, team dynamics, and personal development. Over time, MBTI gained prominence in organizational and industrial psychology (Cohen et al., 2013), as well as career counseling (Savickas, 2019).

However, MBTI also encountered skepticism from scholars and psychometric experts who question its theoretical underpinnings and measurement robustness. Pittenger (1993) argued that MBTI displayed correlations with other personality measures, suggesting challenges in maintaining its distinct assessment qualities. Advocates of MBTI's validity typically cited findings linking one of its four scales to specific behaviors. Nevertheless, the available data did not substantiate the assertion that MBTI effectively measures distinct personality types or that the delineation of the 16 types is crucial for a comprehensive understanding of personality.

Table 4.10: The four dichotomies of the MBTI model (Myers et al., 1998).

Extraversion-Introversion Dichotomy (attitudes or orientations of energy)	
Extraversion (E)	**Introversion (I)**
Directing energy mainly toward the outer world of people and objects	Directing energy mainly toward the inner world of experiences and ideas
Sensing-Intuition Dichotomy (functions or processes of perception)	
Sensing (S)	**Intuition (N)**
Focusing mainly on what can be perceived by the five senses	Focusing mainly on perceiving patterns and interrelationships
Thinking-Feeling Dichotomy (functions or processes of judging)	
Thinking (T)	**Feeling (F)**
Basing conclusions on logical analysis with a focus on objectivity and detachment	Basing conclusions on social values with a focus on understanding and harmony
Judging-Perceiving Dichotomy (attitudes or orientations toward dealing with the outside world)	
Judging (J)	**Perceiving (P)**
Preferring the decisiveness and closure that result from dealing with the outer world through Judging (Thinking or Feeling)	Preferring the flexibility and spontaneity that results from dealing with the outer world through Perceiving (Sensing or Intuition)

4.4.1.3 Personality Psychology

In the area of personality psychology, the research on personality includes personality structure, process (Cervone, 2005), stability and change (Ozer and Benet-Martinez, 2006; Caspi et al., 2005). Cervone (2005) argued that personality structure consists of two aspects: 1) within-person content and organization of personality systems; 2) between-person variation in the population; it is also called intra-individual and inter-individual personality structures, respectively. The intra-individual analysis accounts for the actions and experiences of a person with scientific models (Magnusson and Torestad, 1993; Mischel, 2004; Freud, 1989) for intra-individual mental systems, while inter-individual analysis explains differences between individuals (Borsboom et al., 2003) by classifying inter-individual categories of population variation (Asendorpf, 2003; Ashton et al., 2004; Saucier, 2003). Furthermore, inter-individual personality structure does not capture psychological tendencies or personality changes at the level of the individual case (Cervone, 2005).

Guadagno et al. (2008); Qiu et al. (2012); Johnson and Gardner (2010) believed that personality traits can be externalized through people's behavior, which explores the interaction between personality and the way people use technology. Furthermore, it is demonstrated that the choice of words can be driven by, besides its meaning, psychological phenomena such as relational attitudes, emotions, and personality traits (Tausczik and Pennebaker, 2010). Therefore, it is possible to predict personality traits from texts by employing socioliguistics in techniques for automatic text analysis, which has been proved by several research works (Gill et al., 2009; Luyckx and Daelemans, 2008; Oberlander and Nowson, 2006).

4.4.2 Annotation Schemes

To get the personality trait annotation, the composers or speakers of the corpus (e.g., essays, conversations, etc.) will be scored by themselves or other observers according to given personality questionnaire instruments. As mentioned in section 4.4.1.1, there are a large number of questionnaires, such as TIPI, BFI, NEO-FFI, NEO-PI-R, etc. Since TIPI and BFI contain relatively less items and all the questionnaires share the similar question format, the detailed questionnaires of TIPI (Table 4.13) and BFI (Table 4.11) are token as the examples and shown in this section. For each question in BFI, the participant needs to select a number ranging from 1 to 5 to show the agreement (from disagree strongly to agree strongly) with that statement. Then, the result/ category of each dimension of Big Five is obtained through the scoring of each dimension of the personality traits in Table 4.12. While in TIPI, the participant are required to write a number ranging from 1 to 7 to indicate the extent to which he or she agrees with the statement. The continuous decimal value of each dimension of Big Five is obtained according the scoring rule in Table 4.14.

Table 4.11: BFI. Participants are required to write a number next to each statement to show the degree (1 disagree strongly, 2 disagree a little, 3 neither agree nor disagree, 4 agree a little, 5 agree strongly) to which they agree or disagree with that statement.

I see Myself as Someone Who...		
1. Is talkative	2. Tends to find fault with others	3. Does a thorough job
4. Is depressed, blue	5. Is original, has new ideas	6. Is reserved
7. Is helpful and unselfish with others	8. Can be somewhat careless	9. Is relaxed, handles stress well
10. Is curious about different things	11. Is full of energy	12. Starts quarrels with others
13. Is a reliable worker	14. Can be tense	15. Is ingenious, a deep thinker
16. Generates a lot of enthusiasm	17. Has a forgiving nature	18. Tends to be disorganized
19. Worries a lot	20. Has an active imagination	21. Tends to be quiet
22. Is generally trusting	23. Tends to be lazy	24. Not easily upset
25. Is inventive	26. Has an assertive personality	27. Can be cold and aloof
28. Perseveres until the task is over	29. Can be moody	30. Values aesthetic experiences
31. Is sometimes shy, inhibited	32. Is kind to almost everyone	33. Does things efficiently
34. Remains calm in tense situations	35. Prefers work that is routine	36. Is outgoing, sociable
37. Is sometimes rude to others	38. Makes plans and follows them	39. Gets nervous easily
40. Likes to reflect, play with ideas	41. Has few artistic interests	42. Likes to cooperate with others
43. Is easily distracted	44. Is sophisticated in art or literature	

Table 4.12: BFI scoring. "R" refers to reverse-scored items.

Factor	Corresponding Characteristic Index								
Extraversion	1	6R	11	16	21R	26	31R	36	
Neuroticism	4	9R	14	19	24R	29	34R	39	
Agreeableness	2R	7	12R	17	22	27R	32	37R	42
Conscientiousness	3	8R	13	18R	23R	28	33	38	43R
Openness	5	10	15	20	25	30	35R	40	41R 44

Table 4.13: TIPI. Participants are required to write a number next to each statement to show the degree (from 1 to 7) to which they agree or disagree with that statement.

I see myself as:		
1. Extraverted, enthusiastic.	2. Critical, quarrelsome.	3. Dependable, self-disciplined.
4. Anxious, easily upset.	5. Open to experiences, complex.	6. Reserved, quiet.
7. Sympathetic, warm.	8. Disorganized, careless.	9. Calm, emotionally stable.
10. Conventional, uncreative.		

Table 4.14: TIPI scoring. "R" refers to reverse-scored items.

Factor	Corresponding Characteristic Index	
Extraversion	1	6R
Agreeableness	2R	7
Conscientiousness	3	8R
Emotional Stability	4R	9
Openness	5	10R

4.4.3 Datasets

Many datasets were constructed for personality recognition based on OCEAN and MBTI (Table 4.15). Essay dataset (Pennebaker and King, 1999) contains 2,479 stream-of-consciousness essays written by psychology students who were asked to write down whatever they think off within 20 minutes. The binary personality labels of each essay writer were obtained by asking students to fill in the BFI questionnaire. The following snippet shows an annotated essay sample.

Table 4.15: Personality recognition datasets and statistics.

Reference	Task	Source	# Sample
Pennebaker and King (1999)	Big Five	Essays	2,479 essays
Biel et al. (2013)	Big Five	Vlogs related to diverse domains	408 vlogs
Kerz et al. (2022)	Big Five	Climate change, vaccination, AI	20h speech
Kaggle MBTI	MBTI	Social media posts	8600 instances

```
text: "I haven't chatted on MIRC for a long while. Maybe I'll do that at work, it should be fun. Hmmm, I'm just thinking
like random thoughts, about nothing specific. My gyming is going fine, hope I get a good looking chest/body over the next
month or so, not that I need to show it off or something, just because this way the clothes i wear will I guess look
better on me... I think I'm going to go to my job kind of early and print out some slides for the CS class quiz it should
be easy. and about this whole Coop thing I think I should go check out the companies but not really apply apply I don't
want to as yet I think I can make my GPA higher and then apply what do you say well I can't there's no one really who
might be able to help me out I just have to think about myself."
personality: Yes No Yes No No
```

YouTube dataset (Biel et al., 2013) is built from 408 randomly selected YouTube vlogs with a set of personality traits for each vlogger. The content of the vlogs are relevant to books, movies, politics, personal issues, etc. The personality traits are the Big Five scores annotated by crowdsource workers. the dataset is balanced in gender, with 211 female (52%) and 197 male (48%) vloggers. There are 243k word tokens and 10k unique tokens with an audio of 30h in total in the YouTube dataset.

```
text: "- And here is. I don't know if you can hear me or not, probably or hopefully because this music is really loud.
But uh, I wanted to apologize YouTube, because I've tried
countless time to upload the same videos and for some reason it's just not being
cooperative. I don't know, but um, I've been trying for a week to upload the ones for local bloggers that I did. Sorry
guys, um, my promotion video might be up, might not. I don't know, well- well I mean, like, it might be up eventually.
Um, yeah, ah, right now I'm in North Carolina. Ah, where are we? What town are we in?
- Oh, um, Fairfield?
- No, we're at the Fairfield Marriot but we're in a -
- North Carolina?
- We were just at some first - basketball game, uh, where we played a song and stuff. But um, I just wanted to apologize.
We are going to do another blog with all of my
friends that are in this room. The other ones are downstairs getting our pizza. So,
when they come back -"
personality: 6 4.9 5.1 5.4 5.8
```

SPADE dataset (Kerz et al., 2022) consists of 20h of speech from 220 individuals among which 129 are male, 88 are female and 3 are diverse. The transcribed speech was obtained via the AMT crowdsourcing platform, with 848,827 words in total. The personality traits were collected through the BFI questionnaire.

Kaggle MBTI dataset[23] was gathered from the PersonalityCafe forum, containing over 8600 instances. Each instance includes the last 50 posts of a user and associates an MBTI type of the user.

```
Personality: ENFJ
Post: That sounds like a beautiful relationship already. But don't push too fast as it
seems she needs to be appreciated as a friend. But at the end of the day you can ...
```

4.4.4 Evaluation Metrics

Automatic personality recognition is usually formalized as two tasks: classification and linear regression. Precision, Recall, F_1, and Accuracy are the most commonly used evaluation metrics for personality classification tasks. For the regression tasks, Pearson's r, Adjusted R^2, Mean Absolute Error (MAE), and MSE are the commonly used evaluation metrics. Here, Pearson's r is a linear rank measure of predicted and target values. Adjusted R^2 shows the proportion of variance in the target values that is explained by predicted values. MAE reflects the average of all absolute errors. MSE depicts the average of the squares of errors.

4.4.5 Methods

4.4.5.1 Big Five Personality Recognition

A. Computational Psychology Approaches

There has been a shift from questionnaire approach to computational psychology approaches since the last century, which combines the psychology theories and linguistics models. The early implementation of Pennebaker and King (1999) revealed that the linguistic style was an indicator for the differences of individuals across different contexts and time through leveraging a computational program, i.e., LIWC (Francis and Booth, 1993). LIWC provides a "gold standard" of categories for studying the textual data from the aspects of linguistic, grammatical and psychological dimensions. It has been developed into the current enhanced version (Pennebaker et al., 2015) over the past twenty years. During the same period, several studies (Pennebaker, 2013; Tausczik and Pennebaker, 2010), on the basis of LIWC, found that function words were more informative than content words in terms of showing individual differences. The function words are capable of connecting, shaping and organizing the texts, such as conjunctions, pronouns, and prepositions.

---

[23] https://kaggle.com/datasets/datasnaek/mbti-type?resource=download

The content words refer to those depicting an action or object, e.g., verbs, adjectives and nouns. This means *how*, instead of *what*, people communicate fundamentally reflects more of their psychological situation. However, the approaches based on LIWC had some limitations. They were constrained by the inherent computation mechanism, i.e., statistical features based on word counts, failing to capture some implicit contextual elements such as irony or sarcasm (Pennebaker et al., 2003). Additionally, these methods only explore and analyze the correlations between the LIWC features and the personality, instead of automatically recognizing the personality traits from texts (Štajner and Yenikent, 2020).

B. Computational Linguistic Approaches

With the development of machine learning and computational linguistics, researchers started to think outside the box of previous research framework based on LIWC. Argamon et al. (2005) were the first implementation of automatic personality recognition from texts. They believed personality could be reflected by the writing *style* of the texts and defined the *stylish meaning* of a text as *non-denotational*, i.e., independent of the topic or the content of the text. Accordingly, *function words*, proven useful for stylistic text classification (Argamon et al., 2003), were considered unlikely to be subject to conscious control by the authors and their frequencies varies across different genres and authors. Then, they proposed to utilize the relative frequencies of 675 functional words under three *system networks* as lexical stylistic features, i.e., *cohesion*, *assessment* and *appraisal*, based on the theory of systemic functional grammar. Finally, they carried out binary classification task of recognizing neuroticism (emotional stability) and extraversion from the Essays corpus by using the SMO learning algorithm (Platt, 1999), obtaining an accuracy of 58%.

This work inspired Oberlander and Nowson (2006) to explore what else might be good features for personality recognition tasks besides pre-defined semantic and syntactic categories such as LIWC. They leveraged *n*-gram features (known as *open-vocabulary approach*) to allow the classifier to discover linguistic signals for the four personality traits of Big Five even other signals that cannot be exploited by previous psycholinguistic studies (*closed-vocabulary approach*). Furthermore, besides a binary classification task, they defined personality recognition as a multiple classification task to approximate continuous rating modeling. They experimented on personal weblogs instead of Essays corpus with Naïve Bayes and SVM.

Notably, their study showed that the classification accuracy could be improved by a large amount (from 54% to 93% in terms of agreeableness) as the noise of non-discriminating features was ruled out with manual restrictions such as corpus frequency or relative frequency ratios. It also demonstrated the potential of *open-vocabulary approach* in personality recognition tasks. Mairesse et al. (2007) employed continuous modeling techniques, i.e., regression and ranking models. This work is inspired by the work of Oberlander and Nowson (2006) which defined personality recognition as a fine-grained multiple-classification task. Mairesse et al. (2007) are also the first to recognize personality traits in dialogues and to use both self-reports and observer ratings as the golden labels.

They used a list of features including LIWC, MRC (Coltheart, 1981), utterance features based on speech act categories from Walker and Whittaker (1990), and prosodic features like voice's pitch characterized by Praat (Boersma, 2001). Experimental results came to several interesting conclusions: 1) The reported correlations between features suggested by the psycholinguistic literature and the personality ratings are weak in general; 2) Features selection is of vital importance; 3) More complex models (e.g., SVMs and boosting algorithms) tend to perform better on a large corpus than the simple ones (e.g., Naïve Bayes or regression trees); 4) Ranking models outperformed classifiers in personality recognition task, which can be attributed to that personality traits vary continuously among the population; 5) The observer ratings can be captured more precisely than the self-report scores.

The dramatic performance improvement of deep learning models relax the constraint of feature selection in personality recognition, i.e., from predefined features suggested by the psycholinguistic literature to more dimensions, longer input sequences, or multimodality features. Yuan et al. (2018) employed CNNs to extract the features automatically from the textual content besides LIWC on the MyPersonality dataset (Stillwell and Kosinski, 2015). They discovered the preference for emotion words, social process words, positive words, or negative words varies among different personality traits. This is almost contradictory to the traditional theories mentioned above about the role of function words in personality recognition. It is also notable that myPersonality is based on users' Facebook homepages, i.e., social media content. It also shows the trend of personality recognition from social media data as well as using more powerful deep learning methods.

C. Automatic Personality Recognition from Social Media

As analyzed in the previous section, most of the aforementioned personality recognition research focused on small datasets and/or closed vocabulary approaches. Hence, these methods usually had a poor performance in generalization and regression (or classification) results (Plank and Hovy, 2015; Schwartz et al., 2013; Iacobelli et al., 2011). Kosinski et al. (2013) modeled personality traits on a continuous scale, based on the myPersonality dataset. They utilized singular value decomposition (SVD) to reduce dimensionality of Facebook Likes data and fed them into logistic regression to predict continuous individual personality trait scores. Experimental results showed that, besides personality traits, individual personal attributes such as sexual orientation or intelligence can be inferred through their Facebook Likes automatically and accurately.

Wu et al. (2015) employed LASSO linear regression models in personality recognition to test whether they could surpass human judgements which are observer ratings collected from the participant's family and friends. The results showed the participants' self-reports are more correlated with the employed models ($r = 0.56$) than the human judgements ($r = 0.49$), where r is the Pearson product-moment correlation. However, in the study of Wu et al. (2017), Facebook contents showed better performance than Facebook Likes and self-reports in *homophily* prediction (prediction of the tendency to interact with or be attracted to similar others).

The aforementioned studies showed that social media can provide a good source of data for personality recognition by mitigating the biases coming from questionnaires and predefined psycholinguistic features, as it contains fine-grained and natural descriptions of an individual's personality. The development of deep learning methods exploited the potential of social media data and made it possible to capture information directly from the input sequences, such as n-gram features, besides the suggested features from psycholinguistic literature. Majumder et al. (2017) proposed a novel CNN-based deep learning model for personality recognition. They used the CNN feature extractor to obtain the n-gram sentence representation from the fed sentences, in which each word is vectorized as a word embedding by pretrained word2vec embeddings. The obtained vectors are concatenated with the Mairesse features (Mairesse et al., 2007) to obtain the document-level representation which was fed to a fully connected layer for classification. Experimental results indicated that the was no performance improvement of SVM dealing with the extracted features from CNN compared with a final fully connected layer. It suggested the potential of deep learning models in personality recognition tasks and the research direction of applying LSTM recurrent network to personality recognition.

Su et al. (2016) employed an RNN fed with LIWC-based and grammatical features for personality recognition in conversations. However, both works highly relied on domain expertise as they took predefined linguistic features as inputs and could not exploit the semantic information from the sentences. In the same year when Majumder et al. (2017) presented the CNN-based method, Liu et al. (2017a) developed a hierarchical model, Character to Word to Sentence for Personality Traits (C2W2S4PT), to obtain the vector representations of words and sentences from character level and word level. C2W2S4PT obtained better performance than previous feature-engineering-heavy works and machine learning methods, such as SVM Regression and Random Forest. These works inspired researchers to focus more on explore the capability of deep learning models with engineering-free features in personality recognition.

Sun et al. (2018) proposed to exploit the structures of texts to detect users' personalities by developing a hybrid model, i.e., BiLSTMs concatenated with CNN (2CLSTM). In this work, BiLSTMs are used to encode the context information, and CNN is utilized to capture the structural feature named latent sentence group (LSG) generated from BiLSTMs. According to Ramezani et al. (2022)'s research, recurrent classifiers, e.g., RNN-based, LSTM-based, and BiLSTM-based classifiers, outperformed CNN-based classifiers. They attributed the results to the reason that the inputs are sentences in the form of sequences and recurrent classifiers are more capable of processing temporal information, i.e., exploring the existing interactions among the words in a sentence. BiLSTM was superior to other recurrent classifiers as it does better in remembering information over several timesteps than RNN and focusing more on future information than LSTM, which explains why 2CLSTM employed BiLSTM for context encoding. Additionally, CNN-based classifier was believed to be good at feature extraction, which provides a support for utilizing CNN to capture the structural information in the work of Sun et al. (2018).

Recent studies showed that the researchers are not resigned to only semantic and structural information extraction with lightweight models anymore. They started to pay more attention to incorporating external knowledge graph or LLMs into personality classifiers. Kazemeini et al. (2021) investigated two approaches, a siamese BiLSTM and Sentence-BERT (Reimers and Gurevych, 2019), on distinguishing personality types. Both approaches were of siamese structure, as they were supposed to classify the personality trait type in an interpretable way where the semantics of the statements were captured so that statements with similar personality traits could be embedded close to each other in the semantic vector space. Poria et al. (2013) utilized an SMO classifier, combining commonsense knowledge based features with psycholinguistic features and frequency based features, for personality recognition. The proposed method surpassed other state-of-art methods based on psycholinguistic features and frequency analysis. The commonsense knowledge mainly came from resources such as ConceptNet and SenticNet. However, these methods usually ignored the external knowledge, conventional psycholinguistic features, semantic information, or LLM features and thus failed to fuse them all.

Mehta et al. (2020) developed a BERT-based model integrating traditional psycholinguistic features, commonsense knowledge and language model embeddings for personality prediction. The psycholinguistic features in this work included Mairesse features and Readability. In their fine-tuning setup, they experimented with logistic regression, SVM and a multi-layer perceptron (MLP) with 50 hidden units. Experimental results showed that the language modeling features beat conventional psycholinguistic features with commonsense knowledge features. Ren et al. (2021b) proposed a multi-label model integrating the pretrained BERT model with a neural network, which incorporated semantic and emotional features for personality recognition from social media texts. This work provided several significant implications: 1) External emotion knowledge improved the effectiveness of the personality recognition model; 2) Emotional information was found to be closely related to personality traits and could be applied to the further explanation of personality traits; 3) LLMs such as BERT could extract more contextual information than word embedding methods like word2vec or GloVe (Pennington et al., 2014), and it should the basis for future personality recognition with deep learning. These studies suggested that social media data is a good resource for the assessment of individual personalities in a more objective way than traditional questionnaires (Stachl et al., 2020).

4.4.5.2 MBTI Personality Recognition

A. Automatic Personality Recognition from Social Media

Compared to personality recognition with the Big Five personality traits which has been studied from multiple types of texts and with various approaches, personality recognition with MBTI did not attract the attention of researchers until recently and it exclusively utilizes $\mathbb{X}$ data.

The MBTI model profiled $\mathbb{X}$ users based on their tweets with each personality dimension modeled as a binary classification separately. Plank and Hovy (2015) annotated a corpus of 1.2M English tweets with gender and MBTI type. They trained a binary Logistic Regression classifier for each MBTI dimension (I-E, S-N, T-F, J-P) with the features including gender, n-grams, and count-based meta-features (e.g., the number of tweets, favorites, and followers). The models outperformed the majority-class baseline model on only two personality aspects, Thinking / Feeling and Introversion / Extroversion. Verhoeven et al. (2016) performed MBTI-based personality recognition from Tweets in six European languages with n-gram features. The models got the F_1 score from 0.47 to 0.79. Yamada et al. (2019) built a model for MBTI personality recognition on $\mathbb{X}$ posts in Japanese and behavioral features of $\mathbb{X}$ users. Experimental results indicated that user behaviors are crucial for recognizing the personality of users who have no posts, but it became less important as the number of collected user posts increased. Although all these studies are performed on large training datasets, they barely managed to surpass the majority-based or even random-guessing baselines.

B. Theoretical Analysis on Performance of MBTI Models

As mentioned in the last subsection, MBTI frameworks rarely outperformed the majority-base baselines. On the contrary, Big Five models trained on social media data, e.g., Facebook data, are more capable of prediction personality traits even than untrained humans (Wu et al., 2015). Štajner and Yenikent (2020) attributed the different performance of Big Five models and MBTI models to mainly two causes. First, the Big Five models are data-driven methods. However, MBTI frameworks are relatively more complex, which cover cognitive and information processing hypotheses and require more sophisticated models for the recognition and aggregation of different layers of personality traits. Second, the Big Five models were trained on the Facebook dataset while MBTI models on the $\mathbb{X}$ dataset. It is possible that the Facebook dataset has more content relevant to personal statements, thus it is more suitable for personality recognition tasks.

For example, the Sensing / Intuitive (S-N) typology results from the user's preference for cognitive processing. However, the tweets usually cannot provide sufficient data to figure out whether the user is interested in processing the data in an interpretative and intuitive way, or in a numerical and sensing way. The Judging / Perceiving (J-P) aspect covers the behavioral information of a user, e.g., planning and organization skills. However, the tweets are usually short and thus do not contain content about planning and organization. This had been pointed out and verified through the experiments of both Big Five and MBTI models on $\mathbb{X}$ data in Celli and Lepri (2018)'s research. Furthermore, Štajner and Yenikent (2020) carried out an annotation experiment on randomly selected MBTI-$\mathbb{X}$ data (Plank and Hovy, 2015), in which two annotators, one with psychology knowledge and the other with computational linguistic knowledge, were invited to annotate the users' personality traits from the 16 categories of MBTI.

According to their experiments, both annotators felt confident to label for the J-P and S-N tasks on only 15% and 30% of the samples, respectively. It is lower than that for E-I and T-F tasks (53% and 43%, respectively). The annotators showed low agreements with "gold labels". Moreover, the results raised concern about whether linguistic patterns in textual data could be expected in the MBTI framework constructed from traditional questionnaire. Štajner and Yenikent (2021) also analyzed the poor performance of MBTI models on $\mathbb{X}$ data. They conducted an annotation study to explore theoretical reasons behind this issue. Their research tested and confirmed two hypotheses: 1) $\mathbb{X}$ posts lack sufficient linguistic signals for MBTI personality recognition, even for trained annotators; 2) textual data do not align well with MBTI questionnaire scores. The study revealed three key findings: the style and content of texts can show signals of opposite polarities, individuals in the middle ranges of personality traits may display mixed signals, and the language used in $\mathbb{X}$ posts can make users appear more extroverted due to specific stylistic features like grammar, exclamation marks, sentence structure, and tonality.

C. Automatic Personality Recognition based on Deep Learning Methods

Another reason for the mediocre performance of MBTI models could be that the feature engineering needs to be improved. In the previous studies, the selected features, mainly in the form of n-gram and user behavioral meta-features, were quite simple. In addition, the models employed for personality recognition were machine learning methods, such as SVM, LinearSVC, and logistic regression. These models failed to extract complex features and limited the representation of the fed features as well. The advances of the deep learning methods contributed to the progress of the feature engineering of MBTI personality recognition and better experimental results.

Yang et al. (2021a) categorized existing approaches to combining multiple posts automatically into two groups. First, each document was encoded independently and then averaged to get the user representation (Keh et al., 2019). Second, the posts are concatenated into a long sequence in an arbitrary post order and then encoded hierarchically (Lynn et al., 2020). However, the first approach treated each post of a user equally and failed to distinguish their different importance weights or capture the interactions among them. The second approach overcame the above weakness but introduced the extra post-order bias, affecting the consistency of the users' personalities and the generalization of the models. Additionally, existing methods ignored the association between different post information and different personality aspects. To this end, Yang et al. (2021a) proposed a multi-document Transformer model as a post-order-agnostic encoder to gather the posts of a user without introducing the post-order bias. Moreover, they designed a personality-specific attention mechanism to allow each personality aspect to focus on relevant post information. Their model obtained 0.7047 and 0.6092 in Micro-F_1 on average on Kaggle[24] and Pandora[25] datasets, respectively.

---

[24] https://kaggle.com/datasnaek/mbti-type
[25] https://psy.takelab.fer.hr/datasets/all

However, most of the recent studies focus on applying various deep learning models in a data-driven manner instead of leveraging psycholinguistic knowledge to explore the connections between the user's language use and his or her personality traits. Motivated by the observation, Yang et al. (2021d) designed a psycholinguistic knowledge-based tripartite graph network, TrigNet, consisting of a BERT-based graph initializer and a tripartite graph network. The tripartite graph network injected structural psycholinguistic knowledge from LIWC. A new flow GAT was proposed to transmit information between neighboring parties in the tripartite graph and reduce the computational complexity in graph learning. This work provided a new way to exploit domain knowledge and an effective and efficient model for personality recognition tasks.

Yang et al. (2021b) argued that existing data-driven personality recognition models, capturing personality cues in social media posts implicitly, usually lacked the guidance of psychological knowledge. They believed that the user's posts could contain critical information which would help to answer the psychological questionnaire. Therefore, they presented a novel model, the Psychological Questionnaire Enhanced Network (PQ-Net), to guide personality recognition by linking the texts and questions in the questionnaire. PQ-Net consists of two streams: 1) a context stream to encode each textual input into a contextual text representation; 2) a questionnaire stream to capture question-relevant information in contextual representations and generate candidate answer representations to enhance the contextual representations. According to the experimental results, PQ-Net outperformed the baseline models and generated representations which are more inductive and distinguishable in term of personality dimensions.

Nevertheless, we may notice that there is still room for improvement. In the questionnaire stream, the preferred answer to each question was inferred based on the ground truth personality labels of the user. Then, the inferred answer labels were employed for supervised training PQ-Net. We have known that each personality aspect is based on the accumulation of the score in each question, although the final category of each personality dimension is a binary class. There are cases where people possibly belong to the middle ranges of certain personality aspect. For example, there are two choices "*Quiet and reserved.*" and "*A good mixer.*" corresponding to the Introversion and Extraversion categories, respectively, in the question "*Are you usually a good mixer with groups of people or rather quiet and reserved?*". Even the ground truth personality label of a user is Extraversion, it is still possible that his or her answer to the example question is "*Quiet and reserved.*" with some cues in posts saying "I need time alone sometimes". However, the labeling method in Yang et al. (2021b)'s work would generate a pseudo label "*A good mixer.*" to this question according to the user's final personality label Extraversion. The mismatch between the relevant post information and the inferred label might introduce a personality representation bias to the contextual representation.

4.4.6 Downstream Applications

Although theoretical research about personality traits can date back to the last century, personality recognition is still at a relatively early age. Most of the research focuses were defining the task, building the dataset, feature engineering, and designing methodologies. However, we can find some early applications of personality recognition in different research domains according to existing literature (Vinciarelli and Mohammadi, 2014).

4.4.6.1 Recommendation Systems

There are mainly two types of recommendation systems, i.e., conventional recommendation systems and Personality-Aware recommendation systems according to the research of Dhelim et al. (2022). Conventional recommendation systems usually include three steps: 1) At the rating stage, the users show their interests or evaluations by rating the items; 2) At the filtering stage, the content similarity and/or rating similarity are computed to find out the matched items; 3) At the recommendation stage, the system recommends the systems retrieved by the filtering stage. Compared with conventional recommendation systems, personality-aware recommendation systems include two more stages before the rating stage and also adjust the filtering stage. Firstly, at the personality measurement and matching stage, the system obtains the personality traits of the users through a self-report questionnaire or personality recognition techniques on the users' history data, e.g., social media data. Then, the user's personality type is matched with relevant items. The relevance can be determined by lexical matching or fine-grained rules. Secondly, at the filtering stage, the personality similarity is also calculated between each pair of users, besides the content and/or rating similarity.

Balakrishnan and Arabi (2018) proposed a hybrid movie recommendation system, HyPeRM, enhanced by users' personality traits and demographic information (e.g., age and sex). Specifically, the system employed collaborative filtering to filter the movies on the basis of demographic information and personality. Their work indicated that incorporating users' personality and demographic information improved the results of movie recommendations. Asabere and Acakpovi (2020) designed a TV program recommendation system based on personality and social properties. They generated the group recommendations for TV programs by leveraging personality similarity coupled with normalization procedure and computing tie strength with folksonomy. In the studies mentioned, the users' personality traits were collected through questionnaires, but they proved the effectiveness of incorporating personality traits in recommendation systems. There have been several papers presenting recommendation systems compatible with both questionnaire-based and personality recognition-based techniques. For example, Dhelim et al. (2020) designed a product recommendation system named Meta-Interest to infer users' needs based on their topical interests. They integrated the users' Big Five personalities to enhance the interest mining process and filter the recommended product.

Recently, many research works focused on combining personality recognition with recommendation systems. As some users are not willing to answer questionnaires for privacy issues (Wu, 2017), Lu and Kan (2023) studied whether the automatically detected personality information could improve the performance of recommendation systems through empirical experiments. They designed a Big Five personality-enhanced neural collaborative filtering algorithm, which encoded user, item, and personality into latent vectors and projected them through an MLP to get the probability that the user would be the item. They found that the personality traits of agreeableness and extroversion contributed to the improvement of the recommendation performance. Yakhchi et al. (2020) inferred users' personality traits implicitly by analyzing their online-generated content and then incorporated such traits, users' personal interests, and users' knowledge level into their recommendation system.

Instead, Christodoulou et al. (2022) extracted MBTI personalities from online reviews and employed them with themes obtained from topic modeling to represent restaurant and user features. The two kinds of features with hard-coded review features were combined to enhance the user-item matrix which was utilized to train an XGBoost regression model for restaurant recommendation. The results showed the integration of MBTI personality and topics surpassed the traditional collaborative filtering techniques, such as SVD and non-negative matrix factorization (NMF) models. Yang and Huang (2019) carried out an in-depth study on designing a task-specific personality recognition model for game recommendation systems, instead of using certain existing personality recognition models. They used two methods to recognize the Big Five personality type, i.e., M5' Regression Tree (Wang and Witten, 1996) computing the Pearson's Correlation Coefficients between the Big Five personalities and LIWC, and myPersonality recognizer approach calculating Big Five scores for each keyword of identify the personality traits from texts. Furthermore, they proposed to calculate the personality of a game on top of user personality. Thus, they devised three recommendation schemes, based on the relatedness between personalities: 1) user-based recommendation; 2) game-based recommendation; 3) hybrid recommendation (considering both user- and game-based personality similarity). Experimental results showed promising outcomes, offering a new perspective on personality modeling in recommendation systems.

4.4.6.2 Social Network Analysis

There have been research works that studied the relationship between the personality traits of individuals and different aspects of social media, and how their personality influenced relations with other group members (Maria Balmaceda et al., 2014). According to Dolgova et al. (2010), the user's personality could shape his or her structural position on social networks. Correa et al. (2010) investigated personality traits as important factors for the social media engagement of individuals. They studied the relationship between users' personalities and the use of social media and found that emotional stability is negatively related to social media use while extraversion is positively related.

Maria Balmaceda et al. (2014) studied the association rules among Big Five personality traits of users and provided some insights about how people communicate and who they tend to communicate with based on the personalities on social networks. Their research suggested that users with a certain personality aspect value tended to communicate with others having a similar value in the same personality aspect. Additionally, they also discovered some interaction patterns between different aspects of personality. For example, emotionally stable people tended to communicate with agreeable ones, and agreeable users tended to communicate with extroverted ones. Their research also provided empirical support for friend recommendations, based on the personalities on social networks.

The explosive increase in the user number on social media made it a hard task for users to find interesting people to communicate with. Therefore, many research works on social networks paid more attention to proposing personality-aware recommendation algorithms. Ning et al. (2019) utilized Big Five personality traits obtained from the questionnaire and the users' harmony rating to enhance the hybrid filtering process for the personality-aware friend recommendation system PersoNet. It outperformed the traditional collaborative filtering-based friend recommendation algorithms. Chakrabarty et al. (2020) employed Hellinger-Bhattacharyya Distance (H-B Distance) to compute the similarity of users' Big Five personality traits and used it to recommend friends in their personality-aware friend recommendation systems, named FAFinder (Friend Affinity Finder).

However, the aforementioned methods have some limitations, as users are usually reluctant to filling in the questionnaire even it is for non-commercial objectives such as friend recommendation. Therefore, it is more feasible to use the personality recognition-based friend recommendation algorithm instead of the questionnaire-based. Tommasel et al. (2015) obtained the Big Five personality scores from $\mathbb{X}$ data through the tool of Mairesse et al. (2007) and used them to analyze the influence of personality traits on followee selection. They found that personalities should be considered a distinctive factor in followee selection. Moreover, they also observed the association between different dimensions of personality.

The study by Tommasel et al. (2016) also reached a similar conclusion, suggesting that incorporating personality traits could improve friend recommendation algorithms. Xiao et al. (2018) proposed a personality-aware followee recommendation model based on sentiment analysis, text semantics, and the Big Five personality traits calculated by a Chinese language psychological analysis system TextMind from the Chinese data collected from Sina Weibo. Experimental results showed the personality traits obtained from personality recognition were crucial for followee selection on Chinese social networks as well.

4.4.7 Summary

4.4.7.1 Technical Trends

As shown in Table 4.16, there are mainly two tasks in personality recognition, i.e., Big Five personality recognition and MBTI personality recognition. Due to the distinct research progress of the two tasks, different research trends were observed accordingly. The Big Five personality model has been the subject of extensive research in both psychology and linguistics for a considerable period of time. Therefore, the early research on Big Five-based personality recognition used LIWC (Francis and Booth, 1993) features, a "gold standard" from the aspects of linguistic, grammatical, and psychological dimensions. The advance of machine learning approaches and computational linguistics inspired researchers to attempt on *open-vocabulary approach*, e.g., *n*-gram features (Oberlander and Nowson, 2006), and multimodality features (Mairesse et al., 2007). At the same time, there is a trend to applying more powerful deep learning models such as CNN to extract features automatically from the texts (Yuan et al., 2018) besides LIWC.

With the rapid progress of social media like Facebook and X, there is a growing need to automatically recognize the personality type of people through their posts instead of their self-reports or questionnaire. The development of deep learning models especially the pretrained LLMs enabled the researchers to build powerful models to exploit more complicated structures of texts, and integrate external knowledge and traditional psycholinguistic features for personality recognition (Mehta et al., 2020; Ren et al., 2021b). The MBTI model was developed using an *open-vocabulary approach*, lacking substantial theoretical research background in psycholinguistics. This approach contrasts with Big Five-based personality recognition, where the theoretical discussion precedes empirical research. Furthermore, Štajner and Yenikent (2020, 2021) focused on explaining why MBTI frameworks could not obtain good performance in certain cases. Another major difference between MBTI- and Big Five-based personality recognition is that MBTI frameworks (Plank and Hovy, 2015; Yamada et al., 2019) were initially tailored for processing social media data instead of self-reports. Like Big Five frameworks, MBTI frameworks benefited from the advance of deep learning models (Yang et al., 2021a) and the integration of commonsense and psychological knowledge (Yang et al., 2021d).

4.4.7.2 Application Trends

As shown in Table 4.17, personality recognition has primarily been applied to product recommendation systems and social network analysis. Initially, these fields relied on questionnaires to determine personality types. Over time, they adopted personality recognition frameworks for scoring. However, most applications favored machine learning-based frameworks or existing tools with basic feature engineering over designing more advanced models for downstream tasks.

Table 4.16: Technical trends in personality recognition. SA denotes statistical analysis. W2V denotes word2vec. C2W2S denotes character to word to sentence representations. KG denotes knowledge graph. L&R denotes the number of like and retweets. TPS-E denotes token, position, and segment embeddings. PWC-G denotes post, word and category graphs. PQA denotes posts, questions, and chosen answers. LR denotes logistic regression. TF denotes Transformers. In the metrics of Big Five, the performance is averaged over different classes, where 5 denotes all Big Five (O, C, E, A, and N) classes; 2 denotes N and E classes; 4 denotes C, E, A and N classes. MBTI metrics are measured by the averaged performance on I-E, S-N, T-F, P-J. rRate denotes the ratio of the predicted Pearson product-moment correlation coefficient to the actual Pearson product-moment correlation coefficient.

Task	Reference	Tech	Feature	Framework	Dataset	Score	Metric
	Pennebaker and King (1999)	SA	LIWC	-	Essay	-	-
	Argamon et al. (2005)	ML	Linguistics	SVM	Self-collected	0.58	Acc-2
	Oberlander and Nowson (2006)	ML	LIWC, n-grams	Naïve Bayes	Self-collected	0.92	Acc-4
	Mairesse et al. (2007)	ML	LIWC & MRC	SVM	Self-collected	0.57	Acc-5
	Yuan et al. (2018)	DL	LIWC, W2V	CNN	Alam et al. (2013)	0.62	Acc-5
	Kosinski et al. (2013)	ML	Facebook Likes	SVD, LR	Self-collected	0.53	rRate-5
	Wu et al. (2015)	ML	Facebook Likes	LASSO	Self-collected	0.56	Acc-5
Big 5	Wu et al. (2017)	ML	Facebook content	Linear Reg.	myPersonality	0.33	Corr.-5
	Majumder et al. (2017)	DL	n-grams, W2V	CNN	Essay	0.58	Acc-5
	Su et al. (2016)	DL	LIWC, POS	RNN	Tseng (2004)	0.80	Acc-5
	Liu et al. (2017a)	DL	C2W2S	RNN	Rangel et al. (2015)	0.13	MSE-5
	Sun et al. (2018)	DL	GloVe	BiLSTM, CNN	Essay	0.56	Acc-5
	Ramezani et al. (2022)	DL	KG	BiLSTM	Essay	0.71	Acc-5
	Kazemeini et al. (2021)	DL	RoBERTa	BiLSTM	Essay	0.61	Acc-5
	Mehta et al. (2020)	DL	psycholing., BERT	MLP	Essay	0.61	Acc-5
	Plank and Hovy (2015)	ML	Gender, n-grams	LR	Self-collected	0.67	Acc
	Verhoeven et al. (2016)	ML	n-grams	LinearSVC	Self-collected	0.64	F_1
MBTI	Yamada et al. (2019)	ML	BOW, SVD, L&R	SVM	Self-collected	0.66	AUC
	Yang et al. (2021a)	DL	TPS-E	TF	Kaggle	0.70	F_1
	Yang et al. (2021d)	DL	PWD-G, BERT	Graph network	Kaggle	0.71	F_1
	Yang et al. (2021b)	DL	PQA, BERT	TF, soft gate	Kaggle	0.71	F_1

Instead, they paid more attention to how to combine the obtained personality scores with the product recommendation systems (Yakhchi et al., 2020) or friend recommendation systems (Tommasel et al., 2015; Xiao et al., 2018). As the downstream tasks of personality recognition are still at a relatively early stage, there are a large number of papers studying the effect of incorporating the personalities into the downstream tasks (Lu and Kan, 2023; Xiao et al., 2018; Ning et al., 2019). Their research provided theoretical justification and valuable insights for combining personality recognition with downstream tasks. It is also notable that most of the applications utilized Big Five frameworks instead of MBTI frameworks, which is possible because of the unsatisfying performance of personality recognition models for MBTI.

Table 4.17: Personality recognition downstream task applications.

Reference	Downstream Task	Feature	Parser	Explainability
Balakrishnan and Arabi (2018)	Recommendation Systems	✓	✓	
Asabere and Acakpovi (2020)	Recommendation Systems	✓	✓	
Dhelim et al. (2020)	Recommendation Systems	✓	✓	
Lu and Kan (2023)	Recommendation Systems	✓		
Yakhchi et al. (2020)	Recommendation Systems	✓	✓	
Christodoulou et al. (2022)	Recommendation Systems	✓	✓	
Yang and Huang (2019)	Recommendation Systems	✓	✓	
Correa et al. (2010)	Social Network Analysis		✓	
Maria Balmaceda et al. (2014)	Social Network Analysis		✓	
Ning et al. (2019)	Social Network Analysis	✓		
Chakrabarty et al. (2020)	Social Network Analysis	✓		
Tommasel et al. (2015)	Social Network Analysis		✓	✓
Tommasel et al. (2016)	Social Network Analysis	✓		
Xiao et al. (2018)	Social Network Analysis	✓	✓	

4.4.7.3 Future Works

Automatic personality recognition is to automatically detect and analyze an individual's personality traits through his or her behavior patterns. The personality recognition models are mainly machine learning or deep learning classifiers based on Big Five frameworks or MBTI frameworks. Currently, on the basis of vast knowledge and expertise in computing areas, the personality recognition approaches focused on leveraging machine-detectable cues (e.g., psycholinguistics, affective computing, and social signal processing) to understand and explain social and psychological phenomena.

However, these approaches are more like applications of NLP or machine learning in the psychology domain and seem to pay little attention to the psychological and linguistic theories that set research grounds for modeling personalities. Hence, they ignored the fact that certain personality traits may not be found in textual data (at least social media posts) and tried hard to improve the classification accuracy uncritically. This situation has been found in MBTI frameworks. On the contrary, some psychology researchers seem to use more adequate datasets or well-designed feature engineering for personality recognition, but they may fail to evaluate their frameworks with the same rigorous metrics as the machine learning methods.

In addition, existing downstream applications made much effort to directly incorporate the personality traits into the recommendation systems or friend recommendation algorithms or analyze the impact. However, the personality types or scores in most of the research were collected by questionnaire or predicted by existing *open-vocabulary approaches*-based frameworks. Thus it would be useful for future research work to design both a task-specific personality recognition framework (Zhu et al., 2023) and an effective method to combine the predicted personalities with the application systems (Zhu et al., 2024a).

Finally, there is a desire to combine persona analysis with other NLU tasks. Many NLU tasks involve labels that are subjective to individuals, as the same statement can cause different affective responses between different people. For example, introverted individuals may react negatively to a suggestion to perform in front of audiences, whereas extroverts might welcome such an opportunity. Current NLP research often overlooks personalized prediction, as it tends to rely on majority-defined ground-truth labels. However, this approach contradicts the spirit of human-centric AI, as it neglects user diversity. Therefore, future endeavors could explore the integration of persona analysis with a broader range of NLU tasks.

4.5 Aspect Extraction

Aspect extraction is the process of identifying and extracting specific components, attributes, or features mentioned in text, typically employed to identify opinion targets in the context of ABSA. Besides ABSA, aspect extraction is useful in tasks like text summarization (Tang et al., 2024), financial forecasting (Du et al., 2023a), recommendation systems (Karthik and Ganapathy, 2021), fake review detection (Bathla et al., 2022), and many more. In customer feedback analysis, it pinpoints which aspects of products or services are frequently mentioned and their associated sentiments. Aspect extraction also aids in topic modeling, trend analysis, and competitive analysis by identifying specific subjects or features discussed in texts. In chatbots and virtual assistants, it improves response accuracy to user inquiries. Moreover, it enhances content recommendation and personalization by identifying aspects users engage with, offering more relevant content and insights into market trends.

As mentioned, aspect extraction is key for ABSA and aspect-based opinion mining (ABOM) in order to enable a more nuanced comprehension of specific requirements or opinions related to a product or service. In particular, while ABSA is more focused on analyzing sentiments associated with aspects, ABOM is a broader term that includes not only sentiment analysis but also the extraction and analysis of various types of opinion-related information. Hu and Liu (2004) identified two types of aspects, namely explicit and implicit aspects. Aspect terms that are explicitly specified in given sentences are explicit aspects, while aspects without corresponding words are considered implicit aspects (Dalila et al., 2018).

For instance, consider the statement "it's very light-weight and we can get amazing pix too", where the aspect "weight" is explicitly specified (Ganganwar and Rajalakshmi, 2019). In contrast, in the statement "It is very light. You can carry it everywhere", the aspect "weight" is implicit. Compared with explicit aspects extraction (Tubishat et al., 2021; Behdenna et al., 2022), implicit aspects are more challenging due to their indirect nature and reliance on contextual and common-sense understanding (Verma and Davis, 2021; Zhuang et al., 2022; Ahmed et al., 2022). However, implicit aspects are important linguistic phenomena that cannot be ignored in real-world applications.

4.5.1 Theoretical Research

4.5.1.1 Ontology

Ontology refers to a structured collection of concepts and categories within a specific subject area or domain, representing their properties and relationships. Ontology is important for aspect extraction because it serves to define aspect taxonomies by organizing aspects into classes and specifying their properties and interrelations. While identifying food- and service-related aspects in the restaurant domain may be straightforward, professional contexts such as environmental, social, and governance (ESG)-related aspect extraction require a more systematic approach through ontology engineering. Here, ontology defines the related aspects that need to be extracted, emphasizing the crucial role of ontology engineering in this process.

Noy et al. (2001) defined eight steps for developing an ontology, starting with defining classes to represent concepts or entities in the domain. These classes are then organized into a taxonomic hierarchy, where subclasses inherit properties from their superclasses, aiding in structuring the ontology. Slots are defined to represent properties or attributes of classes, specifying relationships between them. Instances are then created to fill in values for slots, representing specific examples or members of classes. This process is iterative, with the ontology evolving through evaluation and revision based on feedback from applications or experts.

Vrandečić (2009) highlighted key aspects of a good ontology, including conciseness, consistency, and organizational fitness. Conciseness assesses the inclusion of irrelevant or redundant axioms and the ontological commitment imposed. Consistency checks for logical contradictions and consistency between formal and informal descriptions. Organizational fitness evaluates the ease of deployment within the organization and alignment with existing ontologies.

4.5.1.2 Elliptical Constructions

The phenomenon of implicit aspects can be attributed to elliptical constructions in language. These constructions enable speakers and writers to convey information more efficiently by omitting redundant or predictable elements, thereby expediting language production and comprehension (Pickering and Garrod, 2013). From a cognitive point of view, elliptical constructions help conserve mental resources by omitting elements that can be inferred from context, thus reducing processing and memory load (Gibson, 1998). There are several types of elliptical constructions. Verb phrase ellipsis involves the omission of verb phrases, as in the example

(6)   You may like the book, but I don't [like the book].

Noun phrase ellipsis involves the omission of a noun phrase, e.g.,

(7)   My friend likes the first Chinese dish, but I like the second [Chinese dish].

In both cases, the elided elements with square brackets can be inferred from context, but the elided phrase does not necessarily have the same reference as the antecedent, as references are context-dependent.

(8)   Jim likes his new laptop. Jane does too.

In Example (8), Jim likes Jim's new laptop, while Jane likes Jane's laptop. The elided element refers to a different entity. Such a situation also happens to Example (7), where the two Chinese dishes are not the same. Another type of ellipsis is gapping, which involves the omission of non-initial verb phrases in coordinated clauses, where only the first verb phrase is fully expressed (Johnson, 2006).

(9)   For the new laptop, Jim likes the appearance, and Jane [likes] the screen.

Gapping is restricted to coordinations and typically involves the removal of material from the second conjunct, with remnants being major constituents like subjects, and objects. Both verb phrase ellipsis and gapping involve the omission of verb phrases, however, they occur in different contexts. Verb phrase ellipsis often requires the presence of an auxiliary verb in the second clause (e.g., "don't" in Example (6)), while gapping does not have this requirement. Sluicing is another type of ellipsis that occurs when a wh-phrase and a part of the containing clause are omitted, leaving a clause with an interrogative meaning (Chung et al., 1995). For example,

(10)   My friend wrote some reviews for the book, but I don't know [what the reviews are].

In sluicing, the missing information is not explicitly provided, making it challenging for the listener or reader to infer the exact content of the missing elements. Aspect extraction involves identifying and extracting aspects or attributes of entities or topics mentioned in text. Elliptical constructions e.g., verb phrase ellipsis and noun phrase ellipsis can pose challenges for aspect extraction because they involve the omission of key elements (verb phrases or noun phrases) that are relevant for determining aspects. Gapping and sluicing can lead to ambiguity regarding which aspects are associated with which entities or topics in the sentence, further complicating aspect extraction in NLU. Considering the frequent occurrences of elliptical constructions, they cannot be overlooked in developing aspect extraction systems.

4.5.2 Annotation Schemes

Aspect extraction is usually formulated as a sequence-labeling task, in which each token in a sentence should be assigned a label. The three common annotation schemes are shown in Table 4.18. The IO scheme represents aspect extraction as a set of binary classification tasks, where "I" (Inner) indicates that a token is part of an aspect, while "O" (Other) indicates that it is not. In contrast, the BIO scheme introduces three labels: "B" (Beginning), "I" (Inner), and "O" (Other).

The "B" label denotes the beginning of an aspect, the "I" label indicates that the token is inside an aspect, and the "O" label signifies that the token is not part of any aspect. Building upon the BIO scheme, the BIOES scheme incorporates two additional labels: "S" (Single) and "E" (End). These labels are designed to provide clearer annotations of aspect boundaries. The "S" label is used for single-token aspects, while the "E" label marks the end of an aspect entity.

Table 4.18: The three common annotation schemes for aspect extraction.

token	I	like	the	user	interface	design	but	the	price	is	unaffordable
IO	O	O	O	I	I	I	O	O	I	O	O	O
BIO	O	O	O	B	I	I	O	O	B	O	O	O
BIOES	O	O	O	B	I	E	O	O	S	O	O	O

4.5.3 Datasets

Widely used aspect extraction datasets are shown in Table 4.19. The SemEval-2014 Task 4[26] (Pontiki et al., 2014) consists of four subtasks: aspect term extraction, aspect term polarity detection, aspect category detection, and aspect category polarity detection. The task involves two domain-specific datasets focused on laptop and restaurant reviews, which include over 6,000 sentences. These datasets are annotated at a fine-grained aspect level.

Table 4.19: Aspect extraction datasets and statistics. AE denotes aspect extraction; AC denotes aspect category detection; ATP denotes aspect term polarity detection; ACP denotes aspect category polarity detection.

Reference	Task	Source	# Sample
Pontiki et al. (2014)	AE, AC, ATP, ACP	Laptop and restau. reviews	6,000 sentences
Pontiki et al. (2015)	AE, AC, ATP	Laptop and restau. reviews	550 reviews
Pontiki et al. (2016)	AE, AC, ATP	Laptop and restau. reviews	24,897 sentences
Hu and Liu (2004)	AC,ACP	Digital product reviews	4,523 sentences
Orbach et al. (2021)	AE, AC, ATP	Business and product reviews	3,815 sentences
Alturaief et al. (2021)	AE, ATP	App reviews	11,323 sentences
Toledo-Ronen et al. (2022)	AE, ATP	Reviews from multi-domains	2M sentences
Jiang et al. (2019)	AE, AC, ATP, ACP	Restaurant reviews	13,854 instances
Peng et al. (2018)	AE, ATP	Product reviews	2,309 reviews
Vo and Zhang (2015)	AE, ATP	X	6,940 reviews

---

[26] https://alt.qcri.org/semeval2014/task4

```
<sentence id="813">
<text>All the appetizers and salads were fabulous, the steak was mouth watering and
the pasta was delicious!!!</text>
<aspectTerms>
<aspectTerm term="appetizers" polarity="positive" from="8" to="18"/>
<aspectTerm term="salads" polarity="positive" from="23" to="29"/>
<aspectTerm term="steak" polarity="positive" from="49" to="54"/>
<aspectTerm term="pasta" polarity="positive" from="82" to="87"/>
</aspectTerms>
<aspectCategories>
<aspectCategory category="food" polarity="positive"/>
</aspectCategories>
</sentence>
```

The SemEval-2015 Task 12[27] (Pontiki et al., 2015) is a continuation of SemEval-2014 Task 4, which also contains restaurant and laptop reviews. The difference between the two tasks is that the input of the latter dataset contains entire reviews, rather than isolated sentences without context. The SemEval-2015 Task 12 contains in-domain ABSA and out-of-domain ABSA task setups.

```
The food was delicious but do not come here on an empty stomach.
{category= "FOOD#QUALITY", target= "food", from: "4", to: "8", polarity= "positive"},
{category= "FOOD#STYLE", target = "food", from: "4", to: "8", polarity= "negative"}
```

SemEval-2016 Task 5[28] (Pontiki et al., 2016) was also developed for ABSA from laptop and restaurant reviews, but it introduced new test datasets for evaluation and included multilingual datasets encompassing English, Arabic, Dutch, Russian, Spanish, and Turkish at both the sentence and document levels. Hu and Liu (2004) proposed an aspect-level sentiment analysis dataset[29] for mining and summarizing customer reviews of digital products. The instances in the dataset contain aspect category labels and sentiment polarities, where the sentiment polarities range from -3 (strong negative) to +3 (strong positive). In the following example, the numbers in the square brackets denote the sentiment polarities; the terms before the sentiment polarities are aspects; ## denotes the start of a review sentence.

```
Title: great camera
canon g3[+3]##i bought my canon g3 about a month ago and i have to say i am
very satisfied .
photo quality[+2]##i have taken hundreds of photos with it and i continue to be
amazed by their quality .
feature[+2]##the g3 is loaded with many useful features , and unlike many smaller
digital cameras , it is easy to hold steady when using slower shutter speeds .
```

Orbach et al. (2021) proposed a YASO dataset[30] as a solution to the limitation of current ABSA datasets, which are confined to a narrow range of review domains. These datasets cannot accommodate a wide variety of reviews from various domains found on websites such as Amazon or Yelp. The YASO dataset comprises 2215 sentences in English originating from numerous review domains. Aspect terms and their corresponding sentiment are annotated in this dataset.

```
text: "Great food but the service was dreadful !",
target: "food",
from: "6",
to: "10",
sentiment: "positive".
```

AWARE[31] (Alturaief et al., 2021) is an ABSA dataset annotated for App reviews. The 11,323 reviews were collected from 11,323, and were annotated for aspect terms, categories, and sentiment. Reviews were collected from three domains: productivity, social networking, and games.

```
domain: "productivity" ,
app: "things-3" ,
review_id: "c9274c0a-a120-4e09-816b-7a8ba3a16634" ,
sentence_id: "00808934-e8b9-42fa-b37f-cfeac234bbdd" ,
title: "Difficult to update from Things 2" ,
review: "This new version of Things has an entirely different aesthetic from Things 2.
Things 2 is much more minimalist; Things 3 seems to have a lot of UI bloat. Not quite
sure where the design award came from." ,
sentence: "This new version of Things has an entirely different aesthetic from Things
2." ,
rating: "3" ,
is_opinion: "TRUE" ,
category: "usability" ,
term: "new version" ,
from: "6" ,
to: "17" ,
sentiment: "positive".
```

Besides the above datasets, TSA-MD (Toledo-Ronen et al., 2022), MAMS (Jiang et al., 2019), and the studies (Peng et al., 2018; Vo and Zhang, 2015) are also ABSA datasets, which contain aspect extraction as a subtask.

---

[30] https://github.com/IBM/yaso-tsa

[31] https://zenodo.org/record/5528481#.ZE9wXexBz0p

4.5.4 Knowledge Bases

Table 4.20 shows various knowledge bases related to aspect extraction. Besides WordNet (Fellbaum, 1998), SentiWordNet (Baccianella et al., 2010), BabelNet (Navigli and Ponzetto, 2012b), FrameNet (Ruppenhofer et al., 2016), ConceptNet (Speer et al., 2017), SenticNet (Cambria et al., 2024), and Wiktionary (introduced earlier), some other useful knowledge bases for aspect extraction are DBpedia (Auer et al., 2007), YAGO (Suchanek et al., 2007), Freebase (Bollacker et al., 2008), Probase (Wu et al., 2012), Knowledge Vault (Dong et al., 2014), and FinSenticNet (Du et al., 2023b).

In particular, DBpedia is a large-scale multilingual knowledge base extracted from Wikipedia containing millions of entities. YAGO is a large knowledge base with general knowledge about people, cities, countries, movies, and organizations. Freebase is a practical, scalable tuple database that contains structured human knowledge. Probase is a large-scale concept knowledge map proposed by Microsoft, containing entities mapped to different semantic concepts and labeled with corresponding probability labels. Knowledge Vault is a web-scale probabilistic knowledge base that combines extractions from Web content with prior knowledge from existing knowledge bases. FinSenticNet is a lexicon tailored for financial sentiment analysis, including 6,741 concepts, over 65% of which are complex MWEs. Ontologies from professional domains, e.g., medicine (Lipscomb, 2000; Donnelly et al., 2006; Wishart et al., 2018; Hirsch et al., 2016; Johnson et al., 2016; Wheeler et al., 2007) and geology (Ahlers, 2013), can be also helpful for domain-specific aspect extraction.

Table 4.20: Useful knowledge bases for aspect extraction. These knowledge bases can also be helpful for polarity detection in Section 4.6 because they provide either sentiment scores or commonsense knowledge.

Name	Knowledge	# Entities	Structure
Probase	Structured lexicon	12,551,613	Graph
SenticNet	Structured lexicon	400,000	Dictionary
SentiWordNet	Structured lexicon	1,105 WordNet synsets	Graph
FinSenticNet	Structured lexicon	6,741	Dictionary
DBpedia	Structured entities	10M	Graph
Freebase	Structured entities	125M	Graph
YAGO	Structured entities	64M	Graph
BabelNet	Structured entities	22M	Dictionary
Knowledge Vault	Structured entities	45M	Graph

4.5.5 Evaluation Metrics

As a standard sequence labeling task, precision, recall, F_1, and accuracy are the main evaluation metrics in aspect extraction tasks. In particular, precision is used to measure the proportion of correctly identified aspects out of all aspects identified by the model, indicating how accurate the predictions are. Recall assesses the proportion of correctly identified aspects out of all actual relevant aspects present in the data, showing the model's ability to find all relevant aspects. The F_1 score, which is the harmonic mean of precision and recall, balances these two metrics and provides a single measure of a model's performance, especially when there is an uneven distribution between precision and recall. Accuracy measures the overall rate of correct predictions, comparing the total number of correct aspect predictions to the total number of predictions made. Together, these metrics provide a comprehensive evaluation of the model's ability to accurately identify aspects in text data.

4.5.6 Methods

4.5.6.1 Explicit Aspect Extraction

A. Unsupervised Method

Statistics-based methods are the most straightforward solutions for unsupervised aspect extraction. Agrawal et al. (1994) observed that individuals tend to use nouns and noun phrases to represent aspects within a sentence. Thus, they scanned documents for all nouns and noun phrases, categorizing those that exceeded a specified support threshold as frequent aspects. After identifying all the commonly occurring aspects, the closest adjectives were extracted as possible opinion words. This led to the extraction of aspect-opinion pairs. However, Hu and Liu (2004) noted that not all aspects are frequently mentioned. When a sentence had an opinion word but no frequent aspect, they identified the nearest noun as an infrequent aspect. This method produced lists of both frequent and infrequent aspects with their associated opinion words for aspect extraction tasks.

Bafna and Toshniwal (2013) improved the study of Hu and Liu (2004) by introducing probabilistic approach. Aspect extraction was performed using association rule mining. The frequent nouns were extracted as they often represent aspects. However, not all frequent nouns are target aspects, so a probabilistic power equation was utilized to eliminate those that do not represent aspects. Bloom et al. (2007) manually constructed taxonomies of opinion targets for two datasets. Their algorithm utilized a predefined set of dependency tree paths to detect opinion expressions and their associated targets. Due to the absence of annotated datasets containing opinion expressions and targets, they evaluated the accuracy of their algorithm by manually examining a portion of the output in various aspects.

There are also studies that utilized auxiliary information for aspect extraction. Moghaddam and Ester (2010) employed known features mentioned in reviews to identify specific aspects from customer reviews written in free text format. By introducing predefined aspects and their ratings, they improved the accuracy of opinion mining. In addition, Meng and Wang (2009) employed supplementary details about the products available on review websites and correlations between aspects and opinions to pinpoint the relevant aspects. Bancken et al. (2014) proposed an algorithm to automatically detect and rate the product aspects from customer reviews. This algorithm works by matching syntactic dependency path among different words from the sentence. To identify product aspects and their opinion words, ten handcraft dependency paths were defined. Then, the algorithm generates a syntactic dependency tree by using a dependency parser and extracts the basic aspect opinion pair from the sentence. Luo et al. (2019b) proposed an unsupervised model that employs word embeddings for aspect clustering. They utilized Probase and WordNet to identify prominent aspects and emphasized explicit aspects.

B. Semi-supervised Method

We categorized semi-supervised methods for aspect extraction into four groups: bootstrapping-based, parsing-based, statistics-based, and neural-based approaches. Hai et al. (2012) proposed likelihood ratio set and latent semantic analysis bootstrapping methods for computing the pair-wise associations between aspects and opinions. Zhao et al. (2014) employed a combination of a refinement process and a bootstrapping method to improve the aspect and opinion extraction process. The bootstrapping method involves automatic rule refinement to remove false results and update the rules, resulting in a more accurate extraction process. Yu et al. (2011a) proposed a method for automatically identifying significant product aspects from consumer reviews posted online with a shallow dependency parser. Their approach involves using a predetermined list of aspects to determine which ones are frequently ranked highly by users.

Mukherjee and Liu (2012) proposed two new statistical models for automatic extraction and categorization of aspect terms. Based on a Dirichlet forest LDA approach, they jointly trained models for aspects and sentiments with the utilization of distinct seed strategies. This study used the synonym lexicon to expand the aspect list. Neural-based methods have recently dominated the semi-supervised aspect extraction. Li et al. (2020d) introduced an end-to-end attentive framework comprising a general encoding component, a cross-view training component, and a moving-window attention component. Their findings highlighted the significant improvement in aspect prediction achieved by incorporating past nearby information. Zhou et al. (2021a) proposed an adaptive hybrid framework designed for cross-domain ABSA. Their approach involved employing adversarial training to align features through training a domain discriminator. Additionally, they developed an adaptive mean teacher as a semi-supervised component to reduce the impact of noisy pseudo labels.

Bajaj et al. (2021) argued that typical ABSA tasks suffered sentiment inconsistency and colossal search space. To overcome such problems, they propose a similar span-based extract-then-classify framework. An aspect extractor is separately used to extract the multiple possible targets from the sentence. He et al. (2023b) introduced a meta-based self-training method combined with a meta-weighter. They trained a teacher model to produce pseudo-labeled data, which is utilized by a student model for supervised learning. The meta-weighter component was jointly trained with the student model, providing subtask-specific weights to each instance, coordinating their convergence rates, balancing class labels, and reducing the impact of noise introduced during self-training.

C. Supervised Method

Kobayashi et al. (2007) proposed a dictionary-based approach, which identified aspects with the help of syntactic patterns. They found that the combination of contextual clues and a common co-occurrence statistics-based technique can yield a notable enhancement. Later, HMMs were employed to extract aspects by Jin et al. (2009). CRF is another popular method for aspect extraction at the early stage. Chen et al. (2012) utilized CRF-based methods and showed that CRF-based models outperform HMMs in ABOM tasks. Jakob and Gurevych (2010b) further demonstrated the cross-domain capability of CRF-based models in aspect extraction tasks. They introduced domain portability to CRF and evaluated the proposed method in both single and cross-domain settings.

Recently, there has been a trend towards integrating aspect extraction tasks into ABSA or ABOM tasks using neural methods. Two common approaches for ABSA and ABOM tasks are pipeline methods and MTL methods. In pipeline methods (Wang et al., 2017; Xu et al., 2018), aspect terms are first extracted, followed by the identification of sentiment polarities or related opinions. In contrast, MTL methods (He et al., 2019c, 2021) typically involve a shared encoder and multiple linear classifiers built on top of the encoder. These methods prioritize enhancing interactions among subtasks rather than focusing solely on aspect extraction. Mao and Li (2021) integrated the tasks of aspect extraction, opinion extraction, and aspect sentiment polarity classification into an MTL framework. They proposed a novel Gated Bridging Mechanism to facilitate the exchange of relevant information among distinct subtasks, while effectively screening out irrelevant information.

He et al. (2019c) proposed a message passing mechanism that facilitates communication among different tasks by utilizing a shared set of latent variables. This mechanism enables the simultaneous learning of multiple related tasks at both the token and document levels. Chen and Qian (2020) proposed an MTL approach that utilizes a stacked multi-layer network to capture collaborative signals among diverse subtasks. Zheng et al. (2020) utilized transfer learning to address the challenge of cross-domain ABSA. They proposed two approaches – Anchored Model Transfer and Soft Instance Transfer. Both methods are based on MTL and consider both model and instance transfer. The authors found that the proposed methods can mitigate the shortage of labeled data across diverse domains.

Supervised explicit aspect extraction is the most common setting for aspect extraction tasks. In this setting, employing external knowledge is a popular approach for enhancing performance. For example, Sentic LSTM (Ma et al., 2018) enhanced LSTM by integrating a hierarchical attention mechanism that includes target-level attention and sentence-level attention for ABSA. Additionally, they also incorporates commonsense knowledge from SenticNet for sentiment-related aspects. Ghosal et al. (2020) proposed a domain adaptive model to boost the performance of sentiment analysis with specific aspects. They utilized a graph convolutional autoencoder that leverages inter-domain concepts in a domain-invariant manner from ConceptNet.

Liang et al. (2022b) proposed Sentic GCN, building a graph neural network by incorporating affective knowledge from external knowledge bases, which is done to improve the dependency graphs of sentences. The novel approach integrates both the contextual and aspect word dependencies, as well as the affective information between opinion words and aspects, into the graph model, resulting in an enhanced affective graph model. SentiPrompt (Li et al., 2021a) is a method based on prompt tuning, which employs consistency and polarity judgment templates to create prompts related to ground truth aspects. Nonetheless, Mao et al. (2023c) raised concerns about potential biases that could exist in prompt-based methods for sentiment analysis and emotion detection.

4.5.6.2 Implicit Aspect Extraction

A. Unsupervised Method

Zhang et al. (2012) proposed a Weakness Finder method that extracts features and groups them explicitly using a morpheme-based approach and a similarity measure based on HowNet. Implicit features are identified and grouped using a collocation selection method for each aspect. Dependency parsing is a common technique in sentiment analysis that uses predefined rules to extract opinion targets and sentiment words by analyzing the relationships between words.

To enhance the unsupervised dependency parsing, Zainuddin et al. (2016) proposed a hybrid approach that combines rule mining, dependency parsing, and SentiWordNet. This approach enables implicit aspect extraction in an unsupervised manner. Karmaker Santu et al. (2016) utilized a probabilistic topic modeling method to extract implicit features. They modeled the reviews as generative probabilistic feature models, with the reviews represented as associations between sentences and features using hidden variables. Then, they employed the expectation-maximization technique to calculate the model parameters using tagged training data. Finally, they extracted implicit features using hidden variables and the calculated parameter values.

B. Semi-supervised Method

Jiang et al. (2014) employed association rule mining to extract implicit aspects, incorporating indicators like semi-supervised LDA topic models. These were integrated into an improved collocation model, which was then used to extract basic rules. These basic rules were further expanded with new rules generated by the topic model. The combination of these basic and new rules enabled the extraction of implicit aspects. Xu et al. (2015a) combined SVM with topic modeling to extract explicit and implicit aspects. They enhanced the LDA topic model with the proposed cannot-link, must-link, and prior knowledge. The must-link relation provides information about pairs of words that must be present in the same cluster, while the cannot-link relation provides information about pairs of words that cannot be in the same cluster. The LDA explicit topic model, augmented with these enhanced features, was then employed to identify relevant attributes from the dataset. Subsequently, several SVM classifiers were initiated, which were trained on the selected attributes to uncover implicit features.

Yu et al. (2018) argue that many current methods overlook the syntactic relationships between aspect terms and opinion terms, which can result in inconsistencies between the model predictions and syntactic constraints. To address this issue, an MTL framework is first employed to implicitly capture the relations between the aspects and opinions. Then, a global inference method is proposed, which explicitly models several syntactic constraints between aspects and opinions to reveal their intra-task and inter-task relationship. This approach aims to achieve an optimal solution over the neural predictions for both tasks. Maylawati et al. (2020) proposed a method to extract implicit aspect extraction. The proposed method contains feature extraction, feature selection, clustering, and association rule mining to recognize the final implicit aspects.

C. Supervised Method

By utilizing aspect hierarchy and opinion terms, Yu et al. (2011b) identified implicit aspects. They created a hierarchal organization by integrating product specifications and customer reviews, which allowed him to infer implicit aspects within review sentences. Yan et al. (2015) proposed a NodeRank algorithm which first identified all co-occurrences of the opinion words with the explicit aspects. Then, the algorithm calculates the NodeRank value for each potential aspect with the opinion word. The aspect with the highest value was considered as potential implicit aspect.

Liao et al. (2019) identified fact-implied implicit sentiment at the sentence level. To achieve this, they suggested a multi-level semantic fusion method for identifying implicit aspects in sentiment analysis. The corpus is used to learn three distinct features at varying levels, including sentiment target representation at the word level, structure embedded representation at the sentence level, and context semantic background representation at the document level. A fact-implied Chinese implicit sentiment corpus was also developed in this work.

Feng et al. (2019a) utilized a deep CNN with a sequential algorithm to label words in a sentence. Implicit aspects were identified by treating them as topics, and then, determining the degree of sentiment words and aspects that were matched. Initially, They extracted aspects using word vectors, POS vectors, and dependent syntax vectors to train a deep CNN. Next, a sequential algorithm was used to obtain sentiment labels with implicit aspect identification. Rana et al. (2020) argued that prior techniques for implicit aspect extraction have concentrated on certain types of aspects while disregarding the fundamental issue at hand. To deal with such a problem, they utilized co-occurrence and similarity-based techniques to identify implicit aspects in a multi-level approach, which crafted rules to identify clues and used recognized clues to extract implicit aspects. To identify implicit aspects, Xu et al. (2020b) proposed a NMF-based approach. They employed clustering based on the co-occurrence of opinion words and word vectors to gather contextual information about opinion targets from review sentences. Then, they utilized a classifier to forecast the implicit target of users' opinions.

4.5.7 Downstream Applications

4.5.7.1 Aspect-based Sentiment Analysis

ABSA and ABOM are the most popular downstream applications for aspect extraction. To tackle ABSA and ABOM in various settings, numerous tasks have been developed to analyze different sentiment components and their associations, such as aspect terms, aspect categories, opinion terms, and sentiment polarity. Unlike earlier ABSA studies that focused solely on a single sentiment component, recent research has investigated several compound ABSA and ABOM tasks that encompass multiple components to capture more comprehensive aspect-level information. For example, in the aspect-opinion pair extraction task, Chen et al. (2020); Zhao et al. (2020a) aimed to obtain a clearer comprehension of the mentioned opinion target and its associated opinion expression. It is necessary to extract the aspect and opinion terms as a compound, such as the pairs of (pizza, delicious) and (price, expensive).

Dai et al. (2020) tried to deal with aspect extraction and aspect category identification, simultaneously. Recently, finer-grained sentiment analysis tasks, e.g., Aspect Sentiment Quad Prediction (ASQP), Aspect Category Sentiment Analysis (ACSA), Aspect-Category-Sentiment Detection (ACSD), and Aspect Sentiment Triplet Extraction (ASTE) were proposed (Cai et al., 2021; Zhang et al., 2021b; Mao et al., 2022b; Bao et al., 2022; Wan et al., 2020; Wu et al., 2021a; Xu et al., 2020a; Mukherjee et al., 2021; Wu et al., 2020c). These tasks commonly considered aspect extraction as one of the learning tasks, providing necessary information for sentiment analysis. Fundamental aspect extraction tasks along with other advanced tasks can be structured in either an end-to-end fashion or a pipeline approach. Moreover, some tasks rely on explicit aspects, while others depend on implicit aspects.

4.5.7.2 Knowledge Graph Construction

A knowledge graph is a type of semantic network that employs a graph-based data structure comprising nodes and edges. Within a knowledge graph, each node corresponds to a real-world "entity", while each edge denotes a "relationship" linking entities. A knowledge graph is one of the most efficient ways of representing relationships. The typical creation of knowledge graphs involves three principal components: Information Extraction, Information Fusion, and Information Processing. Aspect extraction technology is usually used for extracting properties for given entities in the step of Information Extraction. For example, He et al. (2021) proposed an MTL-based technique for constructing genealogical knowledge graphs. The study collected unstructured online obituary data. Then, they extracted named entities as nodes and categorized the family relationships of recognized individuals as edges to construct genealogical knowledge graphs. For extracted family members, they also extract related aspects (properties), including their age, gender, and residence place to form more complete graphs.

Li et al. (2020c) introduced a systematic method for constructing medical knowledge graphs from electronic medical records. Then, technical experiments were conducted to assess its efficacy. Instead of using the classical triplet (subject, predicate, object), they adopted a novel quadruplet structure that added the property items to represent medical knowledge. The property items facilitated flexible customization for creating knowledge graphs. For instance, their costumed properties include the specific symptom probabilities, specificity, and reliability for given patients. Li et al. (2020a) developed e-commerce knowledge graphs that capture user issues, points of interest, item information, and their corresponding relationships. This approach assists in comprehending user requirements, responding to pre-sales inquiries, and creating explanatory texts. The constructed knowledge graph was implemented in various online business scenarios such as shopping guidance, property-based QA, and recommendation reasoning, all of which yielded favorable outcomes. The concept of item-property-value (IPV) is utilized to record property values of items, e.g., "cleansing foam-ingredient-bisabolol". The "ingredient" is an aspect of the "cleansing foam".

4.5.8 Summary

4.5.8.1 Technical Trends

As shown in Table 4.21, recent research focuses on addressing explicit and implicit aspect extraction tasks. The difference is that explicit aspect extraction identifies aspects explicitly mentioned in text, while implicit aspect extraction aims to uncover aspects that are not explicitly stated but can be inferred from the context. Implicit aspect extraction often involves identifying nuanced or indirect references to aspects, requiring a deeper understanding of the text.

Researchers have developed various techniques for both explicit and implicit aspect extraction, including rule-based methods, machine learning algorithms, and neural network models. These approaches have been applied across different aspect extraction domains, e.g., product, restaurant, and movie reviews, demonstrating the importance of aspect extraction in understanding opinions and sentiments expressed in text. At the early stage, linguistic features played a critical role in this domain. Researchers tried to identify aspects by analyzing the linguistic associations, dependency relationships, POS tags, and customer behaviors. However, such superficial features cannot provide a deeper contextual understanding of aspects.

Table 4.21: Technical trends in aspect extraction. PS denotes product specifications. PB denotes Probase. WN denotes WordNet. SWN denotes SentiWordNet. SN denotes SenticNet. Ling. denotes linguistic features. Dep. denotes dependency features. Cons. denotes consistency features. Some works did not individually report aspect extraction results, as they focused on evaluating models for polarity detection.

Task	Reference	Tech	Feature and KB.	Framework	Dataset	Score	Metric
Explicit	Agrawal et al. (1994)	Rule	Ling., trading	Rule-based	Synthetic data	-	-
	Bafna and Toshniwal (2013)	Rule	Ling., POS	Probabilistic	Self-collected	0.92	Acc.
	Pang and Lee (2004)	Rule	Ling.	Graph-based	Self-collected	0.90	Acc.
	Bloom et al. (2007)*	Rule	Ling., Lexic.	Probabilistic	Pang and Lee (2004)	-	-
	Kobayashi et al. (2007)	Rule	Ling., Stat.	Feature eng.	Self-collected	0.72	Prec.
	Moghaddam and Ester (2010)	Rule	Aspects, ratings	Feature eng.	Self-collected	0.80	Prec.
	Meng and Wang (2009)	Rule	Ling., PS	Feature eng.	Self-collected	0.75	F_1
	Bancken et al. (2014)	Rule	Dep., WN	Feature eng.	Self-collected	0.60	Acc.
	Luo et al. (2019b)	Rule	PB, WN, GloVe	Feature eng.	Self-collected	0.70	Acc.
	Hai et al. (2012)	ML	Ling.	Feature eng.	Self-collected	0.73	F_1
	Jakob and Gurevych (2010b)	ML	Ling., opinion	CRF-based	Self-collected	0.52	F_1
	Zhao et al. (2014)	ML	Depend., opinion	Graph-based	COAE2008	0.79	F_1
	Yu et al. (2011a)	ML	Boolean, Dep.	Probabilistic	Self-collected	0.74	F_1
	Mukherjee and Liu (2012)	ML	Ling.	Probabilistic	Self-collected	-	-
	Jin et al. (2009)	ML	Ling., POS	HMM	Hu and Liu (2004)	0.75	F_1
	Li et al. (2020d)	DL	GloVe	LSTM, att.	SemEval-2014 Task 4	0.72	F_1
	Zhou et al. (2021a)	DL	word2vec	Meta learn.	SemEval-2014 Task 4	-	-
	Bajaj et al. (2021)	DL	BERT	Transformer	SemEval-2014 Task 4	0.76	F_1
	He et al. (2023b)	DL	BERT	Meta learn.	SemEval-2014 Task 4	0.82	F_1
	Chen and Qian (2020)	DL	BERT	MTL	SemEval-2014 Task 4	0.81	F_1
	He et al. (2019c)	DL	BERT	MTL	SemEval-2014 Task 4	0.77	F_1
	Mao and Li (2021)	DL	BERT	MTL	Lapt.14 & Rest.14	0.85	F_1
	Ghosal et al. (2020)	DL	Emb., ConceptNet	Graph-based	Self-collected	-	-
	Liang et al. (2022b)	DL	Dep., SN, BERT	Graph-based	SemEval-2014 Task 4	-	-
	Li et al. (2021a)	DL	Bart, Cons.	Prompt-tun.	SemEval-2014 Task 4	-	-
Implicit	Hu and Liu (2004)	Rule	Ling., POS	Feature eng.	Self-collected	0.56	Prec.
	Zhang et al. (2012)	Rule	Ling., HowNet	Feature eng.	Self-collected	-	-
	Zainuddin et al. (2016)	Rule	Ling., dep., SWN	Feature eng.	Self-collected	-	-
	Jiang et al. (2014)	Rule	Ling.	LDA	Self-collected	0.78	F_1
	Karmaker Santu et al. (2016)	ML	Ling., TF-IDF	EM algorithm	Self-collected	-	-
	Xu et al. (2015a)	ML	Ling., PMI, POS	SVM, LDA	Self-collected	0.78	F_1
	Maylawati et al. (2020)	ML	Ling., TF-IDF	Clustering	SemEval-2014 Task 4	-	-
	Yan et al. (2015)	ML	Ling., NodeRank	Feature eng.	Self-collected	0.84	F_1
	Rana et al. (2020)	ML	Ling., Stat.	Feature eng.	SemEval-2014 Task 4	0.88	F_1
	Xu et al. (2020b)	ML	Ling., dep., Stat.	Matrix factor.	SemEval-2015 Task 12	0.72	F_1
	Yu et al. (2018)	DL	GloVe	BiLSTM	SemEval-2014 Task 4	0.85	F_1
	Liao et al. (2019)	DL	word2vec	CNN	Self-collected	-	-
	Feng et al. (2019a)	DL	Ling., POS, dep.	CNN	Liao et al. (2019)	0.82	F_1

Recently, deep neural networks and PLMs were widely employed. Aspect extraction tasks are learned with other related tasks, e.g., opinion extraction and sentiment polarity prediction. Those related tasks may provide complementary information for each other. Thus, learning these tasks simultaneously can share the useful information, yielding higher accuracy on each task. While implicit aspect extraction is generally considered more challenging than explicit aspect extraction, recent research has shown less interest in this linguistic phenomenon, with a greater focus on achieving state-of-the-art performance on benchmark datasets without considering linguistic properties. Alternatively, another type of effort is to achieve fine-grained analysis of affective expressions, e.g., ASTE aims to extract aspect terms, opinion terms, and their corresponding sentiment polarities. The studies on aspect extraction from different linguistic contexts deserve more research efforts.

Finally, current aspect extraction research commonly focuses on conventional application scenarios, e.g., product or service reviews. This emphasis is likely due to the fact that most research in this field aims to identify sentiment polarities for specific aspect categories, which can hold commercial significance, such as understanding customer preferences for various aspects of a product or service. However, the scope of aspect extraction can be broadened to encompass other scenarios. Unlike NER, which targets specific entities in text, aspect extraction involves identifying various attributes or features of entities, such as food quality, service speed, or restaurant cleanliness. While these are not specific named entities, they contribute to a richer understanding of attributes associated with entities.

Current aspect extraction methods tailored to specific domains may not work well in general domains. Training a model on a domain-specific dataset, like one for restaurant reviews, may not apply to other areas, such as extracting research findings in scientific texts. While aspect extraction is useful in scientific research, there has been limited exploration of domain-agnostic methods.

4.5.8.2 Application Trends

As mentioned before, aspect extraction is usually considered a subtask within fine-grained sentiment analysis and opinion mining tasks. A current application trend involves various task setups that integrate aspect extraction with sentiment analysis and opinion mining, including ABSA, ABOM, ASQP, ACSA, ACSD, and ASTE. Those related works are summarized in Table 4.22. ABSA involves identifying and analyzing sentiments toward specific attributes or aspects of entities, such as products or services, mentioned in text. It seeks to determine whether opinions are positive, negative, or neutral regarding these specific aspects. ABOM is similar to ABSA but has a broader scope, involving the extraction of opinion terms related to aspects. ACSA focuses on detecting the category of an aspect and its sentiment polarity. Categories are broader concepts than aspects. For example, "sushi" is an aspect, and "food" is its associated category in the restaurant domain. ACSD goes further by also detecting the aspect term itself, along with the aspect category and sentiment polarity. ACSD provides additional aspect extraction functionality.

However, ACSA offers flexibility in cases of implicit aspects (e.g., "this is an expensive restaurant") such as when "price" related aspect terms are not explicitly mentioned in the text. ASTE aims to extract sentiment triplets, including an aspect term, an opinion term, and a sentiment polarity, combining elements of both ABSA and ABOM. ASQP builds on ASTE by additionally detecting aspect categories in text. Nevertheless, the aforementioned task setups can be generally named as ABSA, because they all involve analyzing sentiment and aspects. Among these tasks, aspect terms are extracted either explicitly, e.g., ABSA, ABOM, ASTE, ASQP, or implicitly via the identification of aspect categories, e.g., ACSD and ACSD. However, to the best of our knowledge, the retrieval of implicit aspects from elliptical constructions (see Section 4.5.1.2) was rarely studied. In real-world applications, it would be useful if an ABSA model could infer the elliptical aspects from text. The benefit is to achieve more fine-grained sentiment analysis and opinion mining, e.g., ASTE.

Table 4.22: Application trends for aspect extraction tasks.

Reference	Downstream Task	Feature	Parser
Zhang et al. (2015a)	ABSA	✓	✓
Wang et al. (2018a)	ABSA	✓	✓
Li et al. (2019b)	ABSA	✓	✓
Li et al. (2019b)	ABSA	✓	✓
Luo et al. (2019a)	ABSA	✓	✓
He et al. (2019c)	ABSA	✓	✓
Hu et al. (2019)	ABSA	✓	✓
Chen and Qian (2020)	ABSA	✓	✓
Luo et al. (2020)	ABSA	✓	✓
Liang et al. (2020)	ABSA	✓	✓
Yu et al. (2021)	ABSA	✓	✓
Zhao et al. (2020b)	ABOM	✓	✓
Chen et al. (2020)	ABOM	✓	✓
Wu et al. (2020c)	ABOM	✓	✓
Gao et al. (2021a)	ABOM	✓	✓
Wu et al. (2021b)	ABOM	✓	✓
Cai et al. (2021)	ASQP	✓	✓
Zhang et al. (2021b)	ASQP	✓	✓
Mao et al. (2022b)	ASQP	✓	✓
Bao et al. (2022)	ASQP	✓	✓
Zhang et al. (2021b)	ASQP	✓	✓
Schmitt et al. (2018)	ACSA	✓	✓
Cai et al. (2020)	ACSA	✓	✓
Liu et al. (2021a)	ACSA	✓	✓
Wan et al. (2020)	ACSD	✓	✓
Wu et al. (2021a)	ACSD	✓	✓
Zhang et al. (2021c)	ABOM, ABSA, ASTE, ACSD	✓	✓
Peng et al. (2020)	ASTE	✓	✓
Xu et al. (2020a)	ASTE	✓	✓
Wu et al. (2020c)	ASTE	✓	✓
Zhang et al. (2020a)	ASTE	✓	✓
Chen et al. (2021a)	ASTE	✓	✓
Mao et al. (2021b)	ASTE	✓	✓
Yan et al. (2021a)	ASTE	✓	✓
Xu et al. (2021)	ASTE	✓	✓
Fei et al. (2021b)	ASTE	✓	✓
Mukherjee et al. (2021)	ASTE	✓	✓
He et al. (2021)	KGC		✓
Li et al. (2020c)	KGC		✓
Li et al. (2020a)	KGC		✓

4.5.8.3 Future Works

A. Aspect Extraction with Unified Framework

As mentioned above, the main technical trend for aspect extraction is the integration with other sentiment and opinion mining tasks, such as ABSA, ABOM, ASQP, ACSA, ACSD, and ASTE. Different task setups allow users to retrieve different types of information related to sentiment analysis. Then, the natural idea is to design a unified framework to tackle multiple related tasks at the same time. Its practical usefulness lies in the fact that one may prefer not to alter the model architecture and undergo retraining for each instance of new data with varying types of opinion annotations. There are two main paradigms for designing such a unified framework.

The first possible unified framework is to form the tasks as QA tasks. Gao et al. (2021a) divided ABOM into two subtasks: aspect term extraction and aspect-specified opinion extraction. They extract all the candidate aspect terms, followed by extracting the corresponding opinion words given the aspect term. The proposed model employed a span-based tagging scheme and constructed a question-answer-based MRC task to achieve efficient extraction of aspect-opinion pairs. To address the challenges of the ASTE task, Chen et al. (2021a) convert the task into a multi-turn MRC and propose a bidirectional MRC framework. This framework includes three types of queries (non-restrictive extraction, restrictive extraction, and sentiment classification) to establish relationships between various subtasks.

The second possible framework is to develop a Seq2Seq learning framework by directly generating the required sentiment elements in the natural language form. Zhang et al. (2021c) proposed a generative framework comprising two types of modules: annotation-style and extraction-style modeling. The former module adds annotations to a given sentence to include label information in the construction of the target sentence, while the latter module directly adopts the desired natural language label of the input sentence as the target. Yan et al. (2021a) redefined every subtask target as a mixed sequence of pointer indexes and sentiment class indexes, converting all related subtasks into a unified generative formulation. With this unified formulation, they utilized the pretraining Seq2Seq model BART to solve all ABSA subtasks in an end-to-end framework.

B. Multimodal ABSA

The field of sentiment analysis faces a current challenge and opportunity in analyzing emotions and opinions expressed through graphics and videos on social media platforms. This challenge arises from the rapid development of social networks and the increased expressive tendencies of individuals on these platforms. Specifically, the challenge lies in analyzing sentiment in multimodal data, including voice, image, and text. Given that there is often a strong connection between various forms of content, utilizing multimodal information can enhance the analysis of users' sentiments towards different aspects.

Yu et al. (2022) argued that previous approaches either use separately pretrained visual and textual models or use vision-language models pretrained with general pretraining tasks, which are inadequate to identify fine-grained aspects, opinions, and their alignments across modalities. To tackle these limitations, they propose a task-specific Vision-Language Pretraining framework, which is a unified multimodal encoder-decoder architecture for all the pretraining and downstream tasks. They further design three types of task-specific pretraining tasks from the language, vision, and multimodal modalities, respectively. Ling et al. (2022) argue that existing methods did not effectively capture both coarse-grained and fine-grained image-target matching, which includes the relevance between the image and the target, as well as the alignment between visual objects and the target. To address this issue, they proposed a new MTL architecture called the Coarse-to-Fine Grained Image-Target Matching Network. This architecture jointly performs image-target relevance classification, object-target alignment, and targeted sentiment classification.

C. Cross-lingual and Cross-domain ABSA

Currently, much of the research in ABSA is centered around English and Chinese languages, predominantly focusing on commercial domains such as products and services. To address the scarcity of data for languages and domains with limited resources, cross-lingual and cross-domain learning is necessary in the domain of ABSA. This approach requires an understanding of the world knowledge pertaining to relevant aspects, in addition to their linguistic meanings. World knowledge should be independent of languages, because it reflects the properties of entities, cultures, and information of the world. Ontology serves as a formal model of world knowledge, delineating the categories of entities existing in the world and the relationships between them.

The aspects of a concept cannot be naïvely defined by meanings. For example, "dog" may include properties such as "color", "size", "weight" and "kind". While these properties are related to the concept of a dog, they may not be considered semantically central to the definition of a dog. The intra-semantic relevance of this properties is also weak. Therefore, in aspect extraction, without access to world knowledge, identifying all aspects associated with an unseen entity becomes challenging, as different aspects may carry varying meanings and contextual associations. Additionally, transferring knowledge learned from supervised learning across different entities is complex, as different entities may be associated with distinct aspects. For example, a classifier trained on restaurant-related aspects cannot be directly applied to extract aspects related to dogs. Incorporating a systematic ontology can facilitate cross-language and cross-domain transfer learning. In this context, the learning system is not focused on detecting similar aspects from labeled datasets. Instead, its objective is to understand the relationship between entities and their properties based on an ontology and textual data. As ontologies encapsulate world knowledge, which is only loosely dependent on languages, this learning paradigm can aid in cross-language inference as well.

4.6 Downstream Task

The NLU suitcase model proposed in this textbook (Fig. 1.12) can be applied to any downstream task, e.g., polarity detection, emotion detection, text summarization, QA, information retrieval, topic modeling, text classification, machine translation, dialogue systems, and more. Depending on the task, one or more modules may be omitted. For example, if training/test data do not contain any sarcasm, the sarcasm detection module is unnecessary. In this textbook, we pick polarity detection as downstream task. In its simplest form, it can be solved by a binary classifier that merely categorizes text as either "positive" or "negative". Catching all the nuances of how polarity can be expressed (either explicitly or implicitly) in natural language, however, requires a more sophisticated model like the NLU suitcase model.

Polarity detection has received increasing attention from the research community in the past decade mainly because of the rise of massive UGC such as movie reviews and product comments. Unlike formal documents where "factual" and "neutral" content is in the majority, UGC contains a large number of subjective expressions from which valuable information can be extracted. Traditional research analyzes the sentiment polarities, based on textual meanings. The hypothesis is that the words and phrases used in a text carry certain emotional weights that can be classified as positive, negative, or neutral. Recently, Zhu et al. (2024b) proposed personalized sentiment analysis, suggesting that sentiment polarities should be defined by the perceived impressions from the text. This approach acknowledges that individuals may perceive sentiment differently from the same message, such as a proposal to dance in front of a crowd. Therefore, sentiment analysis should incorporate personalized information to differentiate the perceived sentiment accordingly. The research of polarity detection also simulates the interest from the industry as it is used in different application scenarios, such as stock market forecasting (Ma et al., 2023, 2024). In the financial domain, while information that causes stock market fluctuations can be considered positive or negative (Du et al., 2024c), Ma et al. (2023) argued that market sentiment does not always align with the semantic sentiment of text. For example, raising debt may cause different investor reactions depending on the market environment.

Besides text, polarity detection is also applied to images and speech (Fan et al., 2024). The goal of image polarity detection is to differentiate images raising positive emotions in users from images causing negative emotions (Ragusa et al., 2019). An image may contain much information such as facial expressions (Cowen et al., 2021), body posture (Ferres et al., 2022), gestures (Keshari and Palaniswamy, 2019; Atanassov et al., 2021), and scene context information (Kosti et al., 2019; Yang et al., 2021c), which significantly increases the complexity of the task. In recent years, researchers put forward a lot of efforts to deal with difficulties in image polarity detection, e.g., a series of facial expression encoding methods (Li et al., 2009; Soleymani et al., 2015) have been built on the basis of Facial Action Coding System (Ekman and Friesen, 1978) to accurately describe the face and recognize the polarity afterward.

Nowadays, smartphone users interact frequently with voice assistants such as Siri and Google Assistant. It is essential for voice assistants to identify the sentiment of the user and generate empathetic responses accordingly. Compared with text, audio may contain rich tone information which is a good indicator for determining the polarity of the speaker. For instance, a high pitch is usually linked with a positive polarity while low-frequency tones often correspond to negative polarity (Chaturvedi et al., 2022). The main challenge of speech polarity detection is how to learn an intermediate representation of input speech signal without any manual feature engineering (Latif et al., 2021) in an uncontrolled natural environment (Fahad et al., 2021).

4.6.1 Theoretical Research

Despite the fact that most words, idioms, or phrases may occur in both affirmative and negative sentences, there are some which may occur only in affirmative, or only in negative sentences. These words, idioms, or phrases are termed "polarity-sensitive" (Baker, 1970). Here are some examples:

(11)   I *would rather* go to New York.

(12)   I *would rather* not go to New York.

(13)   She did *pretty* well on the project.

(14)   She didn't do *pretty* well on the project.

In this case, "polarity-sensitive" is a strict concept that covers only a small range of polarity-sensitive words, idioms, or phrases. Fauconnier (1975a) described a more general polarity with the extended notion: semantic and pragmatic polarity based on the following observations:

- Some words or phrases are polarized with respect to logical structures.
- Some words or phrases are polarized with respect to context, which means their polarity can therefore vary from context to context.
- The polarity items such as words and phrases have the property of polarity-reversal (Baker, 1970).

4.6.1.1 Logically Polarization

It was noted in the work of Fauconnier (1975b) that grammatical superlatives could yield quantificational effects in some sentences:

(15)   The most stylish suit looks bad on Maxwell ("Any suit looks bad on Maxwell").

(16)   Maxwell cannot solve the simplest question ("Maxwell cannot solve any question").

In the above sentences, superlatives such as "the most stylish" and "the simplest" serve as a universal quantifier in terms of semantics. Nonetheless, there are still many sentences in which the superlative cannot occur even with the semantic value:

(17)   The most stylish suit looks good on Maxwell (≠"Any suit looks good on Maxwell").

(18)   The simplest question is easy for Maxwell (≠"Any question is easy for Maxwell").

The aforementioned examples illustrate that superlatives are not always quantificational. Nonetheless, a *weak polarity principle* can be formulated: a quantifying superlative that is appropriate for use in an affirmative sentence is generally unsuitable for use in the corresponding negative sentence. According to Fauconnier (1975a), quantifying superlatives are polarized in compliance with this principle.

4.6.1.2 Contextually Polarization

As mentioned, the polarity of some words or phrases may change from context to context. Consider the following sentences:

(19)   Even Maxwell doesn't understand "Elements".

(20)   Even the monk is tempted to use contraceptives.

These sentences are suitable only in certain contexts that are compatible with the "premise" of the *even*-phrase. For instance, Example (20) is appropriate only if it is assumed that the monk is highly unlikely to be tempted to use contraceptives; otherwise, it would not be appropriate. It turns out that, in a given context, an *even*-phrase is suitable in an affirmative sentence, but not suitable in the negative counterpart of that sentence:

(21)   Even Maxwell understands "Elements".

(22)   Even the monk is not tempted to use contraceptives.

It is concluded from Sections 4.6.1.1 and 4.6.1.2 that it is impossible to lexically mark words, idioms, or phrases whose polarity highly depends on the logical structure and the context within the sentence. Despite all this, in the 1980s and 1990s, Krifka (1992, 1994); Kadmon and Landman (1993); Lee and Horn (1994) attempted to offer lexical semantics explanations for the distribution of polarity-sensitive items (words, idioms, phrases). The target is to find out general properties to unite a large number of heterogeneous polarity-sensitive items. Israel (1996) argued that polarity sensitivity arises from the interaction of two lexical semantic properties, quantitative and informative values (Kay, 1990), where the quantitative value reflects the fact that a sizable portion of polarity-sensitive items encodes some notion of amount or degree, and the informative value reflects the fact that some propositions are more informative than others in context.

In this case, polarity-sensitive items are lexical expressions combining a high or a low quantitative value with a conventionally emphatic or understanding informative value. Furthermore, the author argued that polarity-sensitive items are scalar operators that require interpretation with respect to a suitable structured scalar model (Fillmore et al., 1988; Kay, 1990). The lexical semantic-pragmatic content of polarity-sensitive items makes them sensitive to scalar inferences. To be specific, scalar operators

> are themselves a special class of what Kay (1990) calls *contextual operators*, that is, expressions whose meanings involve not only what situations they can appropriately describe, but also some notion of the situations in which they can appropriately be used (Fillmore et al., 1988).

contextual operators

> are lexical items or grammatical constructions whose semantic value consists, at least in part, of instructions to find in, or impute to, the context a certain kind of information structure and to locate the information presented by the sentence within that information structure in a specified way (Kay, 1990).

To sum up, a unified and comprehensive account of polarity sensitivity is proposed on the basis of scalar operators and two lexical semantic properties: quantitative and informative values. It cannot be applied to all polarity items but brings insights into the diversity of polarity phenomena (Hoeksema, 1994; Rullmann, 1996).

Szabolcsi (2004) further studied the positive polarity and negative polarity items. It turns out that the distribution of positive polarity items is significantly more complex and shares properties with negative polarity items in certain conditions. Spector (2014) studied two types of positive polarity items that have not been recognized before. For the first type, anti-licensing is a strictly local phenomenon while another type is anti-licensed if and only if their global syntactic environment satisfies a specific semantic property. In recent years, Homer (2021) provided a unified theory of the licensing of negative and positive polarity items *some* and *any*. The author argued that licensing has both a syntactic and a semantic component. Here licensing

> is viewed in terms of a structural relation between a polarity item and an operator equipped with certain features (negativity and downward-entailingness) (Homer, 2021).

The polarity of words, idioms, or phrases has been thoroughly studied from the perspective of linguistics. The polarity of sentences, paragraphs, or documents is defined as an opinion toward a topic, rather than a linguistic concept.

4.6.2 Annotation Schemes

Kim and Hovy (2004) defined opinion as follows:

> The holder believes a claim about the topic, and in many cases associates a sentiment, such as positive or negative.

Opinions are not always associated with the sentiment. For instance, the sentence "I believe Jacky Chan is a famous star" contains no sentiment but a claim. Hence, there are mainly two types of annotation schemes of polarity, i.e., 1) binary classification where $Y \in \{Positive, Negative\}$; 2) 3-class classification where $Y \in \{Positive, Negative, Neutral\}$. Specifically, the annotation schemes of word-level, phrase-level, sentence-level, paragraph-level, and document-level text are slightly different from each other. Besides, when annotating, it is essential to take the domain of the text into consideration.

In the beginning, word-level annotation is conducted in a supervised manner or by experts, e.g., Opinion Lexicon (Hu and Liu, 2004), NTUSD (Ku et al., 2006), SentiWordNet (Baccianella et al., 2010), SO-CAL (Taboada et al., 2011) and AFINN (Nielsen, 2011). The quality of these lexicons is satisfactory. However, it requires huge human resources to build and cannot generalize to certain domains. To this end, researchers developed many data-driven sentiment lexicon construction methods to annotate the polarity of words in a semi-supervised way. Typically, these approaches (Rao and Ravichandran, 2009; Velikovich et al., 2010; Feng et al., 2012; Tang et al., 2014a; Saif et al., 2016; Wu et al., 2016; Li et al., 2018b) employ a limited set of seed words (either through manual annotation or existing sentiment lexicons) to disseminate sentiment labels to a large number of unlabeled words and thus yield a large, domain-specific sentiment lexicon.

Wilson et al. (2005) proposed to identify contextual polarity for a large number of phrase-level sentiment expressions. Annotators are required to tag the polarity of subjective expressions as *positive*, *negative*, *both*, or *neutral*. Here, *both* is applicable to expressions that have both *positive* and *negative* sentiments. In addition to expressions that do not contain polarity, *neutral* is extended to a different type of subjective expression like speculation. Below are the annotation examples of the contextual polarity annotations:

1) Moreover, politicians use the concepts of good and evil (**both**) solely for the purposes of intimidating and exaggerating.
2) Throughout the night, a vast number of individuals who supported the coup celebrated (**positive**) by waving flags and blowing whistles.
3) According to Jerome, the hospital does not feel (**neutral**) any different from a hospital in the United States.

Annotation acquisition is a crucial step in developing supervised classifiers. However, it is time-consuming and cost-intensive to annotate the polarity of texts solely by experts. Therefore, crowdsourcing (Howe, 2008) is introduced to quickly acquire annotations for the purposes of constructing all kinds of predictive models. Hsueh et al. (2009) studied the difference between expert annotators and non-expert annotators (e.g., AMT) in the light of annotation quality. They found that the quality of labels can be improved by eliminating noisy annotators and ambiguous examples. It is also proven that quality measures are useful for selecting annotations that give rise to more accurate classifiers.

4.6.3 Datasets

Initially, polarity detection was performed mostly movie reviews, product reviews, and social media posts in English. Subsequently, many polarity detection datasets for low-resource languages were constructed. Maas et al. (2011) proposed a large movie review dataset[32], including 25,000 training instances and 25,000 testing instances with positive or negative labels sourced from IMDB. Negative reviews are defined as those with a score of 4 or less out of 10, while positive reviews are those with a score of 7 or more out of 10. Neutral reviews are excluded from the dataset.

```
Movie review 1: "This short deals with a severely critical writing teacher whose
undiplomatic criticism extends into his everyday life. When he learns why that's not
a good idea, we learn a bit about the beautiful craft of writing that he's been
lecturing on.",
Label 1: "Positive"
Movie review 2: "I found this movie really hard to sit through, my attention kept
wandering off the tv. As far as romantic movies go..this one is the worst I've seen.
Don't bother with it.",
Label 2: "Negative"
```

Pang and Lee (2004) introduced a movie dataset[33] with 2,000 reviews written before 2002. They extracted at most 20 reviews for each author (312 authors in total). They also released a subjectivity dataset that contains 5,000 subjective and 5,000 objective reviews from Rotten Tomatoes. Only sentences or snippets with more than 10 tokens are selected. Go et al. (2009) introduced a social media posts dataset (Sentiment140)[34] extracted from the X API. There are 160,000 tweets in this dataset annotated using distant supervision.

```
Tweet 1: "Maxwell is my new best friend.",
Label 1: "Positive"
Tweet 2: "just landed at New York.",
Label 2: "Neutral"
Tweet 3: "Math exam studying ugh",
Label 3: "Negative"
```

Socher et al. (2013b) introduced a movie review dataset (Stanford Sentiment Treebank)[35], consisting of 11,855 single sentences extracted from movie reviews. In this dataset, each sentence is labeled as either *negative, somewhat negative, neutral, somewhat positive*, or *positive*. Binary classification experiments (negative $\in$ {somewhat negative, negative}, positive $\in$ {somewhat positive, positive}) on full sentences refer to the dataset SST-2 or SST binary.

---

[32] https://ai.stanford.edu/~amaas/data/sentiment

[33] https://cs.cornell.edu/people/pabo/movie-review-data

[34] https://cs.stanford.edu/people/alecmgo/trainingandtestdata.zip

[35] https://nlp.stanford.edu/sentiment/index.html

Zhang et al. (2015b) built a dataset[36] for binary polarity detection. The authors provided a set of 560,000 highly polar Yelp reviews for training and 38,000 for testing by considering stars 1 and 2 negative, and 3 and 4 positive. Nakov et al. (2013) proposed a message polarity detection task and a corresponding X dataset. The authors extracted tweets from the X API and crowdsourced them on AMT for annotation. Li et al. (2018b) presented a Chinese Tourism Review dataset, consisting of reviews of popular tourism products. The authors manually annotated 30,180 tourism reviews and created a balanced dataset with 7,995 instances and a real-world dataset with the all annotated instances.

4.6.4 Knowledge Bases

Knowledge bases are valuable tools for polarity detection in sentiment analysis because they provide a structured repository of factual information, relationships, and concepts that enhance the understanding of text. Knowledge repositories such as SentiWordNet (Baccianella et al., 2010), SenticNet (Cambria et al., 2024) and FinSenticNet (Du et al., 2023b) (introduced earlier) include dictionaries, ontologies, and databases that encompass various domains, providing context and background knowledge that can be crucial for accurate sentiment interpretation.

As discussed earlier in the textbook, knowledge bases can enable important sub-tasks, e.g., WSD and NER, and hence, increase the accuracy of polarity detection. Moreover, they can offer insights into metaphorical expressions and cultural nuances that might not be evident from the text alone. For example, understanding that the phrase "break the ice" means to alleviate tension can prevent misinterpretation of its literal meaning, which might not directly convey sentiment. By linking such expressions to their underlying meanings, knowledge bases help in correctly identifying polarity. Knowledge bases also facilitate understanding relationships between entities and concepts, which is crucial for detecting sentiment polarity. For instance, if a text mentions a company's stock dropping after a 'scandal', a knowledge base can provide background on the negative connotations of 'scandal' and its typical impact on public perception.

In the era of LLMs, knowledge bases are still important because they provide a reliable source of factual accuracy and consistency that LLMs alone cannot guarantee. LLMs generate responses based on patterns in the data they were trained on, which may not always be up-to-date or factually accurate. Knowledge bases, on the other hand, are curated and regularly updated repositories of structured information, ensuring that the data they contain is accurate and reliable. Furthermore, knowledge bases offer domain-specific expertise that LLMs may lack. While LLMs are generalists, capable of discussing a wide range of topics, they may not possess the depth of knowledge needed for specialized fields like medicine, law, or finance.

---

[36] https://s3.amazonaws.com/fast-ai-nlp/yelp_review_polarity_csv.tgz

The structured nature of knowledge bases also complements the unstructured data processing capabilities of LLMs. Knowledge bases use ontologies and structured formats to organize information, making it easier to understand relationships between different concepts and entities. This structure is particularly useful for complex queries where precise relationships matter, something that LLMs might not inherently handle well due to their training on unstructured data. Additionally, knowledge bases offer a way to maintain temporal consistency, especially for information that frequently changes. Since LLMs are trained on datasets that may quickly become outdated, they can struggle to provide accurate information on recent developments. Knowledge bases, with their regular updates and mechanisms for tracking changes, help ensure that the information remains relevant and up-to-date.

4.6.5 Evaluation Metrics

Polarity detection is a typical classification task. Therefore, accuracy, precision, recall, and F_1 score are the main indicators to evaluate the performance of the models. These different indicators measure different aspects of a model. Alternatively, sentiment polarity can sometimes be represented as sentiment intensity. In such cases, a commonly used evaluation metric is the MAE.

$$\text{MAE} = \frac{1}{n} \sum_{i=1}^{n} |y_i - \hat{y}_i|, \tag{4.1}$$

where n is the total number of observations. y_i is the actual value for the i-th observation. $\hat{y}_i$ is the predicted value for the i-th observation. $|y_i - \hat{y}_i|$ is the absolute difference between the actual and predicted values for the i-th observation.

4.6.6 Methods

This subsection outlines three trends in polarity detection methods: lexicon-based methods, PLMs, and neurosymbolic methods.

4.6.6.1 Lexicon-based Methods

Lexicon-based methods refer to the methods using sentiment lexicons to identify the polarity of sentences. The polarity of a sentence is regarded as the summation of the sentiment scores of positive and negative words within the sentence. To this end, it is essential to build a high-quality sentiment lexicon with a wide coverage of words and phrases. There are three main sentiment lexicon construction methods for polarity detection (Shi et al., 2019).

A. Statistics-based Approach

The main idea of the statistics-based approaches is to design proper statistics rules to extract potential sentiment words from documents. Researchers identified effective statistics-based rules such as pointwise mutual information (PMI) (Church and Hanks, 1990), symmetrical conditional probability (Bu et al., 2010) and enhanced mutual information (Zhang et al., 2009) by observing, analyzing, and summarizing. PMI is an effective statistical indicator that measures statistical relatedness/ similarity between two words. The calculation of PMI is as follows:

$$PMI(w_i, w_j) = \log \frac{n(w_i, w_j)/N}{(n(w_i)/N)(n(w_j)/N)}, \tag{4.2}$$

where $n(w_i, w_j)$ refers to the co-occurrence number of words i and j; N is the total number of documents. Supposing that w_i is a word with polar, if $PMI(w_i, w_j)$ is high, then w_j is likely to share the same polarity with w_i.

Feng et al. (2012) provided an unsupervised method to learn a sentiment lexicon from massive Microblog data with emoticons. In the paper, emoticons acted as noisy labels and are partitioned into positive and negative sets. If a word w_i is more similar to positive emoticons (PE) than negative emoticons (NE), i.e., $score(w_i) = PMI(w_i, PE) - PMI(w_i, NE) > 0$, then the polarity of w_i is regarded as positive, otherwise negative. Emoticon makes it feasible to build a domain-specific sentiment lexicon in an unsupervised manner. However, it cannot detect user-invented words that frequently appear on social media platforms, which may reduce its practical usage. To this end, Wu et al. (2016) extended the previous work by introducing new word detection in Chinese Microblog. The authors also used three types of knowledge to assign sentiment intensities for words. Nevertheless, it did not consider semantic knowledge. Subsequently, Li et al. (2018b) further improved Chinese new word detection in the tourism domain via assembled mutual information (AMI), and introduced semantic similarity knowledge to refine sentiment intensities of words.

In summary, statistics-based methods do not require extra human efforts and can easily extend the coverage of lexicons. Besides, it considers user-invented words and utilizes knowledge from different sources (statistic knowledge, semantic knowledge, prior knowledge) to determine relatively precise sentiment intensities for words. One drawback of this method is that it requires massive data, which may lead to biased sentiment intensity if without enough data.

B. Semantic-based Approach

The idea of semantic-based approaches is to employ paradigm symbols such as sentiment words, emoticons, or any other symbol that delivers polarity as the noisy label and measure the semantic relatedness between paradigm symbols and other words to build sentiment lexicons.

Kamps et al. (2004) put forward an approach calculating semantic similarity between evaluative factors "good" and "bad" and adjectives. If a word is closer to the reference word "good" than "bad" in terms of distance, then it is regarded as positive, otherwise negative. Here, the distance is computed by the shortest path in WordNet between "good" ("bad") and the word, i.e., $EVA(w) = \frac{d(w,bad)-d(w,good)}{d(good-bad)}$. However, WordNet-based measures of distance or similarity mainly focus on taxonomic relations, which makes it almost impossible to calculate the noun and adverb categories in WordNet.

Velikovich et al. (2010) presented a graph propagation method to build polarity lexicons. It starts from a small set of seed polarity words and then propagates information from the seed set to the rest of the graph through the edges. The weights of edges are obtained from the cosine similarity of context vectors of two words. Graph propagation is supposed to overcome the drawbacks of common graph propagation algorithms. However, a destination word may be influenced by a seed word through different paths, resulting in a high polarity. Finally, it is worth noting that the cosine similarity of the context vectors is an indirect way of calculating the semantic similarity of two words, which may be inaccurate or biased.

Viegas et al. (2020) proposed an unsupervised method to expand sentiment lexicons by exploiting semantic relationships. Specifically, the authors employed pretrained word embeddings to directly calculate the semantic similarity between words based on the hypothesis that a closer distance between word vectors indicates a closer polarity and intensity. This unsupervised method provided a simple, yet very effective, technique for improving the coverage of sentiment lexicons. Although the empirical study demonstrated the flexibility and effectiveness of the proposed method, the directly computed semantic similarity between word embeddings cannot avoid a typical error that words that are close in vector space may have opposite semantic meanings, e.g., the semantic similarity between pleasant and unpleasant is 0.7451 according to Spacy. Therefore, a more accurate measurement of polarity similarity should be the future research focus for semantic-based sentiment lexicon construction approaches.

4.6.6.2 Pretrained Language Model-based Methods

The majority of the existing deep learning-based polarity detection models are trained on large labeled data. In general, these supervised methods significantly improve the performance of polarity detection on many benchmark datasets. RNN techniques (Basiri et al., 2021), such as GRU (Cheng et al., 2020) and LSTM (Imran et al., 2020), have been proven to be very effective in the polarity detection task as they naturally support the calculation of sequential text data. In recent years, attention mechanism (Vaswani et al., 2017) and transformer-based techniques, especially PLMs such as BERT, become the standard paradigms and yield state-of-the-art performance on almost all the benchmark datasets. In this subsection, we introduce the trends of PLM-based polarity detection models.

A. PLM as Contextualizer

A typical application scenario is to use PLM as a contextualizer as it is a powerful representation encoder. Satapathy et al. (2022) presented an MTL framework for polarity and subjectivity detection. The authors employed BERT to generate context-aware sentence-level embeddings. NTN (Li et al., 2021c) is designed as a soft parameter-sharing module for both subtasks. An ablation study on a benchmark dataset indicated that BERT representations outperform GloVe (Pennington et al., 2014) embeddings by 2.5% on average. Rahman et al. (2022) proposed a BERT-based model for reputation polarity detection in social media. In the paper, the last BERT hidden state from the first token [CLS] is used to represent the whole input sequence. To address the class imbalance problem, the authors integrated a cost-based strategy (Zhou and Liu, 2005) into the BERT-based model.

Jiang et al. (2022) presented Retweet-BERT, a simple and scalable model for political leaning polarity detection, that encourages profile embeddings to be more similar for users who retweet each other. To this end, BERT is first used to generate profile embedding for each $\mathbb{X}$ user with an objective function to minimize the distance between profile embeddings between retweeted users. Then, such embeddings are used to fine-tune S-BERT (Reimers and Gurevych, 2019). A case study indicated that the most retweeted users by left-leaning users are also left-leaning, and vice versa for right-leaning. As this study mainly focuses on those who reveal their political preference in their profile descriptions, detecting the leaning polarity of users who are more "modest" in vocalizing their opinions remains a challenging task.

B. Fine-tune-based PLM Approaches

Pota et al. (2020) provided an effective BERT-based pipeline for $\mathbb{X}$ polarity detection. The authors converted Italian tweets into plain texts by removing emoticons, emojis, hashtags, etc., and used it to fine-tune the BERT model. It is worth noting that overall positive (opos) and overall negative (oneg) are analyzed by two classification tasks in this paper. Hence, the final polarity can be derived by the combination of opos and oneg, e.g., $oneg(0) + opos(0) = Neutral$, $oneg(1) + opos(0) = Negative$, $oneg(1) + opos(1) = Mixed$. One insightful conclusion is that preprocessing can significantly improve the performance of fine-tuning as BERT is also trained on plain text. Catelli et al. (2022) presented a framework to detect deceptive reviews and identify the sentiment polarity of the review by a multi-label classification mechanism. The label powerset technique is used to transform the multi-label problem into a single-label problem considering the 2^L possible combinations (deceptive positive, deceptive negative, truthful positive, and truthful negative) of the transformed labels, where L is a set of non-disjoint labels. Reviews with deceptive and polarity labels are used to fine-tune BERT to enable dependency modeling between polarity and opinion truthfulness, yielding better performance over baselines. Nevertheless, the transformation is applicable only when the number of non-disjoint labels is low, preventing it from being widely applied to similar problems.

C. Prompt-based PLM Approaches

In recent years, prompt-tuning (Brown et al., 2020) PLM approaches for polarity detection attract more attention from the research community as they yield better performance on a wide range of NLU tasks such as polarity detection (Gao et al., 2021b) and relation extraction (Han et al., 2022b). Li et al. (2021a) put forward a model named SentiPrompt for aspect-based polarity detection (Pontiki et al., 2016) leveraging sentiment knowledge enhanced prompts to tune the PLM in the unified framework. Here, a typical sentiment knowledge enhanced prompt template looks like this: "The A is O? [MASK]. This is [MASK]", where A and O are randomly sampled aspect and opinion terms(e.g., A is sampled from {Sushi, Price}, O is sampled from {Good, High}). [MASK] can be either "yes" (consistent with ground truth) or "no" (inconsistent with ground truth). BART (Lewis et al., 2020) with pointer network (Vinyals et al., 2015b) is used as an Encoder-Decoder module for output (including [MASK] filling) generation. SentiPrompt surpasses the strongest baselines on ABSA by a large margin. Nonetheless, the effect of prompts with different templates remains unclear.

Mao et al. (2023c) conducted a systematic study on prompt-based polarity detection to investigate the influence of prompt templates. The paradigm of the prompt-based PLM is to add a prompt with a [MASK] token upon a sentence (e.g., "I feel [MASK], the movie is very horrible.") and then predict the probabilities of predefined emotion word (e.g., joyful, sad, and frustrated) in the [MASK] position. The polarity of the sentence can be induced by label-word mapping, where each polarity corresponds to a set of sentiment words. One of the advantages of prompt-based approaches over fine-tuning approaches is that it does not require annotating large-scale datasets, which is favored by low-resource languages. This study also observed that different positions of prompts deliver different results, which indicates the PLM is biased on positions as cognitively, different positions do not change the polarity of the original input. Thus, it is essential to design proper prompts to achieve optimal performance.

Du et al. (2024b) introduced an innovative framework called Hierarchical Prompting for Financial Sentiment Analysis. This framework aims to enhance the understanding of crucial factors in financial sentiment analysis for LLMs, focusing on semantic overview, contextual exploration, and influencing variables. It was observed that some LLMs lack the logical reasoning capabilities required for effective financial sentiment analysis. To address this, the hierarchical prompting framework queries additional knowledge from the LLMs, structuring the reasoning process to ultimately improve their performance. Finally, Zhu et al. (2024b) (introduced earlier) also belongs to the group of fine-tune-based PLM approaches to polarity detection. In particular, an LLM was employed for tailored sentiment analysis, based on seven different levels of personalization. Subsequently, the LLM was prompted to evaluate the sentiment perceptions between two interlocutors.

4.6.6.3 Neurosymbolic Methods

Neurosymbolic AI integrates symbolic and statistical paradigms of cognition (Wang and Yang, 2022). It is a promising paradigm that combines the advantages of reasoning and interpretability of symbolic representation and robust learning in neural networks and it might be a crucial step toward the next generation of powerful, trustworthy, and robust AI. Neurosymbolic AI takes inspiration from the concept of System 1 and System 2 types of thinking, introduced by Stanovich and West (2000). System 1 and 2 refer to dual modes of thinking – the former fast and intuitive guided by pattern matching, the latter slow and analytical guided by deliberative reasoning. In this sense, neurosymbolic AI is also brain-inspired as it attempts to emulate the cognitive processes involved in both human learning and reasoning, e.g., bottom-up learning (by examples) and top-down learning (by rules). A recent interpretation of neurosymbolic AI are LLM-Modulo frameworks (Kambhampati et al., 2024), which offer a tight integration between LLMs and symbolic components, extending the scope of model-based planning and reasoning regimes towards more flexible knowledge, problem and preference specifications.

This subsection discusses recent landmark neurosymbolic approaches for polarity detection categorized according to the representation of symbolic knowledge, i.e., lexicon-based, knowledge graph-based, and FOL-based.

A. Lexicon-based Neurosymbolic Approaches

Lexicon-based neurosymbolic approaches refer to frameworks leveraging sentiment lexicon as external symbolic knowledge to improve both accuracy and interpretability of neural network models. Huang et al. (2020b) presented a model to effectively extract features from sentiment words and text utilizing a context-dependent, lexicon-based CNN. The model is first trained with the co-occurrence pattern of polarized words and labels. Subsequently, a context-dependent lexicon is generated and is used to generate sentiment embeddings. These embeddings together with pretrained word embeddings are fed into a 3-channel CNN for sentiment word intensity prediction. Furthermore, the trained model is applied to learning a sentiment strength-specific lexicon. This lexicon has proved to be more effective than baselines in predicting the polarity intensity of documents. However, the neural part (CNN) of this model did not learn sentence- or document-level semantics which may weaken the performance and application scope. Besides, the lexicon-based symbolic knowledge is not fully utilized as it is only treated as an enhancement of word embeddings.

Li et al. (2020f) proposed a lexicon-integrated CNN-LSTM family model for polarity detection in user reviews. In this study, sentiment intensities of words are used as an indicator to augment sentiment-relevant information in an input sentence. Augmented input sentences are then fed into a two-channel CNN-LSTM family model for classification. Extensive experiments on the Chinese Tourism Review dataset demonstrated the effectiveness of the proposed sentiment padding mechanism and the polarity detection framework. Similarly, this approach cannot capture the complicated dependencies in a long sequence.

Polignano et al. (2022) introduced BERT-WMAL, a hybrid approach that integrates knowledge from a neural network model and a polarized lexicon for Italian polarity detection. As the name suggests, WMAL is a weighted version of MAL, a morphologically-inflected affective lexicon (Marco et al., 2019), where different forms relate to the same lemma and share the same polarity as that of the lemma. An input sentence is encoded in two parallel branches, i.e., BERT followed by an LSTM and a dense layer, and WAML representation followed by a dense layer. The outputs of two branches are fed into a dense layer for final prediction. BERT-WMAL achieved better performance on three benchmark Italian polarity detection datasets. Moreover, the authors proposed an explanatory approach using aggregate attention values jointly computed by BERT attention heads and WMAL values to measure how much attention each term receives from the other neighborhoods in the input sentence. The explanatory approach for hybrid models provides an intuitive human-like explanation that makes the model transparent.

Conventional deep neural networks embed word senses with vectorized hidden states, which lack human interpretability. On the other hand, traditional WSD methods frame the task as word sense classification. However, this setup and its outputs are not directly applicable to downstream systems. To this end, as discussed earlier, Zhang et al. (2023b) proposed a neurosymbolic framework that explicitly distinguishes word senses with a paraphrasing paradigm. This is also reflected in the already discussed neurosymbolic framework by Zhang et al. (2024c), who utilized only POS tag sequences and pragmatic representations as input, employing two fully connected layers as the neural classifier. Unlike the conventional practice of assigning a fixed sentiment score to a word, these vector representations account for the subjectivity of annotators, resulting in more robust performance in cross-domain and cross-lingual sentiment analysis tasks.

Finally, SenticNet (introduced earlier) has also been proposing a neurosymbolic approach to polarity detection since its inception back in 2010. The latest version (Cambria et al., 2024) utilizes an ensemble approach that integrates commonsense knowledge representation with HANs (Han et al., 2022a) to improve both performance and transparency of polarity detection in different affective computing domains. In particular, SenticNet leverages a revisited version of the Hourglass of Emotions (Fig. 4.5) to extract emotion-cause pairs from text and, hence, translate different intensities and categories of emotions into polarity values associated to distinct opinion targets in a completely interpretable manner.

SenticNet also aims to reduce the symbol grounding problem via a sort of universal symbolism. For example, it represents $BUY(x)$ as a double transfer of ownership where, at time t_{-1}, agent A owns \$ (a certain amount of money) and agent B owns x while, at time t_0, agent A owns x and agent B owns \$. This semiotic mechanism represents an interesting attempt to recreate language-agnostic representations to refer to concepts in a universal way, the same way as mathematical symbols or musical notes allow anyone to perform mathematical operations or read and write music, no matter what language they speak. Moreover, it uses grounded representations that, unlike words or word embeddings, can better replicate or visualize the current state of affairs of a virtual operating environment on the fly, as narratives unfold.

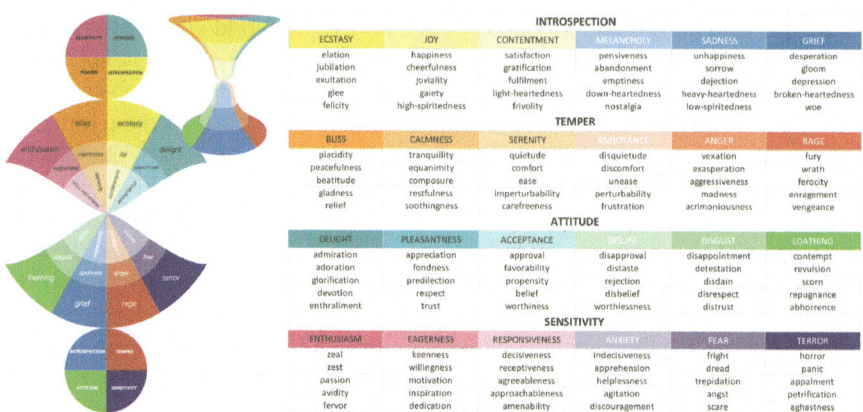

INTROSPECTION					
ECSTASY	**JOY**	**CONTENTMENT**	**MELANCHOLY**	**SADNESS**	**GRIEF**
elation	happiness	satisfaction	pensiveness	unhappiness	desperation
jubilation	cheerfulness	gratification	abandonment	sorrow	gloom
exultation	joviality	fulfilment	emptiness	dejection	depression
glee	gaiety	light-heartedness	down-heartedness	heavy-heartedness	broken-heartedness
felicity	high-spiritedness	frivolity	nostalgia	low-spiritedness	woe
TEMPER					
BLISS	**CALMNESS**	**SERENITY**	**ANNOYANCE**	**ANGER**	**RAGE**
placidity	tranquillity	quietude	disquietude	vexation	fury
peacefulness	equanimity	comfort	discomfort	exasperation	wrath
beatitude	composure	ease	unease	aggressiveness	ferocity
gladness	restfulness	imperturbability	perturbability	madness	enragement
relief	soothingness	carefreeness	frustration	acrimoniousness	vengeance
ATTITUDE					
DELIGHT	**PLEASANTNESS**	**ACCEPTANCE**	**DISLIKE**	**DISGUST**	**LOATHING**
admiration	appreciation	approval	disapproval	disappointment	contempt
adoration	fondness	favorability	distaste	detestation	revulsion
glorification	predilection	propensity	rejection	disdain	scorn
devotion	respect	belief	disbelief	disrespect	repugnance
enthralment	trust	worthiness	worthlessness	distrust	abhorrence
SENSITIVITY					
ENTHUSIASM	**EAGERNESS**	**RESPONSIVENESS**	**ANXIETY**	**FEAR**	**TERROR**
zeal	keenness	decisiveness	indecisiveness	fright	horror
zest	willingness	receptiveness	apprehension	dread	panic
passion	motivation	agreeableness	helplessness	trepidation	appalment
avidity	inspiration	approachableness	agitation	angst	petrification
fervor	dedication	amenability	discouragement	scare	aghastness

Fig. 4.5: The Hourglass of Emotions revisited (Susanto et al., 2020). This emotion categorization model represents affective states both through labels and through four independent but concomitant affective dimensions, which can potentially describe the full range of emotional experiences that are rooted in any of us.

B. Knowledge Graph-based Neurosymbolic Approaches

Graphs are a popular and widely used knowledge representation tool. One of the typical representation forms is the subject-predicate-object (SPO) triplet, where subject and object are entities (e.g., people, places, things), and the predicate (e.g., is_a, is_located_in) is the relation pointing from subject to object. The knowledge recorded in triplets is commonsense knowledge yet essential for people to understand sentences, especially less informative sentences. Researchers developed effective methods to represent graph-based knowledge and combine it with numerical connectionists (neural networks) for a more powerful polarity detection ability.

Liao et al. (2022) presented a framework named KG-MPOA using dynamic commonsense knowledge for Chinese implicit polarity detection. Compared with explicit polarity detection, the implicit one is more challenging due to the lack of sentiment words. To this end, external symbolic knowledge triplets from ConceptNet are fused to the semantic representation of implicit sentences. To embed the graph-based knowledge, the authors designed a pipeline for knowledge distillation, i.e., subjective triplet distillation (step 1), literal-related triplet distillation (step 2), noisy triplet filtering (step 3), and semantic-related triplet distillation (step 4). Specifically, step 1 extracts triplets whose subject or object can be retrieved from an existing sentiment lexicon. Step 2 removes triplets whose subject and object cannot be found in input sentences. Step 3 is used to filter certain triplet relations (e.g., "RelatedTo"). Step 4 leverages BERT to compute the similarity between an input sentence and a triplet-converted expression. The distilled triplets are regarded as a knowledge graph and encoded by a dynamic graph attention layer integrating into the neural module.

Swati et al. (2022) offered an alternative perspective on leveraging graph-based knowledge. They employed a popular neural knowledge model $COMET$ (Hwang et al., 2020) trained on $ATOMIC_{20}^{20}$ knowledge graph to generate the Inferential Commonsense knowledge (IC_Knwl) for each input news headline, e.g., for given news headline H, certain inference type (I) like "xWant, xNeed and xEffect", and the number of returned references (N), the symbolic knowledge can be obtained by $IC\_Knwl = COMET(H, I, N)$. The generated IC_Knwl is further processed to fit the expression of natural language and then combined with the input news headline for classification by PLM. The framework is designed as language-agnostic through the translate-retrieve-translate (TRT) mechanism (Fang et al., 2022b). Experiments conducted on news headline political polarity detection proved the effectiveness of the aforementioned usage of graph-based symbolic knowledge.

C. First-Order Logic-based Neurosymbolic Approaches

FOL, also called quantified logic or predicate logic, is used to express the relationship between objects by allowing for variables in predicates bound by quantifiers (Wang and Yang, 2022). For example, "cat is an animal" can be induced by FOL $\forall x\ Cat(x) \Rightarrow Animal(x)$. Here, $\forall$ is a quantifier and Cat and $Animal$ are variables. A few solutions have emerged to enable the combination of FOL based symbolic knowledge and neural networks. Huang et al. (2022a) presented a novel logic tensor network (Badreddine et al., 2022) with massive rules (LTNMR) for aspect-based polarity detection. Specifically, the authors integrate two types of knowledge into the logic tensor network (LTN), i.e., dependency knowledge and human-defined knowledge rule. A mutual distillation structure knowledge injection (MDSKI) strategy is proposed to transfer dependency knowledge from teacher BERT to student network LTNMR to achieve better performance. Human-defined knowledge is represented in FOL "$R(x, a) \wedge emo_+ \rightarrow P(R(x, a), l_+)$" form integrated into the LTN, where emo_+ is the positive sentiment tag of the word from SenticNet, l_+ is the polarity label, x the input text, and a the aspect term. Intuitively, the logic is to encourage the polarity of the whole sentence to be close to the implicit polarity of aspect-related words.

Zhang et al. (2022) advanced a sentiment interpretable logic tensor network (SILTN) for aspect-term polarity detection. SILTN is interpretable as it is a neurosymbolic formalism that naturally supports learning and reasoning about data with differentiable FOL. FOL provides a flexible declarative language for conveying high-level cognition and representing structured knowledge. Despite the effectiveness of explainability, the performance of SILTN is still unsatisfactory due to its relatively shallow network structure. Therefore, the authors proposed a two-stage syntax knowledge distillation (TSynKD) strategy to improve the inferring accuracy of SILTN. Specifically, a large network BERT serves as teacher one, a big network aspect-specific dynamic GCN (AsDGCN) as teacher two, and SILTN as a student. In the output distillation stage, BERT's logits are used as AsDGCN's learning objective. In the feature distillation stage, SILTN learns dependency knowledge from AsDGCN and also uses BERT's logits as its training objective.

In short, the FOL-based approaches are a promising solution for building trust-worthy, explainable, and powerful AI. However, current approaches rely heavily on the strong representation ability of the subsymbolic system. Hence, improving the predictive performance of the symbolic system without overly leveraging the pretrained subsymbolic system would be a future research focus.

4.6.7 Downstream Applications

Polarity detection supports a variety of downstream tasks, including opinion mining, emotion detection, social media monitoring, social media marketing, customer relationship management, product review summarization, market research, brand positioning, cyber-issue detection, content moderation, political trend analysis, fake news detection, user experience enhancement, competitor analysis, conversational agents, human-robot interaction, recommender systems, stock price prediction, and stock portfolio allocation. In this subsection, we only focus on the last four tasks due to space constraints.

4.6.7.1 Human-robot Interaction

Human-robot interaction refers to the conversation between human and robot agents in this chapter. People often express their opinions over a topic or event in a conversation with a robot agent. Hence, it is essential to identify the polarity or sentiment of the user in this sentiment-aware dialogue system. On one hand, it helps to improve the deficiencies of a product by understanding the user's polarity toward the product. On the other hand, it assists in generating human-like empathetic responses that are appreciated by the users. In recent years, many conversational polarity detection approaches have been presented to better recognize user polarities (Zhong et al., 2019; Li et al., 2021b; Lee and Choi, 2021; Li et al., 2022; Lee and Lee, 2022; Li et al., 2023b). These approaches may act as a pretrained classifier module in the empathetic chatbot to help identify polarity and generate empathetic responses accordingly.

Tahara et al. (2019) proposed an empathetic dialogue system on the basis of extracted $\mathbb{X}$ polarity. The authors assumed that the degree of a user's empathy can be increased if the system can generate an utterance from topic-related tweets that contain the same polarity as the user's utterance. The assumption is confirmed by a preliminary experiment that the degree of empathy can be improved in the case of a polarized user utterance. However, the proposed method cannot increase empathy in the case of a neutral user utterance. Besides, the authors utilized Google API[37] to measure the polarity while fine-grained emotion may provide more insights that are beneficial for an empathetic dialogue system.

---

[37] https://cloud.google.com/speech

Instead of only generating empathetic responses, Liu et al. (2021b) introduced a new paradigm of human-robot interaction, which is to build an emotional support dialogue system grounded on the Helping Skills Theory (Hill, 2009). To this end, the authors defined an emotional support conversation (ESC) task and constructed an associated dataset termed ESConv. ESConv is annotated in a help-seeker (User) and supporter mode (Agent). Specifically, the goal of the robot agent in a conversation is to provide better support to comfort the bad emotional state (e.g., anxiety, shame, depression, and anger) of the user through conversation. Emotional support generally follows the following order: exploration (identify problems), comforting (express empathy and understanding), and action (help the seeker solve the problem). In the conversation of the ESConv dataset, the help-seeker was asked to give feedback every two new utterances to measure the helpfulness of the supporter. The ESC task and ESConv dataset were then measured by both automatic and manual indicators, and the results demonstrated the utility of ESConv in improving the emotional support of the dialogue system.

More recently, Tu et al. (2024) proposed AdaCLF, an adaptive curriculum learning framework to dynamically choose courses of varying complexity according to the learning status of the ESC model. The framework operates within the reinforcement learning paradigm, where a teacher model utilizes feedback from a student model to optimize its teaching strategy, fostering collaborative evolution. Both automatic and human evaluations on benchmark datasets demonstrate that this framework significantly improves existing ESC methods, generating more effective supportive responses.

4.6.7.2 Recommendation System

As discussed in Section 1.3.2, recommender systems gradually spread to most areas of our lives. Polarity detection is crucial for recommender systems because it helps understand and cater to users' preferences and attitudes. For instance, if a user consistently expresses positive emotions towards certain types of content, the system can prioritize similar items. Conversely, it can avoid suggesting items that have received negative feedback. Additionally, understanding polarity allows the system to filter out unwanted content, ensuring that users are only exposed to recommendations that align with their tastes and interests. This level of personalization enhances the user experience and increases user engagement.

Ray et al. (2021) presented an ensemble-based hotel recommender system utilizing polarity detection and aspect categorization of hotel reviews. This study is based on the fact that the decision-making process of choosing a suitable hotel is affected by an ample amount of online reviews. These reviews contain polarities over different aspects (e.g., cleanliness, service, and value) of a hotel. The authors employed BERT and random forests to build an ensemble framework to analyze the polarities of reviews over each pre-defined aspect. Finally, a comprehensive recommender system is built to sort hotels based on the user-preferred aspect (e.g., location, food, room, and facility).

In the context of RRSs, polarity detection can also be applied for analyzing recommender system bias. It is believed that the algorithm of the news recommender system could potentially form filter bubbles, leading users to become more selective in their exposure to news. This could result in increased vulnerability to polarized opinions and fake news. Alam et al. (2022) utilized the users' polarities toward news to analyze and quantify the degree to which the recommendation is biased. The authors paired pre-defined questions with news articles, and categorized the articles into "in favor" or "against" the pre-defined question (e.g., refugee and migration) to transfer German news-related polarity detection to a pretrained classifier. The bias of a recommender system is measured by the average polarity bias score of recommended articles and user history articles. If the score of recommended articles is significantly different from that of user history articles, the recommender system is considered to be biased. Experiments on four news recommender systems indicated that text-based systems tend to amplify user attitudes related to polarity, while knowledge-aware systems are less prone to bias.

4.6.7.3 Stock Price Prediction

The opinions of investors toward a stock, to some extent, reflect the future trend (up or down) of the price of that stock (Du et al., 2024a). Hence, it is valuable to regard investor polarity as an indicator to predict stock price. Derakhshan and Beigy (2019) introduced a human sentiment model to identify investors' opinions. The authors collected comments with explicitly expressed label words such as "*buy*" and "*sell*" and used these label words as handcrafted features for classification. Experiments on both English and Iranian datasets demonstrated that the classifier benefited from investor polarity information.

Colasanto et al. (2022) presented a stock price prediction approach utilizing a fine-tuned BERT-based model named AlBERTino to capture polarity from the market. AlBERTino was fine-tuned on Italian financial news and outperformed its base version AlBERTo (Polignano et al., 2019) by a large margin in terms of polarity detection. The authors exploited the Bayesian inference (Bernardo and Smith, 2001) to obtain a new set of bounded drift and volatility values on the basis of the polarity score from AlBERTino. Finally, the exact future value of the price can be determined on an hourly and daily basis. An empirical study indicated that the proposed approach achieved better results on an hourly basis prediction. This approach brings more insights into the prediction of the exact value of stock price other than only the trend (up or down) prediction.

Ma et al. (2023) proposed a framework for stock price movement prediction. The model incorporated information from multiple sources, e.g., numerical indicators, news, and company relationships. They believed that semantic sentiment is not equal to market sentiment. Thus, they used the stock price movement directions as market sentiment signals, training a sentiment representation generator. The generator is used for embedding news data in the stock price movement classifier.

4.6.7.4 Stock Portfolio Allocation

Stock portfolio allocation refers to the task of allocating funds among the stocks to maximize profit. A profitable yet low-risk allocation strategy would be valued by all investors. Koratamaddi et al. (2021) put forward a market sentiment-aware DRL (Arulkumaran et al., 2017) framework for developing an optimal policy that maximizes returns with minimal risk given historical stock prices, polarity information, and current portfolio. Specifically, observations, i.e., stock price movements and polarity from social media, are fed into the state of the agent. The actor and critic networks are updated based on observations and prediction errors are employed to update Q-value. An empirical study showed that the proposed method achieved better return values than the non-sentiment-aware baseline throughout the experimental period. Ma et al. (2024) investigated the portfolio optimization task by simultaneously learning risk and return. This task is formulated as a ranking problem, where stocks that are highly ranked based on predicted risk and return are included in the portfolio for the next trading day. In a manner similar to (Ma et al., 2023), they also utilized a pretrained news embedding generator, which was trained using market sentiment signals.

4.6.8 Summary

The concept of polarity can be traced back to the 1970s. Researchers studied word-to phrase-level polarity from the perspective of linguistics, which serves as the theoretical base of polarity detection study. As discussed throughout this section, polarity detection can be a downstream task within the NLU suitcase model but also an upstream task for many other applications such as chatbots, recommender systems, and financial forecasting. Thanks to the development of social media platforms and mobile applications, UGC has increased fast in the past decade, which in turn calls for accurate and automatic polarity analysis of users. Real-world polarity detection shows promising commercial value as it can be applied to improving customer service, and product quality contrapuntally. It is worth noting that, while this section mostly surveyed text-based polarity detection in English, multimodal and multilingual polarity detection are also very popular research fields.

4.6.8.1 Technical Trends

As shown in Table 4.23, polarity detection is categorized into three research streams: lexicon-based, PLM model, and neurosymbolic methods. Lexicon-based methods, common in early research, aim to build comprehensive and accurate sentiment lexicons but require significant manual effort and are not easily adaptable to specific domains. Data-driven methods, including statistical and semantic approaches, later automated lexicon creation, particularly for specific domains.

The advent of PLM-based methods improved detection by capturing complex semantics and dependencies in long texts. PLM techniques include using PLMs as contextualizers, fine-tuning them, or employing prompt-based approaches, which are especially useful for low-resource languages and do not require large annotated datasets. A hybrid of subsymbolic and symbolic methods (neurosymbolic approach) is widely recognized as the enabler of the next generation of AI. We roughly categorize neurosymbolic methods into three types based on the representation form of symbolic knowledge, i.e., lexicon-based, knowledge graph-based, and FOL-based neurosymbolic approaches. Neurosymbolic AI combines the learning ability (learn from experience) of the subsymbolic system and the reasoning ability (reason from what has been learned) of the symbolic system. However, the integration of symbolic and neuro parts remains a challenging problem, especially fully integrated systems that directly embed a symbolic reasoning module inside a neural module. Also, defining when, how, and why the subsymbolic components of a neurosymbolic framework interact with the symbolic ones is now an active research area.

Table 4.23: Technical trends in polarity detection. NeSy stands for neurosymbolic AI. SA denotes statistical analysis. DR denotes dynamic rewarding; TRT denotes a translate-retrieve-translate strategy. LTN denotes logic tensor network.

Trend	Reference	Tech	Feature and KB.	Framework	Dataset	Score	Metric
Lexi.	Feng et al. (2012)	SA	Lexicon, emoticons	PMI	Self-collected	0.63	Macro-F_1
	Wu et al. (2016)	SA	Lexicon, WN, syntax	PMI	NLPCC 2012	0.58	Macro-F_1
	Li et al. (2018b)	SA	Lexicon, w2v, semant.	AMI	Self-collected	0.87	Macro-F_1
	Kamps et al. (2004)	SA	Lexicon, WN, syntax	Distance	General Inquirer	0.68	Acc.
	Velikovich et al. (2010)	SA	Lexicon, ling.	Vote-flip	Self-collected	0.81	Avg prec.
	Viegas et al. (2020)	SA	Lexicon, ling., emb.	Cos. sim.	20 datasets	0.75	Macro-F_1
PLM	Satapathy et al. (2022)	DL	BERT	PLM, MTL	Pang and Lee (2005)	0.95	Acc.
	Rahman et al. (2022)	DL	BERT	PLM	Amigó et al. (2013)	0.73	Acc.
	Jiang et al. (2022)	DL	BERT, profile	PLM	Amigó et al. (2013)	0.93	F_1
	Pota et al. (2020)	DL	BERT	PLM	SENTIPOLC 2016	0.75	Macro-F_1
	Catelli et al. (2022)	DL	BERT	PLM	YelpNYC	0.94	Acc.
	Li et al. (2021a)	DL	BART	Prompt	Restaurant 14	0.75	Macro-F_1
	Mao et al. (2023c)	DL	RoBERTa	Prompt	Blitzer et al. (2007)	0.85	Acc.
	Du et al. (2024b)	LLM	Structured prompts, GPT-4	Prompt	Malo et al. (2014)	0.98	Acc.
	Zhu et al. (2024b)	LLM	Persona info., GPT-4	Prompt	Chen et al. (2023b)	0.91	Acc.
NeSy	Huang et al. (2020b)	DL	Lexicon, emb., ling.,	CNN	Thelwall et al. (2012)	0.08	Avg. MAE
	Li et al. (2020f)	DL	Lexicon, w2v	CNN-LSTM	Socher et al. (2013b)	0.5	Acc.
	Polignano et al. (2022)	DL	Lexicon, BERT	LSTM-Att.	SENTIPOLC 2016	0.73	Macro-F_1
	Zhang et al. (2023b)	DL	WN, ALM	HAN, DR	Malo et al. (2014)	0.88	Acc.
	Cambria et al. (2024)	DL	Lexicon, opinion, emo., commonsense	HAN, DR	Go et al. (2009)	0.89	Acc.
	Zhang et al. (2024c)	ML	Lexicon, POS	HAN, FFN	Go et al. (2009)	0.85	Acc.
	Liao et al. (2022)	DL	Lexicon, BERT	Graph, LSTM	SMP2019-ECISA	0.78	Macro-F_1
	Swati et al. (2022)	DL	Commonsense, PLMs	TRT, Att.	Self-collected	0.92	Macro-F_1
	Huang et al. (2022a)	DL	Dep., rules, BERT,	FOL, LTN	Restaurant 14	0.75	Macro-F_1
	Zhang et al. (2022)	DL	Syntax, BERT	FOL, LTN	Restaurant 14	0.76	Macro-F_1

4.6.8.2 Application Trends

As discussed, polarity detection is beneficial for a number of downstream tasks but we only reviewed four (see Table 4.24). The first application is human-robot interaction, where understanding user sentiment polarity enables empathetic responses and supports emotional comfort through conversation. The second application is recommender systems, which use polarity detection to personalize content and assess bias, potentially reducing information cocoons. The third application is stock price prediction, involving the forecasting of stock movements. The fourth application is stock portfolio allocation, where polarity detection helps analyze market sentiment and investor attitudes.

Table 4.24: Polarity detection downstream task applications.

Reference	Downstream Task	Feature Parser	Explainability
Tahara et al. (2019)	Human-robot Interaction	✓	
Liu et al. (2021b)	Human-robot Interaction	✓	
Ray et al. (2021)	Recommendation System	✓	
Alam et al. (2022)	Recommendation System		✓
Derakhshan and Beigy (2019)	Stock Price Prediction	✓	
Colasanto et al. (2022)	Stock Price Prediction	✓	✓
Ma et al. (2023)	Stock Price Prediction	✓	
Koratamaddi et al. (2021)	Stock Portfolio Allocation	✓	
Ma et al. (2024)	Stock Portfolio Allocation	✓	

4.6.8.3 Future Works

Polarity detection research has gained increasing attention due to its importance and promising application value in a number of downstream tasks. Despite the fact that researchers have made great progress in identifying different kinds of (e.g., sentiment and political) polarities of users, several challenges remain unsolved. The first challenge is context-dependent errors. People usually use sarcasm to express their negative sentiments with positive words. For example, "Very good; well done!" is sarcastic when someone does something wrong. One approach to addressing this issue is to incorporate a sarcasm detection module as shown in the NLU suitcase model.

The second challenge is limited training data in low-resource languages. Most of the existing polarity detection datasets are collected from English platforms like X and Amazon. Some low-resource languages have different characteristics (e.g., preprocessing, stopwords, structural differences across languages) that make it difficult to effortlessly transfer English-based models to low-resource languages. Finally, current polarity detection models suffer from bias problems as the training data is collected from human posts which may involve human biases.

Existing studies suggest that polarity detection serves an important role in downstream tasks (Liu et al., 2021b; Alam et al., 2022; Colasanto et al., 2022; Koratamaddi et al., 2021). Apparently, these downstream tasks still suffer from the challenges of polarity detection. Apart from this, dialogue systems are relatively weak at generating controllable responses, which may cause side effects to users in a bad emotional state. The main challenge of the recommender system is the bias problem. People get trapped in similar content so that their attitudes or opinions are strengthened again and again. This is not good for people to have a comprehensive understanding of the world. The stock price is influenced by many factors which cannot be fully reflected in investor comments. Besides, as a financial-related task, stock price prediction calls for a high accuracy as either false positive or false negative results in huge losses.

Amin et al. (2024) found that current expert affective computing systems can exceed LLMs in a wide range of subtasks. Developing robust polarity detection systems is still valuable but challenging. Despite PLMs achieved great accuracy in polarity detection, neurosymbolic approaches significantly improve the reasoning ability without compromising the representation ability of neural models, yielding a trustworthy, powerful and robust AI. In general, the research of neurosymbolic AI is still in its early stages. Hence, there are a series of key problems that remain unsolved. The first problem is how to effectively represent symbolic knowledge. As mentioned, FOL is a promising paradigm but it is hard to model the infinitary FOL in the finiteness neural networks (Bader and Hitzler, 2005). Therefore, it is crucial to explore some other representation paradigms such as programming languages (Nye et al., 2020; Jin et al., 2022) and symbolic expressions (Lample and Charton, 2019; Li et al., 2020e) in future research. On the other hand, subsymbolic and symbolic systems can be combined in different ways and a better combination, especially a fully integrated neurosymbolic system (Wang and Yang, 2022), should be investigated further. Last but not least, the aforementioned four downstream tasks could also benefit from the powerful reasoning ability of the subsymbolic system, e.g., generating controllable empathetic responses with the help of subsymbolic knowledge in human-robot interaction, and reducing the bias of recommender systems.

4.7 Conclusion

In this chapter, we systematically surveyed five important pragmatic processing tasks, namely: metaphor understanding, sarcasm detection, personality recognition, aspect extraction, and polarity detection. We reviewed each task in terms of the relevant theoretical research, datasets, useful knowledge bases, common evaluation metrics, methodology, and downstream applications. By surveying these tasks, we found that most current research has focused on algorithm development, while the scope of the tasks has expanded relatively little over the years. For example, many research tasks have evolved from statistical or rule-based methods to deep learning or PLM-based methods, reflecting the advancement of machine learning algorithms and general NLP research.

These advancements have significantly improved the accuracy of performing pragmatic processing tasks. However, the fundamental definitions of some pragmatic processing tasks have remained largely unchanged, adhering to their original task setups in general. This limitation significantly reduces the usefulness of pragmatic processing tools in downstream applications. LLMs have embedded more general knowledge compared to previous NLP systems or knowledge bases. However, it is unsurprising to see that LLMs were exceeded by expert systems when directly employed or prompted to handle linguistic or pragmatic tasks (Mao et al., 2024a; Amin et al., 2024). Expanding the scope of pragmatic processing tasks, such as introducing new tasks and functions of pragmatic analysis, can likely be achieved by leveraging the knowledge embedded in LLMs, rather than merely focusing on improving accuracy. Relevant theoretical research in pragmatic processing tasks reveals significant opportunities and diverse directions for expanding pragmatic analysis. Future work should emphasize both the linguistic and application aspects of these tasks.

4.8 Learning Resources

A. Reading List

- Rui Mao, Kai He, Claudia Beth Ong, Qian Liu, and Erik Cambria. MetaPro 2.0: Computational Metaphor Processing on the Effectiveness of Anomalous Language Modeling. In Proceedings of ACL, 9891–9908, 2024 (Mao et al., 2024b)

- Bin Liang, Lin Gui, Yulan He, Erik Cambria, and Ruifeng Xu. Fusion and Discrimination: A Multimodal Graph Contrastive Learning Framework for Multimodal Sarcasm Detection. IEEE Transactions on Affective Computing 15, 2024 (Liang et al., 2024)

- Kai He, Rui Mao, Tieliang Gong, Chen Li, and Erik Cambria. Meta-based Self-Training and Re-Weighting for Aspect-based Sentiment Analysis. IEEE Transactions on Affective Computing 14(3): 1731–1742, 2023 (He et al., 2023b)

B. Relevant Videos

- Labcast about Metaphor Understanding: youtu.be/pwSAoR15esw

- Labcast about Persona and Personality Recognition: youtu.be/JVFrhbo9eYU

- Labcast about ABSA: youtu.be/g88JRIkrRkY

C. Related Code

- Github repo about Sarcasm Detection: github.com/SenticNet/Sarcasm-Detection

- Github repo about Personality: github.com/SenticNet/Personality-Recognition

- Github repository about ABSA: github.com/SenticNet/Sentic-GCN

D. Exercises

- **Exercise 1.** Perform metaphor understanding on the sentence "Navigating life's ocean, we face stormy seas of challenges and serene waters of joy; while the sun of hope illuminates our path, we must also beware the shadows of doubt lurking beneath the surface". Take a step-by-step approach whereby you first detect which parts of text are metaphors, then identify source and target for each of the detected metaphors, and finally perform ABSA based on these.

- **Exercise 2.** Choose a set of social media posts and determine if they are sarcastic. Explain the reasoning behind your decisions. Outline a basic approach for detecting sarcasm in text using a rule-based system or machine learning. Discuss potential features, such as word choice, punctuation, sentiment reversal, and context. Discuss the challenges of sarcasm detection in NLU. Consider aspects like lack of tonal clues, cultural differences, and the complexity of language.

- **Exercise 3.** Design a simple system with four independent rule-based classifiers, each responsible for predicting one of the MBTI dichotomies (E/I, S/N, T/F, J/P). Next, develop a second system using a statistical classifier trained on predicting MBTI types. Explain how and why the two systems are different. Compare and discuss results of applying the two systems to a personality recognition dataset.

- **Exercise 4.** Propose a machine learning approach for aspect extraction. Discuss the use of labeled datasets, feature extraction (e.g., POA tagging, word embeddings), and potential algorithms (e.g., CRFs, SVMs, neural networks). Outline the evaluation metrics you would use to assess the performance and accuracy of aspect extraction methods and describe how these metrics can be specifically applied in the context of ABSA.

- **Exercise 5.** Create a simple system able to detect and handle ambivalence in text, where mixed sentiments (both positive and negative) are expressed simultaneously. Discuss possible scenarios where ABSA is an effective solution to the problem and when it might not be suitable.

Chapter 5
Knowledge Representation & Reasoning

Abstract Commonsense knowledge acquisition and representation is a core topic in AI, which is crucial for building more sophisticated and human-like AI systems. Existing commonsense knowledge bases organize facts in an isolated manner, like *bag of facts*, lacking the cognitive-level connections that humans commonly possess. People have the ability to efficiently organize vast amounts of knowledge by linking or generalizing concepts using a limited set of conceptual primitives that serve as the fundamental building blocks of reasoning. Such primitives are basic, foundational elements of thought that humans use to make sense of the world. By combining and recombining these primitives, people can construct complex ideas, solve problems, and understand new concepts. In this chapter, we describe the development of a commonsense knowledge base, termed PrimeNet, to emulate this cognitive mechanism. PrimeNet is organized in a three-layer structure: a small core of conceptual primitives (e.g., FOOD), a bigger set of concepts that connect to such primitives (e.g., `fruit`), and an even larger layer of entities connecting to the concepts (e.g., `banana`).

Key words: Knowledge Representation and Reasoning, Conceptual Primitives, Knowledge Graph Construction, Commonsense Knowledge Acquisition

5.1 Introduction

As covered in previous chapters, syntactic, semantic, and pragmatic processing techniques are crucial for NLU. While it might be tempting to rely solely on machine learning algorithms, given the popularity of subsymbolic AI, it is important to note that LLMs are trained on documents. However, these documents often lack commonsense knowledge. Commonsense knowledge refers to the information about everyday life that humans are expected to know, such as *painters use pencils*, *things fall downwards when dropped*, and *animals do not drive cars*.

This kind of knowledge is usually taken for granted in human communication and reasoning, even though it may not be explicitly stated (Cambria et al., 2009). However, machines lack access to this innate commonsense knowledge, which often results in their inferior performance in simple reasoning tasks. As mentioned by Oren Etzioni, commonsense is *"the dark matter"* of AI: it shapes so much of what we do and what we need to do, and yet it is ineffable. To address this limitation, researchers have dedicated significant effort to construct diverse commonsense knowledge bases like Cyc (Lenat, 1995), FrameNet (Baker et al., 1998), ConceptNet (Speer et al., 2017), ATOMIC (Sap et al., 2019a), TransOMCS (Zhang et al., 2020b), ASER (Zhang et al., 2020c), CSKG (Ilievski et al., 2021b), VoCSK (Liu et al., 2022c), and SenticNet (Cambria et al., 2024). These knowledge bases are compiled from diverse sources (e.g., encyclopedias, crowdsourcing, and expert annotations), aiming to empower machines with access to commonsense knowledge and enhance the reasoning abilities of AI systems.

Despite advancements in existing knowledge bases, the reasoning capabilities of AI systems remain unsatisfactory (Yeo et al., 2024c). One notable limitation is that current knowledge bases often organize facts in a manner resembling a *"millions of facts"*, lacking the cognitive-level connections inherent in human understanding. Humans, on the other hand, exhibit the ability to efficiently organize extensive amounts of knowledge. This capability goes beyond mere accumulation of facts and involves the intricate weaving of cognitive-level connections, enabling a deeper and more nuanced comprehension of the information at hand. We have two observations for human-like knowledge organization.

Firstly, individuals are able to function well in most real-world situations using a much smaller set of *concepts*, as opposed to dealing with an exhaustive list of specific *entities*. For example, humans generally describe commonsense knowledge like *hammer can be used to drive nails into wood*, as illustrated in Fig. 5.1. In this example, the more general concepts such as *hammer*, *nail*, and *wood* are used for the description, rather than getting into overly specific terms like *engineering hammer* or *rubber hammer*. From estimates of effective vocabulary, the number of words that people need in order to understand 95% of everyday texts is around 3000 words, and the average size of American freshman college students' vocabulary has been estimated at about 12,000 words (Zechmeister et al., 1995). This underscores the human ability to distill extensive information into manageable concepts, facilitating a more streamlined expression and understanding of daily experiences.

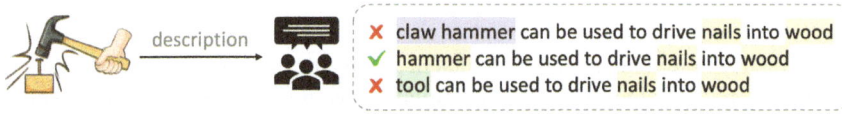

Fig. 5.1: Example of the description of commonsense knowledge with concepts (e.g., *hammer*, *nail*, and *wood*), instead of specific entities (e.g., *claw_hammer*) or abstract primitives (e.g., `tool`).

Secondly, human cognition relies on a small set of fundamental and innate building blocks called *primitives*. In the *conceptual dependency theory* (Jackendoff, 1976; Minsky, 1974; Rumelhart and Ortony, 1977; Schank, 1972; Wierzbicka, 1996), primitives serve as elemental units of information and actions, like COLOR, SHAPE, SIZE, INCREASE, and DECREASE, and forms the foundation for humans to make generalizations, inferences, and predictions, ultimately facilitating efficient reasoning and understanding in a wide range of real-world situations. For example, we generalize concepts with relevant higher-level primitives. Verb concepts such as *eat*, *slurp*, and *munch* could be related to a primitive EAT. Noun concepts like *pasta*, *bread*, and *milk* can be associated with the primitive FOOD. Therefore, *eat pasta* or *slurp milk* can be generalized into a primitive-level description, i.e., EAT FOOD. Hierarchical concept representations have significant applications in diverse domains, e.g., conceptual metaphor understanding (Ge et al., 2022; Mao et al., 2023b) and cognitive analysis (Mao et al., 2023a).

In history, some efforts have been devoted to building knowledge bases more in line with human cognition. For example, VoCSK is designed to exploit concept-level knowledge representation for implicit verb-oriented commonsense knowledge (e.g., *person eats food* instead of *John eats bread*). SenticNet is developed for organizing sentiment knowledge with a core set of primitives. ASER (short for Activities, States, Events, and their Relations) is built to extend the traditional definition of selectional preference to higher-order selectional preference over eventualities. These methods share a common goal of conceptualizing diverse types of commonsense knowledge, mapping them to higher-level cognition, and moving beyond the explicit representation of knowledge as discrete facts. Following this line, we take a further step by constructing a new framework for representing the intricate commonsense knowledge based on conceptual dependency theory.

In this chapter, we propose a new framework for commonsense knowledge representation and reasoning based on conceptual primitives, named PrimeNet. The data and the code used to develop PrimeNet are available on SenticNet github[1]. Additionally, PrimeNet is also available as an API for verb-noun generalization[2] and as a set of embeddings for ABSA available in 80 different languages[3]. The PrimeNet framework consists of three layers, as illustrated in Fig. 5.2:

- **Primitive**: The primitive layer comprises fundamental and universal elements that act as the building blocks of cognition. These primitives form the foundation upon which the entire knowledge representation is constructed. Examples of basic primitives include COLOR, SHAPE, SIZE, OBJECT, TOOL, INCREASE, DECREASE, and others. These primitives are essential for understanding and reasoning about the world.

---

[1] https://github.com/senticnet/primenet

[2] https://sentic.net/api/primenet

[3] https://sentic.net/downloads

- **Concept**: The concept layer is commonly used mental representations of categories or classes of objects, ideas, or events that share common features or characteristics. For example, concepts like *hammer* and *nail* fall into this layer. They allow for efficient information organization and grouping based on shared attributes.

- **Entity**: The entity layer represents specific instances or examples of *concepts*. For example, given the concept *hammer*, specific entities include *brick_hammer*, *rubber_hammer*, and *engineer_hammer*. This layer enables a more specific representation of knowledge, capturing individual objects or instances in the real world.

We begin by gathering extensive commonsense knowledge from diverse sources and integrate it to form a raw knowledge graph (Fig. 5.3). Unlike a simple aggregation of facts, we adopt a gradual expansion approach. Initially, we construct the graph with core concepts and relation types, systematically expanding it by adding more specific entities and incorporating diverse relation types. In the next stage, we establish the conceptual layer of PrimeNet, by assessing the abstractness of all nodes using a new scoring function tailored for conceptualization. We leverage the probabilistic taxonomy Probase (Wu et al., 2012) to identify the abstract concepts. The adopted scoring method centers around core words rather than the peripheral leaves (Liu et al., 2022c; Wang et al., 2015b).

Then, we perform primitive detection on the concepts to build the primitive layer of PrimeNet. Formulating a thorough primitive set demands considerable time and effort. To address this, we design a lexical substitution task to discover the set of primitives. This is grounded in the assumption that within a shared context, the associated concepts under a primitive can be seamlessly interchanged, resulting in grammatically accurate sentences upon substitution. To allocate a representative primitive to each concept cluster, we leverage LLMs to generate the primitive and employ an LLM-based verifier to validate the assignment of the primitive to concepts.

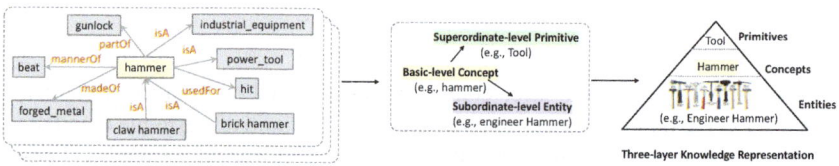

Fig. 5.2: Illustration of three-layer structure in PrimeNet. Given the factual knowledge, a *concept* layer is generated as the basic level, comprising widely recognized mental representations associated with various categories or classes of objects. Its subordinate layer is termed as *entity* layer, which consists of specific entities, and its superordinate layer is defined as *primitive* level, encapsulating overarching and fundamental primitives.

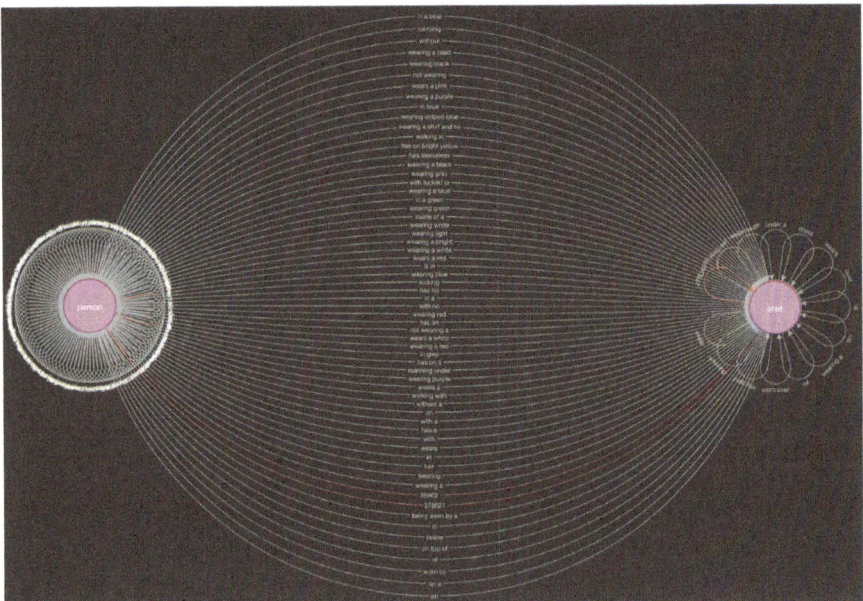

Fig. 5.3: PrimeNet preliminary knowledge graph. The initial knowledge graph collects all natural language relationships (edges) between concepts (nodes) found in the training data. After several rounds of normalization, the final PrimeNet graph only leverages 34 relationships.

Moreover, we manually check the primitives, refine the hierarchy structure of the primitives, and generate the explanation of primitives. For example, DEACTIVATE is defined as *change the status from on to off*, i.e., STATE=ON → STATE=OFF. In Table. 5.1, we present several cases of verb primitives used in PrimeNet. This strategy of constructing a primitive layer balances the need for human hand-coding for accuracy with that for crowdsourcing and machine-based knowledge extraction for coverage.

Table 5.1: Examples of verb primitives in PrimeNet. Given the input string, we illustrate the detected verb primitives, and its primitive-level representation and explanation. Primitives are marked in green.

Input String	Verb Primitives	Primitive-level Representation and Explanation
turn off light	turn off → DEACTIVATE	DEACTIVATE(light) light.STATE=ON → light.STATE = OFF
add salary	add → INCREASE	INCREASE(salary) salary → salary++
cut budget	cut → DECREASE	DECREASE(budget) budget → budget--
drive car	drive → ACCELERATE	ACCELERATE(car) INCREASE(car.SPEED) := car.SPEED++
build house	build → GENERATE	GENERATE(house) ∄ house →∃ house
butcher chickens	butcher → KILL	KILL(chicken) TERMINATE(LIFE(chicken))
revise the manuscript	revise →FIX	Fix(manuscript) manuscript.STATE=BAD → manuscript.STATE=GOOD
illuminate the idea	illuminate →SIMPLIFY	SIMPLIFY(idea) idea.STATE=DIFFICULT → idea.STATE=EASY

The key novelties introduced by PrimeNet are summarized below.

1) **Representation of commonsense knowledge based on conceptual primitives.**
   We propose a multi-layer commonsense knowledge base based on *conceptual primitives* under the hypothesize that commonsense reasoning could depend on a concise core of concepts. To the best of our knowledge, this is the first work incorporating the idea of conceptual primitives into a general-purpose commonsense knowledge base to provide a generalizable, effective representation of commonsense knowledge for AI tasks.

2) **Construction of a new commonsense knowledge base PrimeNet.** Based on the designed multi-layer structure, we construct a brand new commonsense knowledge base. We first collect commonsense knowledge from various sources and perform knowledge integration to build a knowledge graph.

3) **Conceptualization for PrimeNet.** We design a new scoring method to measure the abstractness of a term for conceptualization, according to the conditional probability and connections to core words. Compared with previous methods, the proposed method centers around core words rather than the peripheral leaves, which is effective in measuring the abstractness of concepts.

4) **Primitive Detection for PrimeNet.** We design a new primitive detection method to build the primitive layer. We employ a lexical substitution task to discover related concepts under the assumption that they share a similar context. For the clusters of related concepts, we leverage LLMs to label their primitives and verify the detection process.

The rest of this chapter is organized as follows: Section 5.2 introduces conceptual primitive theory and the challenges of building a commonsense knowledge base; Section 5.3 describes the overall framework; Section 5.4 explains the steps for PrimeNet's initial knowledge graph construction; Section 5.5 illustrates how the concept layer of PrimeNet is built; Section 5.6 introduces the primitive detection algorithm for building the primitive layer of PrimeNet; Section 5.7 reports experiments; Section 5.8 discusses future work; finally, Section 5.9 provides concluding remarks.

5.2 Background

5.2.1 Theory of Conceptual Primitives

In linguistics and cognitive science, *conceptual primitive* commonly refers to a basic, irreducible concept or idea that serves as a foundation for understanding more complex concepts. Conceptual primitives are fundamental elements that are not further defined in terms of other concepts but are instead used to define other, more complex ideas. They are often considered to be the building blocks of thought and language. The exploration of conceptual primitives has a rich history within linguistics.

In the 1950s, Chomsky (1957) introduced the universal grammar theory, positing innate linguistic structures as foundational conceptual primitives. According to this theory, humans inherently possess the capacity to acquire language, with universal linguistic structures serving as fundamental building blocks shared across all languages. The conceptual dependency theory, put forth by Schank (1972), suggested that the basis of natural language is conceptual, forming an interlingual foundation composed of shared concepts and relationships across languages. Jackendoff (1976) delved into explanatory semantic representation, asserting the existence of semantic primitives common to all languages, enabling humans to express a diverse range of semantic information. Wierzbicka (1996) emphasized that "conceptual primitives and semantic universals are the cornerstones of a semantic theory", asserting that this limited set of primitives can determine interpretations for all lexical and grammatical meanings in natural language. These theories collectively aim to identify a core set of fundamental primitives for language, facilitating the description of lexicalized concepts.

In the realm of cognitive science, theoretical studies on commonsense knowledge representation align with similar insights. Jackendoff et al. (1983) highlighted a strong correlation between semantic primitives and cognitive representation. According to Pesina and Solonchak (2015), the primitives studied in linguistics form the basis for the formation of a person's conceptual system, which is both unique and universal in many aspects. In this view, language emerges as a central tool for cognitive functions, including conceptualization and categorization. In the development of knowledge representation theories in cognitive science, many have been based on the idea that humans possess a core set of knowledge connecting a vast array of specific knowledge.

In the early stages, Minsky (1974) studied the framework for knowledge representation and introduced the concept of "frames" as a structured way to organize information about situations or objects. He proposed that humans when encountering new situations, retrieve typical knowledge from their minds. Piaget et al. (1952) introduced the term "schema", representing both the category of knowledge and the process of acquiring that knowledge. The knowledge representation based on schema has also been further researched by Rumelhart and Ortony (1977); Winograd (1976); Bobrow and Norman (1975); Johnson (1989) and others. Spelke and Kinzler (2007) introduced the core knowledge theory, suggesting that infants are born with "core knowledge systems" supporting basic intuitions about the world. West (2011) introduced a data modeling structure divided into primitive and derived concepts, with primitive concepts serving as building blocks for other concepts. These theories collectively underscore that the core primitive set constitutes the fundamental structure of human cognition and provides guidance for knowledge representation.

5.2.2 Challenges

In modern large-scale commonsense knowledge bases, there have been relatively few attempts to build a knowledge base in a way incorporating core primitives based on conceptual dependency theory and linking a vast amount of facts. Cambria et al. (2024) have developed a sentiment analysis system based on primitives such as DECREASE and INCREASE aimed at generalizing sentences into a sort of protolanguage in which it is easier to perform polarity detection, e.g., the sentence "the device's temperature sky rocketed" is first generalized to "INCREASE(device.temperature)" and then later transformed into "device.temperature++" (Fig.5.4). Wachowiak and Gromann (2022) proposed to build on large multilingual PLMs and a small dataset of examples from image schema literature to train a supervised classifier that classifies natural language expressions of varying lengths into image schemas. Liu et al. (2022c) designed conceptualization for verbs and built a knowledge base with conceptual verb-oriented knowledge to represent various instances, e.g., "*John eat apple*" and "*Helen eat bread*" are represented as "people eat food".

The primary challenge hindering progress in this field stems from the complexity of constructing a comprehensive set of core primitives to encompass extensive knowledge across diverse domains. On the one hand, managing large-scale factual data makes manual editing and maintenance of a core primitive set impractical. While it is possible to manually craft a small, high-quality core primitive set, this approach becomes intricate when using primitives to interpret other specific concepts, and its coverage of specific knowledge is limited.

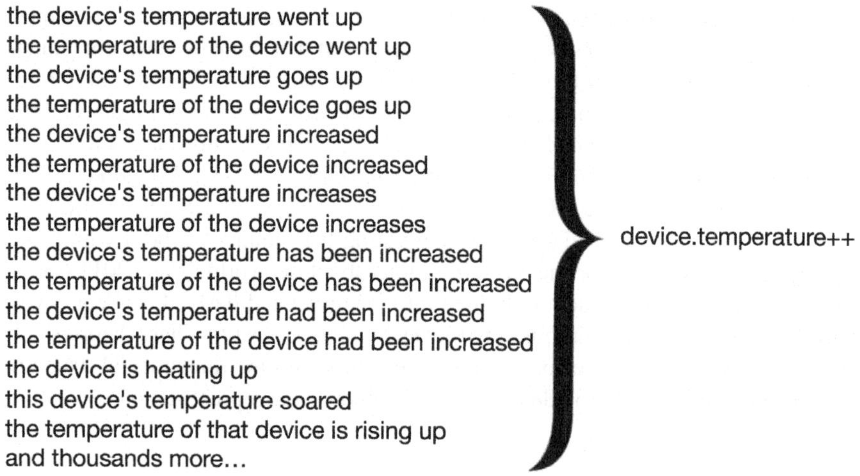

the device's temperature went up
the temperature of the device went up
the device's temperature goes up
the temperature of the device goes up
the device's temperature increased
the temperature of the device increased
the device's temperature increases
the temperature of the device increases
the device's temperature has been increased   device.temperature++
the temperature of the device has been increased
the device's temperature had been increased
the temperature of the device had been increased
the device is heating up
this device's temperature soared
the temperature of that device is rising up
and thousands more…

Fig. 5.4: Generalization of sentences into a more abstract, conceptual form.

On the other hand, primitives are not fixed but rather flexible and adaptable. The core primitives are deeply embedded in the human conceptual system, which is both unique and universal in many aspects. The proposed number of semantic primitives varies significantly, ranging from a few units in some studies (Wierzbicka, 1996; Jackendoff et al., 1983) to several dozens (Wierzbicka, 1996) or even hundreds (Cambria et al., 2024) in others. Pesina and Solonchak (2015) stated that the main concepts of human society remain relatively stable, but their overall volume changes over time.

5.2.3 Related Works

In this section, we conduct a comprehensive literature review on commonsense knowledge acquisition, including crowdsourcing methods, automatic extraction methods, and approaches centered around extracting implicit knowledge from PLMs. Then, we introduce the conceptual primitives theory, which is a pivotal component in the construction of PrimeNet.

5.2.3.1 Commonsense Knowledge Acquisition

Because of its nature, commonsense knowledge is not explicitly stated in documents. It is an inherent understanding of the world that humans possess but machines lack. To narrow the gap between human and machine intelligence, the process of acquiring commonsense knowledge is crucial for improving AI. There are mainly three major approaches to knowledge acquisition: crowdsourcing, automatic extraction, and mining from PLMs.

A. Crowdsourcing

Crowdsourcing is a useful approach for collecting commonsense knowledge from a diverse group of human contributors, such as human experts (Miller, 1995; Lenat, 1995), web users (Singh et al., 2002; Chklovski, 2003), and participants in human computation games (von Ahn et al., 2006; Kuo et al., 2009). By tapping into the collective wisdom of individuals, this approach captures intuitions and insights commonly held by people, thus contributing valuable data to the construction of commonsense knowledge bases. The crowdsourcing approach exhibits high adaptability across diverse tasks and domains. By involving a varied group of contributors, it ensures that multiple viewpoints are considered, leading to the creation of a more comprehensive and balanced knowledge pool. Existing knowledge bases built through crowdsourcing typically encompass the following categories of commonsense knowledge: factual knowledge, lexical knowledge, encyclopedic knowledge, and domain knowledge (discussed below).

Factual knowledge represents concrete and specific details about the world, events, people, places, objects, and other observable phenomena, such as "wheel is part of bicycle", "dog is an animal", and "Los Angeles is located in California". In the early 1980s, the Cyc project (Lenat, 1995) undertook the task of manually constructing a comprehensive knowledge base encompassing the basic facts and rules about the world. After the efforts of its first decade, the Cyc project expanded to include around 100,000 terms. By the time of its release in 2012, known as OpenCyc 4.0, the knowledge base had undergone substantial growth, encompassing over 2 million facts across 239,000 concepts. The DOLCE project (Gangemi et al., 2002) was designed to manually collect the ontological categories underlying natural language and human commonsense with disambiguated concepts and relations. Freebase (Bollacker et al., 2008) is a collaborative knowledge base by gathering data from various sources, including Wikipedia, the Notable Names Database, and contributions from community users. Google Knowledge Graph[4] is powered in part by Freebase, with an extensive collection of billions of facts about people, places, and things. As discussed in earlier sections, ConceptNet (Speer et al., 2017) also leverages crowdsourcing contributions from users to acquire commonsense knowledge. Moreover, ConceptNet is available in 83 languages and can be linked to other knowledge bases, such as WordNet, Wiktionary, OpenCyc, and DBpedia.

As for lexical knowledge, there are several databases manually created by experts, such as WordNet (Miller, 1995), Roget's Thesaurus (Kipfer, 2006), FrameNet (Baker et al., 1998), MetaNet (Dodge et al., 2015), VerbNet (Schuler, 2005), and PropBank (Palmer et al., 2005). As discussed earlier in the textbook, WordNet is also a highly popular lexical knowledge base which captures semantic relations between words. WordNet is now available in over 200 languages, allowing researchers and linguists worldwide to explore the complexities of language and word associations across diverse contexts.

Encyclopedic knowledge is related to a broad understanding of various subjects and topics. For example, Wikidata (Vrandecic and Krötzsch, 2014) is a knowledge graph coupled with Wikipedia, which is a free, open, and multilingual online encyclopedia that is collaboratively edited by volunteers. DBpedia (Auer et al., 2007) extracts structured information from Wikipedia data and converts it into a machine-readable format for use in the Semantic Web and data mining domains. The encyclopedic knowledge resources offer a wide range of information to help people understand various topics and fields.

More recently, commonsense knowledge bases have been specifically developed to cater to particular tasks, such as dialogue systems (Young et al., 2018). For example, SenticNet has been specifically developed for affective computing tasks. Visual Genome (Krishna et al., 2017) contains annotations of concepts and their relations found in a collection of images. ATOMIC (Sap et al., 2019a) is developed to capture inferential commonsense knowledge, such as cause-and-effect relationships. Finally, ATOMIC$^{20}_{20}$ (Hwang et al., 2020) is proposed to unify the triples from ConceptNet and ATOMIC, together with some newly developed relations.

---

[4] https://blog.google/products/search/introducing-knowledge-graph-things-not

B. Automatic Extraction

Despite commonsense knowledge is not explicitly defined, it has been observed that certain types of commonsense knowledge can be extracted through automatic methods, such as text mining and information extraction. Compared with crowdsourcing, these automatic extraction methods can handle large volumes of data efficiently and at a lower cost, making them valuable tools for efficiently capturing and updating commonsense knowledge from various domains.

Firstly, automatic extraction methods generally acquire commonsense knowledge from large-scale text and web pages. For example, Never-Ending Language Learning (NELL) project (Mitchell and Fredkin, 2014) is designed to automatically extract structured information from unstructured Web pages. With hundreds of pre-defined categories and relations, NELL extracts knowledge from more than 500 million web pages, resulting in a large knowledge base comprising over 2.8 million instances. WebChild (Tandon et al., 2014) is constructed through automated extraction and disambiguation from Web contents. It utilizes seeds derived from WordNet and pattern matching techniques on large-scale text collections to gather information, including fine-grained relations like "hasShape," "hasTaste," and "evokesEmotion". As discussed earlier, ASER, SenticNet and Probase also contain elements of automatic extraction. A subsequent version of Probase, named as Microsoft Concept Graph (Ji et al., 2019), harnesses billions of web pages and search logs to build a huge graph of relations between concepts, and has been proven valuable in enhancing search engines, spell-checkers, recommendation engines, and other AI-driven systems.

Secondly, several methods are used to improve existing commonsense knowledge bases. The automatic extraction methods can help fill gaps, update outdated information, and supplement missing commonsense knowledge in existing knowledge bases. For example, BabelNet (Navigli and Ponzetto, 2012a) is a multilingual knowledge base which is automatically created by mapping Wikipedia to English WordNet based on multilingual concept lexicalizations and machine translations. Dense-ATOMIC (Shen et al., 2023a) is designed to overcome the limitations of ATOMIC in knowledge coverage and multi-hop reasoning, by employing a knowledge graph completion approach.

Thirdly, some efforts have been made to automatic integrate diverse commonsense knowledge bases, enhancing the overall coverage and richness of the knowledge base. For example, YAGO (Suchanek et al., 2007) is designed to extract commonsense knowledge from Wikipedia, WordNet, WikiData, GeoNames, and other data sources. Bouraoui et al. (2022) employed Region Connection Calculus to merge open-domain terminological knowledge. CommonSense Knowledge Graph (CSKG) integrates knowledge bases from seven diverse, disjoint sources such as ConceptNet and WordNet. Based on ASER, Zhang et al. (2020b) have developed TransOMCS with an algorithm for discovering patterns from the overlap of existing commonsense and linguistic knowledge bases, and a commonsense knowledge ranking model to select the highest-quality extracted knowledge.

C. Implicit Knowledge in Pretrained Models

Recent advancements in pretrained models have demonstrated significant improvements across various tasks, underscoring their robust representation and generalization capabilities. These models, pretrained on large-scale corpora, have proven adept at encoding diverse forms of knowledge (AlKhamissi et al., 2022; Bhargava and Ng, 2022). For example, BERT uses a masked language model objective in pretraining, where parts of the input are masked, enabling the model to predict concealed words bidirectionally. This process empowers BERT to capture contextualized representations, comprehensively understanding intricate relationships and meanings in different linguistic contexts. Similarly, a GPT (Radford et al., 2019; Brown et al., 2020) follows the generative language model paradigm, predicting the next word based on preceding context. With a unidirectional architecture processing text from left to right during training, it acquires knowledge of grammar, facts, reasoning, and even some degree of commonsense.

Currently, there is a trend to mine commonsense knowledge directly from PLMs, leveraging the rich information embedded in these large models. Several works are designed to probe commonsense knowledge directly from large pretrained models, such as KB-BERT (Yao et al., 2019), KB-BERTSAGE (Fang et al., 2021b), and PseudoReasoner (Fang et al., 2022a). These approaches involve fine-tuning PLMs, such as BERT and BART, on commonsense knowledge bases like ATOMIC, ConceptNet, and ASER, with the tasks typically entails providing the head and relation in a commonsense triple as input, with the tail serving as the expected output.

COMET (Bosselut et al., 2019) (COMmonsEnse Transformers) is designed to leverage GPT to generate rich and diverse commonsense descriptions in natural language. It effectively transforms implicit knowledge from pretrained models into explicit knowledge within commonsense knowledge graphs, and generates novel knowledge that humans rate as high quality. LAMA (Petroni et al., 2019) (LAnguage Model Analysis) is an unsupervised method to leverage BERT to acquire commonsense knowledge. It also serves as a framework[5] for probing and evaluating the factual knowledge encoded in PLMs (Petroni et al., 2020).

West et al. (2022) designed a symbolic knowledge distillation to leverage some seeds from ATOMIC as prompts to acquire commonsense knowledge from GPT-3, resulting a large commonsense knowledge graph $ATOMIC^{10x}$ and a compact commonsense model $COMET_{TIL}^{DIS}$. Their work demonstrates the efficacy of collaborative efforts between humans and language models for curating commonsense knowledge graphs and training efficient, high-performing commonsense models.

---

[5] https://github.com/facebookresearch/LAMA

5.2.3.2 Commonsense Knowledge Representation

Commonsense knowledge representation plays a vital role in AI, as it entails transforming intricate and valuable human commonsense knowledge into machine-readable formats, and enables the facilitation of complex reasoning tasks. Knowledge representation and reasoning are tightly intertwined, as one of the primary objectives of explicitly representing knowledge is to enable the capacity for reasoning, inference drawing, and asserting new knowledge.

Reflecting the complexities of human cognition, commonsense knowledge is represented through a variety of methodologies. Early techniques, like FOL and logic rules, provided structured frameworks for capturing relationships and rules governing the world. These methods encoded knowledge in terms of logical statements and rules, enabling systems to perform deductive reasoning and inference. Beyond logic-based approaches, other methods have emerged to represent commonsense knowledge. For example, semantic networks employ graph structures to depict concepts and their relationships, allowing for intuitive representation and reasoning; frame-based systems organize knowledge into structured frames, capturing attributes, roles, and hierarchies among entities.

More recently, some commonsense knowledge bases have achieved significant success and are widely applied across various AI domains to support different reasoning tasks. Typically, their knowledge representation frameworks adopt a *"millions of facts"* approach. For instance, ConceptNet summarizes millions of facts into a knowledge graph format, where nodes denote entities and edges denote their relationships. Each commonsense knowledge or fact can be represented as a triplet, such as <dog, isa, animal>, forming *millions of triplets* within the knowledge base. Similarly, ATOMIC is a commonsense knowledge graph with 1.33 million everyday inferential knowledge tuples about entities and events. It is represented in the form of IF-THEN statements, like "if X pays Y a compliment, then Y will likely return the compliment", resulting *millions of if-then statements* within the knowledge base.

While these knowledge representation frameworks have been effectively utilized in many applications, one of main limitations of these knowledge representation frameworks is the lack of a cognitive-level connection. When humans store commonsense knowledge, there are often underlying connections that involve the relevance between concepts, underlying reasoning, contextual information, and so on. However, current knowledge representation methods typically only capture surface-level relationships and lack a deep understanding of cognitive processes and underlying thought mechanisms. Consequently, these frameworks may not fully capture human cognitive levels in complex reasoning tasks, limiting their application and effectiveness in some complex reasoning and inference tasks. Moreover, in implementing commonsense knowledge representation, the absence of cognitive-level connections can lead to challenges in identifying meaningful patterns and relationships within data, resulting in suboptimal performance, limited accuracy, and increased risk of erroneous conclusions. This inefficiency may further lead to resource wastage and ultimately diminish effectiveness in addressing complex reasoning tasks across various AI applications.

5.2.3.3 Conceptual Primitives

Conceptual primitives can be defined as concepts that cannot be defined in terms of other concepts in an integration data model which provides an overview of data, thereby forming foundations for definitions of other concepts (West, 2011). Conceptual primitives have been of practical and theoretical interest to researchers in computer science (Minsky, 1974), linguistics (Jackendoff, 1976; Wierzbicka, 1996) and psychology (Rumelhart and Ortony, 1977). Such research reports that the decomposition of meanings into lower-level parts is essential for conceptualization.

We apply the idea of conceptual primitives to construct commonsense knowledge by comprising a small core of primitive commonsense concepts and relations, linked to a much more extensive base of factual knowledge instances. Naturally, humans tend to categorize things, events, and people by identifying common patterns and forms, which is the basis of conceptual dependency theory. Thus, commonsense knowledge bases built upon conceptual primitives possess the greater potential to facilitate reasoning tasks. SenticNet also generalizes words and MWEs into primitives and super-primitives annotated with emotion labels via PLMs, which achieved better performance on various affective tasks and showed the power of conceptual primitives. Unlike SenticNet, which focuses on sentiment knowledge, PrimeNet covers a broader range of general commonsense knowledge based on conceptual primitives.

5.3 Overall Framework

In this section, we first introduce the task definition. Then, we introduce the solution of constructing PrimeNet and the key ideas of each module.

5.3.1 Task Definition

PrimeNet is a hybrid graph $\mathcal{H}$ combining a traditional graph $\mathcal{G}$ where each edge is built among nodes to represent commonsense knowledge in triplets, and a hypergraph $\mathcal{G}^*$ where each edge is built over the nodes to linked their concepts and primitives. For example, in the graph $\mathcal{G}$, its edge is represented as a triplet like (*corgi, isA, dog*), where *dog* and *corgi* are nodes, and *isA* is a relation type. In the hypergraph $\mathcal{G}^*$, *corgi* is linked to *dog* in the concept layer, and *dog* is linked to ANIMAL in the primitive layer. We devise the formal definition of PrimeNet as below.

Definition (PrimeNet) PrimeNet is a hybrid graph $\mathcal{H}$ of a knowledge graph $\mathcal{G}$ and a hypergraph $\mathcal{G}^*$. The knowledge graph is denoted as $\mathcal{G} = \{\mathcal{V}, \mathcal{E}, \mathcal{R}\}$ where $\mathcal{V}$ is a node set, $\mathcal{E}$ is an edge set connecting pairs of nodes, and $\mathcal{R}$ is the set of distinct relation types associated with the edges in $\mathcal{E}$. Each node $v \in \mathcal{V}$ is a term. Each edge $e \in \mathcal{E}$ is a triplet (v_i, r, v_j) where v_i and v_j are the connected nodes, and $r \in \mathcal{R}$ is the relation type.

The hypergraph is denoted as $G^* = \{\mathcal{V}, C, \mathcal{P}, \mathcal{M}\}$, where $\mathcal{V}$ represents the set of entities, C represents the set of concepts, and $\mathcal{P}$ represents the set of primitives. The hyperedge set $\mathcal{M} = \{\mathcal{M}_{v \to c}, \mathcal{M}_{c \to p}\}$ contains two types of hyperedges. The hyperedge $(v, c) \in \mathcal{M}_{v \to c}$ links the entity $v \in \mathcal{V}$ to its concept $c \in C$, and the hyperedge $(c, p) \in \mathcal{M}_{c \to p}$ links the concept $c \in C$ to its primitive $p \in C$. Overall, we have the PrimeNet $\mathcal{H} = \{\mathcal{V}, \mathcal{E}, \mathcal{R}, C, \mathcal{P}, \mathcal{M}\}$. □

5.3.2 Overall Framework of PrimeNet Construction

The solution of PrimeNet mainly consists of three modules. The first module is to construct the knowledge graph $G = \{\mathcal{V}, \mathcal{E}, \mathcal{R}\}$ to organize the large-scale commonsense knowledge. This knowledge graph is designed to cover a wide range of commonsense knowledge, encompassing specific entities and extensive information. We refer to this graph as the entity layer of PrimeNet. The second module is a conceptualization module, which identifies a small set of concepts C on top of the set of specific entities $\mathcal{E}$ in G, as well as the hyperedges $\mathcal{M}_{v \to c}$ to link entities to concepts. We consider this concept set and the mapping between concepts and entities as the concept layer of PrimeNet. The third module is a primitive detection module that constructs the core primitive set $\mathcal{P}$ on top of the concept set C and builds the hyperedges $\mathcal{M}_{c \to p}$ to link concepts to their primitives. This small primitive set and the mapping between primitives and concepts are used as the primitive layer of PrimeNet. In the following, we will provide a more in-depth introduction to each module, along with corresponding examples for illustration.

5.3.2.1 Module-1: Knowledge Graph Construction

Over the course of many years, a vast reservoir of factual knowledge has accumulated, taking on various forms and originating from diverse sources. In order to systematically organize this wealth of knowledge, we have undertaken the construction of a knowledge graph. Drawing inspiration from the theory of cognitive development put forth by Piaget et al. (1952), which posits that human cognitive development occurs in stages, we have adopted a gradual expansion strategy to build the knowledge repository. Rather than merging disparate sources abruptly, our approach is to delicately expand the knowledge base.

The fundamental idea underlying our strategy is that human knowledge acquisition follows a pattern of continuous expansion, rooted in commonly shared and widely accepted information. For instance, individuals typically begin by learning that a "hammer" is a "tool" used for driving "nails", and subsequently delve into more intricate details, such as discerning the differences among various types of hammers like "engineer hammer" and"brick hammer". To emulate this cognitive process, we initially construct a basic graph consisting of widely used concepts and relations as the entity layer of PrimeNet.

Subsequently, we systematically enlarge the graph by incorporating a multitude of facts from diverse sources. This method allows for the gradual incorporation of information, mirroring the incremental nature of human knowledge acquisition. We detail this module in Section 5.4.

5.3.2.2 Module-2: Concept Detection

To construct the concept layer over the knowledge graph $\mathcal{G}$, this module focuses on identifying a suitable concept set C from the node set $\mathcal{V}$ and establishing hyperedges in the set $\mathcal{M}_{v \to c}$ to link entities with their respective concepts. Within PrimeNet, this concept layer encapsulates commonly used mental representations of categories, classes, or ideas that share common features or characteristics. For instance, "hammer" is the concept that represents a category encompassing entities such as "engineering hammer", "brick hammer", and "rubber hammer". Consequently, we initialize the concept set layer using Core WordNet[6], a compilation of about 5000 most commonly used words meticulously curated by experts. Then, we design a concept detection method to discover new concepts and expand the concept set, leveraging Probase and build the edges to link entities to the detected concepts.

Our observation underscores that, for a concept, its hyponyms tend to establish robust connections with diverse concepts in a probabilistic taxonomy, whereas a specific entity is more concentrated in its connection to concepts. To capture this regularity, we introduce a novel scoring function designed to identify whether a term qualifies as a concept. In contrast to alternative conceptualization methods, our approach stands out by centering around core words rather than initiating from the leaves of an extensive taxonomy for concept detection. The pre-defined core words enhance diversity and accuracy, distinguishing our strategy as effective in steering clear of misleading information stemming from isolated graphs or incorrect circles within the large-scale taxonomy.

5.3.2.3 Module-3: Primitive Discovery

This module is dedicated to constructing the primitive layer of PrimeNet, involving the establishment of a core primitive set $\mathcal{P}$ and the creation of the hyperedge set $\mathcal{M}_{c \to p}$ to connect concepts with their higher-level primitives. For instance, the primitive INCREASE is associated with concepts like *ramp up, go up, broaden, step up, elevate, supplement, redouble, pile up, upward spiral, distend*, and more. The manual definition of the primitive set and linking of primitives to their lower-level concepts is impractical. In our approach, an automated method is designed, utilizing concept clustering and subsequent labeling of their primitives using LLMs, followed by error checking to refine both the primitives and concept clusters.

---

[6] https://wordnetcode.princeton.edu/glosstag.shtml

Specifically, it is observed that concepts under the same primitive often share a similar meaning and context. For instance, *elongate* and *stretch* fall under the same primitive GROW and share a similar context. Although intuitive, lexical substitution tends to overlook crucial differences between concepts. For example, verbs such as *stretch* and *compress* belong to opposite primitives, GROW and SHRINK respectively, yet can be identified within similar lexical contexts. To address this issue, we leverage powerful LLMs to filter out incorrect concepts within each cluster, generating a primitive that accurately describes the concept cluster. Manual checks are also employed to ensure the quality of primitives in building the primitive layer. This strategy strikes a balance between human hand-coding for accuracy and crowdsourcing and machine-based knowledge extraction for comprehensive coverage.

5.4 Knowledge Graph Construction

In this section, we detail the construction of the knowledge graph ($\mathcal{G}$) of PrimeNet. It mainly contains four stages. First, *commonsense knowledge acquisition* is to collect high-quality knowledge from diverse sources which are created through manually annotated or crowdsourcing. Then, *knowledge integration* is to map the nodes and relations among different sources. Next, the *graph construction* is to organize the knowledge in a graph. Finally, *exploration* is to define functions to leverage the knowledge graph in the downstream tasks. We detail each stage as follows.

5.4.1 Commonsense Knowledge Acquisition

In constructing a commonsense knowledge base, the acquisition of knowledge stands out as a pivotal initial phase. Collecting commonsense knowledge is a challenging task due to its sheer volume, implicit nature, and diverse forms of expression. With decades of human efforts, a wealth of commonsense knowledge has been amassed and stored in various knowledge bases. To ensure quality, we extract knowledge from expert-crafted databases and crowdsourced repositories, as summarized in Table 5.2, including:

- Lexical knowledge extracted from WordNet, FrameNet, and Roget (Kipfer, 2006);
- Factual knowledge extracted from ConceptNet;
- Structured information in Wikidata and DBpedia. For DBpedia, we extract knowledge from *InfoBoxes* which provide information about a wide variety of topics, e.g., people, places, and organizations, as well as knowledge from *InstanceTypes* which contains instances of 438 types, e.g., book, company, city, and plant.
- Task-specific knowledge, such as inferential knowledge extracted from ATOMIC, which is organized as typed "if-then" relations with variables, and visual knowledge extracted from Visual Genome (Krishna et al., 2017).

Table 5.2: Sources of commonsense knowledge for building the knowledge graph of PrimeNet. *Creation* denotes the construction methods, *# R* denotes the number of relation types, and *Size* denotes the graph scale.

Source	Creation	# R	Size	Example
WordNet	manual	10	155K words, 176K synsets	(denied, morphy, deny)
FrameNet	manual	10	1.2K frames, 12K roles, 1.9K edges	(Criminal_process, Subframe, Arrest)
Roget	manual	2	72k words, 1.4M edges	(explore, Synonym, investigate)
ConceptNet	crowdsourcing	34	8M nodes, 21M edges	(keyboard, part of, computer)
Wikidata	crowdsourcing	6.7K	75M objects, 900M edges	(George Washington, isInstanceOf, human)
DBpedia	crowdsourcing	53.1K	4.8M nodes, 62M edges	(Applied_Artificial_Intelligence, discipline, Artificial_intelligence)
ATOMIC	crowdsourcing	9	300K nodes, 877K edges	(Person X bakes bread, Before X needed to, buy the ingredients)
Visual Genome	crowdsourcing	42.4K	3.8M nodes, 2.3M edges, 2.8M attributes	(man, sit on, bench)

5.4.2 Knowledge Integration

In commonsense KGC, multiple sources can provide complementary knowledge of different types. However, the integration of knowledge from diverse sources is impeded by the varying representation formats. It is noted that many databases provide mappings to other databases, e.g., ConceptNet contains mappings to DBpedia, WordNet, Wikidata, and FrameNet. Yet, these mappings may be incomplete. Recent research endeavors to create high-quality mappings among different knowledge bases, offering a pathway for knowledge integration. For example, CSKG constructs mappings across seven knowledge bases (i.e., ATOMIC, ConceptNet, FrameNet, Roget, Visual Genome, Wikidata, and WordNet). Yang et al. (2023), instead, drew insights from case-based reasoning, which aims to solve a new problem by reasoning with retrieved relevant cases, and investigate the direct application of it to commonsense KGC.

We conduct knowledge integration to build a knowledge graph of PrimeNet using these high-quality mappings, as well as lexical-level and semantic-level matching methods. First, we process the individual sources. More specifically, we keep the initial sets of nodes, edges, and relations in ConceptNet and ATOMIC. For other sources, we extract their nodes and edges and convert their relations to the format of relations in ConceptNet. Then, we conduct mappings between sources for node resolution. On the one hand, we leverage mappings released by Ilievski et al. (2021b)[7] to map nodes from different sources. On the other hand, we represent each node using its label and use exact lexical matching to establish the mappings of nodes from different sources. Moreover, we conduct semantic-level matching to identify the same nodes with different labels. We convert all labels of nodes to embeddings using pretrained Sentence-BERT (Reimers and Gurevych, 2019)[8]. Subsequently, we use node labels from another source as queries for embedding-based semantic search. A link is established between the query and the top-1 similar node if they share the same representation after NLTK lexical tokenization.

---

[7] https://github.com/usc-isi-i2/cskg

[8] https://huggingface.co/sentence-transformers/all-mpnet-base-v2

5.4.3 Graph Construction

Confronted with an extensive dataset of knowledge triplets, creating a graph by incorporating all of them directly is a blunt method. Humans develop core conceptual primitives grounded in the most frequently utilized knowledge. For example, in the realm of *geography*, individuals effortlessly understand fundamental concepts like *country*, *continent*, and *ocean*, forming a foundational understanding without the need to memorize every specific detail, including aspects like the area and visual representation of each country available in DBpedia and Visual Genome, respectively. This insight guides our approach to graph construction through a gradual expansion strategy. We illustrate the construction process in Fig. 5.5.

Initially, we start from core nodes and relations to construct a new knowledge graph. For core nodes, Core WordNet[9], which contains the most frequently used 5,000 words, i.e., 3,300 nouns, 1,000 verbs, and 700 adjectives. We mainly consider knowledge from WordNet and ConceptNet, with a set of core relations: isA, madeOf, partOf, mannerOf, usedFor, and capableOf. Table 5.3 details the core relations and their descriptions and examples. We denote this graph as a basic graph, which contains 488,216 nodes and 962,228 edges. Then, we extract instanceOf and isA relations from DBpedia to expand the core graph with more specific nodes. In this step, we employ an embedding-based semantic similarity method using pretrained Sentence-BERT for mapping. After integration, the graph is expanded to 1.4M nodes and 3M edges.

Finally, we integrate commonsense knowledge from diverse sources into our graph, ensuring a wide-ranging and diverse coverage. To map nodes from other sources to our graph, we employ the mappings developed by CSKG for integration. Moreover, to merge nodes, we use the embedding-based similarity method to identify nodes with the same meaning, and then use the tokenization-based method for verification. After integration, the nodes in PrimeNet are enriched with different kinds of commonsense knowledge, with 2.04M nodes and 6.03M edges.

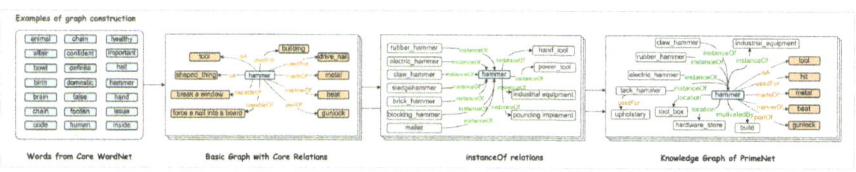

Fig. 5.5: Illustration of graph construction of PrimeNet. Starting with Core WordNet, we first construct a basic graph with core words and relations from WordNet and ConceptNet. Then, we add instanceOf knowledge from DBpedia and Wikipedia. Next, diverse types of knowledge from other knowledge bases are incorporated into the graph of PrimeNet.

---

[9] https://wordnetcode.princeton.edu/standoff-files/core-wordnet.txt

Table 5.3: Core relations of PrimeNet, and their description, example, and mappings to WordNet and ConceptNet.

Relations	Description	Example	Mapping to WordNet	Mapping to ConceptNet
isA	A is a specific instance of B	(car, isA, machine)	hyponym, hypernym	/r/IsA, /r/InstanceOf
madeOf	A is made of B	(car, madeOf, metal)	meronymy	/r/MadeOf
usedFor	A is used for B; the purpose of A is B	(hammer, usedFor, hit)	-	/r/UsedFor
partOf	A is a part of B	(gunlock, partOf, gun)	holonymy	/r/PartOf
mannerOf	A is a specific way of B	(screw, mannerOf, revolve)	-	/r/MannerOf
capableOf	Something that A can typically do is B	(bowl, capableOf, hold_water)	-	/r/CapableOf

Table 5.4: Functions designed for exploring PrimeNet. For each function, we introduce its input, output, and description.

Function	Input	Output	Description
nodes	-	a list of nodes	Return all nodes in PrimeNet.
edges	-	a list of edges	Return all edges in PrimeNet.
get_number_of_nodes	-	an int number	Return the number of nodes in PrimeNet.
get_number_of_edges	-	an int number	Return the number of edges in PrimeNet.
relation_types	a node	a list of relation types	Return all relation types that the node involved.
what_is	a node	a path of the node	Return the first edge of a node.
what_can_be	a node	a list of edges	Return all edges of a node.
relation_exist	a node and a relation type	True or False	If a relation type exists in the node return True, else False.
get_node_with_relation	a node and a relation type	a node	Given a node A and a relation R, return node B if there is an edge (A, R, B).
explain	a node and a relation type	a chain of this node	Return the chain of a node and a relation type.
generalize	a node	a list of edges	Return the root node of each of its relationships.
get_similarity	two nodes	a float score	Return a score that denotes how similar two nodes are, based on the path similarity computed by SequenceMatcher.
get_polarity	a node	Positive or Negative	Return the sentiment polarity of a node.
get_path	start_node and end_node	a path	Return a path from the start_node to the end_node .
find_last_nodes	a node	a list of paths	Return all edges where the end_node is the given node.
find_all_paths	start_node and end_node	a list of paths	Return all paths from start_node to end_node.
get_node_degree	a node	a number	Return the number of edges which connect with the given node.
get_phonetic	a concept	the phonetic information	Return the phonetic information of a concept.
add_node	a node	-	Add a node to PrimeNet if it does not exist in PrimeNet.
add_edge	an edge	-	Add an edge to PrimeNet.
add_primenet_new	a new knowledge graph	-	Add a new knowledge graph to PrimeNet.
print_to_file	a knowledge graph	-	Save a knowledge graph to a file.

5.4.4 Exploration

Then, we design multiple functions for exploring the graph that are capable of:

- Exploring graph structure of PrimeNet. For example, *nodes* and *edges* functions are designed to generate all concepts and relations in PrimeNet, respectively, and *get_number_of_nodes* and *get_number_of_edges* are designed to count the number of nodes and edges in the knowledge graph.
- Exploring commonsense knowledge for specific concepts. For example, given a concept, *what_is* function is designed to get all its relations, *get_polarity* function is used to get its sentiment polarity, and *find_path* function is designed to find a specific path in PrimeNet given a pair of concepts.

- Integrating new knowledge into PrimeNet. For example, the *add_node* and *add_edge* functions are designed to add new concepts and relations into PrimeNet, and the *add_primenet_new* function is able to incorporate a new knowledge base into PrimeNet.

We detail all the designed functions in Table 5.4, including their input, output, and description. These functions make it easy to apply PrimeNet in downstream tasks, as well as update PrimeNet with new commonsense knowledge or domain-specific knowledge.

5.5 Concept Detection

To create the concept layer of PrimeNet, we conduct concept detection to identify concepts that represent categories or classes of objects, ideas, or events based on shared features or characteristics (Cambria et al., 2022). An intuitive approach is to use the *isA* relation to establish mappings between concepts and entities. For example, *(dog, isA, animal)*, *(cat, isA, animal)*, and *(lion, isA, animal)* indicate that *animal* is a concept, and *dog*, *cat*, and *lion* are entities falling under that concept.

Though simple, in practice, it is sub-optimal to identify concepts by checking whether exist entities fall under them. For example, *animal*, *dog*, and *corgi* have specific entities. However, only *animal* and *dog* are widely-used as concepts in human daily reasoning, *corgi* are too specific. In this section, we study how to conduct concept detection with appropriate abstractions.

5.5.1 Preliminaries

When considering the conceptualization, it is important to measure the abstractness of a term. For example, *person* is a more abstract concept compared with *student*. Given a graph with *isA* relation, it is observed that abstract terms are usually located at the higher levels in a graph, while the specific terms tend to be positioned at the lower levels (Liu et al., 2022c). Specifically, the leaf nodes are regarded as the most specific terms, and they are considered as first level. The level of non-leaf nodes defined as the length of the longest path from the leaf nodes to itself. Formally, the *level* of a term is defined as follows.

Definition (Level Score) Given a term c, the level score of c is defined as:

$$level(c) = \begin{cases} \max\limits_{c' \in hypo(c)} level(c') + 1, & \text{if } hypo(c) \neq \phi \\ 1, & \text{otherwise} \end{cases} \tag{5.1}$$

where $hypo(c)$ is a set of hyponyms of c, and ϕ denotes an empty set. □

The abstract words have higher-level scores and specific terms have smaller level scores. For example, the level scores of *dog*, *mammal*, and *animal*, are 72, 89, and 362, respectively. It is also observed that, for an abstract term, its hyponyms are usually positioned at diversified levels, while its hyponyms would be more concentrated for a specific term. Based on it, (Liu et al., 2022c) defined an entropy-based metric for the abstractness measurement. Formally, the entropy score of a term is defined as follows.

Definition (Entropy Score) Given a term c, its entropy score is defined as:

$$entropy(c) = \begin{cases} 0, & \text{if c is a leaf term} \\ -\sum_{i=1}^{l} p_i(c) \cdot \log p_i(c) & \text{otherwise} \end{cases} \tag{5.2}$$

where l is the maximum level, and $p_i(c)$ is the ratio of the number of cs hyponyms at the i-th level to the total number of cs hyponyms.   □

The entropy of abstract terms is often greater than that of specific terms. For example, the entropy scores of *pupil*, *student*, and *people* are 0.563, 0.927, and 1.790, respectively. In general, abstract *concepts* and concrete *entities* are differentiated using these abstractness measure methods by manually-defined thresholds (Liu et al., 2022c). However, these methods are inaccurate and not suitable when applied to complex graphs with large-scale commonsense knowledge. The primary reason is the vast amount of knowledge, inevitably leading to the presence of cycles and isolated subgraphs, significantly reducing the accuracy of the aforementioned methods. Furthermore, some commonly used vocabulary lacks numerous lower-level nodes, e.g., *voice*, *track*, and *driver*, and they have lower scores compared with other words with more hyponyms, e.g., *transport*, *symbol*, and *medicine*. As such, the conceptualization methods which only rely on hierarchical information are not reasonable for such cases.

We perform a probing experiment as illustrated in Fig. 5.6. We assume that words from Core WordNet are concepts, given their fundamental role in describing the world. For all nodes in Core WordNet and our knowledge graph $\mathcal{G}$ of PrimeNet, we show probability distributions of their level scores and entropy scores. It is observed that a considerable number of words in Core WordNet have level scores below 50, and entropy scores under 1. These words are readily excluded from concept sets, by applying previous methods for conceptualization.

5.5.2 Conceptualization

Previous methods employed a *bottom-up* approach to measure abstractness, where a word's score relies on its hyponym set. Leaves without hyponyms are initiated as the seed set and then inferred for the others. We initialize the core concepts and then infer other words accordingly.

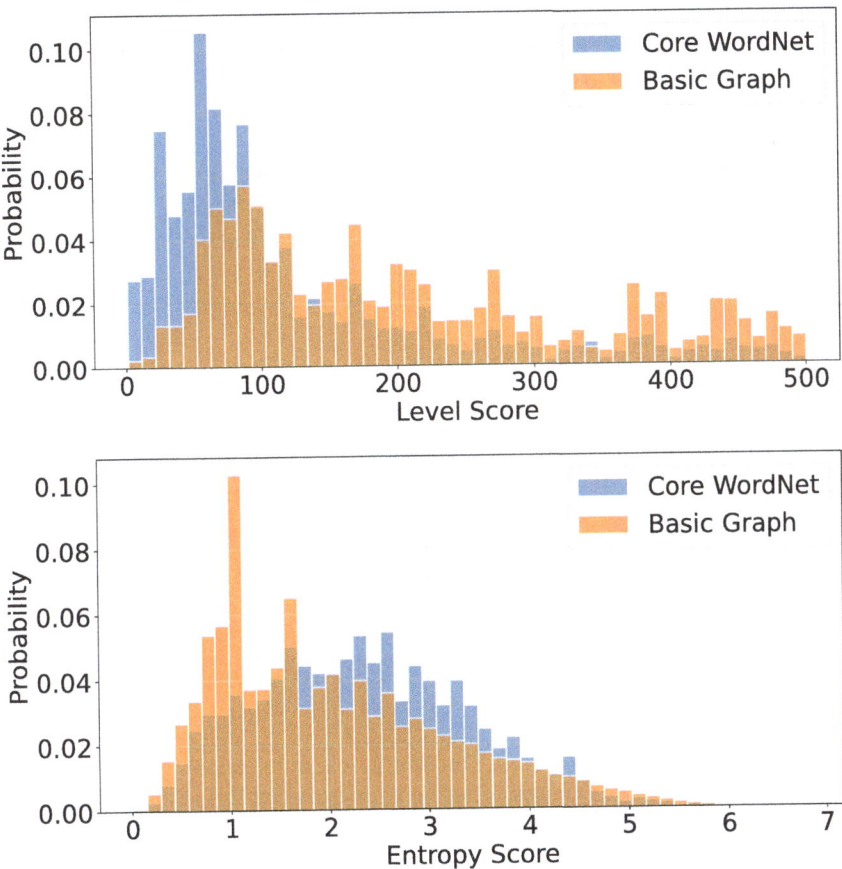

Fig. 5.6: Illustration of data distribution of Core WordNet and the graph of PrimeNet, considering of the level scores and entropy scores of nodes.

Specifically, the initial set of concepts, denoted as $C^0 = \{c_1, c_2, c_3, \ldots\}$, comprises commonly used words from Core WordNet that describe the world in human daily life. In an ideal scenario, the hypernyms of these core words are expected to be more abstract and should be considered as concepts. However, in practical scenario, not all of their hypernyms can be unequivocally regarded as concepts due to the intricate interweaving of commonsense knowledge. For instance, relationships such as *(dog, isA, animal)*, *(dog, isA, pet)*, *(pet, isA, animal)*, and *(dog, isA, species)* are all deemed correct and coexist within the knowledge base. Thus, we need a more accurate method to measure the abstractness of hypernyms. It is observed that not all hypernyms have the same weight when working as the concept of a *dog*. This problem has been deeply studied and Probase has been constructed to provide statistical insights of isA relations.

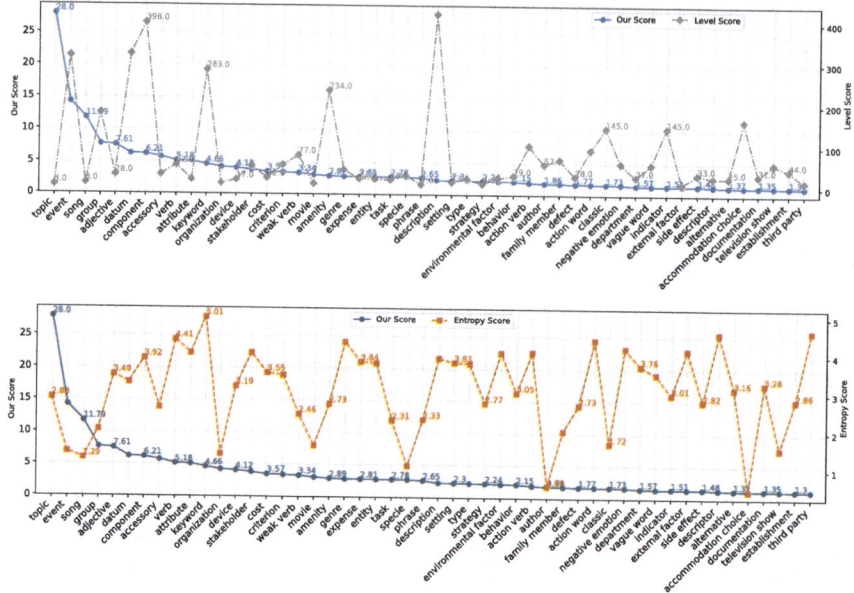

Fig. 5.7: Examples of top-50 words scored by the designed conceptual score function. We compare their level scores and entropy scores with our conceptual scores.

It includes "isA" relations for 2.7 million terms, automatically mined from a corpus of 1.68 billion web pages. That is, each triplet *(t, isA, c)* is linked to a frequency score $frec(t,c)$, providing frequency information computed through a data-driven method based on large-scale corpus. For example, *(dog, isA, animal)* and *(dog, isA, species)* show that both *animal* and *species* are concepts of *dog*, and $freq(dog, animal) > freq(dog, species)$ shows *animal* is a more typical concept for *dog*, compared with *species*.

Given a triplet *(t, isA, c)*, it is associated with a frequency score $freq(t,c)$ in Probase. The frequency score is an important signal to identify whether this relation is typical or not. Based on this observation, Wang et al. (2015b) proposed a *typicality score*, which is defined based on the frequency information to tell how popular a concept *c* is as far as an entity *t* is concerned, and how popular an entity *t* is as far as a concept *c* is concerned:

Definition (Typicality Score) Given a term *t*, the conditional probability $Pr(c|t)$ of a term *c* is defined as:

$$Pr(c|t) = \frac{freq(t,c)}{\sum_{c_i \in hyper(t)} freq(t,c_i)}, \tag{5.3}$$

where $hyper(t) = \{c_1, c_2, c_3, \dots\}$ is the set of hypernyms of *t*.

Given a concept c, the conditional probability $Pr(t|c)$ of an entity t is defined as:

$$Pr(t|c) = \frac{freq(t, c)}{\sum_{t_i \in hypo(c)} freq(t_i, c)}, \qquad (5.4)$$

where $hypo(c) = \{t_1, t_2, t_3, \dots\}$ is the set of hyponyms of c.   □

It is observed that a terms tends to be abstract when it is strongly connected with multiple concepts. Continuing the previous example, the term *animal, pet, species* link to 98, 435, 22 concepts in C^0, respectively. To formalize this regularity, an linking-based metric is designed as follows:

Definition (Conceptual Score) Given a term w and a set of concepts C, the conceptual score of w is defined as:

$$abstract(w) = \sum_{t_i \in hypo(w)} \mathbb{1}(t_i \in C) * \frac{freq(t_i, w)}{\sum_{o_j \in hyper(t_i)} freq(t_i, o_j)} \qquad (5.5)$$

where $hypo(w) = \{t_1, t_2, \dots, t_i, \dots\}$ is the set of hyponyms of w, $hyper(t_i) = \{o_1, o_2, \dots, o_j, \dots\}$ is the set of hypernyms of t_i, and $\mathbb{1}(t_i \in C)$ is set to 1, otherwise 0.   □

This scoring method is designed to quantify the extent to which a term functions as a universal, abstract link across a diverse array of concepts. Utilizing the initial set C^0, we calculate the abstraction scores of their hypernyms, presenting the top 50 terms in Fig. 5.7. According to human analysis, all of them are confirmed as conceptual terms. In addition, we present their *level scores* and *entropy scores*, revealing that these metrics fall short in inferring them as abstract terms. For instance, *topic, song*, and *adjective* exhibit low level scores (i.e., 3, 3, and 28), and *author* and *classic* display low entropy scores (i.e., 0.59 and 1.72), excluding them from being identified as concepts. We employ an iterative approach to augment the concept set by systematically incorporating terms with high abstraction scores. In i-th iteration, we introduce the top-n (e.g., $n = 3$) hypernyms for each concept in C^{i-1}. The constraint imposed is that these hypernyms must surpass a specified threshold T_{abs}. This process results in the construction of an updated concept set, denoted as C^i.

5.6 Primitive Discovery

The primitive discovery is to identify the most basic and essential element of the world knowledge, which provides a way to represent and organize knowledge in a structured and meaningful manner (Schank, 1972; Guarino, 1995). The well-designed primitive set can help to produce more accurate and reusable knowledge bases. However, creating a thorough set of primitives is extremely time-consuming and labor-intensive, hence it is not generally employed in most knowledge bases (Minsky, 1974; Jackendoff, 1976; Schank, 1972; Cambria et al., 2024).

We apply automatically discover a primitive set for commonsense knowledge. The basic idea entails clustering concepts that share similar functions at the cognitive level, then labeling the most representative concept in each cluster as a conceptual primitive. To achieve this goal, initially, we conduct concept clustering to group together related concepts quickly, filtering out those with highly disparate semantic meanings. Subsequently, we conduct a more precise primitive detection process, further refining each cluster to retain only the most consistently coherent concepts at the cognitive level and selecting the most representative concept to serve as the primitive for that specific set of concepts.

5.6.1 Concept Clustering

The concept clustering is designed to group cognitively related concepts while swiftly eliminating highly unrelated ones, thereby simplifying the following task of accurately conducting primitive detection. To achieve this goal, we employ a lexical substitution task to conduct concept clustering. Specifically, this task is to replace a concept in a sentence with a different concept. If the grammatical structure and overall meaning of the sentence are preserved, these two concepts are considered to have similar meanings. For example, in the sentence "the landlord tried to eject the tenants for not paying rent on time", one could substitute the word "eject" with "dispossess", "remove", "oust", or "evict" without changing the overall meaning of the sentence. Inspired by Cambria et al. (2024), we fine-tune PLMs[10] for lexical substitution. More specifically,

1) *Training Data.* We extract all the verb-noun and adjective-noun concepts from ConceptNet 5.5 (Speer et al., 2017) together with a sample sentence for each concept. The collection of concepts is denoted as $\mathcal{E} = \{e_1, e_2, e_3, \ldots, e_n\}$, where each concept $e_i \in \mathcal{E}$ is assigned with a sample sentence s_i. For each concept e_i, we remove it from the sentence s_i and the remaining sentence is denoted as its context c_i. We employ PLMs to represent the concept e_i and its context c_i as fixed-dimensional embeddings, i.e., $\mathbf{e}_i$ and $\mathbf{c}_i$, respectively.

2) *Training Objective.* Then, we fine-tune the PLM with a lexical substitution task. The assumption is that a relevant lexical substitute should be both semantically similar to the target word and have a similar contextual background. Given a concept e_i, its context c_i is regarded as the positive example. We create negative examples by sampling random concepts, which are denoted as $\mathcal{N}(e_i) = \{e_{i,1}^*, e_{i,2}^*, \ldots, e_{i,z}^*\}$.

---

[10] In our experiment, the used pretrained model is *all-mpnet-base-v2*. Having undergone pretraining on over 1 billion sentence pairs, this model is capable of mapping input text to a 768-dimensional vector space, ideal for tasks such as clustering or semantic search. Further details can be found at: https://huggingface.co/sentence-transformers/all-mpnet-base-v2.

The training objective function is defined as:

$$O = \sum_{i=1}^{n} (\log(\sigma(\mathbf{e}_i, \mathbf{c}_i)) + \sum_{e_{i,j}^* \in \mathcal{N}(e_i)} \log(\sigma(-\mathbf{e}_{i,j}^*, \mathbf{c}_i))), \quad (5.6)$$

where n is the number of training examples, z is the number of negative words for each example, and $\mathbf{e}_i^j$ denotes the representation of a negative concept. After fine-tuning, the representation model is expected to map concepts and context into a embedding space, where concepts that are appropriate for a given context are located close to one another.

3) *Semantic Measure.* We design a semantic measure to find the replacement of the concept in the embedding space. Given a concept e_i and its context c_i, we calculate the cosine distance of all the other concepts, e.g., $w \in \mathcal{E}$ in the embedding space as:

$$Sim(\mathbf{w}, (\mathbf{e}_i, \mathbf{c}_i)) = \cos(\mathbf{w}, \mathbf{e}_i) \cdot \cos(\mathbf{w}, \mathbf{c}_i) \cdot \cos(\mathbf{s}_i, \mathbf{s}_i^w), \quad (5.7)$$

where s_i is the original sentence, and s_i^w is a sentence by replacing c_i in s_i with w.

Using pretrained models, we conduct concept clustering through a fast clustering algorithm[11] developed by Sentence-BERT (Reimers and Gurevych, 2019). This algorithm is more efficient than previous hierarchical clustering methods like agglomerative clustering, making it better suited for large-scale, high-dimensional clustering tasks. The clustering process involves two thresholds: the similarity threshold, which determines when two sentences are considered similar, and the min_community_size threshold, specifying the minimum size for a local community. These thresholds allow us to obtain either large, coarse-grained clusters or small, fine-grained ones. In our implementation, based on our experimental observations, we set the similarity threshold to 0.6 and the min_community_size to 10.

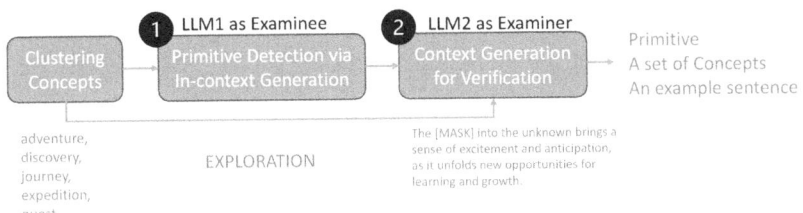

Fig. 5.8: The overall framework for primitive detection. LLM1 is used as an examinee to generate representative primitive for each concept cluster, and LLM2 is used as an examiner to verify the primitive and its related concepts.

---

[11] https://sbert.net/examples/applications/clustering/README.html

5.6.2 Primitive Detection

The primitive detection involves detecting the errors in each cluster, and associating a meaningful and generalizable primitive with a cluster of related concepts. For example, the concepts like *ingest, slurp, munch* are represented by a primitive EAT. It is inherent to human nature to try to categorize things, events and people, finding patterns and forms they have in common. We explore the generative ability of LLMs for primitive detection. To ensure the accuracy, as illustrated in Fig. 5.8, we design a detection-verification framework, where the first LLM works as examinee to generate primitive for the a concept cluster, and another LLM works as examiner to check whether the generated primitive is correct. Specifically,

Step-1: Primitive Detection by Examinee LLM The input of examinee (denoted as LLM1) is a cluster of concepts. The designed prompt is *"Please generate a primitive for the following concepts: C."*, where C is a list of concepts in a cluster.

Step-2: Primitive Verification by Examiner LLM The examiner (denoted as LLM2) is to verify whether the primitive generated by LLM1 is correct or not. To setup LLM2, we input the primitive P and the related concepts C into it, concatenated to the following instructions: *Do you think P is representative for the following concepts: C. Please answer "yes" or "no"*.

Step-3: Explainable context by Examiner LLM For the correct primitive and cluster, we ask the LLM2 to generate a sentence as explainable context. With the primitive P and the related concepts C into it, concatenated to the following instructions: *Please generate a short sentence to describe the primitive P. In this [MASK] can be replaced by the concepts in C.*

5.7 Experiments

In this section, we compare our PrimeNet with other widely-used knowledge bases in terms of coverage, accuracy, and efficiency. Then, we conduct experiments on semantic semantic similarity and commonsense reasoning to verify the accuracy and efficiency of PrimeNet.

5.7.1 Statistics and Analysis

We first evaluate the coverage and accuracy of commonsense knowledge presented in PrimeNet and other widely used knowledge bases. In Table 5.5, we summarize the size of different knowledge bases. Then, we conduct human assessment by utilizing the evaluation method and criteria established by Hwang et al. (2020). Specifically, we randomly select 3,000 triplets from PrimeNet and present each triplet in the format of (head_concept, relation, tail_concept). The evaluation involves three annotators who hold Ph.D. degrees in computer science.

Table 5.5: Accuracy (%) assessed by human annotators. *Size* denotes the number of triplets in different knowledge bases.

Knowledge Bases	Size	Accept	Reject	No Judgment
TransOMCS	18.5M	41.7	53.4	4.9
ATOMIC	877K	88.5	10.0	1.5
ConceptNet	21M	88.6	7.5	3.9
PrimeNet	6M	**92.4**	5.2	2.4

The annotators use four labels to assess each triplet:
1) *always/often*, indicating the triplet is frequently true;
2) *sometimes/likely*, indicating it is occasionally or probably true;
3) *farfetched/never*, indicating it is false or extremely unlikely;
4) *invalid*, indicating it is illogical.

Triplets labeled as *always/often* or *sometimes/likely* are categorized as *Accept*, while others are categorized as *Reject*. To ensure impartial evaluation, annotators are allowed to skip unfamiliar triplets by labeling *No Judgment*. The final results are determined by the majority vote among three annotators.

This experiment assesses PrimeNet's quality and compares it to other common-sense knowledge bases, including:

- **TransOMCS** (Zhang et al., 2020b): This is a knowledge base containing 18.5M triplets that were automatically extracted from syntactic parses of sentences from various web sources, including Wikipedia, Yelp, and Reddit.

- **ATOMIC** (Sap et al., 2019a): It contains 877K textual descriptions of inferential knowledge. It is organized as typed if-then relations with variables, such as "if X pays Y a compliment, then Y will likely return the compliment".

- **ConceptNet** (Speer et al., 2017): This is a large-scale knowledge base that contains relational knowledge collected from resources created by experts, crowdsourcing, and games with a purpose (Von Ahn, 2006).

As shown in Table 5.5[12], it is observed that PrimeNet stands out as the highest quality knowledge base with an acceptance rate of 92.4%, showing that PrimeNet is highly reliable and contains commonsense knowledge that is consistent with human understanding. ConceptNet, $ATOMIC_{20}^{20}$, and ATOMIC also demonstrate high quality, with acceptance rates of 88.6%, 91.3%, and 88.5%, respectively. Although TransOMCS has a vast number of triplets (i.e., 18.5M), it has a lower accuracy compared to the other resources, with an acceptance rate of only 41.7%, indicating it may not be as reliable as the other knowledge bases.

---

[12] Performance of compared knowledge bases is reported by Hwang et al. (2020), evaluated through crowdsourcing on the AMT platform.

5.7.1.1 Task-1: Semantic Similarity

We evaluate the effectiveness of PrimeNet by examining its impact on improving distributional representations on the word semantic similarity task. Following previous works (Faruqui et al., 2015; Speer et al., 2017; Liu et al., 2018c; Hwang et al., 2020), knowledge bases are used as external knowledge to adjust pretrained word embeddings. The resulting refined embeddings, molded by insights from various knowledge bases, undergo systematic evaluation in downstream tasks, such as word semantic similarity assessments. Enhanced performance serves as an indicator of the superior quality of knowledge bases in improving distributional representations.

We employ a retrofitting method[13] designed by Faruqui et al. (2015) to improve pretrained word embeddings with different knowledge bases. It is designed to make words that are known to be related in a given knowledge base have similar representations in embedding space. The training objective is to make the new embedding of a word to be both similar to its initial embedding and nearby words in the knowledge base, by minimizing the following objective function:

$$L = \sum_{i=1}^{n} (\alpha_i ||\mathbf{w}_i - \mathbf{w}_i^*||^2 + \sum_{(w_i, w_j) \in \mathcal{R}} \beta_{i,j} ||\mathbf{w}_i - \mathbf{w}_j||^2), \tag{5.8}$$

where α and β control the relative strengths of associations, $\mathbf{w}_i^*$ is the original embedding of word w_i, and $\mathbf{w}_i$ is its new embedding, $\mathcal{R}$ denotes a set of relations extracted from the knowledge base, and (w_i, w_j) denotes a relation which connects w_i and w_j. We test the retrofitted embeddings with different knowledge bases on two tasks, i.e., semantic similarity and SAT-style analogy.

This task is to measure the degree of similarity between word pairs by calculating the cosine similarities between their embeddings, and then compare the similarities to human judgments. A good method should provide similarities that are strongly correlated with the human judgments evaluated by Spearman correlation coefficient (Myers and Well, 1995). In our experiment, we conduct experiments on eight widely-used word similarity datasets, including

- **YP-130**: A dataset comprising 130 word pairs with similarity ratings provided by human annotators (Yang and Powers, 2005).
- **MenTR-3K**: Consists of 3,000 word pairs with similarity judgments collected from human participants (Bruni et al., 2012).
- **RG-65**: Contains 65 word pairs with similarity ratings obtained through human evaluations (Rubenstein and Goodenough, 1965).
- **MTurk-771**: Comprises 771 word pairs with similarity scores obtained through crowdsourcing on AMT (Halawi et al., 2012).
- **SimLex-999**: Includes 999 word pairs with similarity ratings collected from human subjects, aiming to provide a balanced set for evaluating word similarity models (Hill et al., 2015).

---

[13] https://github.com/mfaruqui/retrofitting

- **SimVerb-3500**: Consists of 3,500 verb pairs annotated with similarity judgments by human raters (Gerz et al., 2016).
- **VERB-143**: Contains 143 verb pairs with similarity ratings collected from human annotators (Baker et al., 2014).
- **WS-353**: Comprises 353 word pairs, including both similarity and relatedness judgments obtained from human raters (Finkelstein et al., 2001).

Two popular pretrained word embeddings are used in our experiments, including word2vec (Mikolov et al., 2013b), which is trained on the first 100M of plain text from Wikipedia[14], and GloVe (Pennington et al., 2014), which are trained on 6 billion words from Wikipedia and English Gigaword[15]. In this task, we compare PrimeNet with FrameNet, WordNet, and ConceptNet, which contain synonyms knowledge.

Table 5.6 presents the overall performance on different word similarity datasets. PrimeNet demonstrated a significant improvement in retrofitting semantic representations, with an average increase of 6.73%, 5.49%, and 5.31% for word2vec (300d), GloVe (50d), and GloVe (300d), respectively. WordNet also achieved notable performance gains, with an average improvement of 4.75%, 3.79%, and 3.98%, benefiting the high-quality synonyms knowledge constructed by experts. While the crowdsourced ConceptNet only slightly outperformed word2vec (300d) and GloVe (50d), and slightly worse than GloVe (300d). The solid performance gain achieved by PrimeNet suggests that it is successful in integrating knowledge from various sources into PrimeNet and creating a robust knowledge base.

Table 5.6: Overall performance on semantic similarity. d denotes the dimension of embeddings. The best performance is marked in bold.

Methods	YP-130	MenTR-3K	RG-65	MTurk-771	SimLex-999	SimVerb-3500	VERB-143	WS-353	Average (Δ)
word2vec (300d)	0.215	0.600	0.633	0.554	0.287	0.155	0.358	0.705	0.438
+FrameNet	0.334	0.589	0.620	0.571	0.295	0.227	0.321	0.651	0.451 (+1.25%)
+WordNet	0.316	0.620	**0.717**	0.598	0.377	0.237	0.318	0.705	0.486 (+4.75%)
+ConceptNet	**0.386**	0.582	0.577	0.533	0.341	0.229	0.302	0.651	0.450 (+1.16%)
+PrimeNet	0.325	**0.638**	0.680	**0.617**	**0.416**	**0.271**	**0.385**	0.715	**0.506 (+6.73%)**
GloVe (50d)	0.377	0.652	0.602	0.554	0.265	0.153	0.250	0.499	0.419
+FrameNet	**0.459**	0.622	0.617	0.568	0.288	0.217	0.240	0.471	0.435 (+1.61%)
+WordNet	0.510	0.649	0.688	0.540	0.342	0.239	0.188	0.500	0.457 (+3.79%)
+ConceptNet	0.427	0.599	0.558	0.493	0.356	0.234	0.236	0.489	0.424 (+0.50%)
+PrimeNet	0.443	**0.674**	**0.707**	**0.597**	**0.376**	**0.236**	**0.273**	0.485	**0.474 (+5.49%)**
GloVe (300d)	0.561	0.737	0.766	0.650	0.371	0.227	0.305	0.605	0.528
+FrameNet	0.589	0.701	0.756	0.639	0.361	0.278	0.274	0.558	0.519 (-0.84%)
+WordNet	**0.610**	0.759	**0.841**	0.679	0.470	0.313	0.256	0.612	0.568 (+3.98%)
+ConceptNet	0.561	0.700	0.747	0.583	0.420	0.288	0.300	0.595	0.524 (-0.34%)
+PrimeNet	0.593	**0.764**	0.818	**0.684**	**0.496**	**0.316**	**0.350**	0.626	**0.581 (+5.31%)**

---

[14] We use the Text8Corpus which is available in Gensim: https://github.com/RaRe-Technologies/gensim-data, and the CBOW model for training: https://code.google.com/archive/p/word2vec

[15] https://nlp.stanford.edu/projects/glove

5.7.1.2 Task-2: Neurosymbolic Commonsense Reasoning

Commonsense knowledge is important to NLU through contextual reasoning. An effective method for assessing this understanding is through commonsense QA tasks, wherein the ability to answer questions often hinges on possessing commonsense knowledge (Liu et al., 2024a). In commonsense QA tasks, PLMs like BERT and RoBERTa have demonstrated their effectiveness in bridging the gap between human and machine performance. Additionally, the incorporation of external knowledge bases has proven crucial for enhancing answer accuracy, providing valuable insights for contextual comprehension and reasoning. Hence, approaches that combine neural PLMs with symbolic knowledge bases, known as *neurosymbolic* methods, have exhibited significant potential for advancing commonsense reasoning.

A. Task Setting

Following previous methods (Ilievski et al., 2021b,a), we use a neurosymbolic method to evaluate the commonsense QA under a zero-shot setting proposed by Ma et al. (2021). Formally, given a natural language question q and a set of possible answers $\mathcal{A} = \{a_1, a_2, \ldots, a_n\}$, the task is to select the most probable answer a^* from $\mathcal{A}$. The wrong answers in $\mathcal{A}$ are denoted as distractors. PLMs are used as backbone. RoBERTa-large is used in our experiments. In a zero-shot setting, the model has no access to the training data. The neurosymbolic solution is to transform knowledge from different knowledge bases into an artificial QA set for pretraining. For example, a triplet *(losing weight, usedFor, being healthier)* is generated as *losing weight is for being healthier*, and several distractors are generated by negative sampling. After pretraining, the model are tested on different datasets. We follow the parameter settings in (Ma et al., 2021). Experiments are tested for five rounds, and the average accuracy of the predicted answers is used as the metric.

B. Baselines

We compare the neurosymbolic methods with the following baselines. *Majority* answers each question with the most frequent option in the entire dataset. Self-Talk (Shwartz et al., 2020) is an unsupervised method. It generates clarification prompts based on a template prefix, which are leveraged to elicit knowledge from another language model, which is used jointly with the original context and question to score each answer candidate. SMLM (Banerjee and Baral, 2020) is designed to pretrain the LM with three representation learning functions which aim to complete a knowledge triple given two of its elements. To show the upper bound, we report the supervised methods on RoBERTa-large model with access to the training data, as well as the human performance. We also include results of a supervised fine-tuned RoBERTa system and of human evaluation. To facilitate the neurosymbolic method for commonsense reasoning, we compare PrimeNet with ATOMIC, ConceptNet, Wikidata, WordNet, and CSKG.

C. Benchmarks

Following the work of Ma et al. (2021), we use five commonsense QA benchmarks for evaluation, including:

- Abductive Natural Language Inference (aNLI) (Bhagavatula et al., 2020) is a binary-classification task, which is to apply abductive reasoning and commonsense to form possible explanations for a given set of observations. Given two observations from narrative contexts, the goal is to pick the most plausible explanatory hypothesis.
- Commonsense Question Answering (CQA) (Talmor et al., 2019) contains 12,247 examples. Each example includes a question and five answer candidates. The questions are sourced from a ConceptNet. Answer candidates are formed by combining ConceptNet nodes with additional distractors gathered through crowdsourcing.
- Physical Interaction Question Answering (PIQA) (Bisk et al., 2020) is a dataset for reasoning about physical commonsense. Each question is associated with two possible solutions. The task is to choose the most appropriate solution, of which exactly one is correct.
- Social Intelligence Question Answering (SIQA) (Sap et al., 2019b) is a dataset for commonsense reasoning about social situations, with 38,000 multiple choice questions. Each example comprises a context, a question, and three answer candidates. The context is derived from ATOMIC, questions are generated based on nine templates corresponding to relations in ATOMIC, and answers are obtained through crowdsourcing.
- WinoGrande (WG) (Sakaguchi et al., 2020) contains 44K problems inspired by pronoun resolution problems in Winograd Schema Challenge (Levesque, 2011). Each example includes a context description featuring an emphasized pronoun, with two options provided as possible references.

D. Performance

The overall performance is shown in Table 5.7. It is observed that pretraining the language models with external knowledge is effectiveness to improve the performance of commonsense QA task. The main reason is that the external knowledge is important supplementary information for implicit knowledge embedding in PLMs. Our PrimeNet achieved the best performance when RoBERTa is used as backbone, with the average performance gains of 1.74%, 2.88%, 0.82% over ATOMIC, ConceptNet+Wikidata+WordNet, and CSKG, respectively. This experiment indicates that PrimeNet has a good quality in organizing commonsense knowledge.

Table 5.7: Performance of neurosymbolic methods across five commonsense QA tasks in a zero-shot setting. RoBERTa-large* denotes the performance of RoBERTa-large under a supervised setting.

Model	Knowledge Base	aNLI	CQA	PIQA	SIQA	WG
Majority	-	50.8	20.9	50.5	33.6	50.4
Self-talk	-	-	32.4	70.2	46.2	54.7
SMLM	-	65.3	38.8	-	48.5	-
RoBERTa-large*	-	85.6	78.5	79.2	76.6	79.3
Human Performance	-	91.4	88.9	94.9	86.9	94.1
	-	65.5	45.0	67.6	47.3	57.5
	ATOMIC	70.8	64.2	72.1	63.1	59.6
RoBERTa-large	ConceptNet, Wikidata, WordNet	70.0	67.9	72.0	54.8	59.4
	CSKG	70.5	67.4	72.4	63.2	60.9
	PrimeNet	**71.2**	**68.3**	**72.4**	**64.5**	**62.1**

5.7.2 Case Studies

In our method, we manually checked the detected primitives. This step is conduct by 5 senior Ph.D. students majors in NLP. We manually code the explainable of primitives. For example, INCREASE is defined as INCREASE(obj) := obj++, which is the basic operation that increments the value of an object and provides a foundation for more complex reasoning. It is observed that some primitives have a hierarchical structure. We show examples of primitives in Fig. 5.9. At *Level-1*, the primitive GROW is defined as GROW(obj) = INCREASE(obj.SIZE) := obj.SIZE++ = obj(l++, h++, w++), which is accomplished by using the INCREASE primitive to increment the object's SIZE attribute, such as length (l), height (h), and width (w).

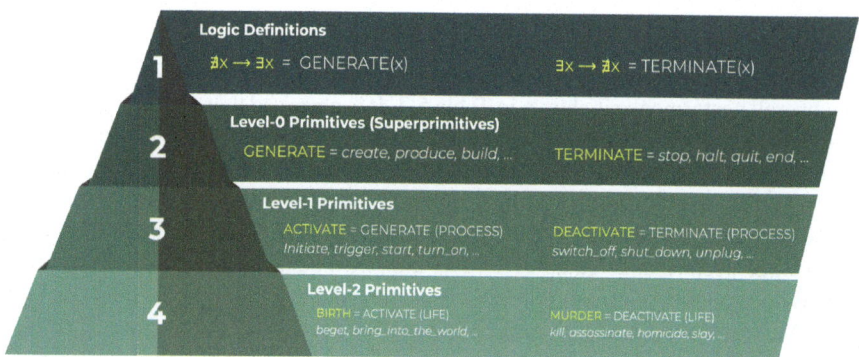

Fig. 5.9: Examples of the hierarchical structure of primitives in PrimeNet.

The *Level-2* primitive LENGTHEN is even more specific, adding only length to an object, and it is defined as LENGTHEN(obj)=INCREASE(obj.SIZE.LENGTH) :=obj.SIZE.LENGTH++ = obj(l++, h, w). We have also performed several experiments on affordances by testing how PrimeNet is able to model human-object interactions in different scenarios, e.g., how to identify and use a liquid container, and in different modalities, e.g., speech processing and computer vision (Fig. 5.10). Finally, we have also carried out preliminary experiments on how PrimeNet can represent and handle different types of domain-specific knowledge, e.g., safety commonsense knowledge (Fig. 5.11). We intend to provide a more detailed account of these experiments and additional ones in our future work.

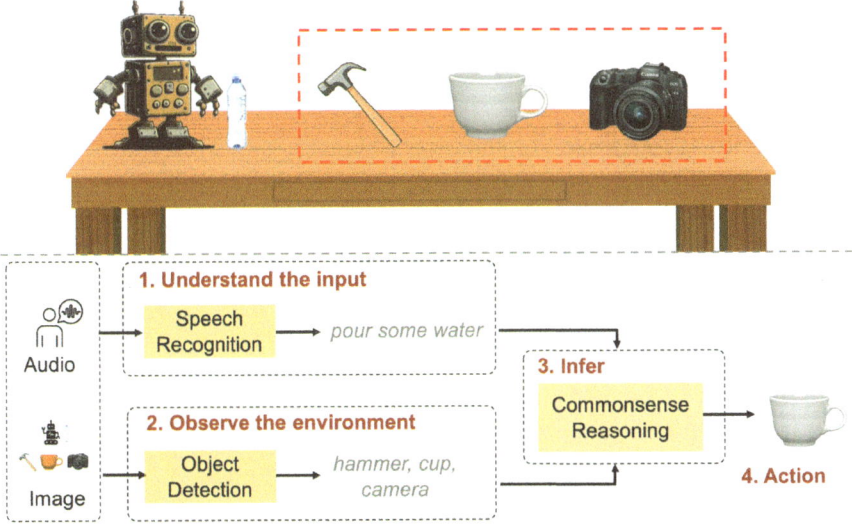

Fig. 5.10: A sample scenario in which PrimeNet's capabilities of understanding and modeling affordances have been tested.

5.8 Future Directions

PrimeNet is grounded in fundamental conceptual principles that are consistent with human reasoning patterns. It has greater potential than current knowledge bases to enhance AI's reasoning capabilities, particularly in response to the growing demand for more intricate reasoning tasks in the era of LLMs. In this section, we discuss several applications where PrimeNet can aid in enhancing AI's reasoning abilities.

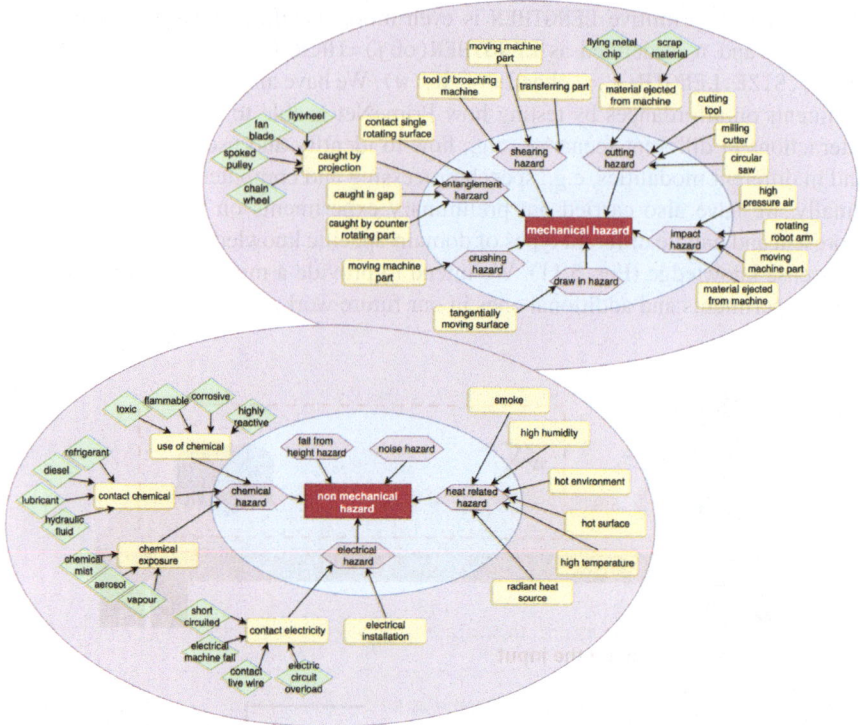

Fig. 5.11: An example of safety commonsense knowledge in PrimeNet.

5.8.1 Logical Reasoning

Logical reasoning is recognized as central to human cognition and intelligence (Goel et al., 2017). It mainly consists of two reasoning types (Salmon, 1989), which are deductive reasoning and inductive reasoning. Previously logical reasoning is mostly investigated in the classic AI field and used formal language as knowledge representation. Recently, there is a trend of research on deductive reasoning (Clark et al., 2020b) and inductive reasoning (Yang et al., 2024) that use natural language as knowledge representation, which has various advantages over the previous paradigm of formal language.

We argue that PrimeNet could have a significant impact on the study of logical reasoning, as LLMs are known to struggle with it. Recently, Xu et al. (2024) demonstrated this with comprehensive evaluations. They selected fifteen logical reasoning datasets and categorized them into deductive, inductive, abductive, and mixed-form reasoning. Then, they evaluated three LLMs across these datasets in zero-shot, one-shot, and three-shot settings. Unlike previous evaluations that focus solely on simple metrics like accuracy, the authors introduced fine-grained evaluations that assess both answers and explanations, covering correctness, completeness, and redundancy.

To identify logical flaws in LLMs, Xu et al. (2024) categorized problematic cases into five error types across two dimensions: evidence selection and reasoning process. Errors in evidence selection include wrong selection and hallucination, while reasoning errors encompass no reasoning, perspective mistakes, and process mistakes. Additionally, they introduce a new dataset with neutral content, consisting of 3,000 samples across deductive, inductive, and abductive settings, to minimize knowledge bias and focus solely on logical reasoning capabilities. Based on these in-depth evaluations, they proposed a general evaluation framework for logical reasoning capabilities, covering six dimensions: Correct, Rigorous, Self-aware, Active, Oriented, and No hallucination (Fig. 5.12). This framework highlights the strengths and weaknesses of LLMs and offers guidance for future research.

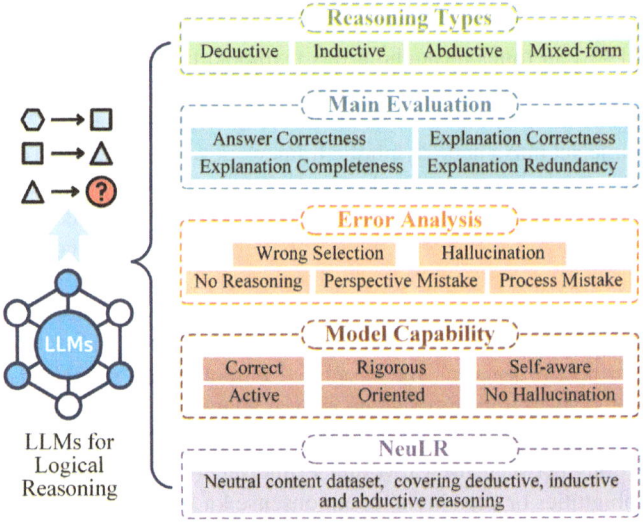

Fig. 5.12: Overall architecture of the LLM evaluation proposed by Xu et al. (2024).

5.8.2 Implicit Reasoning

Implicit reasoning is a challenging task which does not contain explicitly clues for designing reasoning strategies. For example, *"Did Aristotle use a laptop?"* is an implicit question (Geva et al., 2021), and it requires to infer the strategy for answering the implicit question, i.e., *temporal comparison*. Recently, AI systems based on PLMs have achieved impressive performance in answering explicit questions, even surpassing human performance in some datasets, e.g., SQuAD (Rajpurkar et al., 2016) and TriviaQA (Joshi et al., 2017). However, their accuracy on implicit questions is only 66% (Geva et al., 2021).

A key property of implicit reasoning is the diverse strategies. Humans cannot pre-define all of the strategies due to the complexity of scenarios. To conduct implicit reasoning, PrimeNet has the potential to build a finite set of strategies at the primitive level, and apply the primitive-based strategies on concepts and entities. For example, the implicit questions, e.g., "Did Aristotle Use a Laptop?", "Did Shakespeare play guitar?", and "Was NATO involved in World War I?", have the same reasoning strategies in the primitive level, e.g., COMPARE(TIME(Entity-1), TIME(Entity-2)). Primitives can be used to conduct implicit reasoning by providing the basic cognitive processes or mental operations that underlie our ability to reason implicitly.

5.8.3 Neurosymbolic AI

As discussed several times throughout this textbook, the integration of symbolism and connectionism, known as neurosymbolic AI, is widely recognized as a booster for the next generation of AI (Lake et al., 2017; Smolensky et al., 2022). The integration of PrimeNet into a neurosymbolic framework with an LLM can enable better interpretability, improved generalization, and trustworthiness. For example, the LLM can be used first to extract relevant information from an input. This extracted information is then passed to PrimeNet, which applies logical rules and commonsense reasoning to provide structured insights or solutions.

Integration occurs at two main points: pre-processing and post-processing. In pre-processing, the LLM converts the input into a structured format that PrimeNet can utilize. In post-processing, the structured output from PrimeNet is combined with the LLM's capabilities to generate natural language explanations or refine the results. A feedback loop enhances this integration by enabling the system to learn from interactions between the LLM and PrimeNet. For instance, when the LLM identifies ambiguities or gaps in the commonsense knowledge, these can be used to update and improve PrimeNet. This iterative process ensures that the neurosymbolic framework leverages both the advanced language understanding of LLMs and the logical reasoning capabilities of PrimeNet, resulting in a more robust and effective system for tackling complex problems.

5.9 Conclusion

In this chapter, we discussed the importance of knowledge representation and reasoning in the context of NLU and neurosymbolic AI. In particular, we discussed the development of PrimeNet, a commonsense resource based on conceptual dependency theory. Unlike existing knowledge bases, PrimeNet is constructed based on a small core of conceptual primitives and relations, linked to an extensive set of concepts and entities, which is suited for supporting higher-level inference.

Our studies demonstrate that PrimeNet contains high-quality commonsense knowledge that can be used for commonsense reasoning in different downstream tasks thanks to the many intuitive functions developed. In the future, we aim to broaden the scope of commonsense knowledge using generative AI models and will continue to develop additional PrimeNet functions to facilitate commonsense reasoning across a wider range of applications, domains, and languages.

5.10 Learning Resources

A. Reading List

• Qian Liu, Sooji Han, Yang Li, Erik Cambria, and Kenneth Kwok. PrimeNet: A Framework for Commonsense Knowledge Representation and Reasoning based on Conceptual Primitives. Cognitive Computation, 2024 (Liu et al., 2024b)

• Zonglin Yang, Xinya Du, Erik Cambria, Claire Cardie. End-to-end Case-based Reasoning for Commonsense Knowledge Base Completion. Proceedings of EACL, 3509-3522, 2023 (Yang et al., 2023)

• Weijie Yeo, Ranjan Satapathy, Siow Mong Goh, and Erik Cambria. How interpretable are reasoning explanations from prompting large language models? Proceedings of NAACL, 2148–2164, 2024 (Yeo et al., 2024c)

B. Relevant Videos

• Labcast about PrimeNet: youtu.be/KVkcJfispww

• Labcast about Commonsense KGC: youtu.be/8K9TTVj2Woc

• Labcast about Chain of Thought Interpretability: youtu.be/1hM7MZokxz8

C. Related Code

• Github repository about PrimeNet github.com/SenticNet/PrimeNet

• Github repository about Commonsense KGC: github.com/SenticNet/ECBRF

• Github repo about Chain of Thought: github.com/SenticNet/CoT_interpretability

D. Exercises

- **Exercise 1.** Extracting commonsense knowledge is challenging because it is often not explicitly mentioned in written documents. List and describe two potential sources from which commonsense knowledge can be acquired. For each source, explain how it can be used to gather relevant information and what type of commonsense knowledge might be obtained.

- **Exercise 2.** Identify two conceptual primitives, one representing a positive sentiment and one representing a negative sentiment. For each primitive, provide a range of related words and MWEs that are associated with it. Explain how these primitives can be used for enhancing the accuracy and the explainability of sentiment analysis.

- **Exercise 3.** Describe how a commonsense knowledge base can be integrated into an LLM and explain, with examples, why this integration is beneficial. Discuss the methods and strategies for incorporating everyday, intuitive knowledge into the model and illustrate how this integration can enhance the LLM's ability to interpret and generate responses that are contextually accurate and aligned with human expectations.

- **Exercise 4.** Explain with examples how a bottom-up strategy can enable the discovery of conceptual primitives for a specific domain directly from text data, and how a top-down strategy can then be applied on the resulting set of primitives to identify more words and MWEs related to them.

- **Exercise 5.** Describe the development of an AI assistant that helps users with daily tasks. Choose a scenario where the AI assistant must use commonsense knowledge. For example, helping a user decide what to wear based on the weather forecast. List the pieces of commonsense knowledge the AI needs to handle the scenario effectively. Explain how the AI system can represent and use this knowledge to assist the user. Specify whether it would use a rule-based system, machine learning models, or a combination of both.

Chapter 6
Conclusion

Abstract This final chapter provides a comprehensive summary of the textbook's key topics, ranging from the exploration of reliable, responsible, and personalized AI to the detailed description of the NLU suitcase model. It includes implementation insights into syntactic, semantic, and pragmatic processing, and discusses the integration of LLMs within neurosymbolic frameworks. The chapter also provides a short summary on strengths and limitations of LLMs, stressing the importance of combining symbolic and subsymbolic methods for improved reasoning. The chapter concludes with instructions for a group assignment connecting NLU and information retrieval, followed by a quiz.

Key words: Syntactics Processing, Semantics Processing, Pragmatics Processing, Knowledge Representation and Reasoning

In this textbook, we have explored the multifaceted domain of NLU, delving into the complexities and advancements that define this critical area of AI. We have investigated ongoing initiatives aimed at enhancing the reliability, responsibility, and personalization of AI. We have navigated through the essential components of NLU, including syntactic, semantic, and pragmatic processing techniques, and examined the pivotal role of LLMs and their integration into neurosymbolic frameworks. Our journey began with a foundational understanding of how language operates at various levels of abstraction, highlighting the importance of both symbolic and subsymbolic approaches. We then discussed the emergence of LLMs, showcasing their impressive capabilities in handling diverse NLP tasks. However, we also addressed their limitations, particularly in the realm of logical reasoning and commonsense understanding. The introduction of neurosymbolic methods represents a significant stride toward bridging these gaps. By combining the learning prowess of subsymbolic systems with the reasoning capabilities of symbolic systems, we can create more robust and versatile NLU frameworks. The integration of knowledge bases like PrimeNet into these frameworks exemplifies the potential for enhanced reasoning and understanding.

Chapters 2, 3, and 4 provide an in-depth explanation of how the NLU suitcase model operates. They meticulously detail the inner workings and mechanisms of the model, offering a comprehensive overview of its structure, functionality, and the processes involved. It explores how the model integrates various components and techniques to achieve NLU, illustrating its application through practical examples and case studies. In particular, we studied how microtext normalization converts informal or abbreviated text into a standard format, making the text more consistent for further processing. SBD then identifies where sentences begin and end, ensuring that the text is properly segmented for accurate analysis. Following SBD, POS tagging assigns grammatical labels to each word, providing essential structural information that supports text chunking. Text chunking groups words into meaningful phrases, such as noun phrases, which helps in organizing the text for subsequent tasks. Lemmatization standardizes words to their base forms, ensuring consistency and reducing variations that could complicate later stages.

This standardized text aids WSD, which determines the correct meaning of words based on their context. Once WSD has clarified word meanings, NER identifies and classifies named entities like people and locations. This identification relies on the results from WSD and POS tagging, enabling accurate concept extraction. Concept extraction identifies key concepts within the text, utilizing the structured information from NER and WSD to provide a deeper understanding of the text?s content. Anaphora resolution then links pronouns and other referring expressions to their antecedents, using the context provided by POS tagging and concept extraction. Subjectivity detection follows, distinguishing between subjective and objective statements based on the resolved text from previous tasks.

Metaphor understanding interprets metaphorical language, using insights from subjectivity detection and context established earlier. Sarcasm detection identifies sarcastic remarks by analyzing the context and subjectivity established previously. Personality recognition then infers personality traits from the text, leveraging findings from earlier tasks to understand linguistic features that reflect personality. Aspect extraction identifies specific features or aspects of entities within the text, relying on the information from NER and concept extraction to provide detailed sentiment analysis. Finally, polarity detection assesses the sentiment expressed toward these aspects, using the comprehensive context and details gathered from all preceding tasks to determine the overall sentiment.

The remainder of this chapter introduces a group assignment that sits at the intersection of NLU and information retrieval. This assignment aims to engage students in practical applications that combine these two domains, enhancing their comprehension and skills in both areas. Following the assignment, the chapter concludes with a comprehensive quiz consisting of 100 multiple-choice questions covering the contents of the book. This quiz is designed to test students' knowledge and understanding of the key concepts, theories, and methodologies discussed throughout the textbook. Following this, readers can find a comprehensive list of references compiled within the remaining 100 pages.

As we conclude, it is clear that the field of NLU is continuously evolving. The interplay between symbolic and subsymbolic techniques will likely drive future advancements, pushing the boundaries of what machines can comprehend and achieve. Our exploration has underscored the importance of ongoing research and innovation in developing systems that not only process language but also understand and reason with it in ways that mirror human intelligence. The future of NLU holds promise for more intuitive and intelligent interactions between humans and machines. By addressing current challenges and leveraging the strengths of both symbolic and subsymbolic approaches, we can look forward to a new era of AI that truly understands the nuances of human language. This book serves as a testament to the progress made thus far and a guidepost for the exciting developments yet to come.

6.1 Learning Resources

6.1.1 Assignment

In this group assignment, you will build an opinion search engine. Given a specific topic of your choice (e.g., cryptocurrencies), your system should enable users to find relevant opinions about any instance of such topic (e.g., bitcoin) and perform sentiment analysis on the results (e.g., opinions about bitcoin are 70% positive and 30% negative). Once you have chosen a topic, make sure that a) you can find enough data about it (e.g., some topics may be too niche to the point that you would only find a few hundreds data about it) and b) the opinions you get are balanced (e.g., if the topic you chose only has negative opinions associated with it, then it is probably not a good topic). For ideas about interesting topics, you can check our project page at https://sentic.net/projects. You are pretty much free to use anything you want in terms of available tools/libraries. However, your system cannot be just a mashup of existing services. Your final score will depend not only on how you developed your system but also on its novelty and your creativity: in other words, to get a high score you do not only need to implement a system that works, but also a system that is useful and user-friendly. The main tasks of the assignment are crawling (20 points), indexing (40 points) and classification (40 points). A minimum of 60 points is required to pass the assignment. Consult with your course coordinator regarding the procedures and requirements for submitting your assignment.

A. Crawling (20 points)

Crawl text data from any sources which you are interested in and permitted to access, e.g., X API or Reddit API. The crawled corpus should have at least 10,000 records and at least 100,000 words. It is fine to use available datasets for training (e.g., popular sentiment benchmarks), but you still have to at least crawl and label data for testing.

Also, make sure your dataset does not contain duplicates and try your best to make it balanced (e.g., equal number of positive and negative entries). Before crawling any data, carefully consider the questions in this material, e.g., check whether the data have enough details to answer the questions. You can use any third party libraries for the crawling task, e.g.:

- Jsoup: https://jsoup.org
- Twitter4j: https://twitter4j.org
- Facebook marketing: https://developers.facebook.com/docs/marketing-apis
- Instagram: https://instagram.com/developer
- Amazon: https://github.com/ivanpgs/amazon-crawler
- Tinder: https://gist.github.com/rtt/10403467
- Tik Tok: https://developers.tiktok.com

Question 1: Explain and provide the following:

1) How you crawled the corpus (e.g., source, keywords, API, library) and stored it
2) What kind of information users might like to retrieve from your crawled corpus (i.e., applications), with sample queries
3) The numbers of records, words, and types (i.e., unique words) in the corpus

B. Indexing (40 points)

Indexing: You can do this from scratch or use a combination of available tools, e.g., Solr+Lucene+Jetty. Solr runs as a standalone full-text search server within a servlet container such as Jetty and it uses the Lucene search library at its core for text indexing and search. Solr has REST-like HTTP/XML and JSON APIs that make it easy to use from any programming language. Useful documentations include:

- Solr project: https://solr.apache.org
- Solr wiki: https://wiki.apache.org/solr/FrontPage
- Lucene tutorial: https://lucene.apache.org/core/quickstart.html
- Solr with Jetty: https://wiki.apache.org/solr/SolrJetty
- Jetty tutorial: https://jetty.org

You can also choose other inverted-index text search engine open projects, e.g., Sphinx, Nutch, and Lemur. However, you should not simply adopt SQL-based solutions for text search (for example, you cannot solve text search simply using Microsoft Sqlserver or MySql).

Querying: You need to provide a simple but friendly user interface (UI) for querying. It could be either a web-based or mobile app based UI. You could use JSP in Java or Django in Python to develop your UI website. Since Solr provides REST-like APIs to access indexes, one extra JSon or RESTFul library would be enough. Otherwise, you may use any third party library.

The UI must be kept simple: a sophisticated UI is not necessary nor encouraged. Detailed information besides text is allowed to be shown for the query results, e.g., product images on Amazon, ratings on Amazon, and pictures on Instagram. The details should be designed to solve specific problems.

Question 2: Perform the following tasks:

- Design a simple UI (you can design one from scratch or you can tap on an existing one, e.g., Solr UI) to allow users to access your system in a simple way
- Write five queries, get their results, and measure the speed of the querying

Question 3: Explore some innovations for enhancing the indexing and ranking. Explain why they are important to solve specific problems, illustrated with examples. You can list anything that has helped improving your system from the first version to the last one, plus queries that did not work earlier but now work because of the improvements you made. Possible innovations include (but are not limited to) the following:

- Timeline search (e.g., allow user to search within specific time windows)
- Geo-spatial search (e.g., use map information to refine query results)
- Enhanced search (e.g., add histograms, pie charts, word clouds, etc.)
- Interactive search (e.g., refine search results based on users' relevance feedback)
- Multimodal search (e.g., implement image or video retrieval)
- Multilingual search (e.g., enable information retrieval in multiple languages)
- Multifaceted search (e.g., visualize information according to different categories)

C. Classification (40 points)

Choose two or more subtasks from the NLU suitcase model (Fig. 1.12) to perform information extraction on your crawled data. For example, you could choose *subjectivity detection* and *polarity detection* to first categorize your data as *neutral* versus *opinionated* and then classify the resulting opinionated data as *positive* versus *negative*. Different classification approaches can be applied, including:

- knowledge based, e.g., SenticNet
- rule based, e.g., linguistic patterns
- machine learning based, e.g., deep neural networks
- hybrid (a combination of any of the above)

You can tap into any resource or toolkit you like, as long as you motivate your choices and you are able to critically analyze obtained results. Some possible choices include:

- Weka: https://cs.waikato.ac.nz/ml/weka
- Hadoop: https://hadoop.apache.org
- Pylearn2: https://pylearn2.readthedocs.io/en/latest
- SciKit: https://scikit-learn.org
- NLTK: https://nltk.org
- Theano: https://github.com/Theano
- Keras: https://github.com/fchollet/keras
- Tensorflow: https://github.com/tensorflow/tensorflow
- PyTorch: https://pytorch.org
- Huggingface: https://huggingface.co/
- AllenNLP https://github.com/allenai/allennlp

Question 4: Perform the following tasks:

- Justify your classification approach choice in relation to the state of the art
- Discuss whether you had to preprocess data and why
- Build an evaluation dataset by manually labeling at least 1,000 records with an inter-annotator agreement of at least 80%
- Provide metrics such as precision, recall, and F-measure on such dataset
- Perform a random accuracy test on the rest of the data and discuss results
- Discuss performance metrics, e.g., speed, and scalability of the system

Question 5: Explore some innovations for enhancing classification. If you introduce more than one, perform an ablation study to show the contribution of each innovation. For example, if you perform WSD and NER to enhance sentiment analysis, show the increase in accuracy when adding only WSD, the increase in accuracy when adding only NER, and the increase in accuracy when adding both WSD and NER to your system. Explain why they are important to solve specific problems, illustrated with examples. Possible innovations include (but are not limited to) the following:

- Enhanced classification (add another NLU subtask, e.g., sarcasm detection)
- Fine-grained classification (e.g., perform ABSA)
- Hybrid classification (e.g., apply both symbolic and subsymbolic AI)
- Cognitive classification (e.g., use brain-inspired algorithms)
- Multitask classification (e.g., perform two or more NLU tasks jointly)
- Ensemble classification (e.g., use stacked ensemble)

D. Submission

The submission shall consist of one single PDF file. Add some pictures to your report to make it clearer and easier to read. There is no page limit and no special formatting is required. The file shall contain the following five key items:

1) The names of the group members in the first page
2) Your answers to all the above questions

3) A YouTube link to a video presentation of up to 5 minutes: in the video, introduce your group members and their roles, explain the applications and the impact of your work and highlight, if any, the creative parts of your work (note that you do not have to give all the answers in the video presentation)
4) A Dropbox (or similar, e.g., Google Drive or OneDrive) link to a compressed (e.g., zip) file with crawled text data, queries and their results, evaluation dataset, automatic classification results, and any other data for Questions 3 and 5
5) A Dropbox (or similar, e.g., Google Drive or OneDrive) link to a compressed (e.g., zip) file with all your source codes and libraries, with a README file that explains how to compile and run the source codes

6.1.2 Quiz

1) What is the difference between NLP and NLU?
   A) NLP and NLU are synonymous and can be used interchangeably
   B) NLP focuses on text meaning, NLU on processing and generating text
   C) NLU includes more theoretical tasks, NLP is more focused on applications
   D) NLP handles language interaction, NLU focuses on understanding meaning

2) What does task decomposition typically involve?
   A) Splitting a complex NLP task into simpler, manageable subtasks
   B) Translating text from one language to another
   C) Summarizing long documents into shorter versions
   D) Evaluating the accuracy of an NLP model

3) What is the meaning of panalogy?
   A) Comparing machine learning models using parallel datasets
   B) Using parallel processing to speed up computations
   C) Comparing different concepts by examining their parallel structures
   D) Enhancing natural language processing by using parallel corpora

4) What does symbol grounding refer to in the context of AI?
   A) The process of translating symbols into natural language
   B) The ability to create new symbols for unknown concepts
   C) Relating symbols to real-world objects or concepts to give them meaning
   D) The use of symbols to perform logical operations

5) What is a topology-aware similarity measure?
   A) A method that considers the geometric arrangement of data points
   B) A technique for comparing the colors in images
   C) An approach for evaluating the textual similarity between documents
   D) A measure that assesses the similarity of time series based on frequency

6) What does intention awareness refer to in the context of HCI?
    A) The ability of a system to predict user actions based on past behavior
    B) The system's understanding of the user's goals and purposes
    C) The user's ability to comprehend the system's functionalities
    D) The system's ability to execute commands accurately

7) What is the primary difference between explainability and interpretability?
    A) The former focuses on model accuracy, the latter focuses on model complexity
    B) The former is understanding decisions; the latter is explaining model workings
    C) The former is about data transparency, the latter about algorithm transparency
    D) The former deals with system performance, the latter with prediction outcomes

8) Which approach can be used for microtext normalization?
    A) Dictionary-based replacement
    B) Machine translation
    C) Sentiment analysis
    D) Syntax parsing

9) Which of the following techniques is often used in microtext normalization?
    A) TF-IDF
    B) POS tagging
    C) Levenshtein distance
    D) NER

10) Why is microtext normalization important in NLU?
    A) It makes text shorter and more concise
    B) It helps in translating text to other languages
    C) It enhances NLU tasks by standardizing informal text
    D) It helps in generating text summaries

11) Which of the following is a challenge in microtext normalization?
    A) Identifying the language of the text
    B) Converting numeric values to words
    C) Handling the ambiguity of abbreviations and slang
    D) Translating text to multiple languages

12) Which one of these is a popular dataset for microtext normalization?
    A) LexNorm
    B) SenticNet
    C) WordNet
    D) LIWC

13) What challenge does SBD address?
    A) *Recognizing the context of idiomatic expressions*
    B) *Differentiating between abbreviations and sentence endings*
    C) *Identifying the main topic of a paragraph*
    D) *Translating text into different dialects*

14) Which of the following tools can be used for SBD?
    A) *Named entity recognizer*
    B) *Tokenizer*
    C) *POS tagger*
    D) *Lemmatizer*

15) Which of the following is a common dataset used for evaluating SBD systems?
    A) *Penn Treebank*
    B) *OntoNote*
    C) *MIMIC*
    D) *SIGMORPHON*

16) Why is SBD important in NLU?
    A) *It helps in translating text accurately*
    B) *It improves the performance of downstream NLU tasks*
    C) *It reduces the size of text data*
    D) *It helps in detecting the language of the text*

17) In addition to punctuation, what other feature can help in SBD?
    A) *Font size*
    B) *Capitalization*
    C) *Word embeddings*
    D) *Lemmatization*

18) What is POS tagging about?
    A) *Identifying the sentiment of a sentence*
    B) *Labeling words in a sentence as noun, verb, adjective, and more*
    C) *Translating text into another language*
    D) *Parsing the grammatical structure of sentences*

19) Which sentence could be ambiguous for POS tagging?
    A) *"The dog barked loudly"*
    B) *"She will can the peaches"*
    C) *"It is a sunny day"*
    D) *"He enjoys reading books"*

20) Which resource is commonly used for training and evaluating POS taggers?
   A) OntoNote
   B) Penn Treebank
   C) LexNorm
   D) LIWC

21) In addition to words, what other feature can be useful in POS tagging?
   A) Word length
   B) Word frequency
   C) Word embeddings
   D) Lemmatization

22) Text chunking is often referred to as:
   A) Sentence splitting
   B) Shallow parsing
   C) Deep parsing
   D) Text summarization

23) Which machine learning technique is commonly used for text chunking?
   A) K-means clustering
   B) HMM
   C) SVM
   D) Decision trees

24) Which of the following is a key application of text chunking?
   A) Image recognition
   B) Speech synthesis
   C) NER
   D) OCR

25) Which resource is commonly used for training and evaluating chunking systems?
   A) Penn Treebank
   B) OntoNote
   C) LexNorm
   D) MIMIC

26) What challenge does text chunking address in NLU?
   A) Translating text into different dialects
   B) Assigning correct POS tags to each word
   C) Grouping words into meaningful phrases for further analysis
   D) Detecting the boundaries between sentences

27) Which of the following is the lemma of the word "running"?
   A) ran
   B) runs
   C) run
   D) runner

28) How does lemmatization differ from stemming?
   A) It removes only suffixes, stemming removes both prefixes and suffixes
   B) It uses a dictionary to find the base form, stemming uses rules to strip affixes
   C) It is faster than stemming
   D) It is used only for verbs, stemming is used for all POS tags

29) Which of the following tools is commonly used for lemmatization in Python?
   A) NLTK
   B) OpenCV
   C) TensorFlow
   D) Scikit-learn

30) Which library provides the WordNetLemmatizer in Python?
   A) Gensim
   B) Scikit-learn
   C) NLTK
   D) SpaCy

31) Which of the following is the correct lemma for the word "better"?
   A) good
   B) best
   C) well
   D) better

32) In lemmatization, what role does the POS tag play?
   A) It determines the root form of a word
   B) It helps in splitting text into sentences
   C) It is used to identify named entities
   D) It helps in summarizing text

33) Which of the following words has the same lemma and root form in English?
   A) Running
   B) Happier
   C) Cats
   D) Quick

34) Which of the following is a common approach to WSD?
   A) Bag of words
   B) NER
   C) Supervised machine learning
   D) Text summarization

35) Which of the following resources is commonly used in WSD tasks?
   A) WordNet
   B) OntoNote
   C) SIGMORPHON
   D) LexNorm

36) What is the definition of 'cool' language?
   A) A language where a lot of microtext is used
   B) A language with very difficult sentence structure
   C) A language where understanding a sentence relies more on context
   D) A language that uses a lot of stopwords and redundancies

37) Which machine learning model can be used for WSD?
   A) Decision trees
   B) CNN
   C) RNN
   D) K-Means clustering

38) Which of the following is a challenge in WSD?
   A) Identifying the boundaries between sentences
   B) Handling the ambiguity of words with multiple meanings
   C) Translating text to multiple languages
   D) Summarizing long texts

39) In WSD, what role does context play?
   A) It helps in translating text
   B) It is used to identify named entities
   C) It determines the correct meaning of a word
   D) It helps in text summarization

40) Which evaluation metric is commonly used to assess WSD performance?
   A) Accuracy
   B) Euclidean distance
   C) BLEU score
   D) MSE

41) Which of the following is an unsupervised method for WSD?
    A) *Decision trees*
    B) *K-means clustering*
    C) *SVM*
    D) *Naïve Bayes classifier*

42) Which of the following categories is commonly used in NER?
    A) *Sentiment*
    B) *POS*
    C) *Organization*
    D) *Syntax*

43) Which machine learning model is commonly used for NER?
    A) *Decision trees*
    B) *K-means clustering*
    C) *CRF*
    D) *PCA*

44) What is the main challenge in NER?
    A) *Identifying sentence boundaries*
    B) *Disambiguating words with multiple meanings*
    C) *Correctly classifying entities with varying contexts*
    D) *Translating text into different dialects*

45) Which resource is commonly used for training and evaluating NER systems?
    A) *OntoNote*
    B) *Penn Treebank*
    C) *CoNLL-2003 dataset*
    D) *SenticNet*

46) What is a common evaluation metric for assessing NER performance?
    A) *BLEU score*
    B) *F_1 Score*
    C) *MSE*
    D) *Euclidean distance*

47) Which of the following techniques is most useful for concept extraction?
    A) *NER*
    B) *POS tagging*
    C) *TF-IDF*
    D) *OCR*

48) Which of the following functionalities is provided by the Sentic Parser?
   A) *Text summarization*
   B) *Concept extraction*
   C) *NER*
   D) *Shallow parsing*

49) Which NLU task is closely related to concept extraction?
   A) *Text summarization*
   B) *Machine translation*
   C) *NER*
   D) *POS tagging*

50) What is the main challenge in concept extraction?
   A) *Identifying sentence boundaries*
   B) *Disambiguating words with multiple meanings*
   C) *Extracting relevant concepts from unstructured text*
   D) *Translating text into different dialects*

51) Which of the following tools can be used for concept extraction in Python?
   A) *OpenCV*
   B) *NLTK*
   C) *Scikit-learn*
   D) *SpaCy*

52) Which of the following is an example of an anaphoric expression?
   A) *Quickly*
   B) *It*
   C) *Beautiful*
   D) *Running*

53) Which NLU task is most closely related to anaphora resolution?
   A) *NER*
   B) *Text summarization*
   C) *POS tagging*
   D) *Sentiment analysis*

54) What is the primary challenge in anaphora resolution?
   A) *Translating text into different dialects*
   B) *Correctly identifying the antecedent of an anaphoric expression*
   C) *Extracting key ideas or topics from a text*
   D) *Identifying sentence boundaries*

55) Which of the following algorithms can be used for anaphora resolution?
    A) LDA
    B) SVM
    C) CRF
    D) Neural networks

56) Which resource is often used to help with anaphora resolution in NLU?
    A) WordNet
    B) OntoNote
    C) Penn Treebank
    D) SenticNet

57) Which machine learning technique is commonly used for subjectivity detection?
    A) K-means clustering
    B) Decision trees
    C) SVM
    D) PCA

58) What is the difference between ambivalence and neutrality in sentiment analysis?
    A) Ambivalence means mixed feelings, while neutrality means no sentiment
    B) Ambivalence means no sentiment, while neutrality means mixed feelings
    C) Ambivalence and neutrality are the same
    D) Ambivalence means strong feelings, while neutrality means extreme emotions

59) Which of the following features is most useful for subjectivity detection?
    A) Named entities
    B) POS Tags
    C) TF-IDF
    D) OCR

60) Which evaluation metric can be used to assess subjectivity detection?
    A) BLEU score
    B) F_1 score
    C) MSE
    D) Euclidean distance

61) Which of the following techniques can be used for metaphor detection in NLU?
    A) NER
    B) Topic modeling
    C) POS tagging
    D) WSD

62) In metaphor understanding, what is the "source domain"?
   A) *The literal meaning of the metaphorical phrase*
   B) *The figurative context that gives meaning to the metaphor*
   C) *The grammatical structure of the metaphor*
   D) *The direct comparison using "like" or "as"*

63) Which of the following is a common challenge in metaphor understanding?
   A) *Translating text into different languages*
   B) *Detecting metaphors that are culturally specific*
   C) *Summarizing long texts*
   D) *Identifying sentence boundaries*

64) Which of the following NLU techniques is often used to understand metaphors?
   A) *CNN*
   B) *SVM*
   C) *SRL*
   D) *OCR*

65) Which resource can be useful for training metaphor understanding models?
   A) *WordNet*
   B) *OntoNote*
   C) *PropBank*
   D) *Penn Treebank*

66) Which of the following sentences is an example of sarcasm?
   A) *"I absolutely love getting stuck in traffic"*
   B) *"The sky is blue"*
   C) *"I enjoy reading books"*
   D) *"She is very kind"*

67) Which feature can be indicative of sarcasm in text?
   A) *Positive sentiment words in a negative context*
   B) *Frequent use of punctuation*
   C) *Proper noun usage*
   D) *Numerical data*

68) Which of the following is a common challenge in sarcasm detection?
   A) *Identifying sentence boundaries*
   B) *Disambiguating words with multiple meanings*
   C) *Correctly interpreting the intended meaning behind sarcastic statements*
   D) *Translating text into different languages*

69) Which machine learning model can be used for sarcasm detection?
   A) K-means clustering
   B) Decision trees
   C) RNN
   D) Linear regression

70) Which NLU technique can be useful for detecting sarcasm in text?
   A) NER
   B) Sentiment analysis
   C) POS tagging
   D) Text summarization

71) Which of the following resources can help improve sarcasm detection models?
   A) SenticNet
   B) OntoNote
   C) Penn Treebank
   D) SIGMORPHON

72) Which of the following models is commonly used for personality recognition?
   A) CNN
   B) RNN
   C) SVM
   D) Decision trees

73) In personality recognition, what does the "Big Five" model assess?
   A) Cognitive abilities
   B) Openness, conscientiousness, extraversion, agreeableness, and neuroticism
   C) Emotional intelligence
   D) Linguistic diversity

74) Which feature is often used in text-based personality recognition?
   A) NER
   B) TF-IDF
   C) POS tagging
   D) OCR

75) Which approach can assess user personality based on social media activity?
   A) Sentiment analysis
   B) Text classification
   C) Personality prediction models
   D) NER

76) What is a key advantage of using machine learning for personality recognition?
    A) It can provide real-time feedback on personality traits
    B) It can translate text into multiple languages
    C) It can generate new text automatically
    D) It can correct grammatical errors in text

77) Which resource can be useful for evaluating personality recognition models?
    A) WordNet
    B) Personality-labeled text datasets
    C) OntoNote
    D) Penn Treebank

78) Which technique is commonly used for aspect extraction in text?
    A) NER
    B) Sentiment analysis
    C) Topic modeling
    D) POS tagging

79) Which of the following is a common challenge in aspect extraction?
    A) Identifying grammatical errors in text
    B) Accurately extracting relevant aspects from text
    C) Translating text into different languages
    D) Summarizing long texts

80) Which of the following algorithms can be used for aspect extraction?
    A) LDA
    B) K-means clustering
    C) Decision trees
    D) PCA

81) Which NLU tool or library can be useful for aspect extraction?
    A) OpenCV
    B) NLTK
    C) TensorFlow
    D) Keras

82) What is the role of Sentic GCN in sentiment analysis?
    A) It classifies sentiment using word sequences
    B) It uses graph-based models to extract aspects from text
    C) It applies rule-based approaches to sentiment lexicons
    D) It analyzes images to extract sentiment

83) What is polarity detection in the context of NLU?
   A) *Identifying the grammatical structure of a sentence*
   B) *Determining whether the sentiment expressed in text is positive or negative*
   C) *Translating text into another language*
   D) *Splitting text into sentences*

84) Which of the following is a common application of polarity detection?
   A) *Image recognition*
   B) *Spam email filtering*
   C) *Text summarization*
   D) *Speech synthesis*

85) Which of the following methods can be used to perform polarity detection?
   A) *LDA*
   B) *SVM*
   C) *CNN*
   D) *PCA*

86) What is a common challenge in polarity detection?
   A) *Identifying grammatical errors*
   B) *Correctly detecting the sentiment in ambiguous or sarcastic text*
   C) *Summarizing long documents*
   D) *Translating text into multiple languages*

87) Which of the following libraries can be used for polarity detection in Python?
   A) *OpenCV*
   B) *Scikit-learn*
   C) *TensorFlow*
   D) *SpaCy*

88) Which evaluation metric is commonly used to assess polarity detection?
   A) *BLEU score*
   B) *F_1 score*
   C) *MSE*
   D) *Euclidean distance*

89) In polarity detection, what role do sentiment lexicons play?
   A) *They provide grammatical rules for text*
   B) *They contain lists of words with predefined sentiment values*
   C) *They translate text into different languages*
   D) *They summarize the content of the text*

90) Which resource is useful for training a polarity detection model?
   A) SenticNet
   B) OntoNote
   C) Penn Treebank
   D) LexNorm

91) Which of the following is a common technique for knowledge representation?
   A) Gradient descent
   B) Decision trees
   C) Semantic networks
   D) Reinforcement learning

92) What is an ontology in the context of knowledge representation?
   A) A method for reducing dimensionality of data
   B) A formal representation of a set of concepts and their relationships
   C) A type of neural network
   D) A statistical model for predicting outcomes

93) In knowledge representation, what is a frame?
   A) A data structure for dividing a program into separate parts
   B) A network of connected nodes representing knowledge
   C) A collection of attributes and values describing an entity
   D) A technique for visualizing data

94) What is the purpose of a semantic network?
   A) To train machine learning models
   B) To represent knowledge in interconnected nodes and edges
   C) To perform clustering of data
   D) To improve image processing tasks

95) Which of the following best describes a knowledge graph?
   A) A graph used to represent statistical data
   B) A structure that represents entities and their interrelationships
   C) A method for optimizing algorithms
   D) A type of decision tree

96) What is the primary feature of PrimeNet?
   A) It uses a flat structure to organize concepts and entities
   B) It integrates commonsense knowledge via a hierarchical structure
   C) It focuses solely on sentiment analysis without conceptual organization
   D) It employs a rule-based system for automatic text summarization

97) Which aspect of PrimeNet's design helps in emulating human-like reasoning?
   A) A large collection of unrelated facts
   B) A single-layer network with random connections
   C) A three-layer architecture with conceptual primitives, concepts, and entities
   D) Exclusive use of machine translation techniques

98) What is the function of rules in rule-based systems?
   A) To create visualizations from data
   B) To apply logical operations to facts to derive conclusions
   C) To train deep learning models
   D) To perform statistical analysis

99) What is a key limitation of using logic-based knowledge representation?
   A) Inability to handle uncertainty
   B) High computational cost
   C) Lack of interpretability
   D) Difficulty in visualizing data

100) What does a concept hierarchy represent in knowledge representation?
   A) The order of execution in a program
   B) A taxonomy of concepts, from general to specific
   C) The performance metrics of an algorithm
   D) The architecture of a neural network

References

Muhammad Abdul-Mageed and Lyle Ungar. EmoNet: Fine-grained emotion detection with gated recurrent neural networks. In *ACL*, pages 718–728, 2017.

Steven Abney, Robert E Schapire, and Yoram Singer. Boosting applied to tagging and pp attachment. In *1999 joint SIGDAT conference on empirical methods in natural language processing and very large corpora*, 1999.

Waheed Ahmed Abro, Annalena Aicher, Niklas Rach, Stefan Ultes, Wolfgang Minker, and Guilin Qi. Natural language understanding for argumentative dialogue systems in the opinion building domain. *Knowledge-Based Systems*, 242: 108318, 2022.

Rod Adams, Gabriel Nicolae, Cristina Nicolae, and Sanda Harabagiu. Textual entailment through extended lexical overlap and lexico-semantic matching. In *ACL-PASCAL Workshop on Textual Entailment and Paraphrasing*, pages 119–124, 2007.

Gediminas Adomavicius, Bamshad Mobasher, Francesco Ricci, and Alexander Tuzhilin. Context-aware recommender systems. *AI Magazine*, 32(3):67–80, 2011.

Neha Agarwal, Kelley Herndon Ford, and Max Shneider. Sentence boundary detection using a maxEnt classifier. In *MISC*, pages 1–6, 2005.

Eugene Agichtein, Walt Askew, and Yandong Liu. Combining lexical, syntactic, and semantic evidence for textual entailment classification. In *TAC*, pages 1–6, 2008.

Eneko Agirre and Aitor Soroa. Personalizing pagerank for word sense disambiguation. In *12th Conference of the European Chapter of the ACL*, pages 33–41, 2009.

Eneko Agirre, Oier López de Lacalle, and Aitor Soroa. Random walks for knowledge-based word sense disambiguation. *Computational Linguistics*, 40(1):57–84, 2014.

Rakesh Agrawal, Ramakrishnan Srikant, et al. Fast algorithms for mining association rules. In *20th International Conference on Very Large Data Bases*, volume 1215, pages 487–499, 1994.

David W Aha, Dennis Kibler, and Marc K Albert. Instance-based learning algorithms. *Machine learning*, 6(1):37–66, 1991.

Dirk Ahlers. Assessment of the accuracy of GeoNames gazetteer data. In *7th workshop on geographic information retrieval*, pages 74–81, 2013.

Murtadha Ahmed, Shengfeng Pan, Jianlin Su, Xinxin Cao, Wenze Zhang, Bo Wen, and Yunfeng Liu. BERT-ASC: Implicit aspect representation learning through auxiliary-sentence construction for sentiment analysis. *arXiv e-prints*, pages arXiv–2203, 2022.

Naoyoshi Aikawa, Tetsuya Sakai, and Hayato Yamana. Community qa question classification: Is the asker looking for subjective answers or not? *IPSJ Online Transactions*, 4:160–168, 2011.

Alan Akbik, Duncan Blythe, and Roland Vollgraf. Contextual string embeddings for sequence labeling. In *COLING*, pages 1638–1649, 2018.

Iskander Akhmetov, Alexandr Pak, Irina Ualiyeva, and Alexander Gelbukh. Highly language-independent word lemmatization using a machine-learning classifier. *Computación y Sistemas*, 24(3), 2020.

© The Author(s), under exclusive license to Springer Nature Switzerland AG 2025
E. Cambria, *Understanding Natural Language Understanding*,
https://doi.org/10.1007/978-3-031-73974-3

Cem Akkaya, Janyce Wiebe, and Rada Mihalcea. Subjectivity word sense disambiguation. In *EMNLP*, pages 190–199, 2009.

Ahmed Al Hamoud, Amber Hoenig, and Kaushik Roy. Sentence subjectivity analysis of a political and ideological debate dataset using LSTM and BiLSTM with attention and GRU models. *Journal of King Saud University-Computer and Information Sciences*, pages 7975–7987, 2022.

Rabah A. Al-Zaidy, Cornelia Caragea, and C. Lee Giles. Bi-lstm-crf sequence labeling for keyphrase extraction from scholarly documents. In Ling Liu, Ryen W. White, Amin Mantrach, Fabrizio Silvestri, Julian J. McAuley, Ricardo Baeza-Yates, and Leila Zia, editors, *The World Wide Web Conference*, pages 2551–2557. ACM, 2019.

Firoj Alam, Evgeny A Stepanov, and Giuseppe Riccardi. Personality traits recognition on social network-facebook. In *international AAAI conference on web and social media*, pages 6–9, 2013.

Mehwish Alam, Andreea Iana, Alexander Grote, Katharina Ludwig, Philipp Müller, and Heiko Paulheim. Towards analyzing the bias of news recommender systems using sentiment and stance detection. In *Web Conference*, pages 448–457, 2022.

Zakariae Alami Merrouni, Bouchra Frikh, and Brahim Ouhbi. Automatic keyphrase extraction: a survey and trends. *Journal of Intelligent Information Systems*, 54: 391–424, 2020.

Desislava Aleksandrova, François Lareau, and Pierre André Ménard. Multilingual sentence-level bias detection in Wikipedia. In *RANLP*, pages 42–51, 2019.

Badr AlKhamissi, Millicent Li, Asli Celikyilmaz, Mona T. Diab, and Marjan Ghazvininejad. A review on language models as knowledge bases. *CoRR*, abs/2204.06031, 2022.

Abdulrahman Aloraini and Massimo Poesio. Cross-lingual zero pronoun resolution. In *LREC*, pages 90–98, 2020.

Abdulrahman Aloraini and Massimo Poesio. Data augmentation methods for anaphoric zero pronouns. In *Fourth Workshop on Computational Models of Reference, Anaphora and Coreference*, pages 82–93, 2021.

Abdulrahman Aloraini, Sameer Pradhan, and Massimo Poesio. Joint coreference resolution for zeros and non-zeros in Arabic. *arXiv preprint arXiv:2210.12169*, 2022.

Duygu Altinok. An ontology-based dialogue management system for banking and finance dialogue systems. *arXiv preprint arXiv:1804.04838*, 2018.

Nabeela Altrabsheh, Mohamed Medhat Gaber, Mihaela Cocea, et al. SA-E: sentiment analysis for education. *Frontiers in Artificial Intelligence and Applications*, 255: 353–362, 2013.

Yasemin Altun, Ioannis Tsochantaridis, and Thomas Hofmann. Hidden markov support vector machines. In *ICML*, pages 3–10, 2003.

Nouf Alturaief, Hamoud Aljamaan, and Malak Baslyman. Aware: Aspect-based sentiment analysis dataset of apps reviews for requirements elicitation. In *2021 36th IEEE/ACM International Conference on Automated Software Engineering Workshops (ASEW)*, pages 211–218, 2021.

Pooja Alva and Vinay Hegde. Hidden Markov model for POS tagging in word sense disambiguation. In *2016 International Conference on Computation System and Information Technology for Sustainable Solutions (CSITSS)*, pages 279–284, 2016.

Nestor Alvaro, Yusuke Miyao, Nigel Collier, et al. TwiMed: Twitter and PubMed comparable corpus of drugs, diseases, symptoms, and their relations. *JMIR Public Health and Surveillance*, 3(2):e6396, 2017.

Enrique Amigó, Jorge Carrillo de Albornoz, Irina Chugur, Adolfo Corujo, Julio Gonzalo, Tamara Martín, Edgar Meij, Maarten De Rijke, and Damiano Spina. Overview of RepLab 2013: Evaluating online reputation monitoring systems. In *International conference of the cross-language evaluation forum for european languages*, pages 333–352, 2013.

Mostafa Amin, Erik Cambria, and Björn Schuller. Will affective computing emerge from foundation models and General AI? A first evaluation on ChatGPT. *IEEE Intelligent Systems*, 38(2):15–23, 2023.

Mostafa M Amin, Rui Mao, Erik Cambria, and Björn W Schuller. A wide evaluation of ChatGPT on affective computing tasks. *IEEE Transactions on Affective Computing*, 2024.

Ida Amini, Samane Karimi, and Azadeh Shakery. Cross-lingual subjectivity detection for resource lean languages. In *Tenth Workshop on Computational Approaches to Subjectivity, Sentiment and Social Media Analysis*, pages 81–90, 2019.

Henry Anaya-Sánchez, Aurora Pons-Porrata, and Rafael Berlanga-Llavori. Word sense disambiguation based on word sense clustering. In *Advances in Artificial Intelligence-IBERAMIA-SBIA 2006*, pages 472–481. Springer, 2006.

Rie Kubota Ando and Tong Zhang. A framework for learning predictive structures from multiple tasks and unlabeled data. *Journal of Machine Learning Research*, 6:1817–1853, 2005.

Chinatsu Aone and Scott William Bennett. Automated acquisition of anaphora resolution strategies. *AAAI*, 1995.

Ian A Apperly and Stephen A Butterfill. Do humans have two systems to track beliefs and belief-like states? *Psychological Review*, 116(4):953, 2009.

Gor Arakelyan, Karen Hambardzumyan, and Hrant Khachatrian. Towards jointud: Part-of-speech tagging and lemmatization using recurrent neural networks. *arXiv preprint arXiv:1809.03211*, 2018.

Rahul Aralikatte, Heather Lent, Ana Valeria Gonzalez, Daniel Herschcovich, Chen Qiu, Anders Sandholm, Michael Ringaard, and Anders Søgaard. Rewarding coreference resolvers for being consistent with world knowledge. In *EMNLP-IJCNLP*, pages 1229–1235, 2019.

Rahul Aralikatte, Matthew Lamm, Daniel Hardt, and Anders Søgaard. Ellipsis resolution as question answering: An evaluation. In *EACL*, pages 810–817, 2021.

Qazi Mohammad Areeb, Mohammad Nadeem, Shahab Saquib Sohail, Raza Imam, Faiyaz Doctor, Yassine Himeur, Amir Hussain, and Abbes Amira. Filter bubbles in recommender systems: Fact or fallacy—a systematic review. *Wiley Interdisciplinary Reviews: Data Mining and Knowledge Discovery*, 13(6):e1512, 2023.

Shlomo Argamon, Moshe Koppel, Jonathan Fine, and Anat Rachel Shimoni. Gender, genre, and writing style in formal written texts. *Text & talk*, 23(3):321–346, 2003.

Shlomo Argamon, Sushant Dhawle, Moshe Koppel, and James W Pennebaker. Lexical predictors of personality type. In *2005 joint annual meeting of the interface and the classification society of North America*, pages 1–16, 2005.

Jennifer E Arnold. *Reference form and discourse patterns*. Stanford University, 1998.

Ron Artstein and Massimo Poesio. Inter-coder agreement for computational linguistics. *Computational linguistics*, 34(4):555–596, 2008.

Kai Arulkumaran, Marc Peter Deisenroth, Miles Brundage, and Anil Anthony Bharath. Deep reinforcement learning: A brief survey. *IEEE Signal Processing Magazine*, 34(6):26–38, 2017.

Nana Yaw Asabere and Amevi Acakpovi. ROPPSA: TV program recommendation based on personality and social awareness. *Mathematical Problems in Engineering*, 2020:1–15, 2020.

Jens B Asendorpf. Head-to-head comparison of the predictive validity of personality types and dimensions. *European Journal of Personality*, 17(5):327–346, 2003.

Muhammad Zubair Asghar, Aurangzeb Khan, Shakeel Ahmad, and Fazal Masud Kundi. A review of feature extraction in sentiment analysis. *Journal of Basic and Applied Scientific Research*, 4(3):181–186, 2014.

Nicholas Asher and Alex Lascarides. *Logics of Conversation*. Cambridge University Press, 2003.

Michael C Ashton, Kibeom Lee, Marco Perugini, Piotr Szarota, Reinout E De Vries, Lisa Di Blas, Kathleen Boies, and Boele De Raad. A six-factor structure of personality-descriptive adjectives: solutions from psycholexical studies in seven languages. *Journal of personality and social psychology*, 86(2):356, 2004.

Atanas V Atanassov, Dimitar I Pilev, Fani N Tomova, and Vanya D Kuzmanova. Hybrid system for emotion recognition based on facial expressions and body gesture recognition. In *2021 International Conference Automatics and Informatics (ICAI)*, pages 135–140, 2021.

Beryl TS Atkins. Tools for computer-aided corpus lexicography: the hector project. *Acta Linguistica Hungarica*, 41:5–71, 1992.

Sandeep Attree. Gendered ambiguous pronouns shared task: Boosting model confidence by evidence pooling. In *First Workshop on Gender Bias in Natural Language Processing*, pages 134–146, 2019.

Anthony Aue and Michael Gamon. Customizing sentiment classifiers to new domains: A case study. In *RANLP*, 2005.

Sören Auer, Christian Bizer, Georgi Kobilarov, Jens Lehmann, Richard Cyganiak, and Zachary Ives. DBpedia: A nucleus for a web of open data. In *International Semantic Web Conference*, pages 722–735, 2007.

John Langshaw Austin. *How to do things with words*. Oxford University Press, 1975.

Nastaran Babanejad, Heidar Davoudi, Aijun An, and Manos Papagelis. Affective and contextual embedding for sarcasm detection. In *COLING*, pages 225–243, 2020.

Stefano Baccianella, Andrea Esuli, and Fabrizio Sebastiani. SentiWordNet 3.0: An enhanced lexical resource for sentiment analysis and opinion mining. In *Seventh International Conference on Language Resources and Evaluation*, pages 2200–2204, 2010.

Sebastian Bader and Pascal Hitzler. Dimensions of neural-symbolic integration-a structured survey. *arXiv preprint cs/0511042*, 2005.

Samy Badreddine, Artur d'Avila Garcez, Luciano Serafini, and Michael Spranger. Logic tensor networks. *Artificial Intelligence*, 303:103649, 2022.

Kushal Bafna and Durga Toshniwal. Feature based summarization of customers' reviews of online products. *Procedia Computer Science*, 22:142–151, 2013.

Amit Bagga and Breck Baldwin. Algorithms for scoring coreference chains. In *The First International Conference on Language Resources and Evaluation Workshop on Linguistics Coreference*, volume 1, pages 563–566, 1998.

Dzmitry Bahdanau, Kyung Hyun Cho, and Yoshua Bengio. Neural machine translation by jointly learning to align and translate. In *3rd International Conference on Learning Representations*, pages 1–15, 2015.

Vaibhav Bajaj, Kartikey Pant, Ishan Upadhyay, Srinath Nair, and Radhika Mamidi. Teaser: Towards efficient aspect-based sentiment analysis and recognition. In *RANLP*, pages 102–110, 2021.

C Lee Baker. Double negatives. *Linguistic inquiry*, 1(2):169–186, 1970.

Collin F. Baker, Charles J. Fillmore, and John B. Lowe. The berkeley framenet project. In *ACL*, pages 86–90, 1998.

Simon Baker, Roi Reichart, and Anna Korhonen. An unsupervised model for instance level subcategorization acquisition. In *EMNLP*, pages 278–289, 2014.

Vimala Balakrishnan and Hossein Arabi. Hyperm: A hybrid personality-aware recommender for movie. *Malaysian Journal of Computer Science*, 31(1):48–62, 2018.

Vimala Balakrishnan and Ethel Lloyd-Yemoh. Stemming and lemmatization: A comparison of retrieval performances. *Lecture Notes on Software Engineering*, 2, 2014.

Breck Baldwin. CogNIAC: high precision coreference with limited knowledge and linguistic resources. In *Operational Factors in Practical, Robust Anaphora Resolution for Unrestricted Texts*, pages 38–45, 1997.

Timothy Baldwin, Marie Catherine de Marneffe, Bo Han, Young-Bum Kim, Alan Ritter, and Wei Xu. Shared tasks of the 2015 workshop on noisy user-generated text: Twitter lexical normalization and named entity recognition. In *Workshop on Noisy User-generated Text*, pages 126–135, 2015.

David Bamman, Olivia Lewke, and Anya Mansoor. An annotated dataset of coreference in English literature. In *LREC*, pages 44–54, 2020.

Wouter Bancken, Daniele Alfarone, and Jesse Davis. Automatically detecting and rating product aspects from textual customer reviews. In *1st international workshop on interactions between data mining and natural language processing at ECML/PKDD*, volume 1202, pages 1–16, 2014.

Carmen Banea, Rada Mihalcea, and Janyce Wiebe. Multilingual subjectivity: Are more languages better? In *COLING*, pages 28–36, 2010.

Pratyay Banerjee and Chitta Baral. Self-supervised knowledge triplet learning for zero-shot question answering. In *EMNLP*, pages 151–162, 2020.

Satanjeev Banerjee, Ted Pedersen, et al. Extended gloss overlaps as a measure of semantic relatedness. In *IJCAI*, volume 3, pages 805–810, 2003.

Ann Banfield. *Unspeakable Sentences (Routledge Revivals): Narration and Representation in the Language of Fiction*. Routledge, 2014.

Jeesoo Bang, Hyungjong Noh, Yonghee Kim, and Gary Geunbae Lee. Example-based chat-oriented dialogue system with personalized long-term memory. In *2015 International Conference on Big Data and Smart Computing (BIGCOMP)*, pages 238–243, 2015.

Rainer Banse and Klaus R Scherer. Acoustic profiles in vocal emotion expression. *Journal of Personality and Social Psychology*, 70(3):614, 1996.

Hui Bao, Kai He, Xuemeng Yin, Xuanyu Li, Xinrui Bao, Haichuan Zhang, Jialun Wu, and Zeyu Gao. Bert-based meta-learning approach with looking back for sentiment analysis of literary book reviews. In *NLPCC*, pages 235–247, 2021.

Xiaoyi Bao, Z Wang, Xiaotong Jiang, Rong Xiao, and Shoushan Li. Aspect-based sentiment analysis with opinion tree generation. *IJCAI 2022*, pages 4044–4050, 2022.

Roy Bar-Haim, Ido Dagan, Shachar Mirkin, Eyal Shnarch, Idan Szpektor, Jonathan Berant, and Iddo Greental. Efficient semantic deduction and approximate matching over compact parse forests. In *TAC*, pages 1–10, 2008.

Edoardo Barba, Tommaso Pasini, and Roberto Navigli. ESC: Redesigning WSD with extractive sense comprehension. In *NAACL-HLT*, pages 4661–4672, 2021a.

Edoardo Barba, Luigi Procopio, Caterina Lacerra, Tommaso Pasini, and Roberto Navigli. Exemplification modeling: Can you give me an example, please? In *IJCAI*, pages 3779–3785, 2021b.

Luciano Barbosa and Junlan Feng. Robust sentiment detection on twitter from biased and noisy data. In *COLING*, pages 36–44, 2010.

Valentina Bartalesi Lenzi, Giovanni Moretti, and Rachele Sprugnoli. CAT: the CELCT annotation tool. In *Eighth International Conference on Language Resources and Evaluation (LREC'12)*, pages 333–338, 2012.

Fabian Barteld, Ingrid Schröder, and Heike Zinsmeister. Dealing with word-internal modification and spelling variation in data-driven lemmatization. In *10th SIGHUM workshop on language technology for cultural heritage, social sciences, and humanities*, pages 52–62, 2016.

Susan Bartlett, Grzegorz Kondrak, and Colin Cherry. Automatic syllabification with structured SVMs for letter-to-phoneme conversion. In *ACL-08: HLT*, pages 568–576, 2008.

Jon Barwise and John Perry. Situations and attitudes. *The Journal of Philosophy*, 78(11):668–691, 1981.

Pierpaolo Basile, Annalina Caputo, and Giovanni Semeraro. An enhanced lesk word sense disambiguation algorithm through a distributional semantic model. In *COLING*, pages 1591–1600, 2014.

Mohammad Ehsan Basiri, Shahla Nemati, Moloud Abdar, Erik Cambria, and U Rajendra Acharya. Abcdm: An attention-based bidirectional CNN-RNN deep model for sentiment analysis. *Future Generation Computer Systems*, 115:279–294, 2021.

Gourav Bathla, Pardeep Singh, Rahul Kumar Singh, Erik Cambria, and Rajeev Tiwari. Intelligent fake reviews detection based on aspect extraction and analysis using deep learning. *Neural Computing and Applications*, 34:20213–20229, 2022.

Zeynep Batmaz, Ali Yurekli, Alper Bilge, and Cihan Kaleli. A review on deep learning for recommender systems: challenges and remedies. *Artificial Intelligence Review*, 52(1):1–37, 2019.

David I Beaver. The optimization of discourse anaphora. *Linguistics and Philosophy*, 27:3–56, 2004.

Manjot Bedi, Shivani Kumar, Md Shad Akhtar, and Tanmoy Chakraborty. Multimodal sarcasm detection and humor classification in code-mixed conversations. *IEEE Transactions on Affective Computing*, 2021.

Salima Behdenna, Barigou Fatiha, and Ghalem Belalem. Ontology-based approach to enhance explicit aspect extraction in standard arabic reviews. *International Journal of Computing and Digital Systems*, 11(1):277–287, 2022.

Alexander Beider. Beider-morse phonetic matching: An alternative to soundex with fewer false hits. *Avotaynu: the International Review of Jewish Genealogy*, 24(2): 12, 2008.

Ioannis Bekoulis, Johannes Deleu, Thomas Demeester, and Chris Develder. Joint entity recognition and relation extraction as a multi-head selection problem. *Expert Systems With Applications*, 114:34–45, 2018.

Yonatan Belinkov, Lluís Màrquez, Hassan Sajjad, Nadir Durrani, Fahim Dalvi, and James Glass. Evaluating layers of representation in neural machine translation on part-of-speech and semantic tagging tasks. In *IJCNLP*, pages 1–10, 2017.

Farah Benamara, Baptiste Chardon, Yannick Mathieu, and Vladimir Popescu. Towards context-based subjectivity analysis. In *IJCNLP*, pages 1180–1188, 2011.

Emily M Bender and Alexander Koller. Climbing towards nlu: On meaning, form, & understanding in the age of data. In *ACL*, pages 5185–5198, 2020.

Eric Bengtson and Dan Roth. Understanding the value of features for coreference resolution. In *EMNLP*, pages 294–303, 2008.

Luisa Bentivogli and Emanuele Pianta. Exploiting parallel texts in the creation of multilingual semantically annotated resources: the multisemcor corpus. *Natural Language Engineering*, 11(3):247–261, 2005.

Luisa Bentivogli, Peter Clark, Ido Dagan, and Danilo Giampiccolo. The fifth PASCAL recognizing textual entailment challenge. In *TAC*, pages 1–18, 2009.

Gábor Berend. Sparsity makes sense: Word sense disambiguation using sparse contextualized word representations. In *EMNLP*, pages 8498–8508, 2020.

Adam L. Berger, Stephen A. Della Pietra, and Vincent J. Della Pietra. A maximum entropy approach to natural language processing. *Computational Linguistics*, 22 (1):39–71, 1996.

Sabine Bergler, René Witte, Michelle Khalife, Zhuoyan Li, and Frank Rudzicz. Using knowledge-poor coreference resolution for text summarization. In *Workshop on Text Summarization*, pages 1–8, 2003.

Toms Bergmanis and Sharon Goldwater. Context sensitive neural lemmatization with Lematus. In *ACL-HLT*, pages 1391–1400, 2018.

Toms Bergmanis and Sharon Goldwater. Training data augmentation for context-sensitive neural lemmatization using inflection tables and raw text. *arXiv preprint arXiv:1904.01464*, 2019.

José M Bernardo and Adrian FM Smith. Bayesian theory. *Measurement Science and Technology*, 12(2):221–222, 2001.

Steven Bethard, Guergana Savova, Wei-Te Chen, Leon Derczynski, James Pustejovsky, and Marc Verhagen. SemEval-2016 Task 12: Clinical TempEval. In *10th international workshop on semantic evaluation (SemEval-2016)*, pages 1052–1062, 2016.

Michele Bevilacqua and Roberto Navigli. Quasi bidirectional encoder representations from transformers for word sense disambiguation. In *RANLP*, pages 122–131, 2019.

Michele Bevilacqua and Roberto Navigli. Breaking through the 80% glass ceiling: Raising the state of the art in word sense disambiguation by incorporating knowledge graph information. In *ACL*, pages 2854–2864, 2020.

Michele Bevilacqua, Tommaso Pasini, Alessandro Raganato, Roberto Navigli, et al. Recent trends in word sense disambiguation: A survey. In *IJCAI*, pages 4330–4338, 2021.

Chandra Bhagavatula, Ronan Le Bras, Chaitanya Malaviya, Keisuke Sakaguchi, Ari Holtzman, Hannah Rashkin, Doug Downey, Wen-tau Yih, and Yejin Choi. Abductive commonsense reasoning. In *International Conference on Learning Representations, ICLR*, 2020.

Prajjwal Bhargava and Vincent Ng. Commonsense knowledge reasoning and generation with pre-trained language models: A survey. In *AAAI*, pages 12317–12325, 2022.

Joan-Isaac Biel, Vagia Tsiminaki, John Dines, and Daniel Gatica-Perez. Hi YouTube! personality impressions and verbal content in social video. In *15th ACM on International conference on multimodal interaction*, pages 119–126, 2013.

Chris Biemann. Unsupervised part-of-speech tagging employing efficient graph clustering. In *COLING-ACL*, pages 7–12, 2006.

Yonatan Bisk, Rowan Zellers, Ronan Le Bras, Jianfeng Gao, and Yejin Choi. PIQA: reasoning about physical commonsense in natural language. In *AAAI*, pages 7432–7439, 2020.

Prakhar Biyani, Sumit Bhatia, Cornelia Caragea, and Prasenjit Mitra. Using non-lexical features for identifying factual and opinionative threads in online forums. *Knowledge-Based Systems*, 69:170–178, 2014.

Yuri Bizzoni and Shalom Lappin. Predicting human metaphor paraphrase judgments with deep neural networks. In *Workshop on Figurative Language Processing*, pages 45–55, 2018.

William Black, Sabri Elkateb, Horacio Rodriguez, Musa Alkhalifa, Piek Vossen, Adam Pease, Christiane Fellbaum, et al. Introducing the Arabic WordNet project. In *Third International WordNet Conference*, pages 295–300, 2006.

David M Blei, Andrew Y Ng, and Michael I Jordan. Latent Dirichlet allocation. *Journal of Machine Learning Research*, 3:993–1022, 2003.

Terra Blevins and Luke Zettlemoyer. Moving down the long tail of word sense disambiguation with gloss informed bi-encoders. In *ACL*, pages 1006–1017, 2020.

John Blitzer, Mark Dredze, and Fernando Pereira. Biographies, bollywood, boomboxes and blenders: Domain adaptation for sentiment classification. In *45th annual meeting of the association of computational linguistics*, pages 440–447, 2007.

Rexhina Blloshmi, Tommaso Pasini, Niccolò Campolungo, Somnath Banerjee, Roberto Navigli, and Gabriella Pasi. IR like a SIR: Sense-enhanced information retrieval for multiple languages. In *EMNLP*, pages 1030–1041, 2021.

Kenneth Bloom, Navendu Garg, and Shlomo Argamon. Extracting appraisal expressions. In *NAACL-HLT*, pages 308–315, 2007.

Daniel G Bobrow and Donald A Norman. Some principles of memory schemata. In *Representation and understanding*, pages 131–149. Morgan Kaufmann, San Diego, 1975.

Olivier Bodenreider. The unified medical language system (umls): integrating biomedical terminology. *Nucleic acids research*, 32(suppl_1):D267–D270, 2004.

Paul Boersma. Praat, a system for doing phonetics by computer. *Glot. Int.*, 5(9): 341–345, 2001.

Gergo Bogacsovics, Janos Toth, Andras Hajdu, and Balazs Harangi. Enhancing CNNs through the use of hand-crafted features in automated fundus image classification. *Biomedical Signal Processing and Control*, 76:103685, 2022.

Alena Böhmová, Jan Hajic, Eva Hajicová, Barbora Hladká, and Anne Abeillé. The prague dependency treebank: Three-level annotation scenario. *Treebanks: building and using parsed corpora*, 20:103–127, 2003.

Piotr Bojanowski, Edouard Grave, Armand Joulin, and Tomas Mikolov. Enriching word vectors with subword information. *Transactions of the Association for Computational Linguistics*, 5, 2016.

Kurt Bollacker, Colin Evans, Praveen Paritosh, Tim Sturge, and Jamie Taylor. Freebase: A collaboratively created graph database for structuring human knowledge. In *SIGMOD*, pages 1247–1250, 2008.

Rishi Bommasani, Drew A Hudson, Ehsan Adeli, Russ Altman, Simran Arora, Sydney von Arx, Michael S Bernstein, Jeannette Bohg, Antoine Bosselut, Emma Brunskill, et al. On the Opportunities & Risks of Foundation Models. *arXiv preprint arXiv:2108.07258*, 2021.

Francis Bond, Timothy Baldwin, Richard Fothergill, and Kiyotaka Uchimoto. Japanese semcor: A sense-tagged corpus of japanese. In *6th Global WordNet Conference*, pages 56–63, 2012.

Sofia Bonicalzi, Mario De Caro, and Benedetta Giovanola. Artificial intelligence and autonomy: On the ethical dimension of recommender systems. *Topoi*, 2023.

Kalina Bontcheva, Hamish Cunningham, Ian Roberts, Angus Roberts, Valentin Tablan, Niraj Aswani, and Genevieve Gorrell. GATE Teamware: a web-based, collaborative text annotation framework. *Language Resources and Evaluation*, 47 (4):1007–1029, 2013a.

Kalina Bontcheva, Leon Derczynski, Adam Funk, Mark A Greenwood, Diana Maynard, and Niraj Aswani. Twitie: An open-source information extraction pipeline for microblog text. In *RANLP*, pages 83–90, 2013b.

Marco Bonzanini, Miguel Martinez-Alvarez, and Thomas Roelleke. Opinion summarisation through sentence extraction: An investigation with movie reviews. In *35th International ACM SIGIR Conference on Research and Development in Information Retrieval*, pages 1121–1122, 2012.

Ari Bornstein, Arie Cattan, and Ido Dagan. CoRefi: A crowd sourcing suite for coreference annotation. In *EMNLP*, pages 205–215, 2020.

Lera Boroditsky. How language shapes thought. *Scientific American*, 304(2):62–65, 2011.

Oriol Borrega, Mariona Taulé, and M Antø'nia Martı. What do we mean when we speak about named entities. In *Corpus Linguistics*, pages 1–27, 2007.

Denny Borsboom, Gideon J Mellenbergh, and Jaap Van Heerden. The theoretical status of latent variables. *Psychological review*, 110(2):203, 2003.

Johan Bos. Implementing the binding and accommodation theory for anaphora resolution and presupposition projection. *Computational Linguistics*, 29(2):179–210, 2003.

Antoine Bosselut, Hannah Rashkin, Maarten Sap, Chaitanya Malaviya, Asli Celikyilmaz, and Yejin Choi. COMET: commonsense transformers for automatic knowledge graph construction. In *ACL*, pages 4762–4779, 2019.

Mondher Bouazizi and Tomoaki Ohtsuki. Opinion mining in Twitter how to make use of sarcasm to enhance sentiment analysis. In *2015 IEEE/ACM International Conference on Advances in Social Networks Analysis and Mining 2015*, pages 1594–1597, 2015.

Florian Boudin, Stéphane Huet, and Juan-Manuel Torres-Moreno. A graph-based approach to cross-language multi-document summarization. *Polibits*, pages 113–118, 2011.

Gosse Bouma. Finite state methods for hyphenation. *Natural Language Engineering*, 9:5 – 20, 03 2003. doi: 10.1017/S1351324903003073.

Zied Bouraoui, Sébastien Konieczny, Thanh Ma, Nicolas Schwind, and Ivan Varzinczak. Region-based merging of open-domain terminological knowledge. In *International Conference on Principles of Knowledge Representation and Reasoning, KR*, pages 81–90, 2022.

Gordon H Bower. Mood and memory. *American Psychologist*, 36(2):129, 1981.

Sven Branahl. *Das EDGAR (Electronic Data Gathering, Analysis and Retrieval) System der SEC und seine Bedeutung für die Bereitstellung von Rechnungslegungsinformationen.* diplom. de, 1998.

Thorsten Brants. TnT-a statistical part-of-speech tagger. *arXiv preprint cs/0003055*, 2000.

Susan E Brennan. Centering attention in discourse. *Language and Cognitive processes*, 10(2):137–167, 1995.

Susan E. Brennan, Marilyn W. Friedman, and Carl J. Pollard. A centering approach to pronouns. In *ACL*, pages 155–162, 1987.

Eric Brill. Transformation-based error-driven learning and natural language processing: A case study in part-of-speech tagging. *Computational Linguistics*, 21 (4):543–565, 1995.

Eric Brill and Jun Wu. Classifier combination for improved lexical disambiguation. In *ACL*, pages 191–195, 1998.

Sergey Brin and Lawrence Page. The anatomy of a large-scale hypertextual web search engine. *Computer networks and ISDN systems*, 30(1-7):107–117, 1998.

Samuel Brody and Nicholas Diakopoulos. Cooooooooooooooollllllllllll-lll!!!!!!!!!!!!!! using word lengthening to detect sentiment in microblogs. In *EMNLP*, pages 562–570, 2011.

Tom Brown, Benjamin Mann, Nick Ryder, Melanie Subbiah, Jared D Kaplan, Prafulla Dhariwal, Arvind Neelakantan, Pranav Shyam, Girish Sastry, Amanda Askell, et al. Language models are few-shot learners. *NeurIPS*, 33:1877–1901, 2020.

Rebecca Bruce and Janyce Wiebe. Decomposable modeling in natural language processing. *Computational Linguistics*, 25(2):195–207, 1999.

Elia Bruni, Gemma Boleda, Marco Baroni, and Nam-Khanh Tran. Distributional semantics in technicolor. In *ACL*, pages 136–145, 2012.

Fan Bu, Xiaoyan Zhu, and Ming Li. Measuring the non-compositionality of multi-word expressions. In *COLING*, pages 116–124, 2010.

Sébastien Bubeck, Varun Chandrasekaran, Ronen Eldan, Johannes Gehrke, Eric Horvitz, Ece Kamar, Peter Lee, Yin Tat Lee, Yuanzhi Li, Scott Lundberg, Harsha Nori, Hamid Palangi, Marco Tulio Ribeiro, and Yi Zhang. Sparks of artificial general intelligence: Early experiments with GPT-4. *arXiv preprint arXiv:2303.12712*, 2023.

Daniel Büring. *Binding theory*. Cambridge University Press, 2005.

Christopher T Burris, Eddie Harmon-Jones, and W Ryan Tarpley. "by faith alone": Religious agitation and cognitive dissonance. *Basic and Applied Social Psychology*, 19(1):17–31, 1997.

Donna K Byron, Whitney Gegg-Harrison, and Sun-Hee Lee. Resolving zero anaphors and pronouns in korean. *Traitement Automatique des Langues*, 46(1): 91–114, 2006.

Hongjie Cai, Yaofeng Tu, Xiangsheng Zhou, Jianfei Yu, and Rui Xia. Aspect-category based sentiment analysis with hierarchical graph convolutional network. In *COLING*, pages 833–843, 2020.

Hongjie Cai, Rui Xia, and Jianfei Yu. Aspect-category-opinion-sentiment quadruple extraction with implicit aspects and opinions. In *IJCNLP*, pages 340–350, 2021.

Yitao Cai, Huiyu Cai, and Xiaojun Wan. Multi-modal sarcasm detection in Twitter with hierarchical fusion model. In *ACL*, pages 2506–2515, 2019.

Josep Call and Michael Tomasello. Does the chimpanzee have a theory of mind? 30 years later. *Trends in Cognitive Sciences*, 12(5):187–192, 2008.

Francesco Camastra and Gennaro Razi. Italian text categorization with lemmatization and support vector machines. In *Neural approaches to dynamics of signal exchanges*, pages 47–54. Springer, 2020.

Erik Cambria and Bebo White. Jumping NLP curves: A review of natural language processing research. *IEEE Computational Intelligence Magazine*, 9(2):48–57, 2014.

Erik Cambria, Amir Hussain, Catherine Havasi, and Chris Eckl. Common sense computing: From the society of mind to digital intuition and beyond. In Julian Fierrez, Javier Ortega, Anna Esposito, Andrzej Drygajlo, and Marcos Faundez-Zanuy, editors, *Biometric ID Management and Multimodal Communication*, volume 5707 of *Lecture Notes in Computer Science*, pages 252–259. Springer, Berlin Heidelberg, 2009.

Erik Cambria, Daniel Olsher, and Kenneth Kwok. Sentic activation: A two-level affective common sense reasoning framework. In *AAAI*, pages 186–192, Toronto, 2012.

Erik Cambria, Soujanya Poria, Alexander Gelbukh, and Mike Thelwall. Sentiment analysis is a big suitcase. *IEEE Intelligent Systems*, 32(6):74–80, 2017.

Erik Cambria, Rui Mao, Sooji Han, and Qian Liu. Sentic parser: A graph-based approach to concept extraction for sentiment analysis. In *ICDM Workshops*, pages 413–420, 2022.

Erik Cambria, Rui Mao, Melvin Chen, Zhaoxia Wang, and Seng-Beng Ho. Seven pillars for the future of artificial intelligence. *IEEE Intelligent Systems*, 38(6): 62–69, 2023.

Erik Cambria, Xulang Zhang, Rui Mao, Melvin Chen, and Kenneth Kwok. SenticNet 8: Fusing emotion AI and commonsense AI for interpretable, trustworthy, and explainable affective computing. In *International Conference on Human-Computer Interaction (HCII)*, 2024.

Niccolò Campolungo, Federico Martelli, Francesco Saina, and Roberto Navigli. DiBiMT: A novel benchmark for measuring word sense disambiguation biases in machine translation. In *ACL*, pages 4331–4352, 2022.

Olivier Cappé, Simon J Godsill, and Eric Moulines. An overview of existing methods and recent advances in sequential monte carlo. *IEEE*, 95(5):899–924, 2007.

Claire Cardie and Kiri Wagstaff. Noun phrase coreference as clustering. In *1999 Joint SIGDAT Conference on Empirical Methods in Natural Language Processing and Very Large Corpora*, pages 82–89, 1999.

David Carter. *Interpreting Anaphors in Natural Language Texts*. Halsted Press, 1987.

Avshalom Caspi, Brent W Roberts, and Rebecca L Shiner. Personality development: Stability and change. *Annu. Rev. Psychol.*, 56:453–484, 2005.

Giovanni Castiglia, Ayoub El Majjodi, Alain Dominique Starke, Fedelucio Narducci, Yashar Deldjoo, and Federica Calò. Nudging towards health in a conversational food recommender system using multi-modal interactions and nutrition labels. In *Fourth Knowledge-aware and Conversational Recommender Systems Workshop (KaRS)*, volume 3294, pages 29–35, 2022.

Santiago Castro, Devamanyu Hazarika, Verónica Pérez-Rosas, Roger Zimmermann, Rada Mihalcea, and Soujanya Poria. Towards multimodal sarcasm detection (an obviously perfect paper). In *ACL*, pages 4619–4629, Florence, Italy, July 2019.

Rosario Catelli, Hamido Fujita, Giuseppe De Pietro, and Massimo Esposito. Deceptive reviews and sentiment polarity: Effective link by exploiting BERT. *Expert Systems with Applications*, 209:118290, 2022.

Nicolas Cebron and Michael R Berthold. Active learning for object classification: from exploration to exploitation. *Data Mining and Knowledge Discovery*, 18(2): 283–299, 2009.

Giuseppe GA Celano. A gradient boosting-Seq2Seq system for Latin POS tagging and lemmatization. In *LT4HALA 2020-1st Workshop on Language Technologies for Historical and Ancient Languages*, pages 119–123, 2020.

Fabio Celli and Bruno Lepri. Is big five better than MBTI? A personality computing challenge using twitter data. *Computational Linguistics CLiC-it*, 2018:93, 2018.

Daniel Cervone. Personality architecture: Within-person structures and processes. *Annual review of psychology*, 56:423, 2005.

Wallace Chafe. Givenness, contrastiveness, definiteness, subjects, topics, and point of view. *Subject and topic*, 1976.

Haixia Chai and Michael Strube. Incorporating centering theory into neural coreference resolution. In *NAACL-HLT*, pages 2996–3002, 2022.

Abhisek Chakrabarty, Onkar Arun Pandit, and Utpal Garain. Context sensitive lemmatization using two successive bidirectional gated recurrent networks. In *ACL*, pages 1481–1491, 2017.

Abhisek Chakrabarty, Akshay Chaturvedi, and Utpal Garain. CNN-based context sensitive lemmatization. In *ACM India Joint International Conference on Data Science and Management of Data*, pages 334–337, 2019.

Navoneel Chakrabarty, Siddhartha Chowdhury, Sangita D Kanni, and Swarnakeshar Mukherjee. FAFinder: Friend suggestion system for social networking. In *Intelligent Data Communication Technologies and Internet of Things: ICICI 2019*, pages 51–58, 2020.

Craig G. Chambers and Ron Smyth. Structural parallelism and discourse coherence: A test of centering theory. *Journal of Memory and Language*, 39(4):593–608, 1998.

Nathanael Chambers, Daniel Cer, Trond Grenager, David Hall, Chloe Kiddon, Bill MacCartney, Marie-Catherine de Marneffe, Daniel Ramage, Eric Yeh, and Christopher D. Manning. Learning alignments and leveraging natural logic. In *ACL-PASCAL Workshop on Textual Entailment and Paraphrasing*, pages 165–170, 2007.

Jonathan Charteris-Black. Metaphor and political communication. In *Metaphor and Discourse*, pages 97–115. Springer, 2009.

Iti Chaturvedi, Erik Cambria, Soujanya Poria, and Rajiv Bajpai. Bayesian deep convolution belief networks for subjectivity detection. In *ICDM workshops*, pages 916–923, 2016a.

Iti Chaturvedi, Erik Cambria, and David Vilares. Lyapunov filtering of objectivity for spanish sentiment model. In *IJCNN*, pages 4474–4481, 2016b.

Iti Chaturvedi, Edoardo Ragusa, Paolo Gastaldo, Rodolfo Zunino, and Erik Cambria. Bayesian network based extreme learning machine for subjectivity detection. *Journal of The Franklin Institute*, 355(4):1780–1797, 2018.

Iti Chaturvedi, Tim Noel, and Ranjan Satapathy. Speech emotion recognition using audio matching. *Electronics*, 11(23):3943, 2022.

Iti Chaturvedi, Ranjan Satapathy, Curtis Lynch, and Erik Cambria. Predicting word vectors for microtext. *Expert Systems*, page e13589, 2024.

Dushyant Singh Chauhan, Gopendra Vikram Singh, Aseem Arora, Asif Ekbal, and Pushpak Bhattacharyya. An emoji-aware multitask framework for multimodal sarcasm detection. *Knowledge-Based Systems*, 257:109924, 2022.

Chen Chen and Vincent Ng. Chinese zero pronoun resolution: Some recent advances. In *EMNLP*, pages 1360–1365, 2013.

Chen Chen and Vincent Ng. Chinese zero pronoun resolution with deep neural networks. In *ACL*, pages 778–788, 2016.

Danqi Chen and Christopher D Manning. A fast and accurate dependency parser using neural networks. In *EMNLP*, pages 740–750, 2014.

Guanyi Chen, Kees van Deemter, and Chenghua Lin. Modelling pro-drop with the rational speech acts model. In *11th International Conference on Natural Language Generation*, pages 57–66, 2018a.

Hong Chen, Zhenhua Fan, Hao Lu, Alan Yuille, and Shu Rong. PreCo: A large-scale dataset in preschool vocabulary for coreference resolution. In *EMNLP*, pages 172–181, 2018b.

Jun Chen, Xiaoming Zhang, Yu Wu, Zhao Yan, and Zhoujun Li. Keyphrase generation with correlation constraints. In Ellen Riloff, David Chiang, Julia Hockenmaier, and Jun'ichi Tsujii, editors, *EMNLP*, pages 4057–4066, 2018c.

Li Chen, Luole Qi, and Feng Wang. Comparison of feature-level learning methods for mining online consumer reviews. *Expert Systems with Applications*, 39(10): 9588–9601, 2012.

Lingjiao Chen, Matei Zaharia, and James Zou. How is ChatGPT's behavior changing over time? *arXiv preprint arXiv:2307.09009*, 2023a.

Melvin Chen. Trust & trust-engineering in artificial intelligence research: Theory & praxis. *Philosophy & Technology*, 34(4):1429–1447, 2021.

Nuo Chen, Yan Wang, Haiyun Jiang, Deng Cai, Yuhan Li, Ziyang Chen, Longyue Wang, and Jia Li. Large language models meet Harry Potter: A dataset for aligning dialogue agents with characters. In *EMNLP Findings*, pages 8506–8520, 2023b.

Shaowei Chen, Jie Liu, Yu Wang, Wenzheng Zhang, and Ziming Chi. Synchronous double-channel recurrent network for aspect-opinion pair extraction. In *ACL*, pages 6515–6524, Online, July 2020.

Shaowei Chen, Yu Wang, Jie Liu, and Yuelin Wang. Bidirectional machine reading comprehension for aspect sentiment triplet extraction. In *AAAI*, pages 12666–12674, 2021a.

Shisong Chen, Binbin Gu, Jianfeng Qu, Zhixu Li, An Liu, Lei Zhao, and Zhigang Chen. Tackling zero pronoun resolution and non-zero coreference resolution jointly. In *25th Conference on Computational Natural Language Learning*, pages 518–527, 2021b.

Tao Chen and Min-Yen Kan. Creating a live, public short message service corpus: the nus sms corpus. *Language Resources and Evaluation*, 2012.

Tianqi Chen and Carlos Guestrin. Xgboost: A scalable tree boosting system. In *22nd ACM SIGKDD International Conference on Knowledge Discovery and Data Mining*, pages 785–794, 2016.

Wang Chen, Yifan Gao, Jiani Zhang, Irwin King, and Michael R. Lyu. Title-guided encoding for keyphrase generation. In *AAAI*, pages 6268–6275, 2019.

Wangqun Chen, Fuqiang Lin, Xuan Zhang, Guowei Li, and Bo Liu. Jointly learning sentimental clues and context incongruity for sarcasm detection. *IEEE Access*, 10:48292–48300, 2022.

Zhuang Chen and Tieyun Qian. Relation-aware collaborative learning for unified aspect-based sentiment analysis. In *ACL*, pages 3685–3694, 2020.

Yan Cheng, Leibo Yao, Guoxiong Xiang, Guanghe Zhang, Tianwei Tang, and Linhui Zhong. Text sentiment orientation analysis based on multi-channel CNN and bidirectional GRU with attention mechanism. *IEEE Access*, 8:134964–134975, 2020.

Paula Chesley, Bruce Vincent, Li Xu, and Rohini K Srihari. Using verbs and adjectives to automatically classify blog sentiment. *Training*, 580(263):233–241, 2006.

Zheng Lin Chia, Michal Ptaszynski, Fumito Masui, Gniewosz Leliwa, and Michal Wroczynski. Machine learning and feature engineering-based study into sarcasm and irony classification with application to cyberbullying detection. *Information Processing & Management*, 58(4):102600, 2021.

Nancy Chinchor, Lynette Hirschman, and David D Lewis. Evaluating message understanding systems: an analysis of the third message understanding conference (MUC-3). *Computational Linguistics*, 19(3):409–450, 1993.

Nancy A. Chinchor. Overview of MUC-7. In *Seventh Message Understanding Conference (MUC-7)*, pages 1–4, 1998.

Nancy A Chinchor and Beth Sundheim. Message understanding conference (muc) tests of discourse processing. In *Proc. AAAI Spring Symposium on Empirical Methods in Discourse Interpretation and Generation*, pages 21–26, 1995.

Jason PC Chiu and Eric Nichols. Named entity recognition with bidirectional LSTM-CNNs. *Transactions of the Association for Computational Linguistics*, 4:357–370, 2016.

Timothy Chklovski. Learner: a system for acquiring commonsense knowledge by analogy. In John H. Gennari, Bruce W. Porter, and Yolanda Gil, editors, *2nd International Conference on Knowledge Capture (K-CAP 2003)*, pages 4–12, 2003.

Timothy Chklovski and Patrick Pantel. Verbocean: Mining the web for fine-grained semantic verb relations. In *EMNLP*, pages 33–40, 2004.

Kyunghyun Cho, Bart van Merrienboer, Dzmitry Bahdanau, and Yoshua Bengio. On the properties of neural machine translation: Encoder-decoder approaches. In Dekai Wu, Marine Carpuat, Xavier Carreras, and Eva Maria Vecchi, editors, *SSST@EMNLP 2014, Eighth Workshop on Syntax, Semantics and Structure in Statistical Translation*, pages 103–111, 2014.

Noam Chomsky. *Syntactic Structures*. De Gruyter Mouton, Berlin, Boston, 1957. ISBN 9783112316009.

Monojit Choudhury, Rahul Saraf, Vijit Jain, Animesh Mukherjee, Sudeshna Sarkar, and Anupam Basu. Investigation and modeling of the structure of texting language. *International Journal of Document Analysis and Recognition (IJDAR)*, 10(3): 157–174, 2007.

Evripides Christodoulou, Andreas Gregoriades, Maria Pampaka, and Herodotos Herodotou. Personality-informed restaurant recommendation. In *Information Systems and Technologies: WorldCIST 2022, Volume 1*, pages 13–21. Springer, 2022.

Grzegorz Chrupała. Simple data-driven context-sensitive lemmatization. *Proces. del Leng. Natural*, 37, 2006.

Grzegorz Chrupała. Normalizing tweets with edit scripts and recurrent neural embeddings. In *ACL*, pages 680–686, 2014.

Grzegorz Chrupała, Georgiana Dinu, and Josef Genabith. Learning morphology with Morfette. *Chrupała, Grzegorz and Dinu, Georgiana and van Genabith, Josef (2008) Learning morphology with Morfette. In: LREC 2008 - Sixth International Conference on Language Resources and Evaluation, 28-30 May 2008, Marrakech, Morocco.*, 01 2008.

Sandra Chung, William A Ladusaw, and James McCloskey. Sluicing and logical form. *Natural Language Semantics*, 3:239–282, 1995.

Kenneth Church and Patrick Hanks. Word association norms, mutual information, and lexicography. *Computational linguistics*, 16(1):22–29, 1990.

Kenneth Ward Church. A stochastic parts program and noun phrase parser for unrestricted text. In *International Conference on Acoustics, Speech, and Signal Processing,*, pages 695–698, 1989.

Montserrat Civit and Ma Antònia Martí. Building cast3lb: A Spanish treebank. *Research on Language and Computation*, 2(4):549–574, 2004.

Alexander Clark. Combining distributional and morphological information for part of speech induction. In *EACL*, pages 59–66, 2003.

Herbert H Clark. Bridging. In *Theoretical Issues in Natural Language Processing*, pages 169–174, 1975.

Kevin Clark and Christopher D. Manning. Entity-centric coreference resolution with model stacking. In *ACL-IJCNLP*, pages 1405–1415, 2015.

Kevin Clark and Christopher D. Manning. Deep reinforcement learning for mention-ranking coreference models. In *EMNLP*, pages 2256–2262, 2016a.

Kevin Clark and Christopher D. Manning. Improving coreference resolution by learning entity-level distributed representations. In *ACL*, pages 643–653, 2016b.

Kevin Clark, Minh-Thang Luong, Quoc V. Le, and Christopher D. Manning. ELECTRA: pre-training text encoders as discriminators rather than generators. In *8th International Conference on Learning Representations*, pages 1–18, 2020a.

Peter Clark, Oyvind Tafjord, and Kyle Richardson. Transformers as soft reasoners over language. In *IJCAI*, pages 3882–3890, 2020b.

Stephen Clark, James R Curran, and Miles Osborne. Bootstrapping POS-taggers using unlabelled data. In *HLT-NAACL*, pages 49–55, 2003.

Jacob Cohen. A coefficient of agreement for nominal scales. *Educational and Psychological Measurement*, 20(1):37–46, 1960.

K Bretonnel Cohen, Arrick Lanfranchi, Miji Joo-young Choi, Michael Bada, William A Baumgartner, Natalya Panteleyeva, Karin Verspoor, Martha Palmer, and Lawrence E Hunter. Coreference annotation and resolution in the colorado richly annotated full text (craft) corpus of biomedical journal articles. *BMC Bioinformatics*, 18(1):1–14, 2017.

Yuval Cohen, Hana Ornoy, and Baruch Keren. MBTI personality types of project managers and their success: A field survey. *Project Management Journal*, 44(3): 78–87, 2013.

Raphael Cohen-Almagor. Fighting hate and bigotry on the internet. *Policy & Internet*, 3(3):1–26, 2011.

David A Cohn, Zoubin Ghahramani, and Michael I Jordan. Active learning with statistical models. *Journal of artificial intelligence research*, 4:129–145, 1996.

Francesco Colasanto, Luca Grilli, Domenico Santoro, and Giovanni Villani. Al-BERTino for stock price prediction: a Gibbs sampling approach. *Information Sciences*, 597:341–357, 2022.

Michael Collins. Discriminative training methods for hidden markov models: Theory and experiments with perceptron algorithms. In *EMNLP*, pages 1–8, 2002.

Ronan Collobert and Jason Weston. A unified architecture for natural language processing: Deep neural networks with multitask learning. In *ICML*, pages 160–167, 2008.

Ronan Collobert, Jason Weston, Léon Bottou, Michael Karlen, Koray Kavukcuoglu, and Pavel Kuksa. Natural language processing (almost) from scratch. *Journal of Machine Learning Research*, 12:2493–2537, 2011.

Herbert L Colston. On necessary conditions for verbal irony comprehension. *Pragmatics & Cognition*, 8(2):277–324, 2000.

Max Coltheart. The MRC psycholinguistic database. *The Quarterly Journal of Experimental Psychology Section A*, 33(4):497–505, 1981.

Paul Compton, Glenn Edwards, Byeong Kang, Leslie Lazarus, Ron Malor, Phil Preston, and Ashwin Srinivasan. Ripple down rules: Turning knowledge acquisition into knowledge maintenance. *Artificial Intelligence in Medicine*, 4(6):463–475, 1992.

Simone Conia and Roberto Navigli. Framing word sense disambiguation as a multi-label problem for model-agnostic knowledge integration. In *EACL*, pages 3269–3275, 2021.

BNC Consortium. British national corpus. *Oxford Text Archive Core Collection*, 2007.

Teresa Correa, Amber Willard Hinsley, and Homero Gil De Zuniga. Who interacts on the web?: The intersection of users' personality and social media use. *Computers in human behavior*, 26(2):247–253, 2010.

Keith Cortis, André Freitas, Tobias Daudert, Manuela Huerlimann, Manel Zarrouk, Siegfried Handschuh, and Brian Davis. SemEval-2017 task 5: Fine-grained sentiment analysis on financial microblogs and news. In *11th International Workshop on Semantic Evaluation (SemEval-2017)*, pages 519–535, 2017.

Paul T Costa Jr. Still stable after all these years: Personality as a key to some issues in adulthood and old age. *Life span development and behavior*, 3:65–102, 1980.

Paul T Costa Jr and Robert R McCrae. Domains and facets: Hierarchical personality assessment using the revised NEO personality inventory. *Journal of Personality Assessment*, 64(1):21–50, 1995.

Juan M Cotelo, Fermín Cruz, F Javier Ortega, and José A Troyano. Explorando Twitter mediante la integración de información estructurada y no estructurada. *Procesamiento del Lenguaje Natural*, pages 75–82, 2015.

Isaac Councill, Ryan McDonald, and Leonid Velikovich. What's great and what's not: learning to classify the scope of negation for improved sentiment analysis. In *workshop on negation and speculation in natural language processing*, pages 51–59, 2010.

Alan S Cowen, Dacher Keltner, Florian Schroff, Brendan Jou, Hartwig Adam, and Gautam Prasad. Sixteen facial expressions occur in similar contexts worldwide. *Nature*, 589(7841):251–257, 2021.

Michael Crawshaw. Multi-task learning with deep neural networks: A survey. *arXiv preprint arXiv:2009.09796*, pages 1–43, 2020.

William Croft and D Alan Cruse. *Cognitive linguistics*. Cambridge University Press, 2004.

Leyang Cui, Yu Wu, Jian Liu, Sen Yang, and Yue Zhang. Template-based named entity recognition using bart. In *ACL-IJCNLP*, pages 1835–1845, 2021.

James R Curran and Stephen Clark. Investigating GIS and smoothing for maximum entropy taggers. In *EACL*, 2003.

Agata Cybulska and Piek Vossen. Using a sledgehammer to crack a nut? lexical diversity and event coreference resolution. In *Ninth International Conference on Language Resources and Evaluation*, pages 4545–4552, 2014.

W Daelemans, J Zavrel, K van der Sloot, and A van den Bosch. TiMBL: Tilburg memory based learner, version 5.0, reference guide. *Research Group Technical Report Series*, 3, 2003.

Walter Daelemans, Antal Van den Bosch, and Ton Weijters. Igtree: using trees for compression and classification in lazy learning algorithms. In *Lazy learning*, pages 407–423. Springer, 1997.

Walter Daelemans, Sabine Buchholz, and Jorn Veenstra. Memory-based shallow parsing. *arXiv preprint cs/9906005*, 1999a.

Walter Daelemans, Jakub Zavrel, Peter Berck, and Steven Gillis. MBT: A memory-based part of speech tagger-generator. *Fourth Workshop on Very Large Corpora: 1996; Copenhagen, Denmark*, 11 1999b.

Walter Daelemans, Hendrik Johannes Groenewald, and Gerhard B van Huyssteen. Prototype-based active learning for lemmatization. In *International Conference RANLP-2009*, pages 65–70, 2009.

Debishree Dagar, Abir Hudait, Hrudaya Kumar Tripathy, and MN Das. Automatic emotion detection model from facial expression. In *ICACCCT*, pages 77–85, 2016.

Hong-Jie Dai, Po-Ting Lai, Yung-Chun Chang, and Richard Tzong-Han Tsai. Enhancing of chemical compound and drug name recognition using representative tag scheme and fine-grained tokenization. *Journal of cheminformatics*, 7(1):1–10, 2015.

Zehui Dai, Cheng Peng, Huajie Chen, and Yadong Ding. A multi-task incremental learning framework with category name embedding for aspect-category sentiment analysis. In *EMNLP*, pages 6955–6965, 2020.

Bouras Dalila, Amroune Mohamed, and Hakim Bendjanna. A review of recent aspect extraction techniques for opinion mining systems. In *2018 2nd International Conference on Natural Language and Speech Processing (ICNLSP)*, pages 1–6, 2018.

Hayk Danielyan. Sarcasm in social and commercial advertising: A pragmalinguistic perspective. *Armenian Folia Anglistika*, 18(2 (26)):72–84, 2022.

Kareem Darwish, Hamdy Mubarak, Ahmed Abdelali, Mohamed Eldesouki, Younes Samih, Randah Alharbi, Mohammed Attia, Walid Magdy, and Laura Kallmeyer. Multi-dialect Arabic POS tagging: A CRF approach. In *Eleventh International Conference on Language Resources and Evaluation (LREC 2018)*, 2018.

Amitava Das and Sivaji Bandyopadhyay. Theme detection an exploration of opinion subjectivity. In *2009 3rd International Conference on Affective Computing and Intelligent Interaction and Workshops*, pages 1–6, 2009.

Amitava Das and Sivaji Bandyopadhyay. Subjectivity detection using genetic algorithm. *Computational Approaches to Subjectivity and Sentiment Analysis*, pages 14–21, 2010.

Nilanjana Das and Santwana Sagnika. A subjectivity detection-based approach to sentiment analysis. In *Machine Learning and Information Processing*, pages 149–160. Springer, 2020.

Sarkar Snigdha Sarathi Das, Arzoo Katiyar, Rebecca Passonneau, and Rui Zhang. CONTaiNER: Few-shot named entity recognition via contrastive learning. In *ACL*, pages 6338–6353, 2022.

Yann N Dauphin, Angela Fan, Michael Auli, and David Grangier. Language modeling with gated convolutional networks. In *ICML*, pages 933–941, 2017.

Ernest Davis, Leora Morgenstern, and Charles L Ortiz. The first Winograd schema challenge at IJCAI-16. *AI Magazine*, 38(3):97–98, 2017.

Alvise De Biasio, Nicolò Navarin, and Dietmar Jannach. Economic recommender systems – a systematic review. *Electronic Commerce Research and Applications*, 63:101352, 2024.

Matthias De Lange, Rahaf Aljundi, Marc Masana, Sarah Parisot, Xu Jia, Aleš Leonardis, Gregory Slabaugh, and Tinne Tuytelaars. A continual learning survey: Defying forgetting in classification tasks. *IEEE Transactions on Pattern Analysis and Machine Intelligence*, 44(7):3366–3385, 2021.

Kees van Deemter and Rodger Kibble. On coreferring: Coreference in MUC and related annotation schemes. *Computational Linguistics*, 26(4):629–637, 2000.

Dina Demner-Fushman, Kin Wah Fung, Phong Do, Richard D Boyce, and Travis R Goodwin. Overview of the TAC 2018 Drug-Drug Interaction Extraction from Drug Labels Track. *TAC*, November:1–12, 2019.

Yifan Deng, Xinran Xu, Yang Qiu, Jingbo Xia, Wen Zhang, and Shichao Liu. A multimodal deep learning framework for predicting drug–drug interaction events. *Bioinformatics*, 36(15):4316–4322, 2020.

Pascal Denis and Jason Baldridge. Specialized models and ranking for coreference resolution. In *EMNLP*, pages 660–669, 2008.

Pascal Denis and Jason Baldridge. Global joint models for coreference resolution and named entity classification. *Procesamiento del lenguaje natural*, 42, 2009.

Ali Derakhshan and Hamid Beigy. Sentiment analysis on stock social media for stock price movement prediction. *Engineering Applications of Artificial Intelligence*, 85:569–578, 2019.

Leon Derczynski, Eric Nichols, Marieke van Erp, and Nut Limsopatham. Results of the wnut2017 shared task on novel and emerging entity recognition. In *3rd Workshop on Noisy User-generated Text*, pages 140–147, 2017.

Oksana Dereza. Lemmatization for ancient languages: Rules or neural networks? In Dmitry Ustalov, Andrey Filchenkov, Lidia Pivovarova, and Jan Žižka, editors, *Artificial Intelligence and Natural Language*, pages 35–47, Cham, 2018. Springer International Publishing. ISBN 978-3-030-01204-5.

Neelmay Desai and Meera Narvekar. Normalization of noisy text data. *Procedia Computer Science*, 45:127–132, 2015.

Tim Dettmers, Pasquale Minervini, Pontus Stenetorp, and Sebastian Riedel. Convolutional 2d knowledge graph embeddings. In *AAAI*, pages 1811–1818, 2018.

Ann Devitt and Khurshid Ahmad. Sentiment polarity identification in financial news: A cohesion-based approach. In *45th Annual Meeting of the Association of Computational Linguistics*, pages 984–991, 2007.

Jacob Devlin, Ming-Wei Chang, Kenton Lee, and Kristina Toutanova. BERT: Pretraining of deep bidirectional Transformers for language understanding. *NAACL-HLT*, 2018.

Sahraoui Dhelim, Huansheng Ning, Nyothiri Aung, Runhe Huang, and Jianhua Ma. Personality-aware product recommendation system based on user interests mining and metapath discovery. *IEEE Transactions on Computational Social Systems*, 8 (1):86–98, 2020.

Sahraoui Dhelim, Nyothiri Aung, Mohammed Amine Bouras, Huansheng Ning, and Erik Cambria. A survey on personality-aware recommendation systems. *Artificial Intelligence Review*, 55:2409–2454, 2022.

Collins English Dictionary. Collins. *London & Glasgow*, 1982.

Oxford Dictionary. Oxford dictionary of english. *Oxford Dictionary of English, 3rd edn. Oxford University Press. China Translation & Printing Services Ltd, China*, 2010.

Thomas G Dietterich and Ghulum Bakiri. Solving multiclass learning problems via error-correcting output codes. *Journal of artificial intelligence research*, 2: 263–286, 1994.

Eleftherios Dimitrakis, Konstantinos Sgontzos, and Yannis Tzitzikas. A survey on question answering systems over linked data and documents. *Journal of Intelligent Information Systems*, 55(2):233–259, 2020.

Xiaowen Ding and Bing Liu. Resolving object and attribute coreference in opinion mining. In *COLING*, pages 268–276, 2010.

Stefanie Dipper, Anke Lüdeling, and Marc Reznicek. Nosta-d: A corpus of german non-standard varieties. *Non-standard data sources in corpus-based research*, 5: 69–76, 2013.

George R Doddington, Alexis Mitchell, Mark Przybocki, Lance Ramshaw, Stephanie Strassel, and Ralph Weischedel. The automatic content extraction (ACE) program–tasks, data, and evaluation. In *Fourth International Conference on Language Resources and Evaluation*, pages 1–4, 2004.

Ellen K Dodge, Jisup Hong, and Elise Stickles. MetaNet: Deep semantic automatic metaphor analysis. In *Third Workshop on Metaphor in NLP*, pages 40–49, 2015.

Evgenia Dolgova, Woody van Olffen, Frans van den Bosch, and Henk Volberda. The interaction between personality, social network position and involvement in innovation process. Technical report, Erasmus Research Institute of Management, 2010.

Xin Dong, Evgeniy Gabrilovich, Geremy Heitz, Wilko Horn, Ni Lao, Kevin Murphy, Thomas Strohmann, Shaohua Sun, and Wei Zhang. Knowledge vault: A web-scale approach to probabilistic knowledge fusion. In *20th ACM SIGKDD international conference on Knowledge discovery and data mining*, pages 601–610, 2014.

Xin Luna Dong and Gerard De Melo. A helping hand: Transfer learning for deep sentiment analysis. In *ACL*, pages 2524–2534, 2018.

Kevin Donnelly et al. SNOMED-CT: The advanced terminology and coding system for ehealth. *Studies in Health Technology and Informatics*, 121:279, 2006.

Cicero Dos Santos and Bianca Zadrozny. Learning character-level representations for part-of-speech tagging. In *ICML*, pages 1818–1826, 2014.

Alexey Dosovitskiy, Lucas Beyer, Alexander Kolesnikov, Dirk Weissenborn, Xiaohua Zhai, Thomas Unterthiner, Mostafa Dehghani, Matthias Minderer, Georg Heigold, Sylvain Gelly, et al. An image is worth 16x16 words: Transformers for image recognition at scale. In *International Conference on Learning Representations*, pages 1–21, 2021.

Markus Dreyer, Jason Smith, and Jason Eisner. Latent-variable modeling of string transductions with finite-state methods. In *EMNLP*, pages 1080–1089, 2008.

Kelvin Du, Frank Xing, and Erik Cambria. Incorporating multiple knowledge sources for targeted aspect-based financial sentiment analysis. *ACM Transactions on Management Information Systems*, 14(3):23, 2023a.

Kelvin Du, Frank Xing, Rui Mao, and Erik Cambria. FinSenticNet: A concept-level lexicon for financial sentiment analysis. In *2023 IEEE Symposium Series on Computational Intelligence (SSCI)*, pages 109–114, Mexico City, Mexico, 2023b.

Kelvin Du, Rui Mao, Frank Xing, and Erik Cambria. A dynamic dual-graph neural network for stock price movement prediction. In *IJCNN*, Yokohama, Japan, 2024a.

Kelvin Du, Frank Xing, Rui Mao, and Erik Cambria. An evaluation of reasoning capabilities of large language models in financial sentiment analysis. In *IEEE Conference on Artificial Intelligence*, pages 189–194, Singapore, 2024b.

Kelvin Du, Frank Xing, Rui Mao, and Erik Cambria. Financial sentiment analysis: Techniques and applications. *ACM Computing Surveys*, 56(9):1–42, 2024c.

Mark Dunlop and Andrew Crossan. Predictive text entry methods for mobile phones. *Personal Technologies*, 4:2–3, 2000.

Cuc Duong, Vethavikashini Chithrra Raghuram, Amos Lee, Rui Mao, Gianmarco Mengaldo, and Erik Cambria. Neurosymbolic AI for mining public opinions about wildfires. *Cognitive Computation*, 16(4):1531–1553, 2024.

Chris Dyer, Miguel Ballesteros, Wang Ling, Austin Matthews, and Noah A Smith. Transition-based dependency parsing with stack long short-term memory. *arXiv preprint arXiv:1505.08075*, 2015.

Philip Edmonds and Scott Cotton. SensEval-2: Overview. In *SENSEVAL-2 Second International Workshop on Evaluating Word Sense Disambiguation Systems*, pages 1–5, 2001.

Maud Ehrmann, Matteo Romanello, Sven Najem-Meyer, Antoine Doucet, and Simon Clematide. Extended overview of HIPE-2022: Named entity recognition and linking in multilingual historical documents. In Guglielmo Faggioli, Nicola Ferro, Allan Hanbury, and Martin Potthast, editors, *Working Notes of CLEF 2022 - Conference and Labs of the Evaluation Forum*, volume 3180, pages 1038–1063, 2022.

Ben Eisner, Tim Rocktäschel, Isabelle Augenstein, Matko Bošnjak, and Sebastian Riedel. emoji2vec: Learning emoji representations from their description. In *Fourth International Workshop on Natural Language Processing for Social Media*, pages 48–54, 2016.

Christopher Ifeanyi Eke, Azah Anir Norman, Liyana Shuib, and Henry Friday Nweke. Sarcasm identification in textual data: systematic review, research challenges and open directions. *Artificial Intelligence Review*, 53(6):4215–4258, 2020.

Paul Ekman and Wallace V Friesen. Facial action coding system. *Environmental Psychology & Nonverbal Behavior*, 1978.

Abdelkader El Mahdaouy, Abdellah El Mekki, Kabil Essefar, Nabil El Mamoun, Ismail Berrada, and Ahmed Khoumsi. Deep multi-task model for sarcasm detection and sentiment analysis in Arabic language. In *Sixth Arabic Natural Language Processing Workshop*, pages 334–339, 2021.

Jeffrey L Elman. Distributed representations, simple recurrent networks, and grammatical structure. *Machine learning*, 7(2):195–225, 1991.

Ali Emami, Adam Trischler, Kaheer Suleman, and Jackie Chi Kit Cheung. A generalized knowledge hunting framework for the Winograd schema challenge. In *NAACL*, pages 25–31, 2018.

Tomaz Erjavec. The multext-east Slovene lexicon. In *7th Electrotechnical Conference ERK, Volume B*, pages 189–192, 1998.

Tomaž Erjavec and Sasčo Džeroski. Machine learning of morphosyntactic structure: Lemmatizing unknown slovene words. *Applied artificial intelligence*, 18(1):17–41, 2004.

Md Shah Fahad, Ashish Ranjan, Jainath Yadav, and Akshay Deepak. A survey of speech emotion recognition in natural environment. *Digital signal processing*, 110:102951, 2021.

Chunxiao Fan, Jie Lin, Rui Mao, and Erik Cambria. Fusing pairwise modalities for emotion recognition in conversations. *Information Fusion*, 106:102306, 2024.

Rong-En Fan, Kai-Wei Chang, Cho-Jui Hsieh, Xiang-Rui Wang, and Chih-Jen Lin. Liblinear: A library for large linear classification. *the Journal of machine Learning research*, 9:1871–1874, 2008.

Irving E Fang. The "easy listening formula". *Journal of Broadcasting & Electronic Media*, 11(1):63–68, 1966.

Songtao Fang, Zhenya Huang, Ming He, Shiwei Tong, Xiaoqing Huang, Ye Liu, Jie Huang, and Qi Liu. Guided attention network for concept extraction. In Zhi-Hua Zhou, editor, *IJCAI*, pages 1449–1455, 2021a.

Tianqing Fang, Weiqi Wang, Sehyun Choi, Shibo Hao, Hongming Zhang, Yangqiu Song, and Bin He. Benchmarking commonsense knowledge base population with an effective evaluation dataset. In *EMNLP*, pages 8949–8964, 2021b.

Tianqing Fang, Quyet V. Do, Hongming Zhang, Yangqiu Song, Ginny Y. Wong, and Simon See. Pseudoreasoner: Leveraging pseudo labels for commonsense knowledge base population. In *EMNLP*, pages 3379–3394, 2022a.

Yanbo Fang and Yongfeng Zhang. Data-efficient concept extraction from pre-trained language models for commonsense explanation generation. In Yoav Goldberg, Zornitsa Kozareva, and Yue Zhang, editors, *EMNLP*, pages 5883–5893, 2022.

Yuwei Fang, Shuohang Wang, Yichong Xu, Ruochen Xu, Siqi Sun, Chenguang Zhu, and Michael Zeng. Leveraging knowledge in multilingual commonsense reasoning. In *ACL*, pages 3237–3246, 2022b.

Umar Farooq, Tej Prasad Dhamala, Antoine Nongaillard, Yacine Ouzrout, and Muhammad Abdul Qadir. A word sense disambiguation method for feature level sentiment analysis. In *2015 9th International Conference on Software, Knowledge, Information Management and Applications*, pages 1–8, 2015.

Manaal Faruqui, Jesse Dodge, Sujay Kumar Jauhar, Chris Dyer, Eduard H. Hovy, and Noah A. Smith. Retrofitting word vectors to semantic lexicons. In *NAACL-HLT*, pages 1606–1615, 2015.

Mehwish Fatima and Mark-Christoph Mueller. HITS-SBD at the FinSBD task: Machine learning vs. rule-based sentence boundary detection. In *First Workshop on Financial Technology and Natural Language Processing*, pages 115–121, 2019.

Gilles Fauconnier. Polarity and the scale principle. *Chicago*, 1975a.

Gilles Fauconnier. Pragmatic scales and logical structure. *Linguistic Inquiry*, 6(3): 353–375, 1975b.

Gilles Fauconnier and Mark Turner. *The Way We Think: Conceptual Blending and the Mind's Hidden Complexities*. Basic Books, 2008.

William Fedus, Ian Goodfellow, and Andrew M Dai. MaskGAN: Better text generation via filling in the _. In *ICLR*, 2018.

Hao Fei, Donghong Ji, Bobo Li, Yijiang Liu, Yafeng Ren, and Fei Li. Rethinking boundaries: End-to-end recognition of discontinuous mentions with pointer networks. In *AAAI*, pages 12785–12793, 2021a.

Hao Fei, Yafeng Ren, Yue Zhang, and Donghong Ji. Nonautoregressive encoder-decoder neural framework for end-to-end aspect-based sentiment triplet extraction. *IEEE Transactions on Neural Networks and Learning Systems*, 2021b.

Hao Fei, Han Zhang, Bin Wang, Lizi Liao, Qian Liu, and Erik Cambria. Empathyear: An open-source avatar multimodal empathetic chatbot. In *ACL*, pages 61–71, 2024.

Jerome Feldman. *From Molecule to Metaphor: A Neural Theory of Language*. MIT Press, 2008.

Christiane Fellbaum. *WordNet: An Electronic Lexical Database (Language, Speech, and Communication)*. The MIT Press, 1998.

Jinzhan Feng, Shuqin Cai, and Xiaomeng Ma. Enhanced sentiment labeling and implicit aspect identification by integration of deep convolution neural network and sequential algorithm. *Cluster Computing*, 22:5839–5857, 2019a.

Shi Feng, Lin Wang, Weili Xu, Daling Wang, and Ge Yu. Unsupervised learning chinese sentiment lexicon from massive microblog data. In *Advanced Data Mining and Applications: 8th International Conference, ADMA 2012, Nanjing, China, December 15-18, 2012. Proceedings 8*, pages 27–38, 2012.

Xiaocheng Feng, Zhangyin Feng, Wanlong Zhao, Nan Zou, Bing Qin, and Ting Liu. Improved neural machine translation with pos-tagging through joint decoding. In *International Conference on Artificial Intelligence for Communications and Networks*, pages 159–166, 2019b.

Yifan Feng, Haoxuan You, Zizhao Zhang, Rongrong Ji, and Yue Gao. Hypergraph neural networks. In *AAAI*, volume 33, pages 3558–3565, 2019c.

Marco Ferrarotti, Walter Rocchia, and Sergio Decherchi. Finding principal paths in data space. *IEEE Transactions on Neural Networks and Learning Systems*, 30(8): 2449–2462, 2019.

Kim Ferres, Timo Schloesser, and Peter A Gloor. Predicting dog emotions based on posture analysis using deeplabcut. *Future Internet*, 14(4):97, 2022.

David Ferrucci and Adam Lally. UIMA: An architectural approach to unstructured information processing in the corporate research environment. *Natural Language Engineering*, 10(3-4):327–348, 2004.

Leon Festinger. *A Theory of Cognitive Dissonance*. Stanford University Press, 1957. ISBN 9780804709118.

Charles J Fillmore, Paul Kay, Mary C O'connor, et al. Regularity and idiomaticity in grammatical constructions: The case of let alone. *Language*, 64(3):501–538, 1988.

Charles J Fillmore et al. Frame semantics. *Cognitive Linguistics: Basic Readings*, 34:373–400, 2006.

Jenny Rose Finkel and Christopher D Manning. Nested named entity recognition. In *EMNLP*, pages 141–150, 2009.

Jenny Rose Finkel, Trond Grenager, and Christopher D Manning. Incorporating non-local information into information extraction systems by Gibbs sampling. In *ACL*, pages 363–370, 2005.

Lev Finkelstein, Evgeniy Gabrilovich, Yossi Matias, Ehud Rivlin, Zach Solan, Gadi Wolfman, and Eytan Ruppin. Placing search in context: the concept revisited. In *International World Wide Web Conference, WWW*, pages 406–414, 2001.

John Firth. A synopsis of linguistic theory, 1930-1955. *Studies in Linguistic Analysis*, pages 10–32, 1957.

Steve Fligelstone. Developing a scheme for annotating text to show anaphoric relations. *New Directions in English Language Corpora: Methodology, Results, Software Developments*, pages 153–170, 1992.

Luciano Floridi, Josh Cowls, Monica Beltrametti, Raja Chatila, Patrice Chazerand, Virginia Dignum, Christoph Luetge, Robert Madelin, Ugo Pagallo, Francesca Rossi, Burkhard Schafer, Peggy Valcke, and Effy Vayena. Ai4people—an ethical framework for a good ai society: opportunities, risks, principles, and recommendations. *Minds Mach*, 28:689–707, 2018.

Eric N Forsyth. Improving automated lexical and discourse analysis of online chat dialog. Technical report, NAVAL POSTGRADUATE SCHOOL MONTEREY CA, 2007.

Davide Fossati and Barbara Di Eugenio. I saw tree trees in the park: How to correct real-word spelling mistakes. In *LREC*, page 2008, 2008.

ME Francis and Roger J Booth. Linguistic inquiry and word count. *Southern Methodist University: Dallas, TX, USA*, 1993.

W Nelson Francis and Henry Kucera. Brown corpus manual: Manual of information to accompany a standard corpus of present-day edited american english for use with digital computers. *Brown University, Providence, Rhode Island, USA*, 1979.

Dayne Freitag and Andrew McCallum. Information extraction with hmm structures learned by stochastic optimization. *AAAI/IAAI*, 2000:584–589, 2000.

Benoît Frénay and Michel Verleysen. Reinforced extreme learning machines for fast robust regression in the presence of outliers. *IEEE Transactions on Cybernetics*, 46(12):3351–3363, 2015.

Sigmund Freud. The ego and the id (1923). *TACD Journal*, 17(1):5–22, 1989.

Nir Friedman, Kevin P. Murphy, and Stuart Russell. Learning the structure of dynamic probabilistic networks. In Gregory F. Cooper and Serafín Moral, editors, *UAI*, pages 139–147. Morgan Kaufmann, 1998.

A Fritzler, V Logacheva, and M Kretov. Few-shot classification in named entity recognition task. In *ACM Symposium on Applied Computing*, pages 993–1000, 2019.

Chengpeng Fu, Jinqiang Wang, Jitao Sang, Jian Yu, and Changsheng Xu. Beyond literal visual modeling: Understanding image metaphor based on literal-implied concept mapping. In *International Conference on Multimedia Modeling*, pages 111–123, 2020a.

Sunyang Fu, David Chen, Huan He, Sijia Liu, Sungrim Moon, Kevin J Peterson, Feichen Shen, Liwei Wang, Yanshan Wang, Andrew Wen, et al. Clinical concept extraction: a methodology review. *Journal of Biomedical Informatics*, 109: 103526, 2020b.

Zhe Fu, Xi Niu, and Mary Lou Maher. Deep learning models for serendipity recommendations: a survey and new perspectives. *ACM Computing Surveys*, 56 (1):1–26, 2023.

Robert Fullér. *Neural Fuzzy Systems*. Citeseer, 1995.

David C Funder. *The Personality Puzzle*. WW Norton & Co, 1997.

Andrea Gagliano, Emily Paul, Kyle Booten, and Marti A Hearst. Intersecting word vectors to take figurative language to new heights. In *Fifth Workshop on Computational Linguistics for Literature*, pages 20–31, 2016.

Ladislav Gallay and Marián Šimko. Utilizing vector models for automatic text lemmatization. In *International Conference on Current Trends in Theory and Practice of Informatics*, pages 532–543, 2016.

Michel Galley, Kathleen McKeown, Julia Hirschberg, and Elizabeth Shriberg. Identifying agreement and disagreement in conversational speech: use of Bayesian networks to model pragmatic dependencies. In *ACL*, pages 669–677, 2004.

Ankita Gandhi, Kinjal Adhvaryu, Soujanya Poria, Erik Cambria, and Amir Hussain. Multimodal sentiment analysis: A systematic review of history, datasets, multimodal fusion methods, applications, challenges and future directions. *Information Fusion*, 91:424–444, 2023.

Lisa Gandy, Nadji Allan, Mark Atallah, Ophir Frieder, Newton Howard, Sergey Kanareykin, Moshe Koppel, Mark Last, Yair Neuman, and Shlomo Argamon. Automatic identification of conceptual metaphors with limited knowledge. In *AAAI*, pages 328–334, 2013.

Vaishali Ganganwar and R Rajalakshmi. Implicit aspect extraction for sentiment analysis: A survey of recent approaches. *Procedia Computer Science*, 165:485–491, 2019.

Aldo Gangemi, Nicola Guarino, Claudio Masolo, Alessandro Oltramari, and Luc Schneider. Sweetening ontologies with DOLCE. In *Knowledge Engineering and Knowledge Management. Ontologies and the Semantic Web, 13th International Conference, EKAW*, volume 2473 of *Lecture Notes in Computer Science*, pages 166–181, 2002.

Chen Gao, Xiang Wang, Xiangnan He, and Yong Li. Graph neural networks for recommender system. In *Fifteenth ACM International Conference on Web Search and Data Mining*, pages 1623–1625, 2022.

Jie Gao, Sooji Han, Xingyi Song, and Fabio Ciravegna. RP-DNN: A tweet level propagation context based deep neural networks for early rumor detection in social media. In *LREC*, pages 6094–6105, Marseille, France, May 2020. European Language Resources Association. ISBN 979-10-95546-34-4.

Lei Gao, Yulong Wang, Tongcun Liu, Jingyu Wang, Lei Zhang, and Jianxin Liao. Question-driven span labeling model for aspect–opinion pair extraction. In *AAAI*, pages 12875–12883, 2021a.

Tianyu Gao, Adam Fisch, and Danqi Chen. Making pre-trained language models better few-shot learners. In *ACL-IJCNLP*, pages 3816–3830, 2021b.

Peter Gardenfors. *Conceptual Spaces: The Geometry of Thought*. MIT Press, 2004.

Alan Garnham. *Mental Models and the Interpretation of Anaphora*. Psychology Press, 2001.

Mengshi Ge, Rui Mao, and Erik Cambria. Explainable metaphor identification inspired by conceptual metaphor theory. *AAAI*, 36(10):10681–10689, 2022.

Mengshi Ge, Rui Mao, and Erik Cambria. A survey on computational metaphor processing techniques: From identification, interpretation, generation to application. *Artificial Intelligence Review*, 56:1829–1895, 2023.

Niyu Ge, John Hale, and Eugene Charniak. A statistical approach to anaphora resolution. In *Sixth workshop on very large corpora*, pages 161–170, 1998.

Sebastian Gehrmann, Franck Dernoncourt, Yeran Li, Eric T Carlson, Joy T Wu, Jonathan Welt, John Foote Jr, Edward T Moseley, David W Grant, Patrick D Tyler, et al. Comparing deep learning and concept extraction based methods for patient phenotyping from clinical narratives. *PloS one*, 13(2):e0192360, 2018.

Daniela Gerz, Ivan Vulic, Felix Hill, Roi Reichart, and Anna Korhonen. Simverb-3500: A large-scale evaluation set of verb similarity. In *EMNLP*, pages 2173–2182, 2016.

Andrea Gesmundo and Tanja Samardzic. Lemmatisation as a tagging task. In *ACL*, pages 368–372, 2012.

Mor Geva, Daniel Khashabi, Elad Segal, Tushar Khot, Dan Roth, and Jonathan Berant. Did aristotle use a laptop? A question answering benchmark with implicit reasoning strategies. *Transactions of the Association for Computational Linguistics*, 9:346–361, 2021.

Abbas Ghaddar and Philippe Langlais. WikiCoref: An English coreference-annotated corpus of wikipedia articles. In *Tenth International Conference on Language Resources and Evaluation*, pages 136–142, 2016.

Deepanway Ghosal, Devamanyu Hazarika, Abhinaba Roy, Navonil Majumder, Rada Mihalcea, and Soujanya Poria. Kingdom: Knowledge-guided domain adaptation for sentiment analysis. In *ACL*, pages 3198–3210, 2020.

Debanjan Ghosh, Alexander Richard Fabbri, and Smaranda Muresan. The role of conversation context for sarcasm detection in online interactions. *arXiv preprint arXiv:1707.06226*, 2017.

Raymond W Gibbs Jr. A new look at literal meaning in understanding what is said and implicated. *Journal of pragmatics*, 34(4):457–486, 2002.

Edward Gibson. Linguistic complexity: Locality of syntactic dependencies. *Cognition*, 68(1):1–76, 1998.

Alastair Gill, Scott Nowson, and Jon Oberlander. What are they blogging about? personality, topic and motivation in blogs. In *International AAAI Conference on Web and Social Media*, pages 18–25, 2009.

Dan Gillick. Sentence boundary detection and the problem with the US. In *NAACL-HLT*, pages 241–244, 2009.

Jesús Giménez and Lluís Màrquez. SVMTool: A general POS tagger generator based on support vector machines. In *Fourth International Conference on Language Resources and Evaluation (LREC'04)*, 2004.

Jesús Giménez and Lluis Marquez. Fast and accurate part-of-speech tagging: The SVM approach revisited. *RANLP*, pages 153–162, 2004.

Kevin Gimpel, Nathan Schneider, Brendan O'Connor, Dipanjan Das, Daniel Mills, Jacob Eisenstein, Michael Heilman, Dani Yogatama, Jeffrey Flanigan, and Noah A Smith. Part-of-speech tagging for twitter: Annotation, features, and experiments. Technical report, Carnegie-Mellon Univ Pittsburgh Pa School of Computer Science, 2010.

Christian Girardi, Manuela Speranza, Rachele Sprugnoli, and Sara Tonelli. Cromer: A tool for cross-document event and entity coreference. In *LREC*, pages 3204–3208, 2014.

Njagi Dennis Gitari, Zhang Zuping, Hanyurwimfura Damien, and Jun Long. A lexicon-based approach for hate speech detection. *International Journal of Multimedia and Ubiquitous Engineering*, 10(4):215–230, 2015.

Talmy Givón. Topic continuity in discourse: The functional domain of switch reference. *Switch reference and universal grammar*, 51:82, 1983.

Alec Go, Richa Bhayani, and Lei Huang. Twitter sentiment classification using distant supervision. *CS224N project report, Stanford*, 1(12):2009, 2009.

Andrew Goatly. *Washing the brain: Metaphor and hidden ideology*, volume 23. John Benjamins Publishing, 2007.

Vinod Goel, Gorka Navarrete, Ira A Noveck, and Jérôme Prado. The reasoning brain: The interplay between cognitive neuroscience and theories of reasoning, 2017.

Adele Goldberg and Laura Suttle. Construction grammar. *Wiley Interdisciplinary Reviews: Cognitive Science*, 1(4):468–477, 2010.

Lewis R Goldberg. Language and individual differences: The search for universals in personality lexicons. *Review of personality and social psychology*, 2(1):141–165, 1981.

Lewis R Goldberg. Standard markers of the Big-Five factor structure. In *First International Workshop on Personality Language*, 1989.

Julio Gonzalo, Felisa Verdejo, Irina Chugur, and Juan Cigarran. Indexing with WordNet synsets can improve text retrieval. In *Usage of WordNet in Natural Language Processing Systems*, pages 38–44, 1998.

Julio Gonzalo, Anselmo Penas, and Felisa Verdejo. Lexical ambiguity and information retrieval revisited. In *EMNLP-VLC*, pages 195–202, 1999.

Samuel D Gosling, Peter J Rentfrow, and William B Swann Jr. A very brief measure of the big-five personality domains. *Journal of Research in personality*, 37(6):504–528, 2003.

Yoshihiko Gotoh and Steve Renals. Sentence boundary detection in broadcast speech transcripts. In *ASR2000-Automatic Speech Recognition*, 2000.

Alex Graves and Jürgen Schmidhuber. Framewise phoneme classification with bidirectional LSTM and other neural network architectures. *Neural Networks*, 18 (5-6):602–610, 2005.

Alex Graves, Abdel-rahman Mohamed, and Geoffrey Hinton. Speech recognition with deep recurrent neural networks. In *2013 IEEE international conference on acoustics, speech and signal processing*, pages 6645–6649, 2013.

Gregory Grefenstette and Pasi Tapanainen. What is a word, what is a sentence?: Problems of tokenisation. Report, Grenoble Laboratory, 1994.

Denis Griffis, Chaitanya Shivade, Eric Fosler-Lussier, and Albert M. Lai. A quantitative and qualitative evaluation of sentence boundary detection for the clinical domain. *AMIA Joint Summits on Translational Science*, pages 88–97, 2016.

Ralph Grishman and Beth M Sundheim. Message understanding conference-6: A brief history. In *COLING*, pages 466–471, 1996.

Barbara J. Grosz, Aravind K. Joshi, and Scott Weinstein. Providing a unified account of definite noun phrases in discourse. In *ACL*, pages 44–50, 1983.

Barbara J. Grosz, Aravind K. Joshi, and Scott Weinstein. Centering: A framework for modeling the local coherence of discourse. *Computational Linguistics*, 21(2): 203–225, 1995.

Barbara Jean Grosz. *The Representation and Use of Focus in Dialogue Understanding*. University of California, Berkeley, 1977.

Adam J Grove and Dan Roth. Linear concepts and hidden variables. *Machine Learning*, 42(1-2):123–141, 2001.

Rosanna E Guadagno, Bradley M Okdie, and Cassie A Eno. Who blogs? Personality predictors of blogging. *Computers in human behavior*, 24(5):1993–2004, 2008.

Nicola Guarino. Formal ontology, conceptual analysis and knowledge representation. *International journal of human-computer studies*, 43(5-6):625–640, 1995.

Tao Gui, Qi Zhang, Haoran Huang, Minlong Peng, and Xuanjing Huang. Part-of-speech tagging for twitter with adversarial neural networks. In *EMNLP*, pages 2411–2420, 2017.

Liane Guillou. Improving pronoun translation for statistical machine translation. In *EACL*, pages 1–10, 2012.

Jeanette K Gundel, Nancy Hedberg, and Ron Zacharski. Cognitive status and the form of referring expressions in discourse. *Language*, pages 274–307, 1993.

Honey Gupta, Aveena Kottwani, Soniya Gogia, and Sheetal Chaudhari. Text analysis and information retrieval of text data. In *WiSPNET*, pages 788–792, 2016a.

Pankaj Gupta, Hinrich Schütze, and Bernt Andrassy. Table filling multi-task recurrent neural network for joint entity and relation extraction. In *COLING*, pages 2537–2547, 2016b.

Harsha Gurulingappa, Abdul Mateen Rajput, Angus Roberts, Juliane Fluck, Martin Hofmann-Apitius, and Luca Toldo. Development of a benchmark corpus to support the automatic extraction of drug-related adverse effects from medical case reports. *Journal of Biomedical Informatics*, 45(5):885–892, 2012.

Dario Gutierrez, Ekaterina Shutova, Tyler Marghetis, and Benjamin Bergen. Literal and metaphorical senses in compositional distributional semantic models. In *ACL*, pages 183–193, 2016.

Christian Hadiwinoto, Hwee Tou Ng, and Wee Chung Gan. Improved word sense disambiguation using pre-trained contextualized word representations. In *EMNLP-IJCNLP*, pages 5297–5306, 2019.

Aria Haghighi and Dan Klein. Simple coreference resolution with rich syntactic and semantic features. In *EMNLP*, pages 1152–1161, 2009.

Zhen Hai, Kuiyu Chang, and Gao Cong. One seed to find them all: Mining opinion features via association. In *CIKM*, pages 255–264, 2012.

Jan Hajič, Massimiliano Ciaramita, Richard Johansson, Daisuke Kawahara, Maria Antònia Martí, Lluís Màrquez, Adam Meyers, Joakim Nivre, Sebastian Padó, Jan Štěpánek, Pavel Straňák, Mihai Surdeanu, Nianwen Xue, and Yi Zhang. The CoNLL-2009 shared task: Syntactic and semantic dependencies in multiple languages. In *CoNLL*, pages 1–18, 2009.

Péter Halácsy and V Trón. Benefits of deep NLP-based lemmatization for information retrieval. In *CLEF (Working Notes)*, 2006.

Guy Halawi, Gideon Dror, Evgeniy Gabrilovich, and Yehuda Koren. Large-scale learning of word relatedness with constraints. In *ACM SIGKDD International Conference on Knowledge Discovery and Data Mining*, pages 1406–1414, 2012.

Jeffrey A Hall, Jess Dominguez, and Teodora Mihailova. Interpersonal media and face-to-face communication: Relationship with life satisfaction and loneliness. *Journal of Happiness Studies*, 24(1):331–350, 2023.

Bo Han and Timothy Baldwin. Lexical normalisation of short text messages: Makn sens a# twitter. In *ACL-HLT*, pages 368–378, 2011.

Na-Rae Han. *Korean zero pronouns: analysis and resolution*. University of Pennsylvania, 2006.

Sooji Han, Jie Gao, and Fabio Ciravegna. Neural language model based training data augmentation for weakly supervised early rumor detection. In *2019 IEEE/ACM International Conference on Advances in Social Networks Analysis and Mining*, pages 105–112, 2019.

Sooji Han, Rui Mao, and Erik Cambria. Hierarchical attention network for explainable depression detection on Twitter aided by metaphor concept mappings. In *COLING*, pages 94–104, 2022a.

Xu Han, Weilin Zhao, Ning Ding, Zhiyuan Liu, and Maosong Sun. PTR: Prompt tuning with rules for text classification. *AI Open*, 3:182–192, 2022b.

Sanda M. Harabagiu and Steven J. Maiorano. Knowledge-lean coreference resolution and its relation to textual cohesion and coherence. In *The Relation of Discourse/Dialogue Structure and Reference*, pages 29–38, 1999.

Sanda M. Harabagiu, Razvan C. Bunescu, and Steven J. Maiorano. Text and knowledge mining for coreference resolution. In *NAACL*, pages 1–8, 2001.

Jaron Harambam, Dimitrios Bountouridis, Mykola Makhortykh, and Joris Van Hoboken. Designing for the better by taking users into account: A qualitative evaluation of user control mechanisms in (news) recommender systems. In *13th ACM conference on recommender systems*, pages 69–77, 2019.

Christian Hardmeier and Marcello Federico. Modelling pronominal anaphora in statistical machine translation. In *IWSLT (International Workshop on Spoken Language Translation); Paris, France; December 2nd and 3rd, 2010.*, pages 283–289, 2010.

Peter E Hart, Nils J Nilsson, and Bertram Raphael. A formal basis for the heuristic determination of minimum cost paths. *IEEE transactions on Systems Science and Cybernetics*, 4(2):100–107, 1968.

Md Rajibul Hasan, Ashish Kumar Jha, and Yi Liu. Excessive use of online video streaming services: Impact of recommender system use, psychological factors, and motives. *Computers in Human Behavior*, 80:220–228, 2018.

Laura Hasler, Constantin Orăsan, and Karin Naumann. NPs for events: Experiments in coreference annotation. In *Fifth International Conference on Language Resources and Evaluation (LREC'06)*, pages 1167–1172, 2006.

Catherine Havasi and Robert Speer. ConceptNet 3: A flexible, multilingual semantic network for common sense knowledge. In *RANLP*, pages 27–29, 2007.

Taher H Haveliwala. Topic-sensitive PageRank. In *11th International Conference on World Wide Web*, pages 517–526, 2002.

Kai He, Na Hong, Samuel Lapalme-Remis, Yangyang Lan, Ming Huang, Chen Li, and Lixia Yao. Understanding the patient perspective of epilepsy treatment through text mining of online patient support groups. *Epilepsy & Behavior*, 94: 65–71, 2019a.

Kai He, Jialun Wu, Xiaoyong Ma, Chong Zhang, Ming Huang, Chen Li, and Lixia Yao. Extracting kinship from obituary to enhance electronic health records for genetic research. In *Fourth social media mining for health applications (# SMM4H) workshop & shared task*, pages 1–10, 2019b.

Kai He, Lixia Yao, JiaWei Zhang, Yufei Li, Chen Li, et al. Construction of genealogical knowledge graphs from obituaries: Multitask neural network extraction system. *Journal of Medical Internet Research*, 23(8):e25670, 2021.

Kai He, Bing Mao, Xiangyu Zhou, Yufei Li, Tieliang Gong, Chen Li, and Jialun Wu. Knowledge enhanced coreference resolution via gated attention. In *2022 IEEE International Conference on Bioinformatics and Biomedicine (BIBM)*, pages 2287–2293, 2022a.

Kai He, Rui Mao, Tieliang Gong, Erik Cambria, and Chen Li. JCBIE: a joint continual learning neural network for biomedical information extraction. *BMC Bioinformatics*, 23(1):1–20, 2022b.

Kai He, Yucheng Huang, Rui Mao, Tieliang Gong, Chen Li, and Erik Cambria. Virtual prompt pre-training for prototype-based few-shot relation extraction. *Expert Systems With Applications*, 213:118927, 2023a.

Kai He, Rui Mao, Tieliang Gong, Chen Li, and Erik Cambria. Meta-based self-training and re-weighting for aspect-based sentiment analysis. *IEEE Transactions on Affective Computing*, 14(3):1731–1742, 2023b.

Kai He, Rui Mao, Qika Lin, Yucheng Ruan, Xiang Lan, Mengling Feng, and Erik Cambria. A survey of large language models for healthcare: From data, technology, and applications to accountability and ethics. *arXiv preprint arXiv:2310.05694*, 2023c.

Kai He, Rui Mao, Yucheng Huang, Tieliang Gong, Chen Li, and Erik Cambria. Template-free prompting for few-shot named entity recognition via semantic-enhanced contrastive learning. *IEEE Transactions on Neural Networks and Learning Systems*, 2024a.

Kaiming He, Xiangyu Zhang, Shaoqing Ren, and Jian Sun. Deep residual learning for image recognition. In *IEEE conference on computer vision and pattern recognition*, pages 770–778, 2016.

Ruidan He, Wee Sun Lee, Hwee Tou Ng, and Daniel Dahlmeier. An interactive multi-task learning network for end-to-end aspect-based sentiment analysis. In *ACL*, pages 504–515, 2019c.

Xinyue He, Qi Liu, and Sunho Jung. The impact of recommendation system on user satisfaction: A moderated mediation approach. *Journal of Theoretical and Applied Electronic Commerce Research*, 19(1):448–466, 2024b.

Yulan He. Bayesian models for sentence-level subjectivity detection. Technical report, Knowledge Media Institute, 2010.

Irene Roswitha Heim. *The semantics of definite and indefinite noun phrases.* University of Massachusetts Amherst, 1982.

Sigrún Helgadóttir. Icelandic frequency dictionary 2012.11-training/testing sets, 2012.

Mohammad Helmi and Seyed Mohammad T AlModarresi. Human activity recognition using a fuzzy inference system. In *2009 IEEE International Conference on Fuzzy Systems*, pages 1897–1902, 2009.

Iris Hendrickx, Gosse Bouma, Frederik Coppens, Walter Daelemans, Veronique Hoste, Geert Kloosterman, Anne-Marie Mineur, Joeri Van Der Vloet, and Jean-Luc Verschelde. A coreference corpus and resolution system for Dutch. In *Sixth International Conference on Language Resources and Evaluation*, pages 1–6, 2008.

Sam Henry, Kevin Buchan, Michele Filannino, Amber Stubbs, and Ozlem Uzuner. 2018 n2c2 shared task on adverse drug events and medication extraction in electronic health records. *Journal of the American Medical Informatics Association*, 27(1):3–12, 2020.

María Herrero-Zazo, Isabel Segura-Bedmar, Paloma Martínez, and Thierry Declerck. The DDI corpus: An annotated corpus with pharmacological substances and drug-drug interactions. *Journal of Biomedical Informatics*, 46(5):914–920, 2013.

Clara E Hill. *Helping skills: Facilitating, exploration, insight, and action.* American Psychological Association, 2009.

Felix Hill, Roi Reichart, and Anna Korhonen. Simlex-999: Evaluating semantic models with (genuine) similarity estimation. *Computational Linguistics*, 41(4): 665–695, 2015.

Dustin Hillard, Mari Ostendorf, and Elizabeth Shriberg. Detection of agreement vs. disagreement in meetings: Training with unlabeled data. In *HLT-NAACL*, pages 34–36, 2003.

Geoffrey E Hinton, Nitish Srivastava, Alex Krizhevsky, Ilya Sutskever, and Ruslan R Salakhutdinov. Improving neural networks by preventing co-adaptation of feature detectors. *arXiv preprint arXiv:1207.0580*, pages 1–18, 2012.

JA Hirsch, G Nicola, G McGinty, RW Liu, RM Barr, MD Chittle, and L Manchikanti. ICD-10: history and context. *American Journal of Neuroradiology*, 37(4):596–599, 2016.

Lynette Hirschman and Nancy Chinchor. Appendix F: MUC-7 coreference task definition (version 3.0). In *Seventh Message Understanding Conference (MUC-7)*, pages 1–17, 1998.

Lynette Hirschman, Patricia Robinson, John Burger, and Marc Vilain. Automating coreference: The role of annotated training data. In *AAAI Spring Symposium on Applying Machine Learning to Discourse Processing*, pages 118–121, 1997.

Janet Hitzeman, Alan W Black, Paul Taylor, Chris Mellish, and Jon Oberlander. On the use of automatically generated discourse-level information in a concept-to-speech synthesis system. In *5th International Conference on Spoken Language Processing*, pages 2763–2766, 1998.

Seng-Beng Ho, Zhaoxia Wang, Boon-Kiat Quek, and Erik Cambria. Knowledge representation for conceptual, motivational, and affective processes in natural language communication. In *BICS*, volume 14374, pages 14–30, 2023.

Shuk Ying Ho and Kai H Lim. Nudging moods to induce unplanned purchases in imperfect mobile personalization contexts. *Mis Quarterly*, 42(3):757–A13, 2018.

Jerry R Hobbs. Resolving pronoun references. *Lingua*, 44(4):311–338, 1978.

Jerry R. Hobbs, Mark Stickel, Paul Martin, and Douglas Edwards. Interpretation as abduction. In *ACL*, pages 95–103, 1988.

Sepp Hochreiter and Jürgen Schmidhuber. Long short-term memory. *Neural computation*, 9(8):1735–1780, 1997.

Jack Hoeksema. On the grammaticalization of negative polarity items. In *Annual Meeting of the Berkeley Linguistics Society*, pages 273–282, 1994.

Patrick Hohenecker, Frank Mtumbuka, Vid Kocijan, and Thomas Lukasiewicz. Systematic comparison of neural architectures and training approaches for open information extraction. In *EMNLP*, pages 8554–8565, 2020.

John H Holland. *Adaptation in Natural and Artificial Systems: An Introductory Analysis with Applications to Biology, Control, and Artificial Intelligence*. MIT press, 1992.

Janusz A Hołyst, Philipp Mayr, Michael Thelwall, Ingo Frommholz, Shlomo Havlin, Alon Sela, Yoed N Kenett, Denis Helic, Aljoša Rehar, Sebastijan R Maček, Przemysław Kazienko, Tomasz Kajdanowicz, Przemysław Biecek, Boleslaw K. Szymanski, and Julian Sienkiewicz. Protect our environment from information overload. *Nature Human Behaviour*, pages 1–2, 2024.

Vincent Homer. Domains of polarity items. *Journal of Semantics*, 38(1):1–48, 2021.

Albert Sydney Hornby and Anthony P Cowie. Oxford advanced learner's dictionary of current english. *Paperback,*, 1974.

David W. Hosmer and Stanley Lemeshow. *Applied Logistic Regression*. Wiley, 2 edition, 2000.

Yufang Hou, Katja Markert, and Michael Strube. Unrestricted bridging resolution. *Computational Linguistics*, 44(2):237–284, 2018.

Eduard Hovy, Mitch Marcus, Martha Palmer, Lance Ramshaw, and Ralph Weischedel. Ontonotes: the 90% solution. In *human language technology conference of the NAACL, Companion Volume: Short Papers*, pages 57–60, 2006.

Newton Howard and Erik Cambria. Intention awareness: Improving upon situation awareness in human-centric environments. *Human-centric Computing and Information Sciences*, 3(9), 2013.

Jeff Howe. *Crowdsourcing: How the Power of the Crowd is Driving the Future of Business*. Random House, 2008.

Pei-Yun Hsueh, Prem Melville, and Vikas Sindhwani. Data quality from crowdsourcing: A study of annotation selection criteria. In *NAACL HLT 2009 Workshop on Active Learning for Natural Language Processing*, pages 27–35, 2009.

Jiawei Hu, Zheng Li, and Bin Xu. An approach of ontology based knowledge base construction for chinese k12 education. In *2016 First International Conference on Multimedia and Image Processing (ICMIP)*, pages 83–88, 2016.

Minghao Hu, Yuxing Peng, Zhen Huang, Dongsheng Li, and Yiwei Lv. Open-domain targeted sentiment analysis via span-based extraction and classification. *arXiv preprint arXiv:1906.03820*, 2019.

Minqing Hu and Bing Liu. Mining and summarizing customer reviews. In *tenth ACM SIGKDD international conference on Knowledge discovery and data mining*, pages 168–177, 2004.

Renqing Hu and Xue Wang. A cognitive pragmatic analysis of conceptual metaphor in political discourse based on text data mining. In *2021 4th International Conference on Information Systems and Computer Aided Education*, pages 235–238, 2021.

Hu Huang, Bowen Zhang, Liwen Jing, Xianghua Fu, Xiaojun Chen, and Jianyang Shi. Logic tensor network with massive learned knowledge for aspect-based sentiment analysis. *Knowledge-Based Systems*, 257:109943, 2022a.

James Huang. On the distribution and reference of empty pronouns. *Linguistic Inquiry*, pages 531–574, 1984.

Lishan Huang, Zheng Ye, Jinghui Qin, Liang Lin, and Xiaodan Liang. GRADE: Automatic graph-enhanced coherence metric for evaluating open-domain dialogue systems. In *EMNLP*, pages 9230–9240, 2020a.

Luyao Huang, Chi Sun, Xipeng Qiu, and Xuan-Jing Huang. GlossBERT: BERT for word sense disambiguation with gloss knowledge. In *EMNLP-IJCNLP*, pages 3509–3514, 2019a.

Minghui Huang, Haoran Xie, Yanghui Rao, Jingrong Feng, and Fu Lee Wang. Sentiment strength detection with a context-dependent lexicon-based convolutional neural network. *Information Sciences*, 520:389–399, 2020b.

Xiaoqing Huang, Qi Liu, Chao Wang, Haoyu Han, Jianhui Ma, Enhong Chen, Yu Su, and Shijin Wang. Constructing educational concept maps with multiple relationships from multi-source data. In Jianyong Wang, Kyuseok Shim, and Xindong Wu, editors, *ICDM*, pages 1108–1113. IEEE, 2019b.

Yucheng Huang, Kai He, Yige Wang, Xianli Zhang, Tieliang Gong, Rui Mao, and Chen Li. COPNER: Contrastive learning with prompt guiding for few-shot named entity recognition. In *COLING*, pages 2515–2527, 2022b.

Zhiheng Huang, Wei Xu, and Kai Yu. Bidirectional LSTM-CRF models for sequence tagging. *arXiv preprint arXiv:1508.01991*, 2015.

Christoph Hube and Besnik Fetahu. Detecting biased statements in Wikipedia. In *Web Conference*, pages 1779–1786, 2018.

Christoph Hube and Besnik Fetahu. Neural based statement classification for biased language. In *Twelfth ACM International Conference on Web Search and Data Mining*, pages 195–203, 2019.

Anette Hulth. Improved automatic keyword extraction given more linguistic knowledge. In *EMNLP*, pages 1–8, 2003.

Chihli Hung and Shiuan-Jeng Chen. Word sense disambiguation based sentiment lexicons for sentiment classification. *Knowledge-Based Systems*, 110:224–232, 2016.

Chihli Hung and Hao-Kai Lin. Using objective words in SentiWordNet to improve word-of-mouth sentiment classification. *IEEE Intelligent Systems*, 28(02):47–54, 2013.

Hairong Huo and Mizuho Iwaihara. Utilizing BERT pretrained models with various fine-tune methods for subjectivity detection. In *Asia-Pacific Web (APWeb) and Web-Age Information Management (WAIM) Joint International Conference on Web and Big Data*, pages 270–284, 2020.

Jena D. Hwang, Chandra Bhagavatula, Ronan Le Bras, Jeff Da, Keisuke Sakaguchi, Antoine Bosselut, and Yejin Choi. Comet-atomic 2020: On symbolic and neural commonsense knowledge graphs. In *AAAI*, 2020.

Ignacio Iacobacci, Mohammad Taher Pilehvar, and Roberto Navigli. Embeddings for word sense disambiguation: An evaluation study. In *ACL*, pages 897–907, 2016.

Francisco Iacobelli, Alastair J Gill, Scott Nowson, and Jon Oberlander. Large scale personality classification of bloggers. In *Affective Computing and Intelligent Interaction: Fourth International Conference, ACII 2011, Memphis, TN, USA, October 9–12, 2011, Proceedings, Part II*, pages 568–577, 2011.

Nancy Ide and Jean Véronis. Multext: Multilingual text tools and corpora. In *COLING*, 1994.

Ryu Iida and Massimo Poesio. A cross-lingual ilp solution to zero anaphora resolution. In *ACL-HLT*, pages 804–813, 2011.

Ryu Iida, Kentaro Inui, and Yuji Matsumoto. Zero-anaphora resolution by learning rich syntactic pattern features. *ACM Transactions on Asian Language Information Processing*, 6(4):1–22, 2007a.

Ryu Iida, Mamoru Komachi, Kentaro Inui, and Yuji Matsumoto. Annotating a Japanese text corpus with predicate-argument and coreference relations. In *Linguistic Annotation Workshop*, pages 132–139, 2007b.

Filip Ilievski, Alessandro Oltramari, Kaixin Ma, Bin Zhang, Deborah L. McGuinness, and Pedro A. Szekely. Dimensions of commonsense knowledge. *Knowledge-Based Systems*, 229:107347, 2021a.

Filip Ilievski, Pedro A. Szekely, and Bin Zhang. CSKG: the commonsense knowledge graph. In *Semantic Web - 18th International Conference, ESWC*, volume 12731 of *Lecture Notes in Computer Science*, pages 680–696, 2021b.

Ali Shariq Imran, Sher Muhammad Daudpota, Zenun Kastrati, and Rakhi Batra. Cross-cultural polarity and emotion detection using sentiment analysis and deep learning on COVID-19 related tweets. *Ieee Access*, 8:181074–181090, 2020.

Andrea Iovine, Fedelucio Narducci, and Giovanni Semeraro. Conversational recommender systems and natural language:: A study through the converse framework. *Decision Support Systems*, 131:113250, 2020.

Hideki Isozaki and Tsutomu Hirao. Japanese zero pronoun resolution based on ranking rules and machine learning. In *EMNLP*, pages 184–191, 2003.

Michael Israel. Polarity sensitivity as lexical semantics. *Linguistics and Philosophy*, pages 619–666, 1996.

Ray Jackendoff. Toward an explanatory semantic representation. *Linguistic inquiry*, 7(1):89–150, 1976.

Ray S Jackendoff et al. *Semantics and Cognition*. The MIT Press, Cambridge, Massachusetts, 1983.

Vincent Jahjah, Richard Khoury, and Luc Lamontagne. Word normalization using phonetic signatures. In *Canadian Conference on Artificial Intelligence*, pages 180–185, 2016.

Niklas Jakob and Iryna Gurevych. Using anaphora resolution to improve opinion target identification in movie reviews. In *ACL*, pages 263–268, 2010a.

Niklas Jakob and Iryna Gurevych. Extracting opinion targets in a single and cross-domain setting with conditional random fields. In *EMNLP*, pages 1035–1045, 2010b.

Bernard J Jansen. *Understanding user-web interactions via web analytics*. Springer Nature, 2022.

Bernard J Jansen, Soon-Gyo Jung, Lene Nielsen, Kathleen W Guan, and Joni Salminen. How to create personas: Three persona creation methodologies with implications for practical employment. *Pacific Asia Journal of the Association for Information Systems*, 14(3):1, 2022a.

Bernard J Jansen, Joni Salminen, Soon-gyo Jung, and Kathleen Guan. *Data-driven personas*. Springer Nature, 2022b.

Bernard J Jansen, Soon-gyo Jung, and Joni Salminen. Employing large language models in survey research. *Natural Language Processing Journal*, 4:100020, 2023.

Lei Ji, Yujing Wang, Botian Shi, Dawei Zhang, Zhongyuan Wang, and Jun Yan. Microsoft concept graph: Mining semantic concepts for short text understanding. *Data Intelligence*, 1(3):238–270, 2019.

Shaoxiong Ji, Xue Li, Zi Huang, and Erik Cambria. Suicidal ideation and mental disorder detection with attentive relation networks. *Neural Computing and Applications*, 34(13):10309–10319, 2022.

Sittichai Jiampojamarn, Grzegorz Kondrak, and Tarek Sherif. Applying many-to-many alignments and hidden markov models to letter-to-phoneme conversion. In *ACL-HLT*, pages 372–379, 2007.

Sittichai Jiampojamarn, Colin Cherry, and Grzegorz Kondrak. Integrating joint n-gram features into a discriminative training framework. In *HLT-ACL*, pages 697–700, 2010.

Julie Jiang, Xiang Ren, and Emilio Ferrara. Retweet-BERT: Political leaning detection using language features and information diffusion on social networks. *arXiv preprint arXiv:2207.08349*, 2022.

Qingnan Jiang, Lei Chen, Ruifeng Xu, Xiang Ao, and Min Yang. A challenge dataset and effective models for aspect-based sentiment analysis. In *EMNLP-IJCNLP*, pages 6280–6285, 2019.

Tianwen Jiang, Qingkai Zeng, Tong Zhao, Bing Qin, Ting Liu, Nitesh V Chawla, and Meng Jiang. Biomedical knowledge graphs construction from conditional statements. *IEEE/ACM Transactions on Computational Biology and Bioinformatics*, 18(3):823–835, 2020.

Wei Jiang, Hao Pan, and Qing Ye. An improved association rule mining approach to identification of implicit product aspects. *The Open Cybernetics & Systemics Journal*, 8(1), 2014.

Mu Jin, Zhihao Ma, Kebing Jin, Hankz Hankui Zhuo, Chen Chen, and Chao Yu. Creativity of AI: Automatic symbolic option discovery for facilitating deep reinforcement learning. In *AAAI*, pages 7042–7050, 2022.

Wei Jin, Hung Hay Ho, and Rohini K Srihari. OpinionMiner: A novel machine learning system for web opinion mining and extraction. In *15th ACM SIGKDD International Conference on Knowledge Discovery and Data Mining*, pages 1195–1204, 2009.

Hongyan Jing, Daniel Lopresti, and Chilin Shih. Summarization of noisy documents: a pilot study. In *HLT-NAACL 03 Text Summarization Workshop*, pages 25–32, 2003.

Ni Jinjie, Pandelea Vlad, Young Tom, Zhou Haicang, and Cambria Erik. HiTKG: Towards goal-oriented conversations via multi-hierarchy learning. In *AAAI*, pages 11112–11120, 2022.

Yohan Jo and Alice H Oh. Aspect and sentiment unification model for online review analysis. In *fourth ACM International Conference on Web Search and Data Mining*, pages 815–824, 2011.

Thorsten Joachims. Optimizing search engines using clickthrough data. In *eighth ACM SIGKDD international conference on Knowledge discovery and data mining*, pages 133–142, 2002.

O John, E Donahue, and R Kentle. The Big Five inventory-versions 4a and 54. Berkeley, CA: University of California. *Berkeley, Institute of Personality and Social Research*, 1991.

Alistair Johnson, Tom J Pollard, Lu Shen, H Lehman Li-Wei, Mengling Feng, Mohammad Ghassemi, Benjamin Moody, Peter Szolovits, Leo Anthony Celi, and Roger G Mark. Mimic-iii, a freely accessible critical care database. *Scientific data*, 3(1):1–9, 2016.

Daniel Johnson and John Gardner. Personality, motivation and video games. In *22nd Conference of the Computer-Human Interaction Special Interest Group of Australia on Computer-Human Interaction*, pages 276–279, 2010.

Kyle Johnson. Gapping. *The Blackwell companion to syntax*, pages 407–435, 2006.

Mark Johnson. The body in the mind: The bodily basis of meaning, imagination, and reason. *Journal of Aesthetics and Art Criticism*, 47(4), 1989.

Rie Johnson and Tong Zhang. A high-performance semi-supervised learning method for text chunking. In *ACL*, pages 1–9, 2005.

Karen Sparck Jones. A statistical interpretation of term specificity and its application in retrieval. *Journal of documentation*, 1972.

Bart Jongejan and Hercules Dalianis. Automatic training of lemmatization rules that handle morphological changes in pre-, in-and suffixes alike. In *ACL-IJCNLP*, pages 145–153, 2009.

Greety Jose and Nisha S Raj. Lexico-syntactic normalization model for noisy SMS text. In *2014 International Conference on Electronics, Communication and Computational Engineering (ICECCE)*, pages 163–168, 2014.

Aditya Joshi, Vaibhav Tripathi, Pushpak Bhattacharyya, and Mark James Carman. Harnessing sequence labeling for sarcasm detection in dialogue from TV series 'friends'. In *CoNLL*, pages 146–155, 2016.

Aravind K Joshi and Steve Kuhn. Centered logic: The role of entity centered sentence representation in natural language inferencing. In *IJCAI*, pages 435–439, 1979.

Mandar Joshi, Eunsol Choi, Daniel Weld, and Luke Zettlemoyer. TriviaQA: A large scale distantly supervised challenge dataset for reading comprehension. In *ACL*, pages 1601–1611, 2017.

Mandar Joshi, Omer Levy, Luke Zettlemoyer, and Daniel Weld. BERT for coreference resolution: Baselines and analysis. In *EMNLP-IJCNLP*, pages 5803–5808, 2019.

Mandar Joshi, Danqi Chen, Yinhan Liu, Daniel S. Weld, Luke Zettlemoyer, and Omer Levy. SpanBERT: Improving pre-training by representing and predicting spans. *Transactions of the Association for Computational Linguistics*, 8:64–77, 2020.

Armand Joulin, Edouard Grave, Piotr Bojanowski, Matthijs Douze, Hérve Jégou, and Tomas Mikolov. FastText.zip: Compressing text classification models. *arXiv preprint arXiv:1612.03651*, 2016.

Armand Joulin, Edouard Grave, Piotr Bojanowski, and Tomás Mikolov. Bag of tricks for efficient text classification. In Mirella Lapata, Phil Blunsom, and Alexander Koller, editors, *EACL*, pages 427–431, 2017.

John Judge, Aoife Cahill, and Josef van Genabith. QuestionBank: Creating a corpus of parse-annotated questions. In *ACL*, pages 497–504, 2006.

Carl Jung and John Beebe. *Psychological types*. Routledge, 2016.

Matjaž Juršič, Igor Mozetič, and Nada Lavrač. Learning ripple down rules for efficient lemmatization. In *10th international multiconference information society, IS*, pages 206–209, 2007.

Nirit Kadmon and Fred Landman. Any. *Linguistics and Philosophy*, pages 353–422, 1993.

Mikael Kågebäck and Hans Salomonsson. Word sense disambiguation using a bidirectional LSTM. In *5th Workshop on Cognitive Aspects of the Lexicon (CogALex-V)*, pages 51–56, 2016.

Nal Kalchbrenner, Edward Grefenstette, and Phil Blunsom. A convolutional neural network for modelling sentences. In *ACL*, pages 655–665, 2014.

Ahmad Kamal. Subjectivity classification using machine learning techniques for mining feature-opinion pairs from web opinion sources. *International Journal of Computer Science Issues*, 10(5):191, 2013.

Ashraf Kamal and Muhammad Abulaish. CAT-BiGRU: Convolution and attention with bi-directional gated recurrent unit for self-deprecating sarcasm detection. *Cognitive Computation*, 14(1):91–109, 2022.

Subbarao Kambhampati, Karthik Valmeekam, Lin Guan, Mudit Verma, Kaya Stechly, Siddhant Bhambri, Lucas Saldyt, and Anil Murthy. LLMs can't plan, but can help planning in LLM-Modulo frameworks. *arXiv preprint arXiv:2402.01817*, 2024.

Megumi Kameyama. *Zero Anaphora: The Case of Japanese (Discourse Aanalysis, Pronouns, Sytax, Computational Llinguistics, Typology)*. Stanford University, 1985.

Megumi Kameyama. Recognizing referential links: an information extraction prespective. In *Operational Factors in Practical, Robust Anaphora Resolution for Unrestricted Texts*, pages 46–53, 1997.

Aida-zade Kamil, Samir Rustamov, Mark A Clements, and Elshan Mustafayev. Adaptive neuro-fuzzy inference system for classification of texts. In *Recent Developments and the New Direction in Soft-Computing Foundations and Applications*, pages 63–70. Springer, 2018.

Hans Kamp and Uwe Reyle. *From discourse to logic: Introduction to modeltheoretic semantics of natural language, formal logic and discourse representation theory*, volume 42. Springer Science & Business Media, 2013.

Jaap Kamps, Maarten Marx, Robert J Mokken, Maarten De Rijke, et al. Using WordNet to measure semantic orientations of adjectives. In *Lrec*, volume 4, pages 1115–1118, 2004.

Suvarna G Kanakaraddi and Suvarna S Nandyal. Survey on parts of speech tagger techniques. In *2018 International Conference on Current Trends towards Converging Technologies (ICCTCT)*, pages 1–6, 2018.

Jakub Kanis and Luděk Müller. Automatic lemmatizer construction with focus on OOV words lemmatization. In *International Conference on Text, Speech and Dialogue*, pages 132–139, 2005.

Jakub Kanis and Lucie Skorkovská. Comparison of different lemmatization approaches through the means of information retrieval performance. In Petr Sojka, Aleš Horák, Ivan Kopeček, and Karel Pala, editors, *Text, Speech and Dialogue*, pages 93–100, Berlin, Heidelberg, 2010. Springer Berlin Heidelberg. ISBN 978-3-642-15760-8.

Samaneh Karimi and Azadeh Shakery. A language-model-based approach for subjectivity detection. *Journal of Information Science*, 43(3):356–377, 2017.

Shubhra Kanti Karmaker Santu, Parikshit Sondhi, and ChengXiang Zhai. Generative feature language models for mining implicit features from customer reviews. In *CIKM*, pages 929–938, 2016.

RV Karthik and Sannasi Ganapathy. A fuzzy recommendation system for predicting the customers interests using sentiment analysis and ontology in e-commerce. *Applied Soft Computing*, 108:107396, 2021.

Brijesh S. Bhatt Kartik Detroja, C.K. Bhensdadia. A survey on relation extraction. *Intelligent Systems with Applications*, 2023.

Gabriele Kasper and Richard Schmidt. Developmental issues in interlanguage pragmatics. *Studies in Second Language Acquisition*, 18(2):149–169, 1996.

Arzoo Katiyar and Claire Cardie. Nested named entity recognition revisited. In *NAACL-HLT*, pages 861–871, 2018.

Albert N Katz and Christopher J Lee. The role of authorial intent in determining verbal irony and metaphor. *Metaphor and Symbol*, 8(4):257–279, 1993.

Max Kaufmann and Jugal Kalita. Syntactic normalization of twitter messages. In *International conference on natural language processing, Kharagpur, India*, volume 16, 2010.

Vishal Kaushal and Manasi Patwardhan. Emerging trends in personality identification using online social networks—a literature survey. *ACM Transactions on Knowledge Discovery from Data (TKDD)*, 12(2):1–30, 2018.

Paul Kay. Even. *Linguistics and Philosophy*, 13:59–111, 1990.

Mahmut Kaya and Hasan Şakir Bilge. Deep metric learning: A survey. *Symmetry*, 11(9):1066, 2019.

Amirmohammad Kazemeini, Sudipta Singha Roy, Robert E Mercer, and Erik Cambria. Interpretable representation learning for personality detection. In *ICDM Workshops*, pages 158–165, 2021.

Przemysław Kazienko and Erik Cambria. Towards responsible recommender systems. *IEEE Intelligent Systems*, 39(3):5–12, 2024.

Guolin Ke, Qi Meng, Thomas Finley, Taifeng Wang, Wei Chen, Weidong Ma, Qiwei Ye, and Tie-Yan Liu. Lightgbm: A highly efficient gradient boosting decision tree. *NeurIPS*, 30:3146–3154, 2017.

Sedrick Scott Keh, I Cheng, et al. Myers-briggs personality classification and personality-specific language generation using pre-trained language models. *arXiv preprint arXiv:1907.06333*, 2019.

Andrew Kehler. Current theories of centering for pronoun interpretation: A critical evaluation. *Computational linguistics*, 23(3):467–475, 1997.

Elma Kerz, Yu Qiao, Sourabh Zanwar, and Daniel Wiechmann. SPADE: A Big Five-mturk dataset of argumentative speech enriched with socio-demographics for personality detection. In *LREC*, pages 6405–6419, 2022.

Tanya Keshari and Suja Palaniswamy. Emotion recognition using feature-level fusion of facial expressions and body gestures. In *2019 International Conference on Communication and Electronics Systems (ICCES)*, pages 1184–1189, 2019.

HR Keshavarz and M Saniee Abadeh. MHSubLex: Using metaheuristic methods for subjectivity classification of microblogs. *Journal of AI and data mining*, 6(2): 341–353, 2018.

Mike Kestemont, Guy De Pauw, Renske van Nie, and Walter Daelemans. Lemmatization for variation-rich languages using deep learning. *Digital Scholarship in the Humanities*, 32(4):797–815, 2017.

Mitesh Khapra, Anup Kulkarni, Saurabh Sohoney, and Pushpak Bhattacharyya. All words domain adapted WSD: Finding a middle ground between supervision and unsupervision. In *ACL*, pages 1532–1541, 2010.

Aparup Khatua, Erik Cambria, Shirley S Ho, and Jin Cheon Na. Deciphering public opinion of nuclear energy on Twitter. In *IJCNN*, pages 1–8, 2020.

Mikhail Khodak, Nikunj Saunshi, and Kiran Vodrahalli. A large self-annotated corpus for sarcasm. In *Eleventh International Conference on Language Resources and Evaluation (LREC 2018)*, pages 641–646, 2018.

Richard Khoury. Phonetic normalization of microtext. In *2015 IEEE/ACM International Conference on Advances in Social Networks Analysis and Mining 2015*, pages 1600–1601, 2015.

Adam Kilgarriff and Martha Palmer. Introduction to the special issue on senseval. *Computers and the Humanities*, 34(1):1–13, 2000.

A Kim, Hyun-Je Song, Seong-Bae Park, et al. A two-step neural dialog state tracker for task-oriented dialog processing. *Computational Intelligence and Neuroscience*, 2018:1–12, 2018.

J-D Kim, Tomoko Ohta, Yuka Tateisi, and Jun'ichi Tsujii. Genia corpus—a semantically annotated corpus for bio-textmining. *Bioinformatics*, 19(suppl_1): i180–i182, 2003.

Jeong-Dong Kim, Jiseong Son, and Doo-Kwon Baik. CA 5W1H onto: ontological context-aware model based on 5W1H. *International Journal of Distributed Sensor Networks*, 8(3):247346, 2012.

Sang-Bum Kim, Hee-Cheol Seo, and Hae-Chang Rim. Information retrieval using word senses: root sense tagging approach. In *27th Annual International ACM SIGIR Conference on Research and Development in Information Retrieval*, pages 258–265, 2004.

Soo-Min Kim and Eduard Hovy. Determining the sentiment of opinions. In *COLING*, pages 1367–1373, 2004.

Soo-Min Kim and Eduard Hovy. Automatic detection of opinion bearing words and sentences. In *Companion Volume to the Proceedings of Conference including Posters/Demos and tutorial abstracts*, pages 61–66, 2005.

Soo-Min Kim and Eduard Hovy. Identifying and analyzing judgment opinions. In *NAACL-HLT*, pages 200–207, 2006.

Su Nam Kim, Olena Medelyan, Min-Yen Kan, and Timothy Baldwin. Semeval-2010 task 5: Automatic keyphrase extraction from scientific articles. In Katrin Erk and Carlo Strapparava, editors, *5th International Workshop on Semantic Evaluation*, pages 21–26, 2010.

Yoon Kim. Convolutional neural networks for sentence classification. In *EMNLP*, pages 1746–1751, 2014.

Yoon Kim, Yacine Jernite, David Sontag, and Alexander M Rush. Character-aware neural language models. In *AAAI*, 2016.

Barbara Ann Kipfer. *Roget's 21st century thesaurus in dictionary form (Third Edition)*. Bantam Dell, New York, NY, 2006.

Christo Kirov, Ryan Cotterell, John Sylak-Glassman, Géraldine Walther, Ekaterina Vylomova, Patrick Xia, Manaal Faruqui, Sabrina J Mielke, Arya D McCarthy, Sandra Kübler, et al. Unimorph 2.0: universal morphology. *arXiv preprint arXiv:1810.11101*, 2018.

Yuval Kirstain, Ori Ram, and Omer Levy. Coreference resolution without span representations. In *ACL-IJCNLP*, pages 14–19, 2021.

Tibor Kiss and Jan Strunk. Viewing sentence boundary detection as collocation identification. In *KONVENS*, volume 2002, pages 75–82, 2002.

Tibor Kiss and Jan Strunk. Unsupervised multilingual sentence boundary detection. *Computational Linguistics*, 32(4):485–525, 2006.

Günter Klambauer, Thomas Unterthiner, Andreas Mayr, and Sepp Hochreiter. Self-normalizing neural networks. In *31st international conference on neural information processing systems*, pages 972–981, 2017.

Jan-Christoph Klie, Michael Bugert, Beto Boullosa, Richard Eckart de Castilho, and Iryna Gurevych. The INCEpTION platform: Machine-assisted and knowledge-oriented interactive annotation. In *COLING*, pages 5–9, 2018.

Benjamin C. Knoll, Elizabeth A. Lindemann, Arian L. Albert, Genevieve B. Melton, and Serguei V. S. Pakhomov. Recurrent deep network models for clinical nlp tasks: Use case with sentence boundary disambiguation. *Studies in health technology and informatics*, 264(31437913):198–202, 2019.

Nozomi Kobayashi, Kentaro Inui, and Yuji Matsumoto. Extracting aspect-evaluation and aspect-of relations in opinion mining. In *EMNLP-CoNLL*, pages 1065–1074, 2007.

Vid Kocijan, Oana-Maria Camburu, Ana-Maria Cretu, Yordan Yordanov, Phil Blunsom, and Thomas Lukasiewicz. WikiCREM: A large unsupervised corpus for coreference resolution. In *EMNLP-IJCNLP*, pages 4303–4312, 2019.

Philipp Koehn, Hieu Hoang, Alexandra Birch, Chris Callison-Burch, Marcello Federico, Nicola Bertoldi, Brooke Cowan, Wade Shen, Christine Moran, Richard Zens, et al. Moses: Open source toolkit for statistical machine translation. In *ACL*, pages 177–180, 2007.

Rob Koeling. Chunking with maximum entropy models. In *Fourth Conference on Computational Natural Language Learning and the Second Learning Language in Logic Workshop*, 2000.

Keshav Kolluru, Vaibhav Adlakha, Samarth Aggarwal, Soumen Chakrabarti, et al. OpenIE6: Iterative grid labeling and coordination analysis for open information extraction. In *EMNLP*, pages 3748–3761, 2020.

Dan Kondratyuk. Cross-lingual lemmatization and morphology tagging with two-stage multilingual BERT fine-tuning. In *16th Workshop on Computational Research in Phonetics, Phonology, and Morphology*, pages 12–18, 2019.

Dan Kondratyuk and Milan Straka. 75 languages, 1 model: Parsing universal dependencies universally. *arXiv preprint arXiv:1904.02099*, 2019.

Daniel Kondratyuk, Tomáš Gavenčiak, Milan Straka, and Jan Hajič. Lemmatag: Jointly tagging and lemmatizing for morphologically-rich languages with BRNNs. *arXiv preprint arXiv:1808.03703*, 2018.

Fang Kong and Guodong Zhou. A tree kernel-based unified framework for Chinese zero anaphora resolution. In *EMNLP*, pages 882–891, 2010.

Prahlad Koratamaddi, Karan Wadhwani, Mridul Gupta, and Sriram G Sanjeevi. Market sentiment-aware deep reinforcement learning approach for stock portfolio allocation. *Engineering Science and Technology, an International Journal*, 24(4): 848–859, 2021.

Michal Kosinski, David Stillwell, and Thore Graepel. Private traits and attributes are predictable from digital records of human behavior. *National Academy of Sciences*, 110(15):5802–5805, 2013.

Ronak Kosti, Jose M Alvarez, Adria Recasens, and Agata Lapedriza. Context based emotion recognition using emotic dataset. *IEEE Transactions on Pattern Analysis and Machine Intelligence*, 42(11):2755–2766, 2019.

Martin Krallinger, Obdulia Rabal, Anália Lourenço, Julen Oyarzabal, and Alfonso Valencia. Information Retrieval and Text Mining Technologies for Chemistry. *Chemical Reviews*, 117(12):7673–7761, 2017.

Mikalai Krapivin, Aliaksandr Autaeu, and Maurizio Marchese. Large dataset for keyphrases extraction. Technical report, University of Trento, 2009.

Roger J Kreuz and Sam Glucksberg. How to be sarcastic: The echoic reminder theory of verbal irony. *Journal of Experimental Psychology: General*, 118(4): 374, 1989.

Manfred Krifka. Some remarks on polarity items. *Semantic Universals and Universal Semantics*, pages 150–189, 1992.

Manfred Krifka. The semantics and pragmatics of weak and strong polarity items in assertions. In *Semantics and Linguistic Theory*, volume 4, pages 195–219, 1994.

Saul A Kripke. Naming and necessity. In *Semantics of natural language*, pages 253–355. Springer, 1972.

Ranjay Krishna, Yuke Zhu, Oliver Groth, Justin Johnson, Kenji Hata, Joshua Kravitz, Stephanie Chen, Yannis Kalantidis, Li-Jia Li, David A. Shamma, Michael S. Bernstein, and Li Fei-Fei. Visual genome: Connecting language and vision using crowdsourced dense image annotations. *International Journal of Computer Vision*, 123(1):32–73, 2017.

Paul R Kroeger. *Analyzing Meaning: An Introduction to Semantics and Pragmatics*. Language Science Press, 2023.

Joshua Krook and Jan Blockx. Recommender systems, autonomy and user engagement. In *First International Symposium on Trustworthy Autonomous Systems*, pages 1–9, 2023.

Robert Krovetz and W Bruce Croft. Lexical ambiguity and information retrieval. *ACM Transactions on Information Systems (TOIS)*, 10(2):115–141, 1992.

Lun-Wei Ku, Yu-Ting Liang, and Hsin-Hsi Chen. Opinion extraction, summarization and tracking in news and blog corpora. In *AAAI*, pages 100–107, 2006.

Sandra Kübler and Desislava Zhekova. Singletons and coreference resolution evaluation. In *RANLP*, pages 261–267, 2011.

Taku Kudo and Yuji Matsumoto. Use of support vector learning for chunk identification. In *Fourth Conference on Computational Natural Language Learning and the Second Learning Language in Logic Workshop*, 2000.

Brian Kulis et al. Metric learning: A survey. *Foundations and Trends® in Machine Learning*, 5(4):287–364, 2013.

Akshi Kumar, Shubham Dikshit, and Victor Hugo C Albuquerque. Explainable artificial intelligence for sarcasm detection in dialogues. *Wireless Communications and Mobile Computing*, 2021, 2021.

Sawan Kumar, Sharmistha Jat, Karan Saxena, and Partha Talukdar. Zero-shot word sense disambiguation using sense definition embeddings. In *ACL*, pages 5670–5681, 2019a.

Sudhanshu Kumar, Mahendra Yadava, and Partha Pratim Roy. Fusion of EEG response and sentiment analysis of products review to predict customer satisfaction. *Information Fusion*, 52:41–52, 2019b.

Ziva Kunda. The case for motivated reasoning. *Psychological Bulletin*, 108(3):480, 1990.

Florian Kunneman, Christine Liebrecht, Margot Van Mulken, and Antal Van den Bosch. Signaling sarcasm: From hyperbole to hashtag. *Information Processing & Management*, 51(4):500–509, 2015.

Yen-Ling Kuo, Jong-Chuan Lee, Kai-yang Chiang, Rex Wang, Edward Shen, Cheng-wei Chan, and Jane Yung-jen Hsu. Community-based game design: experiments on social games for commonsense data collection. In *ACM SIGKDD Workshop on Human Computation*, pages 15–22, 2009.

Julian Kupiec. Robust part-of-speech tagging using a hidden markov model. *Computer Speech & Language*, 6(3):225–242, 1992.

Onur Kuru, Ozan Arkan Can, and Deniz Yuret. CharNER: Character-level named entity recognition. In *COLING*, pages 911–921, 2016.

John Lafferty and Chengxiang Zhai. Document language models, query models, and risk minimization for information retrieval. In *24th Annual International ACM SIGIR Conference on Research and Development in Information Retrieval*, pages 111–119, 2001.

John D. Lafferty, Andrew McCallum, and Fernando C. N. Pereira. Conditional random fields: Probabilistic models for segmenting and labeling sequence data. In *ICML*, ICML '01, pages 282–289, San Francisco, CA, USA, 2001. Morgan Kaufmann Publishers Inc. ISBN 1558607781.

Brenden M Lake, Tomer D Ullman, Joshua B Tenenbaum, and Samuel J Gershman. Building machines that learn and think like people. *Behavioral and brain sciences*, 40:e253, 2017.

George Lakoff. *Master Metaphor List*. University of California, 1994.

George Lakoff and Mark Johnson. *Metaphors We Live by*. University of Chicago press, 1980.

Guillaume Lample and François Charton. Deep learning for symbolic mathematics. In *International Conference on Learning Representations*, 2019.

Zhenzhong Lan, Mingda Chen, Sebastian Goodman, Kevin Gimpel, Piyush Sharma, and Radu Soricut. Albert: A lite bert for self-supervised learning of language representations. In *International Conference on Learning Representations*, pages 1–17, 2020.

Mark J Landau, Brian P Meier, and Lucas A Keefer. A metaphor-enriched social cognition. *Psychological Bulletin*, 136(6):1045, 2010.

Lukas Lange, Heike Adel, and Jannik Strötgen. Closing the gap: Joint de-identification and concept extraction in the clinical domain. In Dan Jurafsky, Joyce Chai, Natalie Schluter, and Joel R. Tetreault, editors, *ACL*, pages 6945–6952, 2020.

Joseph LaPorte. Rigid designators for properties. *Philosophical Studies*, 130(2): 321–336, 2006.

Shalom Lappin and Herbert J Leass. An algorithm for pronominal anaphora resolution. *Computational Linguistics*, 20(4):535–561, 1994.

Christine Largeron, Christophe Moulin, and Mathias Géry. Entropy based feature selection for text categorization. In *2011 ACM Symposium on Applied Computing*, pages 924–928, 2011.

Ida Unmack Larsen, Tua Vinther-Jensen, Anders Gade, Jørgen Erik Nielsen, and Asmus Vogel. Do I misconstrue? Sarcasm detection, emotion recognition, and theory of mind in Huntington disease. *Neuropsychology*, 30(2):181, 2016.

Siddique Latif, Rajib Rana, Sara Khalifa, Raja Jurdak, Junaid Qadir, and Bjoern W Schuller. Survey of deep representation learning for speech emotion recognition. *IEEE Transactions on Affective Computing*, 2021.

Minh Le, Marten Postma, Jacopo Urbani, and Piek Vossen. A deep dive into word sense disambiguation with lstm. In *COLING*, pages 354–365, 2018.

Thi Thuy Le, Thanh Hung Vo, Duc Trung Mai, Than Tho Quan, and Tuoi Thi Phan. Sentiment analysis using anaphoric coreference resolution and ontology inference. In *International Workshop on Multi-disciplinary Trends in Artificial Intelligence*, pages 297–303, 2016.

Ronan Le Nagard and Philipp Koehn. Aiding pronoun translation with co-reference resolution. In *Joint Fifth Workshop on Statistical Machine Translation and MetricsMATR*, pages 252–261, 2010.

Claudia Leacock, Geoffrey Towell, and Ellen M Voorhees. Corpus-based statistical sense resolution. In *HLT*, pages 260–265, 1993.

Yann LeCun, Léon Bottou, Yoshua Bengio, and Patrick Haffner. Gradient-based learning applied to document recognition. *IEEE*, 86(11):2278–2324, 1998.

Bongseok Lee and Yong Suk Choi. Graph based network with contextualized representations of turns in dialogue. In *EMNLP*, pages 443–455, 2021.

Heeyoung Lee, Angel Chang, Yves Peirsman, Nathanael Chambers, Mihai Surdeanu, and Dan Jurafsky. Deterministic coreference resolution based on entity-centric, precision-ranked rules. *Computational Linguistics*, 39(4):885–916, 2013.

Heeyoung Lee, Mihai Surdeanu, and Dan Jurafsky. A scaffolding approach to coreference resolution integrating statistical and rule-based models. *Natural Language Engineering*, 23(5):733–762, 2017a.

Joosung Lee and Wooin Lee. CoMPM: Context modeling with speaker's pre-trained memory tracking for emotion recognition in conversation. In *NAACL-HLT*, pages 5669–5679, 2022.

Kenton Lee, Luheng He, Mike Lewis, and Luke Zettlemoyer. End-to-end neural coreference resolution. In *EMNLP*, pages 188–197, 2017b.

Kenton Lee, Luheng He, and Luke Zettlemoyer. Higher-order coreference resolution with coarse-to-fine inference. In *NAACL-HLT*, pages 687–692, 2018.

Young-Suk Lee and Laurence Horn. Any as indefinite plus even. *Manuscript, Yale University*, 1994.

Yue-Shi Lee and Yu-Chieh Wu. A robust multilingual portable phrase chunking system. *Expert Systems with Applications*, 33(3):590–599, 2007.

Samuel Leeman-Munk, James Lester, and James Cox. Ncsu_sas_sam: Deep encoding and reconstruction for normalization of noisy text. In *Workshop on Noisy User-generated Text*, pages 154–161, 2015.

Douglas B. Lenat. CYC: A large-scale investment in knowledge infrastructure. *Communications of The ACM*, 38(11):32–38, 1995.

Michael Lesk. Automatic sense disambiguation using machine readable dictionaries: how to tell a pine cone from an ice cream cone. In *5th Annual International Conference on Systems Documentation*, pages 24–26, 1986.

Hector J. Levesque. The winograd schema challenge. In *Logical Formalizations of Commonsense Reasoning, Papers from the 2011 AAAI Spring Symposium, Technical Report SS-11-06*, pages 1–1, 2011.

Hector J Levesque, Ernest Davis, and Leora Morgenstern. The Winograd schema challenge. In *Thirteenth International Conference on Principles of Knowledge Representation and Reasoning*, pages 552–561, 2012.

Mike Lewis, Yinhan Liu, Naman Goyal, Marjan Ghazvininejad, Abdelrahman Mohamed, Omer Levy, Veselin Stoyanov, and Luke Zettlemoyer. BART: Denoising sequence-to-sequence pre-training for natural language generation, translation, and comprehension. In *ACL*, pages 7871–7880, 2020.

Baoli Li, Yandong Liu, and Eugene Agichtein. CoCQA: Co-training over questions and answers with an application to predicting question subjectivity orientation. In *EMNLP*, pages 937–946, 2008a.

Baoli Li, Yandong Liu, Ashwin Ram, Ernest V Garcia, and Eugene Agichtein. Exploring question subjectivity prediction in community qa. In *31st Annual International ACM SIGIR Conference on Research and Development in Information Retrieval*, pages 735–736, 2008b.

Chen Li and Yang Liu. Normalization of text messages using character-and phone-based machine translation approaches. In *Thirteenth Annual Conference of the International Speech Communication Association*, 2012.

Chen Li, Marco Donizelli, Nicolas Rodriguez, Harish Dharuri, Lukas Endler, Vijayalakshmi Chelliah, Lu Li, Enuo He, Arnaud Henry, Melanie I Stefan, et al. BioModels database: An enhanced, curated and annotated resource for published quantitative kinetic models. *BMC Systems Biology*, 4(1):1–14, 2010.

Chen Li, Xiangdong Xu, Guanghua Zhou, Kai He, Tianliang Qi, Wei Zhang, Feng Tian, Qinghua Zheng, Jianping Hu, et al. Implementation of national health informatization in China: Survey about the status quo. *JMIR Medical Informatics*, 7(1):e12238, 2019a.

Chengxi Li, Feiyu Gao, Jiajun Bu, Lu Xu, Xiang Chen, Yu Gu, Zirui Shao, Qi Zheng, Ningyu Zhang, Yongpan Wang, et al. Sentiprompt: Sentiment knowledge enhanced prompt-tuning for aspect-based sentiment analysis. *arXiv preprint arXiv:2109.08306*, 2021a.

Chuang Li, Hengchang Hu, Yan Zhang, Min-Yen Kan, and Haizhou Li. A conversation is worth a thousand recommendations: A survey of holistic conversational recommender systems. *Knowledge-aware and Conversational Recommender Systems*, 3560:7–20, 2023a.

Fei Li, Meishan Zhang, Guohong Fu, and Donghong Ji. A neural joint model for entity and relation extraction from biomedical text. *BMC Bioinformatics*, 18(1): 1–11, 2017a.

Feng-Lin Li, Hehong Chen, Guohai Xu, Tian Qiu, Feng Ji, Ji Zhang, and Haiqing Chen. AliMeKG: Domain knowledge graph construction and application in e-commerce. In *29th ACM International Conference on Information & Knowledge Management*, pages 2581–2588, 2020a.

Hongsong Li, Kenny Q Zhu, and Haixun Wang. Data-driven metaphor recognition and explanation. *Transactions of the Association for Computational Linguistics*, 1:379–390, 2013.

Jiangnan Li, Zheng Lin, Peng Fu, and Weiping Wang. Past, present, and future: Conversational emotion recognition through structural modeling of psychological knowledge. In *EMNLP*, pages 1204–1214, 2021b.

Jing Li, Aixin Sun, Jianglei Han, and Chenliang Li. A survey on deep learning for named entity recognition. *IEEE Transactions on Knowledge and Data Engineering*, 34(1):50–70, 2020b.

Keqian Li, Hanwen Zha, Yu Su, and Xifeng Yan. Concept mining via embedding. In *ICDM*, pages 267–276, 2018a.

Linfeng Li, Peng Wang, Jun Yan, Yao Wang, Simin Li, Jinpeng Jiang, Zhe Sun, Buzhou Tang, Tsung-Hui Chang, Shenghui Wang, et al. Real-world data medical knowledge graph: construction and applications. *Artificial Intelligence in Medicine*, 103:101817, 2020c.

Ning Li, Chi-Yin Chow, and Jia-Dong Zhang. Emova: A semi-supervised end-to-end moving-window attentive framework for aspect mining. In *Advances in Knowledge Discovery and Data Mining: 24th Pacific-Asia Conference, PAKDD 2020, Singapore, May 11–14, 2020, Proceedings, Part II 24*, pages 811–823, 2020d.

Peng Li and Heng Huang. UTA DLNLP at semeval-2016 task 12: Deep learning based natural language processing system for clinical information identification from clinical notes and pathology reports. In Steven Bethard, Daniel M. Cer, Marine Carpuat, David Jurgens, Preslav Nakov, and Torsten Zesch, editors, *10th International Workshop on Semantic Evaluation*, pages 1268–1273, 2016.

Qing Li, Siyuan Huang, Yining Hong, Yixin Chen, Ying Nian Wu, and Song-Chun Zhu. Closed loop neural-symbolic learning via integrating neural perception, grammar parsing, and symbolic reasoning. In *ICML*, pages 5884–5894, 2020e.

Wei Li, Kun Guo, Yong Shi, Luyao Zhu, and Yuanchun Zheng. Dwwp: Domain-specific new words detection and word propagation system for sentiment analysis in the tourism domain. *Knowledge-Based Systems*, 146:203–214, 2018b.

Wei Li, Luyao Zhu, Yong Shi, Kun Guo, and Erik Cambria. User reviews: Sentiment analysis using lexicon integrated two-channel CNN-LSTM family models. *Applied Soft Computing*, 94:106435, 2020f.

Wei Li, Luyao Zhu, and Erik Cambria. Taylor's theorem: A new perspective for neural tensor networks. *Knowledge-Based Systems*, 228:107258, 2021c.

Wei Li, Wei Shao, Shaoxiong Ji, and Erik Cambria. BiERU: Bidirectional emotional recurrent unit for conversational sentiment analysis. *Neurocomputing*, 467:73–82, 2022.

Wei Li, Luyao Zhu, Rui Mao, and Erik Cambria. SKIER: A symbolic knowledge integrated model for conversational emotion recognition. In *AAAI*, pages 13121–13129, 2023b.

Xiaoya Li, Jingrong Feng, Yuxian Meng, Qinghong Han, Fei Wu, and Jiwei Li. A unified MRC framework for named entity recognition. In *ACL*, pages 5849–5859, 2020g.

Xin Li, Lidong Bing, Piji Li, and Wai Lam. A unified model for opinion target extraction and target sentiment prediction. In *AAAI*, pages 6714–6721, 2019b.

Xiujun Li, Yun-Nung Chen, Lihong Li, Jianfeng Gao, and Asli Celikyilmaz. End-to-end task-completion neural dialogue systems. In *IJCNLP*, pages 733–743, 2017b.

Yang Li, Kangbo Liu, Ranjan Satapathy, Suhang Wang, and Erik Cambria. Recent developments in recommender systems: A survey. *IEEE Computational Intelligence Magazine*, 19(2):78–95, 2024.

Yufei Li, Xiaoyong Ma, Xiangyu Zhou, Pengzhen Cheng, Kai He, and Chen Li. Knowledge enhanced lstm for coreference resolution on biomedical texts. *Bioinformatics*, 37(17):2699–2705, 2021d.

Zisheng Li, Jun-ichi Imai, and Masahide Kaneko. Facial-component-based bag of words and phog descriptor for facial expression recognition. In *2009 IEEE International Conference on Systems, Man and Cybernetics*, pages 1353–1358, 2009.

Bin Liang, Chenwei Lou, Xiang Li, Min Yang, Lin Gui, Yulan He, Wenjie Pei, and Ruifeng Xu. Multi-modal sarcasm detection via cross-modal graph convolutional network. In *ACL*, pages 1767–1777, 2022a.

Bin Liang, Hang Su, Lin Gui, Erik Cambria, and Ruifeng Xu. Aspect-based sentiment analysis via affective knowledge enhanced graph convolutional networks. *Knowledge-Based Systems*, 235:107643, 2022b.

Bin Liang, Lin Gui, Yulan He, Erik Cambria, and Ruifeng Xu. Fusion and discrimination: A multimodal graph contrastive learning framework for multimodal sarcasm detection. *IEEE Transactions on Affective Computing*, 15, 2024.

Tyne Liang and Dian-Song Wu. Automatic pronominal anaphora resolution in English texts. In *Research on Computational Linguistics Conference*, pages 111–127, 2003.

Yu Liang and Martijn C Willemsen. Promoting music exploration through personalized nudging in a genre exploration recommender. *International Journal of Human–Computer Interaction*, 39(7):1495–1518, 2023.

Yunlong Liang, Fandong Meng, Jinchao Zhang, Yufeng Chen, Jinan Xu, and Jie Zhou. An iterative multi-knowledge transfer network for aspect-based sentiment analysis. *arXiv preprint arXiv:2004.01935*, 2020.

Jian Liao, Suge Wang, and Deyu Li. Identification of fact-implied implicit sentiment based on multi-level semantic fused representation. *Knowledge-Based Systems*, 165:197–207, 2019.

Jian Liao, Min Wang, Xin Chen, Suge Wang, and Kai Zhang. Dynamic commonsense knowledge fused method for Chinese implicit sentiment analysis. *Information Processing & Management*, 59(3):102934, 2022.

Rensis Likert. A technique for the measurement of attitudes. *Archives of Psychology*, 1932.

Chenghua Lin, Yulan He, and Richard Everson. Sentence subjectivity detection with weakly-supervised learning. In *IJCNLP*, pages 1153–1161, 2011.

Jerry Chun-Wei Lin, Yinan Shao, Ji Zhang, and Unil Yun. Enhanced sequence labeling based on latent variable conditional random fields. *Neurocomputing*, 403:431–440, 2020.

Wang Ling, Chris Dyer, Alan W Black, and Isabel Trancoso. Two/too simple adaptations of word2vec for syntax problems. In *ACL-HLT*, pages 1299–1304, 2015a.

Wang Ling, Tiago Luís, Luís Marujo, Ramón Fernandez Astudillo, Silvio Amir, Chris Dyer, Alan W Black, and Isabel Trancoso. Finding function in form: Compositional character models for open vocabulary word representation. *arXiv preprint arXiv:1508.02096*, 2015b.

Yan Ling, Jianfei Yu, and Rui Xia. Vision-language pre-training for multimodal aspect-based sentiment analysis. In *ACL*, pages 2149–2159, 2022.

Carolyn E Lipscomb. Medical subject headings (MeSH). *Bulletin of the Medical Library Association*, 88(3):265, 2000.

Seymour Martin Lipset. *The Confidence Gap: Business, Labor, and Government in the Public Mind*. The Johns Hopkins University Press, 1987.

Ken Litkowski. SensEval-3 task: Word sense disambiguation of wordnet glosses. In *SENSEVAL-3, the Third International Workshop on the Evaluation of Systems for the Semantic Analysis of Text*, pages 13–16, 2004.

Nick Littlestone. Learning quickly when irrelevant attributes abound: A new linear-threshold algorithm. *Machine learning*, 2(4):285–318, 1988.

Bing Liu et al. Sentiment analysis and subjectivity. *Handbook of Natural Language Processing*, 2(2010):627–666, 2010.

Chia-Wei Liu, Ryan Lowe, Iulian Vlad Serban, Mike Noseworthy, Laurent Charlin, and Joelle Pineau. How not to evaluate your dialogue system: An empirical study of unsupervised evaluation metrics for dialogue response generation. In *PEMNLP*, pages 2122–2132, 2016.

Emmy Liu, Chen Cui, Kenneth Zheng, and Graham Neubig. Testing the ability of language models to interpret figurative language. *arXiv e-prints*, pages arXiv–2204, 2022a.

Fei Liu, Fuliang Weng, Bingqing Wang, and Yang Liu. Insertion, deletion, or substitution? normalizing text messages without pre-categorization nor supervision. In *ACL-HLT*, pages 71–76, 2011.

Fei Liu, Fuliang Weng, and Xiao Jiang. A broad-coverage normalization system for social media language. In *ACL*, pages 1035–1044, 2012.

Fei Liu, Julien Perez, and Scott Nowson. A language-independent and compositional model for personality trait recognition from short texts. In *EACL*, pages 754–764, 2017a.

Frederick Liu, Han Lu, and Graham Neubig. Handling homographs in neural machine translation. In *NAACL-HLT*, pages 1336–1345, 2018a.

Haochen Liu, Jamell Dacon, Wenqi Fan, Hui Liu, Zitao Liu, and Jiliang Tang. Does gender matter? towards fairness in dialogue systems. *arXiv preprint arXiv:1910.10486*, 2019a.

Jian Liu, Zhiyang Teng, Leyang Cui, Hanmeng Liu, and Yue Zhang. Solving aspect category sentiment analysis as a text generation task. *arXiv preprint arXiv:2110.07310*, 2021a.

Jing Liu, Yue Wang, Lihua Huang, Chenghong Zhang, and Songzheng Zhao. Identifying adverse drug reaction-related text from social media: A multi-view active learning approach with various document representations. *Information*, 13(4): 189, 2022b.

Jingping Liu, Tao Chen, Chao Wang, Jiaqing Liang, Lihan Chen, Yanghua Xiao, Yunwen Chen, and Ke Jin. Vocsk: Verb-oriented commonsense knowledge mining with taxonomy-guided induction. *Artificial Intelligence*, 310:103744, 2022c.

Pan Liu, Yanming Guo, Fenglei Wang, and Guohui Li. Chinese named entity recognition: The state of the art. *Neurocomputing*, 473:37–53, 2022d.

Peng Liu, Wei Chen, Gaoyan Ou, Tengjiao Wang, Dongqing Yang, and Kai Lei. Sarcasm detection in social media based on imbalanced classification. In *International Conference on Web-Age Information Management*, pages 459–471, 2014.

Qi Liu, Zai Huang, Zhenya Huang, Chuanren Liu, Enhong Chen, Yu Su, and Guoping Hu. Finding similar exercises in online education systems. In Yike Guo and Faisal Farooq, editors, *SIGKDD*, pages 1821–1830, 2018b.

Qian Liu, Heyan Huang, Guangquan Zhang, Yang Gao, Junyu Xuan, and Jie Lu. Semantic structure-based word embedding by incorporating concept convergence and word divergence. In *AAAI*, pages 5261–5268, 2018c.

Qian Liu, Rui Mao, Xiubo Geng, and Erik Cambria. Semantic matching in machine reading comprehension: An empirical study. *Information Processing & Management*, 60(2):103145, 2023a.

Qian Liu, Xiubo Geng, Yu Wang, Erik Cambria, and Daxin Jiang. Disentangled retrieval and reasoning for implicit question answering. *IEEE Transactions on Neural Networks and Learning Systems*, 35(6):7804–7815, 2024a.

Qian Liu, Sooji Han, Yang Li, Erik Cambria, and Kenneth Kwok. PrimeNet: A framework for commonsense knowledge representation and reasoning based on conceptual primitives. *Cognitive Computation*, 2024b.

Ruicheng Liu, Guanyi Chen, Rui Mao, and Erik Cambria. A multi-task learning model for gold-two-mention co-reference resolution. In *IJCNN*, pages 1–8, 2023b.

Ruicheng Liu, Rui Mao, Anh Tuan Luu, and Erik Cambria. A brief survey on advances in coreference resolution. *Artificial Intelligence Review*, 56:14439–14481, 2023c.

Siyang Liu, Chujie Zheng, Orianna Demasi, Sahand Sabour, Yu Li, Zhou Yu, Yong Jiang, and Minlie Huang. Towards emotional support dialog systems. In *ACL-IJCNLP*, pages 3469–3483, 2021b.

Ting Liu, Yiming Cui, Qingyu Yin, Wei-Nan Zhang, Shijin Wang, and Guoping Hu. Generating and exploiting large-scale pseudo training data for zero pronoun resolution. In *ACL*, pages 102–111, 2017b.

Xiao Liu, Yanan Zheng, Zhengxiao Du, Ming Ding, Yujie Qian, Zhilin Yang, and Jie Tang. GPT understands, too. *arXiv preprint arXiv:2103.10385*, 2021c.

Xiao Liu, Kaixuan Ji, Yicheng Fu, Weng Tam, Zhengxiao Du, Zhilin Yang, and Jie Tang. P-tuning: Prompt tuning can be comparable to fine-tuning across scales and tasks. In *ACL*, pages 61–68, 2022e.

Xiaodong Liu, Pengcheng He, Weizhu Chen, and Jianfeng Gao. Multi-task deep neural networks for natural language understanding. In *ACL*, pages 4487–4496, 2019b.

Yang Liu, Andreas Stolcke, Elizabeth Shriberg, and Mary Harper. Comparing and combining generative and posterior probability models: Some advances in sentence boundary detection in speech. In *EMNLP*, pages 64–71, 2004.

Yang Liu, Andreas Stolcke, Elizabeth Shriberg, and Mary Harper. Using conditional random fields for sentence boundary detection in speech. In *ACL*, pages 451–458, 2005.

Yang Liu, Nitesh V. Chawla, Mary P. Harper, Elizabeth Shriberg, and Andreas Stolcke. A study in machine learning from imbalanced data for sentence boundary detection in speech. *Computer Speech & Language*, 20(4):468–494, 2006.

Ye Liu, Han Wu, Zhenya Huang, Hao Wang, Jianhui Ma, Qi Liu, Enhong Chen, Hanqing Tao, and Ke Rui. Technical phrase extraction for patent mining: A multi-level approach. In *ICDM*, pages 1142–1147, 2020a.

Yinhan Liu, Myle Ott, Naman Goyal, Jingfei Du, Mandar Joshi, Danqi Chen, Omer Levy, Mike Lewis, Luke Zettlemoyer, and Veselin Stoyanov. RoBERTa: A robustly optimized BERT pretraining approach. *arXiv e-prints*, pages arXiv–1907, 2019c.

Yupeng Liu, Guodong Li, and Xiaochen Zhang. Semi-Markov CRF model based on stacked neural Bi-LSTM for sequence labeling. In *2020 IEEE 3rd International Conference of Safe Production and Informatization (IICSPI)*, pages 19–23, 2020b. doi: 10.1109/IICSPI51290.2020.9332321.

Zengjian Liu, Ming Yang, Xiaolong Wang, Qingcai Chen, Buzhou Tang, Zhe Wang, and Hua Xu. Entity recognition from clinical texts via recurrent neural network. *BMC Medical Informatics Decis. Mak.*, 17(2):53–61, 2017c.

Zhengyuan Liu, Ke Shi, and Nancy Chen. Coreference-aware dialogue summarization. In *22nd Annual Meeting of the Special Interest Group on Discourse and Dialogue*, pages 509–519, 2021d.

Guido Löhr. What are abstract concepts? on lexical ambiguity and concreteness ratings. *Review of Philosophy and Psychology*, 13(3):549–566, 2022.

Adam Lopez. Statistical machine translation. *ACM Computing Surveys*, 40(3):1–49, 2008.

Daniel Loureiro and Alipio Jorge. Language modelling makes sense: Propagating representations through wordnet for full-coverage word sense disambiguation. In *ACL*, pages 5682–5691, 2019.

Ismini Lourentzou, Kabir Manghnani, and ChengXiang Zhai. Adapting sequence to sequence models for text normalization in social media. In *International AAAI Conference on Web and Social Media*, volume 13, pages 335–345, 2019.

Xinyuan Lu and Min-Yen Kan. Improving recommendation systems with user personality inferred from product reviews. *arXiv preprint arXiv:2303.05039*, 2023.

Zhiyun Lu, Liangliang Cao, Yu Zhang, Chung-Cheng Chiu, and James Fan. Speech sentiment analysis via pre-trained features from end-to-end ASR models. In *ICASSP*, pages 7149–7153, 2020.

Hans Peter Luhn. A statistical approach to mechanized encoding and searching of literary information. *IBM Journal of research and development*, 1(4):309–317, 1957.

Edwin Lunando and Ayu Purwarianti. Indonesian social media sentiment analysis with sarcasm detection. In *2013 International Conference on Advanced Computer Science and Information Systems (ICACSIS)*, pages 195–198, 2013.

Hongyin Luo and Jim Glass. Learning word representations with cross-sentence dependency for end-to-end co-reference resolution. In *EMNLP*, pages 4829–4833, 2018.

Huaishao Luo, Tianrui Li, Bing Liu, and Junbo Zhang. DOER: Dual cross-shared rnn for aspect term-polarity co-extraction. *arXiv preprint arXiv:1906.01794*, 2019a.

Huaishao Luo, Lei Ji, Tianrui Li, Nan Duan, and Daxin Jiang. GRACE: Gradient harmonized and cascaded labeling for aspect-based sentiment analysis. *arXiv preprint arXiv:2009.10557*, 2020.

Xiaoqiang Luo. On coreference resolution performance metrics. In *EMNLP-HLT*, pages 25–32, 2005.

Xiaoqiang Luo and Sameer Pradhan. Evaluation metrics. In *Anaphora Resolution*, pages 141–163. Springer, 2016.

Zhiyi Luo, Shanshan Huang, and Kenny Q Zhu. Knowledge empowered prominent aspect extraction from product reviews. *Information Processing & Management*, 56(3):408–423, 2019b.

Minh-Thang Luong, Hieu Pham, and Christopher D Manning. Effective approaches to attention-based neural machine translation. *arXiv preprint arXiv:1508.04025*, 2015.

M. Lusetti, T. Ruzsics, A. Göhring, T. Samardzic, and E. Stark. Encoder-decoder methods for text normalization. In *VarDial@COLING 2018*, 2018.

Kim Luyckx and Walter Daelemans. Using syntactic features to predict author personality from text. *Digital Humanities*, 2008:146–9, 2008.

Veronica Lynn, Niranjan Balasubramanian, and H Andrew Schwartz. Hierarchical modeling for user personality prediction: The role of message-level attention. In *ACL*, pages 5306–5316, 2020.

Dimitrios P Lyras, Kyriakos N Sgarbas, and Nikolaos D Fakotakis. Using the Levenshtein edit distance for automatic lemmatization: A case study for modern Greek and English. In *19th IEEE International Conference on Tools with Artificial Intelligence (ICTAI 2007)*, volume 2, pages 428–435, 2007.

Ji Ma, Jingbo Zhu, Tong Xiao, and Nan Yang. Easy-first POS tagging and dependency parsing with beam search. In *ACL*, pages 110–114, 2013.

Kaixin Ma, Filip Ilievski, Jonathan Francis, Yonatan Bisk, Eric Nyberg, and Alessandro Oltramari. Knowledge-driven data construction for zero-shot evaluation in commonsense question answering. In *AAAI*, pages 13507–13515, 2021.

Ruotian Ma, Xin Zhou, Tao Gui, Yiding Tan, Linyang Li, Qi Zhang, and Xuanjing Huang. Template-free prompt tuning for few-shot NER. In *NAACL-HLT*, pages 5721–5732, 2022a.

Xuezhe Ma and Eduard Hovy. End-to-end sequence labeling via bi-directional LSTM-CNNs-CRF. *arXiv preprint arXiv:1603.01354*, 2016.

Youmi Ma, Tatsuya Hiraoka, and Naoaki Okazaki. Named entity recognition and relation extraction using enhanced table filling by contextualized representations. *Journal of Natural Language Processing*, 29(1):187–223, 2022b.

Yu Ma, Rui Mao, Qika Lin, Peng Wu, and Erik Cambria. Multi-source aggregated classification for stock price movement prediction. *Information Fusion*, 91:515–528, 2023.

Yu Ma, Rui Mao, Qika Lin, Peng Wu, and Erik Cambria. Quantitative stock portfolio optimization by multi-task learning risk and return. *Information Fusion*, 104: 102165, 2024.

Yukun Ma, Haiyun Peng, and Erik Cambria. Targeted aspect-based sentiment analysis via embedding commonsense knowledge into an attentive LSTM. In *AAAI*, pages 5876–5883, 2018.

Mohamed Maamouri, Ann Bies, Tim Buckwalter, and Wigdan Mekki. The Penn Arabic Treebank: Building a large-scale annotated Arabic corpus. In *NEMLAR conference on Arabic language resources and tools*, volume 27, pages 466–467, 2004.

Mohamed Maamouri, Sondos Krouna, Dalila Tabessi, Nadia Hamrouni, and Nizar Habash. Egyptian Arabic morphological annotation guidelines, 2012.

Andrew Maas, Raymond E Daly, Peter T Pham, Dan Huang, Andrew Y Ng, and Christopher Potts. Learning word vectors for sentiment analysis. In *ACL-HLT*, pages 142–150, 2011.

David Magnusson and Bertil Torestad. A holistic view of personality: A model revisited. *Annual Review of Psychology*, 44:427, 1993.

Ahsan Mahmood, Hikmat Ullah Khan, Zahoor ur Rehman, and Wahab Khan. Query based information retrieval and knowledge extraction using hadith datasets. In *2017 13th International Conference on Emerging Technologies (ICET)*, pages 1–6, 2017. doi: 10.1109/ICET.2017.8281714.

François Mairesse, Marilyn A Walker, Matthias R Mehl, and Roger K Moore. Using linguistic cues for the automatic recognition of personality in conversation and text. *Journal of Artificial Intelligence Research*, 30:457–500, 2007.

Navonil Majumder, Soujanya Poria, Alexander Gelbukh, and Erik Cambria. Deep learning-based document modeling for personality detection from text. *IEEE Intelligent Systems*, 32(2):74–79, 2017.

Chaitanya Malaviya, Shijie Wu, and Ryan Cotterell. A simple joint model for improved contextual neural lemmatization. *arXiv preprint arXiv:1904.02306*, 2019.

Pekka Malo, Ankur Sinha, Pekka Korhonen, Jyrki Wallenius, and Pyry Takala. Good debt or bad debt: Detecting semantic orientations in economic texts. *Journal of the Association for Information Science and Technology*, 65(4):782–796, 2014.

Suresh Manandhar, Sašo Džeroski, and Tomaž Erjavec. Learning multilingual morphology with clog. In *International Conference on Inductive Logic Programming*, pages 135–144, 1998.

Enrique Manjavacas, Ákos Kádár, and Mike Kestemont. Improving lemmatization of non-standard languages with joint learning. *arXiv preprint arXiv:1903.06939*, 2019.

Christopher D. Manning. Part-of-speech tagging from 97% to 100%: Is it time for some linguistics? In Alexander F. Gelbukh, editor, *Computational Linguistics and Intelligent Text Processing*, pages 171–189, Berlin, Heidelberg, 2011. Springer Berlin Heidelberg. ISBN 978-3-642-19400-9.

Christopher D Manning, Mihai Surdeanu, John Bauer, Jenny Rose Finkel, Steven Bethard, and David McClosky. The Stanford CoreNLP natural language processing toolkit. In *ACL*, pages 55–60, 2014.

Rui Mao and Xiao Li. Bridging towers of multi-task learning with a gating mechanism for aspect-based sentiment analysis and sequential metaphor identification. In *AAAI*, volume 35, pages 13534–13542, 2021.

Rui Mao, Chenghua Lin, and Frank Guerin. Word embedding and WordNet based metaphor identification and interpretation. In *ACL*, pages 1222–1231, 2018.

Rui Mao, Chenghua Lin, and Frank Guerin. End-to-end sequential metaphor identification inspired by linguistic theories. In *ACL*, pages 3888–3898, 2019.

Rui Mao, Chenghua Lin, and Frank Guerin. Combining pre-trained word embeddings and linguistic features for sequential metaphor identification. *arXiv preprint arXiv:2104.03285*, 2021a.

Rui Mao, Xiao Li, Mengshi Ge, and Erik Cambria. MetaPro: A computational metaphor processing model for text pre-processing. *Information Fusion*, 86-87: 30–43, 2022a.

Rui Mao, Kelvin Du, Yu Ma, Luyao Zhu, and Erik Cambria. Discovering the cognition behind language: Financial metaphor analysis with MetaPro. In *ICDM*, pages 1211–1216, 2023a.

Rui Mao, Xiao Li, Kai He, Mengshi Ge, and Erik Cambria. MetaPro Online: A computational metaphor processing online system. In *ACL*, pages 127–135, 2023b.

Rui Mao, Qian Liu, Kai He, Wei Li, and Erik Cambria. The biases of pre-trained language models: An empirical study on prompt-based sentiment analysis and emotion detection. *IEEE Transactions on Affective Computing*, 14(3):1743–1753, 2023c.

Rui Mao, Guanyi Chen, Xulang Zhang, Frank Guerin, and Erik Cambria. GPTEval: A survey on assessments of ChatGPT and GPT-4. In *LREC-COLING*, pages 7844–7866, 2024a.

Rui Mao, Kai He, Claudia Beth Ong, Qian Liu, and Erik Cambria. MetaPro 2.0: Computational metaphor processing on the effectiveness of anomalous language modeling. In *ACL*, pages 9891–9908, 2024b.

Rui Mao, Kai He, Xulang Zhang, Guanyi Chen, Jinjie Ni, Zonglin Yang, and Erik Cambria. A survey on semantic processing techniques. *Information Fusion*, 101 (101988), 2024c.

Rui Mao, Qika Lin, Qiawen Liu, Gianmarco Mengaldo, and Erik Cambria. Understanding public perception towards weather disasters through the lens of metaphor. In *IJCAI*, 2024d.

Rui Mao, Tianwei Zhang, Qian Liu, Amir Hussain, and Erik Cambria. Unveiling diplomatic narratives: Analyzing United Nations Security Council debates through metaphorical cognition. In *Annual Meeting of the Cognitive Science Society (CogSci)*, pages 1709–1716, 2024e.

Yue Mao, Yi Shen, Chao Yu, and Longjun Cai. A joint training dual-MRC framework for aspect based sentiment analysis. In *AAAI*, pages 13543–13551, 2021b.

Yue Mao, Yi Shen, Jingchao Yang, Xiaoying Zhu, and Longjun Cai. Seq2Path: Generating sentiment tuples as paths of a tree. In *ACL*, pages 2215–2225, Dublin, Ireland, May 2022b.

Vassallo Marco, Gabrieli Giuliano, Valerio Basile, Cristina Bosco, et al. The tenuousness of lemmatization in lexicon-based sentiment analysis. In *Italian Conference on Computational Linguistics*, volume 2481, pages 1–6, 2019.

Mitchell P. Marcus, Beatrice Santorini, and Mary Ann Marcinkiewicz. Building a large annotated corpus of English: The Penn Treebank. *Computational Linguistics*, 19(2):313–330, 1993.

Jose Maria Balmaceda, Silvia Schiaffino, and Daniela Godoy. How do personality traits affect communication among users in online social networks? *Online Information Review*, 38(1):136–153, 2014.

Hazel Rose Markus and Shinobu Kitayama. Culture and the self: Implications for cognition, emotion, and motivation. In *College student development and academic life*, pages 264–293. Routledge, 2014.

Mónica Marrero, Julián Urbano, Sonia Sánchez-Cuadrado, Jorge Morato, and Juan Miguel Gómez-Berbís. Named entity recognition: fallacies, challenges and opportunities. *Computer Standards & Interfaces*, 35(5):482–489, 2013.

Edison Marrese-Taylor, Juan D Velásquez, and Felipe Bravo-Marquez. A novel deterministic approach for aspect-based opinion mining in tourism products reviews. *Expert Systems with Applications*, 41(17):7764–7775, 2014.

M Antonia Martı, Mariona Taulé, Lluıs Márquez, and Manuel Bertran. CESS-ECE: A multilingual and multilevel annotated corpus, 2007. URL http://lsi.upc.edu/~mbertran/cess-ece.

James H Martin. *A Computational Model of Metaphor Interpretation*. Academic Press Professional, Inc., 1990.

James R Martin and Peter R White. *The Language of Evaluation*, volume 2. Springer, 2003.

Scott Martin. The role of salience ranking in anaphora resolution, 2015.

Sebastian Martschat and Michael Strube. Recall error analysis for coreference resolution. In *EMNLP*, pages 2070–2081, 2014.

Marco Maru, Federico Scozzafava, Federico Martelli, and Roberto Navigli. Syntagnet: Challenging supervised word sense disambiguation with lexical-semantic combinations. In *EMNLP-IJCNLP*, pages 3534–3540, 2019.

Rebecca Marvin and Philipp Koehn. Exploring word sense disambiguation abilities of neural machine translation systems. In *13th Conference of the Association for Machine Translation in the Americas (Volume 1: Research Track)*, pages 125–131. Association for Machine Translation in the Americas, 2018.

Zachary J Mason. CorMet: A computational, corpus-based conventional metaphor extraction system. *Computational Linguistics*, 30(1):23–44, 2004.

Spyros Matsoukas, Ivan Bulyko, Bing Xiang, Kham Nguyen, Richard Schwartz, and John Makhoul. Integrating speech recognition and machine translation. In *2007 IEEE International Conference on Acoustics, Speech and Signal Processing-ICASSP'07*, volume 4, pages IV–1281, 2007.

Tomoko Matsui, Tagiru Nakamura, Akira Utsumi, Akihiro T Sasaki, Takahiko Koike, Yumiko Yoshida, Tokiko Harada, Hiroki C Tanabe, and Norihiro Sadato. The role of prosody and context in sarcasm comprehension: Behavioral and fMRI evidence. *Neuropsychologia*, 87:74–84, 2016.

Chanda Maurya, T Muhammad, Preeti Dhillon, and Priya Maurya. The effects of cyberbullying victimization on depression and suicidal ideation among adolescents and young adults: A three year cohort study from India. *BMC Psychiatry*, 22(1):1–14, 2022.

Diah Hevyka Maylawati, Warih Maharani, and Ibnu Asror. Implicit aspect extraction in product reviews using FIN algorithm. In *2020 8th International Conference on Information and Communication Technology (ICoICT)*, pages 1–5, 2020.

Diana Maynard and Mark A Greenwood. Who cares about sarcastic Tweets? Investigating the impact of sarcasm on sentiment analysis. In *Ninth International Conference on Language Resources and Evaluation*, pages 4238–4243, 2014.

Michael Mayor. *Longman dictionary of contemporary English*. Pearson Education India, 2009.

Andrew McCallum and Ben Wellner. Object consolidation by graph partitioning with a conditionally-trained distance metric. In *KDD Workshop on Data Cleaning, Record Linkage and Object Consolidation*, pages 1–6, 2003.

Andrew McCallum and Ben Wellner. Conditional models of identity uncertainty with application to noun coreference. *NeurIPS*, 17:1–8, 2004.

Andrew McCallum, Dayne Freitag, and Fernando CN Pereira. Maximum entropy markov models for information extraction and segmentation. In *Icml*, volume 17, pages 591–598, 2000.

Arya D McCarthy, Ekaterina Vylomova, Shijie Wu, Chaitanya Malaviya, Lawrence Wolf-Sonkin, Garrett Nicolai, Christo Kirov, Miikka Silfverberg, Sabrina J Mielke, Jeffrey Heinz, et al. The SIGMORPHON 2019 shared task: Morphological analysis in context and cross-lingual transfer for inflection. *arXiv preprint arXiv:1910.11493*, 2019.

Joseph F McCarthy and Wendy G Lehnert. Using decision trees for conference resolution. In *IJCAI*, pages 1050–1055, 1995.

Michael C McCord. Slot grammar. In *Natural language and logic*, pages 118–145. Springer, 1990.

Iain McCowan, Jean Carletta, Wessel Kraaij, Simone Ashby, Sebastien Bourban, Mike Flynn, Mael Guillemot, Thomas Hain, Jaroslav Kadlec, Vasilis Karaiskos, et al. The AMI meeting corpus. In *5th International Conference on Methods and Techniques in Behavioral Research*, volume 88, page 100, 2005.

Kathleen E McCoy and Michael Strube. Generating anaphoric expressions: pronoun or definite description? *The Relation of Discourse/Dialogue Structure and Reference*, 1999.

Robert R McCrae and Paul T Costa. Updating norman's "adequacy taxonomy"': Intelligence and personality dimensions in natural language and in questionnaires. *Journal of Personality and Social Psychology*, 49(3):710, 1985.

Robert R McCrae and Paul T Costa. Validation of the five-factor model of personality across instruments and observers. *Journal of Personality and Social Psychology*, 52(1):81, 1987.

Robert R McCrae and Paul T Costa Jr. A contemplated revision of the NEO five-factor inventory. *Personality and Individual Differences*, 36(3):587–596, 2004.

Robert R McCrae and Oliver P John. An introduction to the five-factor model and its applications. *Journal of Personality*, 60(2):175–215, 1992.

R.R. McCrae and P.T. Costa. *Personality in Adulthood*. Guilford Publications, 1990. ISBN 9780898625288.

Ryan McDonald, Koby Crammer, and Fernando Pereira. Flexible text segmentation with structured multilabel classification. In *HLT-EMNLP*, pages 987–994, 2005.

Skye McDonald and Samantha Pearce. Clinical insights into pragmatic theory: Frontal lobe deficits and sarcasm. *Brain and Language*, 53(1):81–104, 1996.

Skye McDonald, Sharon Flanagan, Jennifer Rollins, and Julianne Kinch. TASIT: A new clinical tool for assessing social perception after traumatic brain injury. *The Journal of Head Trauma Rehabilitation*, 18(3):219–238, 2003.

Brian McFee, Colin Raffel, Dawen Liang, Daniel P Ellis, Matt McVicar, Eric Battenberg, and Oriol Nieto. librosa: Audio and music signal analysis in Python. In *14th Python in Science Conference*, volume 8, pages 18–25, 2015.

Walaa Medhat, Ahmed Hassan, and Hoda Korashy. Sentiment analysis algorithms and applications: A survey. *Ain Shams Engineering Journal*, 5(4):1093–1113, 2014.

Douglas L Medin and Marguerite M Schaffer. Context theory of classification learning. *Psychological Review*, 85(3):207, 1978.

Raveesh Meena, Gabriel Skantze, and Joakim Gustafson. Data-driven models for timing feedback responses in a map task dialogue system. *Computer Speech & Language*, 28(4):903–922, 2014.

Sara Meftah and Nasredine Semmar. A neural network model for part-of-speech tagging of social media texts. In *Eleventh International Conference on Language Resources and Evaluation (LREC 2018)*, Miyazaki, Japan, May 2018. European Language Resources Association (ELRA).

Yash Mehta, Samin Fatehi, Amirmohammad Kazameini, Clemens Stachl, Erik Cambria, and Sauleh Eetemadi. Bottom-up and top-down: Predicting personality with psycholinguistic and language model features. In *ICDM*, pages 1184–1189, 2020.

Rui Meng, Sanqiang Zhao, Shuguang Han, Daqing He, Peter Brusilovsky, and Yu Chi. Deep keyphrase generation. In Regina Barzilay and Min-Yen Kan, editors, *ACL*, pages 582–592, 2017.

Xinfan Meng and Houfeng Wang. Mining user reviews: from specification to summarization. In *ACL-IJCNLP 2009 Conference Short Papers*, pages 177–180, 2009.

Amina Ben Meriem, Lobna Hlaoua, and Lotfi Ben Romdhane. A fuzzy approach for sarcasm detection in social networks. *Procedia Computer Science*, 192:602–611, 2021.

Andreas Messalas, Yiannis Kanellopoulos, and Christos Makris. Model-agnostic interpretability with shapley values. In *2019 10th International Conference on Information, Intelligence, Systems and Applications (IISA)*, pages 1–7, 2019.

Mayuri Mhatre, Dakshata Phondekar, Pranali Kadam, Anushka Chawathe, and Kranti Ghag. Dimensionality reduction for sentiment analysis using preprocessing techniques. In *2017 International Conference on Computing Methodologies and Communication (ICCMC)*, pages 16–21, 2017.

Lesly Miculicich Miculicich and Andrei Popescu-Belis. Using coreference links to improve Spanish-to-English machine translation. In *2nd Workshop on Coreference Resolution Beyond OntoNotes (CORBON 2017)*, pages 30–40, 2017.

Rada Mihalcea, Carmen Banea, and Janyce Wiebe. Learning multilingual subjective language via cross-lingual projections. In *45th annual meeting of the association of computational linguistics*, pages 976–983, 2007.

Andrei Mikheev. Tagging sentence boundaries. In *NAACL*, 2000.

Andrei Mikheev. Periods, capitalized words, etc. *Computational Linguistics*, 28(3): 289–318, 2002.

Tomas Mikolov, Kai Chen, G.s Corrado, and Jeffrey Dean. Efficient estimation of word representations in vector space. *Workshop at ICLR*, 2013, 01 2013a.

Tomas Mikolov, Ilya Sutskever, Kai Chen, Greg S Corrado, and Jeff Dean. Distributed representations of words and phrases and their compositionality. In *NIPS*, pages 3111–3119, 2013b.

Kirill Milintsevich and Kairit Sirts. Enhancing sequence-to-sequence neural lemmatization with external resources. *arXiv preprint arXiv:2101.12056*, 2021.

George A Miller. WordNet: A lexical database for English. *Communications of the ACM*, 38(11):39–41, 1995.

George A Miller, Claudia Leacock, Randee Tengi, and Ross T Bunker. A semantic concordance. In *HLT*, pages 303–308, 1993.

Gary Miner, John Elder IV, Andrew Fast, Thomas Hill, Robert Nisbet, and Dursun Delen. *Practical text mining and statistical analysis for non-structured text data applications*. Academic Press, 2012.

Marvin Minsky. A framework for representing knowledge, 1974.

Marvin Minsky. *The Emotion Machine: Commonsense Thinking, Artificial Intelligence, and the Future of the Human Mind*. Simon & Schuster, New York, 2006.

Shachar Mirkin, Ido Dagan, and Sebastian Padó. Assessing the role of discourse references in entailment inference. In *ACL*, pages 1209–1219, 2010.

Walter Mischel. Toward an integrative science. *Annual Review of Psychology*, 55: 1–22, 2004.

Abhijit Mishra, Diptesh Kanojia, Seema Nagar, Kuntal Dey, and Pushpak Bhattacharyya. Harnessing cognitive features for sarcasm detection. In *ACL*, pages 1095–1104, Berlin, Germany, 2016.

Rishabh Misra and Prahal Arora. Sarcasm detection using hybrid neural network. *arXiv preprint arXiv:1908.07414*, 2019.

Tom Mitchell and E. Fredkin. Never-ending language learning. In *2014 IEEE International Conference on Big Data (Big Data)*, pages 1–1, 2014.

Ruslan Mitkov. *Anaphora Resolution*. Routledge, 2014.

Ruslan Mitkov. *The Oxford Handbook of Computational Linguistics*. Oxford University Press, 2022.

Ruslan Mitkov, Richard Evans, Constantin Orăsan, Le An Ha, and Viktor Pekar. Anaphora resolution: To what extent does it help nlp applications? In *Anaphora: Analysis, Algorithms and Applications: 6th Discourse Anaphora and Anaphor Resolution Colloquium*, pages 179–190, 2007.

Ankush Mittal, Pooja Bhatt, and Padam Kumar. Phonetic matching and syntactic tree similarity based QA system for SMS queries. In *2014 International Conference on Green Computing Communication and Electrical Engineering (ICGCCEE)*, pages 1–6, 2014.

Makoto Miwa and Mohit Bansal. End-to-end relation extraction using lstms on sequences and tree structures. In *ACL*, pages 1105–1116, 2016.

Makoto Miwa and Yutaka Sasaki. Modeling joint entity and relation extraction with table representation. In *EMNLP*, pages 1858–1869, 2014.

Dunja Mladenic. Automatic word lemmatization. In *5th International Multi-Conference Information Society, IS-2002 B*, pages 153–159, 2002.

Aditya Mogadala and Vasudeva Varma. Language independent sentence-level subjectivity analysis with feature selection. In *26th Pacific Asia Conference on Language, Information, and Computation*, pages 171–180, 2012.

Samaneh Moghaddam and Martin Ester. Opinion digger: An unsupervised opinion miner from unstructured product reviews. In *CIKM*, pages 1825–1828, 2010.

Saif Mohammad, Ekaterina Shutova, and Peter Turney. Metaphor as a medium for emotion: An empirical study. In *Fifth Joint Conference on Lexical and Computational Semantics*, pages 23–33, 2016.

Michael Mohler, Bryan Rink, David Bracewell, and Marc Tomlinson. A novel distributional approach to multilingual conceptual metaphor recognition. In *COLING*, pages 1752–1763, 2014.

Michael Mohler, Mary Brunson, Bryan Rink, and Marc Tomlinson. Introducing the LCC metaphor datasets. In *Tenth International Conference on Language Resources and Evaluation (LREC'16)*, pages 4221–4227, 2016.

Antonio Molina and Ferran Pla. Shallow parsing using specialized HMMs. *Journal of Machine Learning Research*, 2(Mar):595–613, 2002.

Natawut Monaikul, Giuseppe Castellucci, Simone Filice, and Oleg Rokhlenko. Continual learning for named entity recognition. In *AAAI*, pages 13570–13577, 2021.

Christine A. Montgomery. Concept extraction. *American Journal of Computational Linguistics*, 8(2):70–73, 1982.

Andrés Montoyo, Patricio Martínez-Barco, and Alexandra Balahur. Subjectivity and sentiment analysis: An overview of the current state of the area and envisaged developments. *Decision Support Systems*, 53(4):675–679, 2012.

Nafise Sadat Moosavi and Michael Strube. Which coreference evaluation metric do you trust? a proposal for a link-based entity aware metric. In *ACL*, pages 632–642, 2016.

Louis-Philippe Morency, Ariadna Quattoni, and Trevor Darrell. Latent-dynamic discriminative models for continuous gesture recognition. In *2007 IEEE Conference on Computer Vision and Pattern Recognition*, pages 1–8, 2007. doi: 10.1109/CVPR.2007.383299.

Louis-Philippe Morency, Rada Mihalcea, and Payal Doshi. Towards multimodal sentiment analysis: Harvesting opinions from the web. In *13th international conference on multimodal interfaces*, pages 169–176, 2011.

Andrea Moro and Roberto Navigli. SemEval-2015 task 13: Multilingual all-words sense disambiguation and entity linking. In *9th International Workshop on Semantic Evaluation*, pages 288–297, 2015.

Andrea Moro, Francesco Cecconi, and Roberto Navigli. Multilingual word sense disambiguation and entity linking for everybody. In *2014 International Conference on Posters & Demonstrations Track-Volume 1272*, pages 25–28, 2014a.

Andrea Moro, Alessandro Raganato, and Roberto Navigli. Entity linking meets word sense disambiguation: a unified approach. *Transactions of the Association for Computational Linguistics*, 2:231–244, 2014b.

Guanyi Mou and Kyumin Lee. Malicious bot detection in online social networks: Arming handcrafted features with deep learning. In *International Conference on Social Informatics*, pages 220–236, 2020.

Mohamad Syahrul Mubarok, Adiwijaya, and Muhammad Dwi Aldhi. Aspect-based sentiment analysis to review products using Naïve Bayes. In *AIP Conference Proceedings*, volume 1867, page 020060, 2017.

Andrius Mudinas, Dell Zhang, and Mark Levene. Combining lexicon and learning based approaches for concept-level sentiment analysis. In *ISDOMW*, 2012.

Aldrian Obaja Muis and Wei Lu. Weak semi-Markov CRFs for noun phrase chunking in informal text. In *NAACL-HLT*, pages 714–719, 2016.

Aldrian Obaja Muis and Wei Lu. Labeling gaps between words: Recognizing overlapping mentions with mention separators. In *EMNLP*, pages 2608–2618, 2017.

Arjun Mukherjee and Bing Liu. Aspect extraction through semi-supervised modeling. In *ACL*, pages 339–348, 2012.

Rajdeep Mukherjee, Tapas Nayak, Yash Butala, Sourangshu Bhattacharya, and Pawan Goyal. Paste: A tagging-free decoding framework using pointer networks for aspect sentiment triplet extraction. *arXiv preprint arXiv:2110.04794*, 2021.

Thomas Müller, Helmut Schmid, and Hinrich Schütze. Efficient higher-order crfs for morphological tagging. In *EMNLP*, pages 322–332, 2013.

Thomas Müller, Ryan Cotterell, Alexander Fraser, and Hinrich Schütze. Joint lemmatization and morphological tagging with Lemming. In *EMNLP*, pages 2268–2274, 2015.

Gabriel Murray and Giuseppe Carenini. Predicting subjectivity in multimodal conversations. In *EMNLP*, pages 1348–1357, 2009.

Gabriel Murray and Giuseppe Carenini. Subjectivity detection in spoken and written conversations. *Natural Language Engineering*, 17(3):397–418, 2011.

Sumiya Mushtaq and Neerendra Kumar. Text-based automatic personality recognition: Recent developments. In *Third International Conference on Computing, Communications, and Cyber-Security*, pages 537–549, 2023.

Judith Muzerelle, Anaïs Lefeuvre, Jean-Yves Antoine, Emmanuel Schang, Denis Maurel, Jeanne Villaneau, and Iris Eshkol. ANCOR, the first large French speaking corpus of conversational speech annotated in coreference to be freely available. In *TALN*, pages 555–563, 2013.

Isabel Briggs Myers, Mary H. McCaulley, Naomi L. Quenk, and Allen L. Hammer. *MBTI Manual: A Guide to the Development and Use of the Myers - Briggs Type Indicator*. Consulting Psychologists Press, third edition, 1998.

Jerome L Myers and Arnold D Well. *Research Design & Statistical Analysis*. Routledge, New York, 1995.

Tetsuji Nakagawa, Taku Kudo, and Yuji Matsumoto. Unknown word guessing and part-of-speech tagging using support vector machines. In *NLPRS*, pages 325–331, 2001.

Hiromi Nakaiwa and Satoru Ikehara. Zero pronoun resolution in a machine translation system by using japanese to english verbal semantic attributes. In *Third Conference on Applied Natural Language Processing*, pages 201–208, 1992.

Hiromi Nakaiwa and Satoshi Shirai. Anaphora resolution of Japanese zero pronouns with deictic reference. In *COLING*, pages 812–817, 1996.

Hiromi Nakaiwa, Satoshi Shirai, Satoru Ikehara, and Tsukasa Kawaoka. Extrasentential resolution of Japanese zero pronouns using semantic and pragmatic constraints. In *AAAI 1995 Spring Symposium Series*, pages 99–105, 1995.

Preslav Nakov, Sara Rosenthal, Zornitsa Kozareva, Veselin Stoyanov, Alan Ritter, and Theresa Wilson. SemEval-2013 task 2: Sentiment analysis in Twitter. In *SemEval*, pages 312–320, 2013.

Sri Nandhini and JI Sheeba. Cyberbullying detection and classification using information retrieval algorithm. In *ICARCSET*, 2015.

Zara Nasar, Syed Waqar Jaffry, and Muhammad Kamran Malik. Named entity recognition and relation extraction: State-of-the-art. *ACM Computing Surveys*, 54 (1):1–39, 2021.

Arman Khadjeh Nassirtoussi, Saeed Aghabozorgi, Teh Ying Wah, and David Chek Ling Ngo. Text mining of news-headlines for FOREX market prediction: A multi-layer dimension reduction algorithm with semantics and sentiment. *Expert Systems with Applications*, 42(1):306–324, 2015.

Vivi Nastase, Michael Strube, Benjamin Boerschinger, Caecilia Zirn, and Anas Elghafari. WikiNet: A very large scale multi-lingual concept network. In *LREC*, pages 1015–1022, 2010.

Roberto Navigli. Word sense disambiguation: A survey. *ACM Computing Surveys*, 41(2):1–69, 2009.

Roberto Navigli and Mirella Lapata. Graph connectivity measures for unsupervised word sense disambiguation. In *IJCAI*, pages 1683–1688, 2007.

Roberto Navigli and Simone Paolo Ponzetto. Babelnet: The automatic construction, evaluation and application of a wide-coverage multilingual semantic network. *Artificial Intelligence*, 193:217–250, 2012a.

Roberto Navigli and Simone Paolo Ponzetto. BabelNet: The automatic construction, evaluation and application of a wide-coverage multilingual semantic network. *Artificial Intelligence*, 193:217–250, 2012b.

Roberto Navigli, David Jurgens, and Daniele Vannella. SemEval-2013 task 12: Multilingual word sense disambiguation. In *SemEval*, pages 222–231, 2013.

Steven Neale, Joao Silva, and António Branco. A flexible tool for manual word sense annotation. In *11th Joint ACL-ISO Workshop on Interoperable Semantic Annotation (ISA-11)*, pages 1–5, 2015.

A Nedoluzhko, J Mírovskỳ, E Fučíková, and J Pergler. Annotation of coreference in Prague Czech-English dependency treebank. Technical report, Technical report 2014/57. Prague: ÚFAL MFF UK, 2014.

Anna Nedoluzhko, Michal Novák, Martin Popel, Zdeněk Žabokrtskỳ, Amir Zeldes, and Daniel Zeman. CorefUD 1.0: Coreference meets universal dependencies. In *LREC*, pages 4859–4872, 2022.

Daniel Neiberg, Kjell Elenius, Inger Karlsson, and Kornel Laskowski. Emotion recognition in spontaneous speech. In *Fonetik*, pages 101–104, 2006.

Mariana Neves and Jurica Ševa. An extensive review of tools for manual annotation of documents. *Briefings in Bioinformatics*, 22(1):146–163, 2021.

Hwee Tou Ng. Getting serious about word sense disambiguation. In *Tagging Text with Lexical Semantics: Why, What, and How?*, pages 1–7, 1997.

Hwee Tou Ng and Hian Beng Lee. Integrating multiple knowledge sources to disambiguate word sense: an exemplar-based approach. In *ACL*, pages 40–47, 1996.

Vincent Ng. Supervised noun phrase coreference research: The first fifteen years. In *ACL*, pages 1396–1411, 2010.

Vincent Ng and Claire Cardie. Combining sample selection and error-driven pruning for machine learning of coreference rules. In *EMNLP*, pages 55–62, 2002a.

Vincent Ng and Claire Cardie. Improving machine learning approaches to coreference resolution. In *ACL*, pages 104–111, 2002b.

Grace Ngai and Radu Florian. Transformation based learning in the fast lane. In *NAACL*, 2001.

Dat Quoc Nguyen, Thanh Vu, Dai Quoc Nguyen, Mark Dras, and Mark Johnson. From word segmentation to POS tagging for Vietnamese. *arXiv preprint arXiv:1711.04951*, 2017.

Thuy Dung Nguyen and Min-Yen Kan. Keyphrase extraction in scientific publications. In Dion Hoe-Lian Goh, Tru Hoang Cao, Ingeborg Sølvberg, and Edie M. Rasmussen, editors, *10th International Conference on Asian Digital Libraries*, volume 4822 of *Lecture Notes in Computer Science*, pages 317–326, 2007.

Jinjie Ni, Tom Young, Vlad Pandelea, Fuzhao Xue, and Erik Cambria. Recent advances in deep learning based dialogue systems: A systematic survey. *Artificial Intelligence Review*, pages 3055–3155, 2022.

Cristina Nicolae and Gabriel Nicolae. BESTCUT: A graph algorithm for coreference resolution. In *EMNLP*, pages 275–283, 2006.

Garrett Nicolai and Grzegorz Kondrak. Leveraging inflection tables for stemming and lemmatization. In *ACL*, pages 1138–1147, 2016.

Nicolas Nicolov, Franco Salvetti, and Steliana Ivanova. Sentiment analysis: Does coreference matter. In *AISB 2008 Convention Communication, Interaction and Social Intelligence*, volume 1, page 37, 2008.

Jan Niehues and Eunah Cho. Exploiting linguistic resources for neural machine translation using multi-task learning. *arXiv preprint arXiv:1708.00993*, 2017.

Finn Årup Nielsen. A new anew: Evaluation of a word list for sentiment analysis in microblogs. In *Workshop on'Making Sense of Microposts: Big Things Come in Small Packages*, pages 93–98, 2011.

Huansheng Ning, Sahraoui Dhelim, and Nyothiri Aung. PersoNet: Friend recommendation system based on big-five personality traits and hybrid filtering. *IEEE Transactions on Computational Social Systems*, 6(3):394–402, 2019.

Joakim Nivre, Johan Hall, Sandra Kübler, Ryan McDonald, Jens Nilsson, Sebastian Riedel, and Deniz Yuret. The CoNLL 2007 shared task on dependency parsing. In *EMNLP-CoNLL*, pages 915–932, Prague, Czech Republic, June 2007.

Joakim Nivre, Marie-Catherine De Marneffe, Filip Ginter, Yoav Goldberg, Jan Hajic, Christopher D Manning, Ryan McDonald, Slav Petrov, Sampo Pyysalo, Natalia Silveira, et al. Universal dependencies v1: A multilingual treebank collection. In *Tenth International Conference on Language Resources and Evaluation (LREC'16)*, pages 1659–1666, 2016.

Jorge Nocedal and Stephen Wright. *Numerical optimization*. Springer Science & Business Media, 2006.

Warren T Norman. Toward an adequate taxonomy of personality attributes: Replicated factor structure in peer nomination personality ratings. *The Journal of Abnormal and Social Psychology*, 66(6):574, 1963.

Warren T. Norman. "To see ourselves as others see us!": Relations among self-perceptions, peer-perceptions, and expected peer-perceptions of personality. *Multivariate Behavioral Research*, 4(4):417–443, 1969.

Natalya F Noy, Deborah L McGuinness, et al. Ontology development 101: A guide to creating your first ontology, 2001.

Gertrude E Noyes. The first English dictionary, Cawdrey's table alphabeticall. *Modern Language Notes*, 58(8):600–605, 1943.

Maxwell Nye, Armando Solar-Lezama, Josh Tenenbaum, and Brenden M Lake. Learning compositional rules via neural program synthesis. *NeurIPS*, 33:10832–10842, 2020.

Jon Oberlander and Scott Nowson. Whose thumb is it anyway? Classifying author personality from weblog text. In *COLING-ACL*, pages 627–634, 2006.

Bruno Oberle. SACR: A Drag-and-Drop Based Tool for Coreference Annotation. In *Eleventh International Conference on Language Resources and Evaluation*, pages 389–394, 2018.

Brendan O'Connor, Michel Krieger, and David Ahn. Tweetmotif: Exploratory search and topic summarization for twitter. In *Fourth International AAAI Conference on Weblogs and Social Media*, 2010.

Maciej Ogrodniczuk, Katarzyna Głowińska, Mateusz Kopeć, Agata Savary, and Magdalena Zawisławska. Polish coreference corpus. In *Language and Technology Conference*, pages 215–226, 2013.

Bruno Ohana and Brendan Tierney. Sentiment classification of reviews using SentiWordNet. *IT&T*, 8, 2009.

Tom O'Hara, Rebecca Bruce, Jeff Donner, and Janyce Wiebe. Class-based collocations for word sense disambiguation. In *SENSEVAL-3, the Third International Workshop on the Evaluation of Systems for the Semantic Analysis of Text*, pages 199–202, 2004.

Manabu Okumura and Kouji Tamura. Zero pronoun resolution in Japanese discourse based on centering theory. In *COLING*, pages 871–876, 1996.

Silviu Oprea and Walid Magdy. iSarcasm: A dataset of intended sarcasm. In *ACL*, pages 1279–1289, Online, July 2020.

Shereen Oraby, Vrindavan Harrison, Lena Reed, Ernesto Hernandez, Ellen Riloff, and Marilyn Walker. Creating and characterizing a diverse corpus of sarcasm in dialogue. In *17th Annual Meeting of the Special Interest Group on Discourse and Dialogue*, pages 31–41, Los Angeles, September 2016.

Constantin Orasan. The influence of pronominal anaphora resolution on term-based summarisation. *RANLP*, pages 291–300, 2007.

Matan Orbach, Orith Toledo-Ronen, Artem Spector, Ranit Aharonov, Yoav Katz, and Noam Slonim. YASO: A targeted sentiment analysis evaluation dataset for open-domain reviews. In *EMNLP*, pages 9154–9173, 2021.

Naho Orita, Naomi Feldman, Jordan Boyd-Graber, and Eliana Vornov. Quantifying the role of discourse topicality in speakers' choices of referring expressions. In *Fifth Workshop on Cognitive Modeling and Computational Linguistics*, pages 63–70, 2014.

Naho Orita, Eliana Vornov, Naomi Feldman, and Hal Daumé III. Why discourse affects speakers' choice of referring expressions. In *ACL-IJCNLP*, volume 1, pages 1639–1649, 2015.

Reynier Ortega, Adrian Fonseca, Yoan Gutiérrez, and Andrés Montoyo. Improving subjectivity detection using unsupervised subjectivity word sense disambiguation. *Procesamiento del Lenguaje Natural*, 51:179–186, 2013.

Daniel W Otter, Julian R Medina, and Jugal K Kalita. A survey of the usages of deep learning for natural language processing. *IEEE transactions on neural networks and learning systems*, 32(2):604–624, 2020.

Olutobi Owoputi, Brendan O'Connor, Chris Dyer, Kevin Gimpel, Nathan Schneider, and Noah A Smith. Improved part-of-speech tagging for online conversational text with word clusters. In *NAACL-HLT*, pages 380–390, 2013.

Daniel J Ozer and Veronica Benet-Martinez. Personality and the prediction of consequential outcomes. *Annual Review of Psychology*, 57:401, 2006.

Allan Paivio. Abstractness, imagery, and meaningfulness in paired-associate learning. *Journal of Verbal Learning and Verbal Behavior*, 4(1):32–38, 1965.

David D Palmer and Marti A Hearst. Adaptive sentence boundary disambiguation. *arXiv preprint cmp-lg/9411022*, 1994.

David D. Palmer and Marti A. Hearst. Adaptive multilingual sentence boundary disambiguation. *Comput. Linguist.*, 23(2):241–267, June 1997.

Frank Robert Palmer and Palmer Frank Robert. *Semantics*. Cambridge university press, 1981.

Martha Palmer, Paul R. Kingsbury, and Daniel Gildea. The proposition bank: An annotated corpus of semantic roles. *Computational Linguistics*, 31(1):71–106, 2005.

Georgios Paltoglou and Anastasia Giachanou. Opinion retrieval: Searching for opinions in social media. In *Professional Search in the Modern World*, pages 193–214. Springer, 2014.

Hongliang Pan, Zheng Lin, Peng Fu, Yatao Qi, and Weiping Wang. Modeling intra and inter-modality incongruity for multi-modal sarcasm detection. In *EMNLP*, pages 1383–1392, 2020.

Liangming Pan, Xiaochen Wang, Chengjiang Li, Juanzi Li, and Jie Tang. Course concept extraction in MOOCs via embedding-based graph propagation. In *IJC-NLP*, pages 875–884, 2017.

Rajnish Pandey, Abhinav Kumar, Jyoti Prakash Singh, and Sudhakar Tripathi. Hybrid attention-based long short-term memory network for sarcasm identification. *Applied Soft Computing*, 106:107348, 2021.

Bo Pang and Lillian Lee. A sentimental education: Sentiment analysis using subjectivity summarization based on minimum cuts. In *ACL*, pages 271–278, 2004.

Bo Pang and Lillian Lee. Seeing stars: exploiting class relationships for sentiment categorization with respect to rating scales. In *ACL*, pages 115–124, 2005.

Kartikey Pant, Tanvi Dadu, and Radhika Mamidi. Towards detection of subjective bias using contextualized word embeddings. In *Web Conference*, pages 75–76, 2020.

Kishore Papineni, Salim Roukos, Todd Ward, and Wei-Jing Zhu. BLEU: A method for automatic evaluation of machine translation. In *ACL*, pages 311–318, 2002.

Tommaso Pasini and Roberto Navigli. Train-o-matic: Large-scale supervised word sense disambiguation in multiple languages without manual training data. In *EMNLP*, pages 78–88, 2017.

Tommaso Pasini, Francesco Elia, and Roberto Navigli. Huge automatically extracted training-sets for multilingual word sense disambiguation. In *Eleventh International Conference on Language Resources and Evaluation*, pages 1694–1698, 2018.

Rebecca Passonneau. Instructions for applying discourse reference annotation for multiple applications (drama). *Unpublished Manuscript*, page 46, 1997.

Aneta Pavlenko. Emotion and emotion-laden words in the bilingual lexicon. *Bilingualism: Language and cognition*, 11(2):147–164, 2008.

Haiyun Peng, Yukun Ma, Yang Li, and Erik Cambria. Learning multi-grained aspect target sequence for chinese sentiment analysis. *Knowledge-Based Systems*, 148: 167–176, 2018.

Haiyun Peng, Lu Xu, Lidong Bing, Fei Huang, Wei Lu, and Luo Si. Knowing what, how and why: A near complete solution for aspect-based sentiment analysis. In *AAAI*, pages 8600–8607, 2020.

Yifan Peng, Shankai Yan, and Zhiyong Lu. Transfer learning in biomedical natural language processing: An evaluation of bert and elmo on ten benchmarking datasets. *BioNLP 2019*, page 58, 2019.

James W Pennebaker. *The Secret Life of Pronouns: What Our Words Say About Us.* Bloomsbury Publishing USA, 2013.

James W Pennebaker and Laura A King. Linguistic styles: Language use as an individual difference. *Journal of Personality and Social Psychology*, 77(6):1296, 1999.

James W Pennebaker, Martha E Francis, and Roger J Booth. Linguistic inquiry and word count: Liwc 2001. *Mahway: Lawrence Erlbaum Associates*, 71(2001):2001, 2001.

James W Pennebaker, Matthias R Mehl, and Kate G Niederhoffer. Psychological aspects of natural language use: Our words, our selves. *Annual Review of Psychology*, 54(1):547–577, 2003.

James W Pennebaker, Ryan L Boyd, Kayla Jordan, and Kate Blackburn. The development and psychometric properties of LIWC2015. Technical report, The University of Texas at Austin, 2015.

Deana Pennell and Yang Liu. Normalization of text messages for text-to-speech. In *2010 IEEE International Conference on Acoustics, Speech and Signal Processing*, pages 4842–4845, 2010.

Deana Pennell and Yang Liu. A character-level machine translation approach for normalization of SMS abbreviations. In *IJCNLP*, pages 974–982, 2011.

Jeffrey Pennington, Richard Socher, and Christopher D Manning. GloVe: Global vectors for word representation. In *EMNLP*, pages 1532–1543, 2014.

Iryna Pentina, Tyler Hancock, and Tianling Xie. Exploring relationship development with social chatbots: A mixed-method study of replika. *Computers in Human Behavior*, 140:107600, 2023.

S Pesina and T Solonchak. Semantic primitives and conceptual focus. *Procedia-Social and Behavioral Sciences*, 192:339–345, 2015.

Robert Peters and Norbert Nagel. Das digitale, referenzkorpus mittelniederdeutsch/niederrheinisch (ReN)'. *Jahrbuch für Germanistische Sprachgeschichte*, 5(1):165–175, 2014.

Fabio Petroni, Tim Rocktäschel, Sebastian Riedel, Patrick Lewis, Anton Bakhtin, Yuxiang Wu, and Alexander Miller. Language models as knowledge bases? In *EMNLP-IJCNLP*, pages 2463–2473, 2019.

Fabio Petroni, Patrick S. H. Lewis, Aleksandra Piktus, Tim Rocktäschel, Yuxiang Wu, Alexander H. Miller, and Sebastian Riedel. How context affects language models' factual predictions. In *Conference on Automated Knowledge Base Construction, AKBC*, 2020.

Slav Petrov, Dipanjan Das, and Ryan McDonald. A universal part-of-speech tagset. *arXiv preprint arXiv:1104.2086*, 2011.

Saša Petrović, Miles Osborne, and Victor Lavrenko. The Edinburgh twitter corpus. In *NAACL HLT 2010 workshop on computational linguistics in a world of social media*, pages 25–26, 2010.

Miriam RL Petruck. Frame semantics. *Handbook of pragmatics*, 2, 1996.

Lawrence Philips. Hanging on the metaphone. *Computer Language*, 7(12):39–43, 1990.

Lawrence Philips. The double metaphone search algorithm. *C/C++ users journal*, 18(6):38–43, 2000.

Jean Piaget, Margaret Cook, et al. *The origins of intelligence in children*, volume 8. International Universities Press, New York, 1952.

Emanuele Pianta, Luisa Bentivogli, and Christian Girardi. MultiWordNet: developing an aligned multilingual database. In *First International Conference on Global WordNet*, pages 293–302, 2002.

Martin J Pickering and Simon Garrod. An integrated theory of language production and comprehension. *Behavioral and Brain Sciences*, 36(4):329–347, 2013.

David J Pittenger. The utility of the Myers-Briggs Type Indicator. *Review of educational research*, 63(4):467–488, 1993.

Barbara Plank and Dirk Hovy. Personality traits on Twitter—or—how to get 1,500 personality tests in a week. In *6th Workshop on Computational Approaches to Subjectivity, Sentiment and Social Media Analysis*, pages 92–98, 2015.

John C Platt. Fast training of support vector machines using sequential minimal optimization, advances in kernel methods. *Support Vector Learning*, pages 185–208, 1999.

Joël Plisson, Nada Lavrac, Dunja Mladenic, et al. A rule based approach to word lemmatization. In *IS*, volume 3, pages 83–86, 2004.

Massimo Poesio. Annotating a corpus to develop and evaluate discourse entity realization algorithms: Issues and preliminary results. In *Second International Conference on Language Resources and Evaluation*, pages 1–8, 2000.

Massimo Poesio. Associative descriptions and salience: A preliminary investigation. In *2003 EACL Workshop on The Computational Treatment of Anaphora*, pages 31–38, 2003.

Massimo Poesio. The MATE/GNOME proposals for anaphoric annotation, revisited. In *HLT-NAACL*, pages 154–162, 2004.

Massimo Poesio and M Alexandrov-Kabadjov. A general-purpose, off the shelf anaphoric resolver. In *LREC*, pages 653–656, 2004.

Massimo Poesio and Ron Artstein. Anaphoric annotation in the ARRAU corpus. In *Sixth International Conference on Language Resources and Evaluation*, pages 1–5, 2008.

Massimo Poesio, Florence Bruneseaux, and Laurent Romary. The MATE meta-scheme for coreference in dialogues in multiple languages. In *ACL'99 Workshop Towards Standards and Tools for Discourse Tagging*, pages 65–74, 1999a.

Massimo Poesio, Renate Henschel, Janet Hitzeman, Rodger Kibble, Shane Montague, and Kees van Deemter. Towards an annotation scheme for noun phrase generation. Technical report, EACL, 1999b.

Massimo Poesio, Juntao Yu, Silviu Paun, Abdulrahman Aloraini, Pengcheng Lu, Janosch Haber, and Derya Cokal. Computational models of anaphora. *Annual Review of Linguistics*, 9:561–587, 2023.

Livia Polanyi and Annie Zaenen. Contextual valence shifters. In *Computing Attitude and Affect in Text: Theory and Applications*, pages 1–10. Springer, 2006.

Marco Polignano, Pierpaolo Basile, Marco De Gemmis, Giovanni Semeraro, Valerio Basile, et al. Alberto: Italian BERT language understanding model for NLP challenging tasks based on tweets. In *CEUR Workshop Proceedings*, volume 2481, pages 1–6, 2019.

Marco Polignano, Valerio Basile, Pierpaolo Basile, Giuliano Gabrieli, Marco Vassallo, and Cristina Bosco. A hybrid lexicon-based and neural approach for explainable polarity detection. *Information Processing & Management*, 59(5):103058, 2022.

Maria Pontiki, Dimitris Galanis, John Pavlopoulos, Harris Papageorgiou, Ion Androutsopoulos, and Suresh Manandhar. SemEval-2014 task 4: Aspect based sentiment analysis. In *8th International Workshop on Semantic Evaluation (SemEval 2014)*, pages 27–35, 2014.

Maria Pontiki, Dimitris Galanis, Haris Papageorgiou, Suresh Manandhar, and Ion Androutsopoulos. SemEval-2015 task 12: Aspect based sentiment analysis. In *9th International Workshop on Semantic Evaluation (SemEval 2015)*, pages 486–495, 2015.

Maria Pontiki, Dimitris Galanis, Haris Papageorgiou, Ion Androutsopoulos, Suresh Manandhar, Mohammed AL-Smadi, Mahmoud Al-Ayyoub, Yanyan Zhao, Bing Qin, Orphée De Clercq, et al. Semeval-2016 task 5: Aspect based sentiment analysis. In *ProWorkshop on Semantic Evaluation (SemEval-2016)*, pages 19–30, 2016.

Simone Paolo Ponzetto and Michael Strube. Exploiting semantic role labeling, WordNet and Wikipedia for coreference resolution. In *Human Language Technology Conference of the NAACL, Main Conference*, pages 192–199, 2006.

Alexander Popov. Word sense disambiguation with recurrent neural networks. In *Student Research Workshop associated with RANLP 2017*, pages 25–34, 2017.

Soujanya Poria, Alexandar Gelbukh, Basant Agarwal, Erik Cambria, and Newton Howard. Common sense knowledge based personality recognition from text. In *Mexican International Conference on Artificial Intelligence*, pages 484–496, 2013.

Marten Postma, Emiel Van Miltenburg, Roxane Segers, Anneleen Schoen, and Piek Vossen. Open Dutch WordNet. In *8th Global WordNet Conference (GWC)*, pages 302–310, 2016.

Marco Pota, Mirko Ventura, Rosario Catelli, and Massimo Esposito. An effective BERT-based pipeline for twitter sentiment analysis: A case study in italian. *Sensors*, 21(1):133, 2020.

Rolandos-Alexandros Potamias, Georgios Siolas, and Andreas Stafylopatis. A robust deep ensemble classifier for figurative language detection. In *Engineering Applications of Neural Networks*, pages 164–175. Springer International Publishing, 2019. ISBN 978-3-030-20257-6.

Rolandos Alexandros Potamias, Georgios Siolas, and Andreas-Georgios Stafylopatis. A transformer-based approach to irony and sarcasm detection. *Neural Computing and Applications*, 32(23):17309–17320, 2020.

Sameer Pradhan, Edward Loper, Dmitriy Dligach, and Martha Palmer. SemEval-2007 task-17: English lexical sample, SRL and all words. In *Fourth International Workshop on Semantic Evaluations (SemEval-2007)*, pages 87–92, 2007.

Sameer Pradhan, Alessandro Moschitti, Nianwen Xue, Olga Uryupina, and Yuchen Zhang. CoNLL-2012 shared task: Modeling multilingual unrestricted coreference in ontonotes. In *Joint Conference on EMNLP and CoNLL-Shared Task*, pages 1–40, 2012.

Group Pragglejaz. MIP: A method for identifying metaphorically used words in discourse. *Metaphor and Symbol*, 22(1):1–39, 2007.

Judita Preiss and David Yarowsky. Senseval-2 second international workshop on evaluating word sense disambiguation systems. In *SENSEVAL-2 Second International Workshop on Evaluating Word Sense Disambiguation Systems*, pages 1–163, 2001.

Erich Prem, Julia Neidhardt, Peter Knees, Stefan Woltran, and Hannes Werthner. Digital humanism and norms in recommender systems. In *First Workshop on Normative Design and Evaluation of Recommender Systems*, volume 3639, 2023.

David Premack and Guy Woodruff. Does the chimpanzee have a theory of mind? *Behavioral and Brain Sciences*, 1(4):515–526, 1978.

Aman Priyanshu, Vedant Rishi Das, Shashank Rajiv Moghe, Harsh Rathod, Sai Sravan Medicherla, Mini Shail Chhabra, and Sarthak Shastri. Stance classification with improved elementary classifiers using lemmatization (grand challenge). In *2020 IEEE Sixth International Conference on Multimedia Big Data (BigMM)*, pages 466–470, 2020. doi: 10.1109/BigMM50055.2020.00077.

Sven-Oliver Proksch, Will Lowe, Jens Wäckerle, and Stuart Soroka. Multilingual sentiment analysis: A new approach to measuring conflict in legislative speeches. *Legislative Studies Quarterly*, 44(1):97–131, 2019.

Reid Pryzant, Richard Diehl Martinez, Nathan Dass, Sadao Kurohashi, Dan Jurafsky, and Diyi Yang. Automatically neutralizing subjective bias in text. In *AAAI*, volume 34, pages 480–489, 2020.

Tomáš Ptáček, Ivan Habernal, and Jun Hong. Sarcasm detection on Czech and English Twitter. In *COLING*, pages 213–223, 2014.

Vasin Punyakanok and Dan Roth. The use of classifiers in sequential inference. *NeurIPS*, 13:995–1001, 2000.

Tobias Pütz, Daniël De Kok, Sebastian Pütz, and Erhard Hinrichs. Seq2seq or perceptrons for robust lemmatization. an empirical examination. In *17th International Workshop on Treebanks and Linguistic Theories (TLT 2018)*, pages 193–207, 2018.

Peng Qi, Timothy Dozat, Yuhao Zhang, and Christopher D. Manning. Universal Dependency parsing from scratch. In *CoNLL 2018 Shared Task: Multilingual Parsing from Raw Text to Universal Dependencies*, pages 160–170, Brussels, Belgium, October 2018. doi: 10.18653/v1/K18-2016.

Peng Qi, Yuhao Zhang, Yuhui Zhang, Jason Bolton, and Christopher D Manning. Stanza: A python natural language processing toolkit for many human languages. *arXiv preprint arXiv:2003.07082*, 2020.

Lin Qiu, Han Lin, Jonathan Ramsay, and Fang Yang. You are what you tweet: Personality expression and perception on Twitter. *Journal of research in personality*, 46(6):710–718, 2012.

Stephan Raaijmakers, Khiet P Truong, and Theresa Wilson. Multimodal subjectivity analysis of multiparty conversation. In *EMNLP*, pages 466–474, 2008.

Alec Radford, Karthik Narasimhan, Tim Salimans, Ilya Sutskever, et al. Improving language understanding by generative pre-training. Technical report, OpenAI, 2018.

Alec Radford, Jeffrey Wu, Rewon Child, David Luan, Dario Amodei, Ilya Sutskever, et al. Language models are unsupervised multitask learners. *OpenAI blog*, 1(8), 2019.

Colin Raffel, Minh-Thang Luong, Peter J Liu, Ron J Weiss, and Douglas Eck. Online and linear-time attention by enforcing monotonic alignments. In *ICML*, pages 2837–2846, 2017.

Alessandro Raganato, Claudio Delli Bovi, and Roberto Navigli. Neural sequence learning models for word sense disambiguation. In *EMNLP*, pages 1156–1167, 2017a.

Alessandro Raganato, Jose Camacho-Collados, and Roberto Navigli. Word sense disambiguation: A unified evaluation framework and empirical comparison. In *EACL*, pages 99–110, 2017b.

Alessandro Raganato, Yves Scherrer, and Jörg Tiedemann. The MuCoW test suite at WMT 2019: Automatically harvested multilingual contrastive word sense disambiguation test sets for machine translation. In *Fourth Conference on Machine Translation (Volume 2: Shared Task Papers, Day 1)*, pages 470–480, 2019.

Karthik Raghunathan, Heeyoung Lee, Sudarshan Rangarajan, Nathanael Chambers, Mihai Surdeanu, Dan Jurafsky, and Christopher Manning. A multi-pass sieve for coreference resolution. In *EMNLP*, pages 492–501, 2010.

Edoardo Ragusa, Erik Cambria, Rodolfo Zunino, and Paolo Gastaldo. A survey on deep learning in image polarity detection: Balancing generalization performances and computational costs. *Electronics*, 8(7):783, 2019.

Afshin Rahimi, Yuan Li, and Trevor Cohn. Massively multilingual transfer for NER. In *ACL*, pages 151–164, 2019.

Altaf Rahman and Vincent Ng. Narrowing the modeling gap: A cluster-ranking approach to coreference resolution. *Journal of Artificial Intelligence Research*, 40:469–521, 2011.

Altaf Rahman and Vincent Ng. Resolving complex cases of definite pronouns: The Winograd schema challenge. In *EMNLP-CoNLL*, pages 777–789, 2012.

Mohammad Wali Ur Rahman, Sicong Shao, Pratik Satam, Salim Hariri, Chris Padilla, Zoe Taylor, and Carlos Nevarez. A BERT-based deep learning approach for reputation analysis in social media. In *2022 IEEE/ACS 19th International Conference on Computer Systems and Applications (AICCSA)*, pages 1–8, 2022.

Sunny Rai and Shampa Chakraverty. Metaphor detection using fuzzy rough sets. In *International Joint Conference on Rough Sets*, pages 271–279, 2017.

Sunny Rai, Shampa Chakraverty, Devendra K Tayal, Divyanshu Sharma, and Ayush Garg. Understanding metaphors using emotions. *New Generation Computing*, 37 (1):5–27, 2019.

Pranav Rajpurkar, Jian Zhang, Konstantin Lopyrev, and Percy Liang. SQuAD: 100,000+ questions for machine comprehension of text. In *EMNLP*, pages 2383–2392, 2016.

Majid Ramezani, Mohammad-Reza Feizi-Derakhshi, and Mohammad-Ali Balafar. Knowledge graph-enabled text-based automatic personality prediction. *Computational Intelligence and Neuroscience*, 2022, 2022.

Lance A Ramshaw and Mitchell P Marcus. Text chunking using transformation-based learning. In *Natural language processing using very large corpora*, pages 157–176. Springer, 1999.

Toqir A Rana, Yu-N Cheah, and Tauseef Rana. Multi-level knowledge-based approach for implicit aspect identification. *Applied Intelligence*, 50:4616–4630, 2020.

Francisco Rangel, Fabio Celli, Paolo Rosso, Potthast Martin, Benno Stein, Walter Daelemans, et al. Overview of the 3rd author profiling task at PAN 2015. In *CLEF2015 Working Notes. Working Notes of CLEF 2015-Conference and Labs of the Evaluation forum.*, 2015.

Rasika Ransing and Archana Gulati. A survey of different approaches for word sense disambiguation. In *ICT Analysis and Applications*, pages 435–445. Springer, 2022.

Delip Rao and Deepak Ravichandran. Semi-supervised polarity lexicon induction. In *EACL*, pages 675–682, 2009.

David Rapaport, Merton Gill, and Roy Schafer. *Diagnostic psychological testing: The theory, statistical evaluation, and diagnostic application of a battery of tests: Volume II*. The Year Book, 1946.

Pushpendre Rastogi, Ryan Cotterell, and Jason Eisner. Weighting finite-state transductions with neural context. In *NAACL-HLT*, pages 623–633, 2016.

Lev Ratinov and Dan Roth. Design challenges and misconceptions in named entity recognition. In *CoNLL*, pages 147–155, 2009.

Adwait Ratnaparkhi. A maximum entropy model for part-of-speech tagging. In *EMNLP*, 1996.

Biswarup Ray, Avishek Garain, and Ram Sarkar. An ensemble-based hotel recommender system using sentiment analysis and aspect categorization of hotel reviews. *Applied Soft Computing*, 98:106935, 2021.

Jonathon Read, Rebecca Dridan, Stephan Oepen, and Lars Jørgen Solberg. Sentence boundary detection: A long solved problem? In *COLING 2012: Posters*, pages 985–994, 2012.

Marta Recasens and Eduard Hovy. BLANC: Implementing the rand index for coreference evaluation. *Natural Language Engineering*, 17(4):485–510, 2011.

Marta Recasens, Cristian Danescu-Niculescu-Mizil, and Dan Jurafsky. Linguistic models for analyzing and detecting biased language. In *ACL*, pages 1650–1659, 2013.

Marek Rei. Semi-supervised multitask learning for sequence labeling. *arXiv preprint arXiv:1704.07156*, 2017.

Nils Reimers and Iryna Gurevych. Sentence-BERT: Sentence embeddings using siamese BERT-networks. In *EMNLP-IJCNLP*, pages 3982–3992, 2019.

Tanya Reinhart. Coreference and bound anaphora: A restatement of the anaphora questions. *Linguistics and Philosophy*, pages 47–88, 1983.

Nils Reiter. CorefAnnotator - a new annotation tool for entity references. In *Abstracts of EADH: Data in the Digital Humanities*, pages 1–4, 2018.

Ahmed Remaida, Benyoussef Abdellaoui, Aniss Moumen, and Younes El Bouzekri El Idrissi. Personality traits analysis using artificial neural networks: A literature survey. In *2020 1st international conference on innovative research in applied science, engineering and technology (IRASET)*, pages 1–6, 2020.

Robert Remus. Improving sentence-level subjectivity classification through readability measurement. In *18th Nordic Conference of Computational Linguistics*, pages 168–174, 2011.

Feiliang Ren, Longhui Zhang, Shujuan Yin, Xiaofeng Zhao, Shilei Liu, Bochao Li, and Yaduo Liu. A novel global feature-oriented relational triple extraction model based on table filling. In *EMNLP*, pages 2646–2656, 2021a.

Lu Ren, Bo Xu, Hongfei Lin, Xikai Liu, and Liang Yang. Sarcasm detection with sentiment semantics enhanced multi-level memory network. *Neurocomputing*, 401:320–326, 2020.

Zhancheng Ren, Qiang Shen, Xiaolei Diao, and Hao Xu. A sentiment-aware deep learning approach for personality detection from text. *Information Processing & Management*, 58(3):102532, 2021b.

Philip Resnik and David Yarowsky. Distinguishing systems and distinguishing senses: New evaluation methods for word sense disambiguation. *Natural Language Engineering*, 5(2):113–133, 1999.

Jeffrey C Reynar and Adwait Ratnaparkhi. A maximum entropy approach to identifying sentence boundaries. *arXiv preprint cmp-lg/9704002*, 1997.

Marco Tulio Ribeiro, Sameer Singh, and Carlos Guestrin. "why should I trust you?" Explaining the predictions of any classifier. In *22nd ACM SIGKDD international conference on knowledge discovery and data mining*, pages 1135–1144, 2016.

Michael Riley. Some applications of tree-based modelling to speech and language. In *Proceedings of Speech and Natural Language*, 1989.

Ellen Riloff. Automatically generating extraction patterns from untagged text. In *AAAI*, pages 1044–1049, 1996.

Ellen Riloff and Janyce Wiebe. Learning extraction patterns for subjective expressions. In *EMNLP*, pages 105–112, 2003.

Ellen Riloff, Ashequl Qadir, Prafulla Surve, Lalindra De Silva, Nathan Gilbert, and Ruihong Huang. Sarcasm as contrast between a positive sentiment and negative situation. In *EMNLP*, pages 704–714, 2013.

Kyeongmin Rim. MAE2: Portable annotation tool for general natural language use. In *Proc 12th Joint ACL-ISO Workshop on Interoperable Semantic Annotation*, pages 75–80, 2016.

Nicky Ringland, Xiang Dai, Ben Hachey, Sarvnaz Karimi, Cecile Paris, and James R. Curran. NNE: A dataset for nested named entity recognition in english newswire. In *ACL*, pages 5176–5181, 2019a.

Nicky Ringland, Xiang Dai, Ben Hachey, Sarvnaz Karimi, Cecile Paris, and James R Curran. Nne: A dataset for nested named entity recognition in english newswire. In *ACL*, pages 5176–5181, 2019b.

Annette Rios Gonzales, Laura Mascarell, and Rico Sennrich. Improving word sense disambiguation in neural machine translation with sense embeddings. In *Second Conference on Machine Translation*, pages 11–19, 2017.

David Ritchie. Metaphors in conversational context: Toward a connectivity theory of metaphor interpretation. *Metaphor and Symbol*, 19(4):265–287, 2004.

Alan Ritter, Sam Clark, Oren Etzioni, et al. Named entity recognition in tweets: an experimental study. In *EMNLP*, pages 1524–1534, 2011.

Leah Roberts. *Syntactic processing*, pages 227–247. Cambridge Handbooks in Language and Linguistics. Cambridge University Press, 2016. doi: 10.1017/CBO9781107425965.011.

Kevin Dela Rosa and Jeffrey Ellen. Text classification methodologies applied to micro-text in military chat. In *ICML*, pages 710–714, 2009. doi: 10.1109/ICMLA.2009.49.

Rudolf Rosa and Zdeněk Žabokrtský. Unsupervised lemmatization as embeddings-based word clustering. *arXiv preprint arXiv:1908.08528*, 2019.

Eleanor Rosch and Carolyn B Mervis. Family resemblances: Studies in the internal structure of categories. *Cognitive Psychology*, 7(4):573–605, 1975.

Eleanor Rosch, Carolyn B Mervis, Wayne D Gray, David M Johnson, and Penny Boyes-Braem. Basic objects in natural categories. *Cognitive Psychology*, 8(3):382–439, 1976.

Eleanor H Rosch. Natural categories. *Cognitive psychology*, 4(3):328–350, 1973.

Zachary Rosen. Computationally constructed concepts: A machine learning approach to metaphor interpretation using usage-based construction grammatical cues. In *Workshop on Figurative Language Processing*, pages 102–109, 2018.

Frank Rosenblatt. The perceptron: a probabilistic model for information storage and organization in the brain. *Psychological review*, 65(6):386, 1958.

Sascha Rothe and Hinrich Schütze. Autoextend: Extending word embeddings to embeddings for synsets and lexemes. In *ACL-IJCNLP*, pages 1793–1803, 2015.

Kathrin Rothermich, Ayotola Ogunlana, and Natalia Jaworska. Change in humor and sarcasm use based on anxiety and depression symptom severity during the COVID-19 pandemic. *Journal of Psychiatric Research*, 140:95–100, 2021.

Herbert Rubenstein and John B. Goodenough. Contextual correlates of synonymy. *Communications of the ACM*, 8(10):627–633, 1965.

Rachel Rudinger, Jason Naradowsky, Brian Leonard, and Benjamin Van Durme. Gender bias in coreference resolution. In *NAACL-HLT*, pages 8–14, 2018.

Dwijen Rudrapal, Anupam Jamatia, Kunal Chakma, Amitava Das, and Björn Gambäck. Sentence boundary detection for social media text. In *12th International Conference on Natural Language Processing*, pages 254–260, Trivandrum, India, December 2015. NLP Association of India.

Hotze Rullmann. Two types of negative polarity items. In *North East Linguistics Society*, 1996.

David E Rumelhart and Andrew Ortony. The representation of knowledge in memory. *Schooling and the acquisition of knowledge*, 99:135, 1977.

Josef Ruppenhofer, Michael Ellsworth, Myriam Schwarzer-Petruck, Christopher R Johnson, and Jan Scheffczyk. FrameNet II: Extended theory and practice. Technical report, International Computer Science Institute, 2016.

Alexander M Rush, Roi Reichart, Michael Collins, and Amir Globerson. Improved parsing and POS tagging using inter-sentence consistency constraints. In *EMNLP-CoNLL*, pages 1434–1444, 2012.

Samir Rustamov. A hybrid system for subjectivity analysis. *Advances in Fuzzy Systems*, 2018, 2018.

Samir Rustamov, Elshan Mustafayev, and Mark A Clements. Sentence-level subjectivity detection using neuro-fuzzy models. In *4th Workshop on Computational Approaches to Subjectivity, Sentiment and Social Media Analysis*, pages 108–114, 2013.

Tatyana Ruzsics and Tanja Samardzic. Neural sequence-to-sequence learning of internal word structure. In *CoNLL*, pages 184–194, 2017.

Nipun Sadvilkar and Mark Neumann. PySBD: Pragmatic sentence boundary disambiguation. *arXiv preprint arXiv:2010.09657*, 2020.

Horacio Saggionα and Adam Funk. Interpreting SentiWordNet for opinion classification. In *Seventh Conference on International Language Resources and Evaluation*, pages 1129–1133, 2010.

Santwana Sagnika, Bhabani Shankar Prasad Mishra, and Saroj K Meher. Improved method of word embedding for efficient analysis of human sentiments. *Multimedia Tools and Applications*, 79(43):32389–32413, 2020.

Santwana Sagnika, Bhabani Shankar Prasad Mishra, and Saroj K Meher. An attention-based CNN-LSTM model for subjectivity detection in opinion-mining. *Neural Computing and Applications*, 33(24):17425–17438, 2021.

Hassan Saif, Yulan He, Miriam Fernandez, and Harith Alani. Contextual semantics for sentiment analysis of Twitter. *Information Processing & Management*, 52(1): 5–19, 2016.

Keisuke Sakaguchi, Ronan Le Bras, Chandra Bhagavatula, and Yejin Choi. Winogrande: An adversarial winograd schema challenge at scale. In *AAAI*, pages 8732–8740, 2020.

Keisuke Sakaguchi, Ronan Le Bras, Chandra Bhagavatula, and Yejin Choi. Winogrande: An adversarial winograd schema challenge at scale. *Communications of the ACM*, 64(9):99–106, 2021.

David Salgado, Martin Krallinger, Marc Depaule, Elodie Drula, Ashish V Tendulkar, Florian Leitner, Alfonso Valencia, and Christophe Marcelle. MyMiner: a web application for computer-assisted biocuration and text annotation. *Bioinformatics*, 28(17):2285–2287, 2012.

Said A Salloum, Rehan Khan, and Khaled Shaalan. A survey of semantic analysis approaches. In *International Conference on Artificial Intelligence and Computer Vision*, pages 61–70, 2020.

Merrilee H Salmon. *Introduction to Logic and Critical Thinking*. Matthew J. Van Cleave, 1989.

Steven L. Salzberg. C4.5: Programs for machine learning by j. ross quinlan. morgan kaufmann publishers, inc., 1993. *Machine Learning*, 16(3):235–240, 1994.

Fahime Same, Guanyi Chen, and Kees Van Deemter. Non-neural models matter: a re-evaluation of neural referring expression generation systems. In *ACL*, pages 5554–5567, Dublin, Ireland, May 2022. doi: 10.18653/v1/2022.acl-long.380.

George Sanchez. Sentence boundary detection in legal text. In *Natural Legal Language Processing Workshop 2019*, pages 31–38, 2019.

Mark Sanderson. Word sense disambiguation and information retrieval. In *SIGIR*, pages 142–151, 1994.

Erik F Sang and Sabine Buchholz. Introduction to the CoNLL-2000 shared task: Chunking. *arXiv preprint cs/0009008*, 2000.

Erik F Sang and Jorn Veenstra. Representing text chunks. *arXiv preprint cs/9907006*, 1999.

Erik Tjong Kim Sang. Text chunking by system combination. In *Fourth Conference on Computational Natural Language Learning and the Second Learning Language in Logic Workshop*, 2000.

Erik Tjong Kim Sang and Fien De Meulder. Introduction to the CoNLL-2003 shared task: Language-independent named entity recognition. In *HLT-NAACL*, pages 142–147, 2003.

Cicero Nogueira dos Santos and Victor Guimaraes. Boosting named entity recognition with neural character embeddings. *arXiv preprint arXiv:1505.05008*, 2015.

Maarten Sap, Ronan Le Bras, Emily Allaway, Chandra Bhagavatula, Nicholas Lourie, Hannah Rashkin, Brendan Roof, Noah A Smith, and Yejin Choi. ATOMIC: An atlas of machine commonsense for if-then reasoning. In *AAAI*, volume 33, pages 3027–3035, 2019a.

Maarten Sap, Hannah Rashkin, Derek Chen, Ronan Le Bras, and Yejin Choi. Social iqa: Commonsense reasoning about social interactions. In *EMNLP-IJCNLP*, pages 4462–4472, 2019b.

Sunita Sarawagi and William W Cohen. Semi-Markov conditional random fields for information extraction. *NeurIPS*, 17:1185–1192, 2004.

Ryohei Sasano and Sadao Kurohashi. A discriminative approach to Japanese zero anaphora resolution with large-scale lexicalized case frames. In *IJCNLP*, pages 758–766, 2011.

Ryohei Sasano, Daisuke Kawahara, and Sadao Kurohashi. A fully-lexicalized probabilistic model for Japanese zero anaphora resolution. In *COLING*, pages 769–776, 2008.

Kumara Sastry, David Goldberg, and Graham Kendall. Genetic algorithms. In *Search Methodologies*, pages 97–125. Springer, 2005.

Ranjan Satapathy, Iti Chaturvedi, Erik Cambria, Shirley S Ho, and Jin Cheon Na. Subjectivity detection in nuclear energy tweets. *Computación y Sistemas*, 21(4): 657–664, 2017a.

Ranjan Satapathy, Claudia Guerreiro, Iti Chaturvedi, and Erik Cambria. Phonetic-based microtext normalization for twitter sentiment analysis. In *ICDM Workshops*, pages 407–413, 2017b.

Ranjan Satapathy, Yang Li, Sandro Cavallari, and Erik Cambria. Seq2seq deep learning models for microtext normalization. In *IJCNN*, pages 1–8, 2019a.

Ranjan Satapathy, Aalind Singh, and Erik Cambria. PhonSenticNet: A cognitive approach to microtext normalization for concept-level sentiment analysis. In *International Conference on Computational Data and Social Networks*, pages 177–188, 2019b.

Ranjan Satapathy, Erik Cambria, Andrea Nanetti, and Amir Hussain. A review of shorthand systems: From brachygraphy to microtext and beyond. *Cognitive Computation*, 12(4):778–792, 2020.

Ranjan Satapathy, Shweta Rajesh Pardeshi, and Erik Cambria. Polarity and subjectivity detection with multitask learning and BERT embedding. *Future Internet*, 14(7):191, 2022.

Ville Satopaa, Jeannie Albrecht, David Irwin, and Barath Raghavan. Finding a "kneedle" in a haystack: Detecting knee points in system behavior. In *2011 31st International Conference on Distributed Computing Systems Workshops*, pages 166–171, 2011.

Gerard Saucier. An alternative multi-language structure for personality attributes. *European Journal of Personality*, 17(3):179–205, 2003.

Danielle Saunders, Rosie Sallis, and Bill Byrne. Neural machine translation doesn't translate gender coreference right unless you make it. In *Second Workshop on Gender Bias in Natural Language Processing*, pages 35–43, 2020.

Agata Savary, Bartosz Zaborowski, Aleksandra Krawczyk-Wieczorek, and Filip Makowiecki. Sejfek - a lexicon and a shallow grammar of polish economic multi-word units. In *3rd Workshop on Cognitive Aspects of the Lexicon*, pages 195–214, 2012.

Mark Savickas. *Career Counseling*. American Psychological Association Washington, DC, 2019.

Ramit Sawhney, Shivam Agarwal, Atula Tejaswi Neerkaje, Nikolaos Aletras, Preslav Nakov, and Lucie Flek. Towards suicide ideation detection through online conversational context. In *45th International ACM SIGIR Conference on Research and Development in Information Retrieval*, pages 1716–1727, 2022.

Prateek Saxena and Soma Paul. EPIE dataset: A corpus for possible idiomatic expressions. In *International Conference on Text, Speech, and Dialogue*, pages 87–94, 2020.

Bianca Scarlini, Tommaso Pasini, and Roberto Navigli. Just "onesec" for producing multilingual sense-annotated data. In *ACL*, pages 699–709, 2019.

Bianca Scarlini, Tommaso Pasini, and Roberto Navigli. With more contexts comes better performance: Contextualized sense embeddings for all-round word sense disambiguation. In *EMNLP*, pages 3528–3539, 2020a.

Bianca Scarlini, Tommaso Pasini, and Roberto Navigli. SensEmBERT: Context-enhanced sense embeddings for multilingual word sense disambiguation. In *AAAI*, pages 8758–8765, 2020b.

Roger C Schank. Conceptual dependency: A theory of natural language understanding. *Cognitive psychology*, 3(4):552–631, 1972.

Robert E Schapire and Yoram Singer. Improved boosting algorithms using confidence-rated predictions. *Machine learning*, 37(3):297–336, 1999.

Robert E Schapire and Yoram Singer. BoosTexter: A boosting-based system for text categorization. *Machine Learning*, 39(2):135–168, 2000.

Timo Schick and Hinrich Schutze. It's not just size that matters: Small language models are also few-shot learners. In *NAACL-HLT*, pages 2339–2352, 2021.

Helmut Schmid. Unsupervised learning of period disambiguation for tokenisation. *Internal Report, IMS-CL*, 2000.

Marine Schmitt and Matthieu Constant. Neural lemmatization of multiword expressions. In *Joint Workshop on Multiword Expressions and WordNet (MWE-WN 2019)*, pages 142–148, 2019.

Martin Schmitt, Simon Steinheber, Konrad Schreiber, and Benjamin Roth. Joint aspect and polarity classification for aspect-based sentiment analysis with end-to-end neural networks. *arXiv preprint arXiv:1808.09238*, 2018.

Kaitlyn B Schodt, Selena I Quiroz, Brittany Wheeler, Deborah L Hall, and Yasin N Silva. Cyberbullying and mental health in adults: The moderating role of social media use and gender. *Frontiers in Psychiatry*, page 954, 2021.

Karin Kipper Schuler. *VerbNet: A broad-coverage, comprehensive verb lexicon.* University of Pennsylvania, Philadelphia, PA,United States, 2005.

H Andrew Schwartz, Johannes C Eichstaedt, Margaret L Kern, Lukasz Dziurzynski, Stephanie M Ramones, Megha Agrawal, Achal Shah, Michal Kosinski, David Stillwell, Martin EP Seligman, et al. Personality, gender, and age in the language of social media: The open-vocabulary approach. *PloS One*, 8(9):e73791, 2013.

Federico Scozzafava, Marco Maru, Fabrizio Brignone, Giovanni Torrisi, and Roberto Navigli. Personalized pagerank with syntagmatic information for multilingual word sense disambiguation. In *ACL*, pages 37–46, 2020.

Djamé Seddah, Reut Tsarfaty, Sandra Kübler, Marie Candito, Jinho Choi, Richárd Farkas, Jennifer Foster, Iakes Goenaga, Koldo Gojenola, Yoav Goldberg, et al. Overview of the SPMRL 2013 shared task: cross-framework evaluation of parsing morphologically rich languages. In *Fourth Workshop on Statistical Parsing of Morphologically-Rich Languages*, 2013.

Norbert M Seel. *Encyclopedia of the Sciences of Learning.* Springer Science & Business Media, 2011.

Kazuhiro Seki, Atsushi Fujii, and Tetsuya Ishikawa. A probabilistic model for japanese zero pronoun resolution integrating syntactic and semantic features. In *NLPRS*, pages 403–410, 2001.

Kazuhiro Seki, Atsushi Fujii, and Tetsuya Ishikawa. A probabilistic method for analyzing japanese anaphora integrating zero pronoun detection and resolution. In *COLING*, pages 1–7, 2002.

Rico Sennrich, Orhan Firat, Kyunghyun Cho, Alexandra Birch, Barry Haddow, Julian Hitschler, Marcin Junczys-Dowmunt, Samuel Läubli, Antonio Valerio Miceli Barone, Jozef Mokry, et al. Nematus: a toolkit for neural machine translation. *arXiv preprint arXiv:1703.04357*, 2017.

Jesus Serrano-Guerrero, Jose A Olivas, Francisco P Romero, and Enrique Herrera-Viedma. Sentiment analysis: A review and comparative analysis of web services. *Information Sciences*, 311:18–38, 2015.

Fei Sha and Fernando Pereira. Shallow parsing with conditional random fields. In *NAACL-HLT*, pages 213–220, 2003.

Sarah Shafqat, Hammad Majeed, Qaisar Javaid, and Hafiz Farooq Ahmad. Standard NER tagging scheme for big data healthcare analytics built on unified medical corpora. *Journal of Artificial Intelligence and Technology*, 2(4):152–157, 2022.

Yan Shao, Christian Hardmeier, Jörg Tiedemann, and Joakim Nivre. Character-based joint segmentation and pos tagging for chinese using bidirectional rnn-crf. *arXiv preprint arXiv:1704.01314*, 2017.

Guangyao Shen, Jia Jia, Liqiang Nie, Fuli Feng, Cunjun Zhang, Tianrui Hu, Tat-Seng Chua, and Wenwu Zhu. Depression detection via harvesting social media: A multimodal dictionary learning solution. In *IJCAI*, pages 3838–3844, 2017.

Libin Shen, Giorgio Satta, and Aravind Joshi. Guided learning for bidirectional sequence classification. In *45th Annual Meeting of the Association of Computational Linguistics*, pages 760–767, Prague, Czech Republic, June 2007.

Xiangqing Shen, Siwei Wu, and Rui Xia. Dense-atomic: Towards densely-connected ATOMIC with high knowledge coverage and massive multi-hop paths. In *ACL*, pages 13292–13305, 2023a.

Yiqiu Shen, Laura Heacock, Jonathan Elias, Keith D Hentel, Beatriu Reig, George Shih, and Linda Moy. Chatgpt & other large language models are double-edged swords. *Radiology*, 307(2):e230163, 2023b.

Jonathan Richard Shewchuk et al. An introduction to the conjugate gradient method without the agonizing pain, 1994.

Yong Shi, Luyao Zhu, Wei Li, Kun Guo, and Yuanchun Zheng. Survey on classic and latest textual sentiment analysis articles and techniques. *International Journal of Information Technology & Decision Making*, 18(04):1243–1287, 2019.

Takashi Shibuya and Eduard Hovy. Nested named entity recognition via second-best sequence learning and decoding. *Transactions of the Association for Computational Linguistics*, 8:605–620, 2020.

Mayank Shrivastava and Shishir Kumar. A pragmatic and intelligent model for sarcasm detection in social media text. *Technology in Society*, 64:101489, 2021.

Ekaterina Shutova. Automatic metaphor interpretation as a paraphrasing task. In *NAACL-HLT*, pages 1029–1037, 2010.

Ekaterina Shutova. Design and evaluation of metaphor processing systems. *Computational Linguistics*, 41(4):579–623, 2015.

Ekaterina Shutova and Simone Teufel. Metaphor corpus annotated for source-target domain mappings. In *Seventh International Conference on Language Resources and Evaluation (LREC'10)*, pages 3255–3261, 2010.

Ekaterina Shutova, Douwe Kiela, and Jean Maillard. Black holes and white rabbits: Metaphor identification with visual features. In *NAACL-HLT*, pages 160–170, 2016.

Vered Shwartz, Peter West, Ronan Le Bras, Chandra Bhagavatula, and Yejin Choi. Unsupervised commonsense question answering with self-talk. In *EMNLP*, pages 4615–4629, 2020.

Candace Lee Sidner. Towards a computational theory of definite anaphora comprehension in english discourse. Technical report, Massachusetts Inst of Tech Cambridge Artificial Intelligence lab, 1979.

Natalia Silveira, Timothy Dozat, Marie-Catherine De Marneffe, Samuel R Bowman, Miriam Connor, John Bauer, and Christopher D Manning. A gold standard dependency corpus for English. In *LREC*, pages 2897–2904, 2014.

Stefano Silvestri, Francesco Gargiulo, and Mario Ciampi. Iterative annotation of biomedical ner corpora with deep neural networks and knowledge bases. *Applied Sciences*, 12(12):5775, 2022.

John Simpson and Edmund Weiner. *The Oxford English Dictionary*. Oxford University Press, 2 edition, 1989.

Push Singh, Thomas Lin, Erik T. Mueller, Grace Lim, Travell Perkins, and Wan Li Zhu. Open mind common sense: Knowledge acquisition from the general public. In *On the Move to Meaningful Internet Systems*, volume 2519 of *Lecture Notes in Computer Science*, pages 1223–1237, 2002.

Richard Laishram Singh, Krishnendu Ghosh, Kishorjit Nongmeikapam, and Sivaji Bandyopadhyay. A decision tree based word sense disambiguation system in manipuri language. *Advanced Computing*, 5(4):17, 2014.

Juan Sixto, Aitor Almeida, and Diego López-de Ipiña. An approach to subjectivity detection on Twitter using the structured information. In *International Conference on Computational Collective Intelligence*, pages 121–130, 2016.

Sara Skilbred-Fjeld, Silje Endresen Reme, and Svein Mossige. Cyberbullying involvement and mental health problems among late adolescents. *Cyberpsychology: Journal of Psychosocial Research on Cyberspace*, 14(1), 2020.

Wojciech Skut, Brigitte Krenn, Thorsten Brants, and Hans Uszkoreit. An annotation scheme for free word order languages. *5th Conference on Applied Natural Language Processing*, 05 2002. doi: 10.3115/974557.974571.

Stavroula Skylaki, Ali Oskooei, Omar Bari, Nadja Herger, and Zac Kriegman. Named entity recognition in the legal domain using a pointer generator network. *arXiv preprint arXiv:2012.09936*, 2020.

Edgar A Smith. Devereux readability index. *The Journal of Educational Research*, 54(8):298–303, 1961.

Paul Smolensky, Richard McCoy, Roland Fernandez, Matthew Goldrick, and Jianfeng Gao. Neurocompositional computing: From the central paradox of cognition to a new generation of ai systems. *AI Magazine*, 43(3):308–322, 2022.

Jake Snell, Kevin Swersky, and Richard Zemel. Prototypical networks for few-shot learning. In *31st International Conference on Neural Information Processing Systems*, pages 4080–4090, 2017. ISBN 9781510860964.

Rion Snow, Dan Jurafsky, and Andrew Y Ng. Semantic taxonomy induction from heterogenous evidence. In *ACL*, pages 801–808, 2006.

Benjamin Snyder and Martha Palmer. The English all-words task. In *SENSEVAL-3, the Third International Workshop on the Evaluation of Systems for the Semantic Analysis of Text*, pages 41–43, 2004.

Richard Socher, Danqi Chen, Christopher D Manning, and Andrew Ng. Reasoning with neural tensor networks for knowledge base completion. In *NIPS*, pages 926–934, 2013a.

Richard Socher, Alex Perelygin, Jean Wu, Jason Chuang, Christopher D Manning, Andrew Y Ng, and Christopher Potts. Recursive deep models for semantic compositionality over a sentiment treebank. In *EMNLP*, pages 1631–1642, 2013b.

Anders Søgaard. Simple semi-supervised training of part-of-speech taggers. In *ACL 2010 Conference Short Papers*, pages 205–208, Uppsala, Sweden, July 2010.

Mohammad Soleymani, Sadjad Asghari-Esfeden, Yun Fu, and Maja Pantic. Analysis of eeg signals and facial expressions for continuous emotion detection. *IEEE Transactions on Affective Computing*, 7(1):17–28, 2015.

Swapna Somasundaran and Janyce Wiebe. Recognizing stances in ideological online debates. In *NAACL HLT 2010 workshop on computational approaches to analysis and generation of emotion in text*, pages 116–124, 2010.

Swapna Somasundaran, Josef Ruppenhofer, and Janyce Wiebe. Detecting arguing and sentiment in meetings. In *8th SIGdial Workshop on Discourse and Dialogue*, pages 26–34, 2007.

Sheetal Sonawane and Parag Kulkarni. The role of coreference resolution in extractive summarization. In *2016 International Conference on Computing, Analytics and Security Trends (CAST)*, pages 351–356, 2016.

Bosheng Song, Fen Li, Yuansheng Liu, and Xiangxiang Zeng. Deep learning methods for biomedical named entity recognition: a survey and qualitative comparison. *Briefings in Bioinformatics*, 22(6):bbab282, 2021.

Linfeng Song, Kun Xu, Yue Zhang, Jianshu Chen, and Dong Yu. ZPR2: Joint zero pronoun recovery and resolution using multi-task learning and BERT. In *ACL*, pages 5429–5434, 2020a.

Min Song, Il-Yeol Song, Xiaohua Hu, and Robert B Allen. Integrating text chunking with mixture hidden Markov models for effective biomedical information extraction. In *International Conference on Computational Science*, pages 976–984, 2005.

Wei Song, Jingjin Guo, Ruiji Fu, Ting Liu, and Lizhen Liu. A knowledge graph embedding approach for metaphor processing. *IEEE/ACM Transactions on Audio, Speech, and Language Processing*, 29:406–420, 2020b.

Wee Meng Soon, Hwee Tou Ng, and Daniel Chung Yong Lim. A machine learning approach to coreference resolution of noun phrases. *Computational Linguistics*, 27(4):521–544, 2001.

Hoong-Cheng Soong, Norazira Binti A Jalil, Ramesh Kumar Ayyasamy, and Rehan Akbar. The essential of sentiment analysis and opinion mining in social media: Introduction and survey of the recent approaches and techniques. In *2019 IEEE 9th Symposium on Computer Applications & Industrial Electronics*, pages 272–277, 2019.

Vibeke Sorensen, John Stephen Lansing, Nagaraju Thummanapalli, and Erik Cambria. Mood of the planet: Challenging visions of big data in the arts. *Cognitive Computation*, 14(1):310–321, 2022.

Benjamin Spector. Global positive polarity items and obligatory exhaustivity. *Semantics and Pragmatics*, 7:11–1, 2014.

Robyn Speer, Joshua Chin, and Catherine Havasi. ConceptNet 5.5: An open multilingual graph of general knowledge. In *AAAI*, pages 4444–4451, 2017.

Elizabeth S Spelke and Katherine D Kinzler. Core knowledge. *Developmental science*, 10(1):89–96, 2007.

Drahomíra Spoustová, Jan Hajič, Jan Raab, and Miroslav Spousta. Semi-supervised training for the averaged perceptron POS tagger. In *12th Conference of the European Chapter of the ACL (EACL 2009)*, pages 763–771, Athens, Greece, March 2009.

Clemens Stachl, Florian Pargent, Sven Hilbert, Gabriella M Harari, Ramona Schoedel, Sumer Vaid, Samuel D Gosling, and Markus Bühner. Personality research and assessment in the era of machine learning. *European Journal of Personality*, 34(5):613–631, 2020.

Sanja Štajner and Seren Yenikent. A survey of automatic personality detection from texts. In *COLING*, pages 6284–6295, 2020.

Sanja Štajner and Seren Yenikent. Why is MBTI personality detection from texts a difficult task? In *EACL*, pages 3580–3589, 2021.

Efstathios Stamatatos, Nikos Fakotakis, and George Kokkinakis. Automatic extraction of rules for sentence boundary disambiguation. In *Workshop on Machine Learning in Human Language Technology*, pages 88–92, 1999.

Keith Stanovich and Richard West. Advancing the rationality debate. *Behavioral and brain sciences*, 23(5):701–717, 2000.

Harald Steck, Chaitanya Ekanadham, and Nathan Kallus. Is cosine-similarity of embeddings really about similarity? In *Web Conference*, WWW '24. ACM, May 2024. doi: 10.1145/3589335.3651526.

Gerard Steen, Lettie Dorst, Berenike Herrmann, Anna Kaal, Tina Krennmayr, and Trijntje Pasma. A method for linguistic metaphor identification from MIP to MIPVU preface. *Method For Linguistic Metaphor Identification: From MIP To MIPVU*, 14:IX–+, 2010.

Josef Steinberger, Massimo Poesio, Mijail A Kabadjov, and Karel Ježek. Two uses of anaphora resolution in summarization. *Information Processing & Management*, 43(6):1663–1680, 2007.

Pontus Stenetorp, Sampo Pyysalo, Goran Topić, Tomoko Ohta, Sophia Ananiadou, and Jun'ichi Tsujii. Brat: a web-based tool for nlp-assisted text annotation. In *EACL*, pages 102–107, 2012.

Mark Stevenson and Robert Gaizauskas. Experiments on sentence boundary detection. In *Sixth Applied Natural Language Processing Conference*, pages 84–89, 2000.

Rosemary J. Stevenson, Rosalind A. Crawley, and David Kleinman. Thematic roles, focus and the representation of events. *Language and Cognitive Processes*, 9(4): 519–548, 1994.

DJ Stillwell and M Kosinski. myPersonality project website, 2015. URL https://sites.google.com/michalkosinski.com/mypersonality.

Dario Stojanovski and Alexander Fraser. Coreference and coherence in neural machine translation: A study using oracle experiments. In *Third Conference on Machine Translation: Research Papers*, pages 49–60, 2018.

Christopher Stokoe, Michael P Oakes, and John Tait. Word sense disambiguation in information retrieval revisited. In *26th Annual International ACM SIGIR Conference on Research and Development in Informaion Retrieval*, pages 159–166, 2003.

Philip J Stone, Dexter C Dunphy, and Marshall S Smith. The general inquirer: A computer approach to content analysis. *MIT press*, 1966.

Veselin Stoyanov, Claire Cardie, and Janyce Wiebe. Multi-perspective question answering using the opqa corpus. In *EMNLP-HLT*, pages 923–930, 2005.

Veselin Stoyanov, Nathan Gilbert, Claire Cardie, and Ellen Riloff. Conundrums in noun phrase coreference resolution: Making sense of the state-of-the-art. In *ACL-IJCNLP*, pages 656–664, 2009.

Jana Straková, Milan Straka, and Jan Hajic. Neural architectures for nested NER through linearization. In *ACL*, pages 5326–5331, 2019.

Carlo Strapparava, Alessandro Valitutti, et al. Wordnet affect: an affective extension of wordnet. In *LREC*, pages 1083–1086, 2004.

Stephanie Strassel. Simple metadata annotation specification version 5.0–may 14, 2003, 2003.

Michael Strube, Stefan Rapp, and Christoph Müller. The influence of minimum edit distance on reference resolution. In *EMNLP*, pages 312–319, 2002.

Tomek Strzalkowski, George Aaron Broadwell, Sarah Taylor, Laurie Feldman, Samira Shaikh, Ting Liu, Boris Yamrom, Kit Cho, Umit Boz, Ignacio Cases, et al. Robust extraction of metaphor from novel data. In *First Workshop on Metaphor in NLP*, pages 67–76, 2013.

Amber Stubbs and Özlem Uzuner. Annotating longitudinal clinical narratives for de-identification: The 2014 I2B2/uthealth corpus. *Journal of Biomedical Informatics*, 58:S20–S29, 2015.

Chang Su, Shuman Huang, and Yijiang Chen. Context-dependent metaphor interpretation based on semantic relatedness. In *Natural Language Processing and Chinese Computing*, pages 182–193. Springer, 2015.

Chang Su, Shuman Huang, and Yijiang Chen. Automatic detection and interpretation of nominal metaphor based on the theory of meaning. *Neurocomputing*, 219:300–311, 2017.

Chang Su, Ying Peng, Shuman Huang, and Yijiang Chen. A metaphor comprehension method based on culture-related hierarchical semantic model. *Neural Processing Letters*, 51(3):2807–2826, 2020.

Jianlin Su, Ahmed Murtadha, Shengfeng Pan, Jing Hou, Jun Sun, Wanwei Huang, Bo Wen, and Yunfeng Liu. Global pointer: Novel efficient span-based approach for named entity recognition. *arXiv preprint arXiv:2208.03054*, 2022.

Ming-Hsiang Su, Chung-Hsien Wu, and Yu-Ting Zheng. Exploiting turn-taking temporal evolution for personality trait perception in dyadic conversations. *IEEE/ACM Transactions on Audio, Speech, and Language Processing*, 24(4):733–744, 2016.

Niranjan Subrahmanya and Yung C Shin. Sparse multiple kernel learning for signal processing applications. *IEEE Transactions on Pattern Analysis and Machine Intelligence*, 32(5):788–798, 2009.

Amarnag Subramanya, Slav Petrov, and Fernando Pereira. Efficient graph-based semi-supervised learning of structured tagging models. In *EMNLP*, EMNLP '10, pages 167–176, USA, 2010.

Fabian M. Suchanek, Gjergji Kasneci, and Gerhard Weikum. YAGO: A core of semantic knowledge. In *International Conference on the World Wide Web*, pages 697–706, 2007.

Rhea Sukthanker, Soujanya Poria, Erik Cambria, and Ramkumar Thirunavukarasu. Anaphora and coreference resolution: A review. *Information Fusion*, 59:139–162, 2020.

Chi Sun, Xipeng Qiu, Yige Xu, and Xuanjing Huang. How to fine-tune BERT for text classification? In *China National Conference on Chinese Computational Linguistics*, pages 194–206, 2019.

Xiangguo Sun, Bo Liu, Jiuxin Cao, Junzhou Luo, and Xiaojun Shen. Who am I? Personality detection based on deep learning for texts. In *2018 IEEE International Conference on Communications (ICC)*, pages 1–6, 2018.

Xu Sun, Louis-Philippe Morency, Daisuke Okanohara, Yoshimasa Tsuruoka, and Jun'ichi Tsujii. Modeling latent-dynamic in shallow parsing: A latent conditional model with improved inference. In *COLING*, pages 841–848, 2008.

Yu Sun, Shuohuan Wang, Yukun Li, Shikun Feng, Hao Tian, Hua Wu, and Haifeng Wang. ERNIE 2.0: A continual pre-training framework for language understanding. In *AAAI*, pages 8968–8975, 2020.

Yosephine Susanto, Andrew Livingstone, Bee Chin Ng, and Erik Cambria. The hourglass model revisited. *IEEE Intelligent Systems*, 35(5):96–102, 2020.

Charles Sutton, Andrew McCallum, and Khashayar Rohanimanesh. Dynamic conditional random fields: Factorized probabilistic models for labeling and segmenting sequence data. *Journal of Machine Learning Research*, 8(3), 2007.

Jun Suzuki and Hideki Isozaki. Semi-supervised sequential labeling and segmentation using giga-word scale unlabeled data. In *ACL-08: HLT*, pages 665–673, 2008.

Swati Swati, Adrian Mladenić Grobelnik, Dunja Mladenić, and Marko Grobelnik. A commonsense-infused language-agnostic learning framework for enhancing prediction of political polarity in multilingual news headlines. *arXiv e-prints*, pages arXiv–2212, 2022.

Afraz Z Syed, Muhammad Aslam, and Ana Maria Martinez-Enriquez. Associating targets with sentiunits: a step forward in sentiment analysis of urdu text. *Artificial intelligence review*, 41(4):535–561, 2014.

Symeon Symeonidis, Dimitrios Effrosynidis, and Avi Arampatzis. A comparative evaluation of pre-processing techniques and their interactions for twitter sentiment analysis. *Expert Systems with Applications*, 110:298–310, 2018.

Anna Szabolcsi. Positive polarity–negative polarity. *Natural Language & Linguistic Theory*, 22(2):409–452, 2004.

Charles S Taber and Milton Lodge. Motivated skepticism in the evaluation of political beliefs. *American Journal of Political Science*, 50(3):755–769, 2006.

Maite Taboada, Julian Brooke, Milan Tofiloski, Kimberly Voll, and Manfred Stede. Lexicon-based methods for sentiment analysis. *Computational Linguistics*, 37(2): 267–307, 2011.

Kaveh Taghipour and Hwee Tou Ng. Semi-supervised word sense disambiguation using word embeddings in general and specific domains. In *NAACL-HLT*, pages 314–323, 2015a.

Kaveh Taghipour and Hwee Tou Ng. One million sense-tagged instances for word sense disambiguation and induction. In *Nineteenth Conference on Computational Natural Language Learning*, pages 338–344, 2015b.

Shunichi Tahara, Kazushi Ikeda, and Keiichiro Hoashi. Empathic dialogue system based on emotions extracted from tweets. In *24th International Conference on Intelligent User Interfaces*, pages 52–56, 2019.

Henri Tajfel, John C Turner, William G Austin, and Stephen Worchel. An integrative theory of intergroup conflict. *Organizational Identity: A Reader*, 56(65): 9780203505984–16, 1979.

Alon Talmor, Jonathan Herzig, Nicholas Lourie, and Jonathan Berant. Commonsenseqa: A question answering challenge targeting commonsense knowledge. In *NAACL-HLT*, pages 4149–4158, 2019.

Partha Pratim Talukdar and Koby Crammer. New regularized algorithms for transductive learning. In *Machine Learning and Knowledge Discovery in Databases: European Conference*, pages 442–457, 2009.

Xin Tan, Shaohui Kuang, and Deyi Xiong. Detecting and translating dropped pronouns in neural machine translation. In *NLPCC*, pages 343–354, 2019.

Niket Tandon, Gerard de Melo, Fabian M. Suchanek, and Gerhard Weikum. Webchild: harvesting and organizing commonsense knowledge from the web. In *7th ACM International Conference on Web Search and Data Mining, WSDM*, pages 523–532, 2014.

An Quang Tang, Xiuzhen Zhang, Minh Ngoc Dinh, and Erik Cambria. Prompted aspect key point analysis for quantitative review summarization. In *ACL*, 2024.

Duyu Tang, Furu Wei, Bing Qin, Ming Zhou, and Ting Liu. Building large-scale Twitter-specific sentiment lexicon: A representation learning approach. In *COLING*, pages 172–182, 2014a.

Duyu Tang, Furu Wei, Nan Yang, Ming Zhou, Ting Liu, and Bing Qin. Learning sentiment-specific word embedding for twitter sentiment classification. In *ACL*, volume 1, pages 1555–1565, 2014b.

Gongbo Tang, Rico Sennrich, and Joakim Nivre. Encoders help you disambiguate word senses in neural machine translation. In *EMNLP-IJCNLP*, pages 1429–1435, 2019.

Mariona Taulé, M Antònia Martí, and Marta Recasens. AnCora: Multilevel annotated corpora for catalan and spanish. In *Sixth International Conference on Language Resources and Evaluation (LREC'08)*, pages 1–6, 2008.

Yla R Tausczik and James W Pennebaker. The psychological meaning of words: LIWC and computerized text analysis methods. *Journal of Language and Social Psychology*, 29(1):24–54, 2010.

Paul Taylor, Alan W Black, and Richard Caley. The architecture of the Festival speech synthesis system. In *The Third ESCA/COCOSDA Workshop (ETRW) on Speech Synthesis*, 1998.

Simone Tedeschi and Roberto Navigli. MultiNERD: A multilingual, multi-genre and fine-grained dataset for named entity recognition (and disambiguation). In *NAACL*, pages 801–812, 2022.

Simone Tedeschi, Simone Conia, Francesco Cecconi, and Roberto Navigli. Named entity recognition for entity linking: What works and what's next. In *EMNLP*, pages 2584–2596, 2021a.

Simone Tedeschi, Valentino Maiorca, Niccolò Campolungo, Francesco Cecconi, and Roberto Navigli. WikiNEuRal: Combined neural and knowledge-based silver data creation for multilingual ner. In *EMNLP*, pages 2521–2533, 2021b.

Heike Telljohann, Erhard Hinrichs, Sandra Kübler, and Ra Kübler. The TüBa-D/Z treebank: Annotating german with a context-free backbone. In *LREC*, pages 2229–2232, 2004.

Puneshkumar U Tembhare, Ritesh Hiware, Shrey Ojha, Abhisheik Nimpure, and Faiz raza. Content recommender system based on users reviews. In *International Conference on ICT for Sustainable Development*, pages 441–451, 2023.

Joel R. Tetreault. A corpus-based evaluation of centering and pronoun resolution. *Computational Linguistics*, 27(4):507–520, 2001.

Mike Thelwall, Kevan Buckley, and Georgios Paltoglou. Sentiment strength detection for the social web. *Journal of the American Society for Information Science and Technology*, 63(1):163–173, 2012.

Yuan Tian, Nan Xu, Wenji Mao, and Daniel Zeng. Modeling conceptual attribute likeness and domain inconsistency for metaphor detection. In *EMNLP*, pages 7736–7752, 2023.

Orith Toledo-Ronen, Matan Orbach, Yoav Katz, and Noam Slonim. Multi-domain targeted sentiment analysis. In *NAACL-HLT*, pages 2751–2762, Seattle, United States, July 2022.

Antonela Tommasel, Alejandro Corbellini, Daniela Godoy, and Silvia Schiaffino. Exploring the role of personality traits in followee recommendation. *Online Information Review*, 39(6):812–830, 2015.

Antonela Tommasel, Alejandro Corbellini, Daniela Godoy, and Silvia Schiaffino. Personality-aware followee recommendation algorithms: An empirical analysis. *Engineering Applications of Artificial Intelligence*, 51:24–36, 2016.

Hanghang Tong, Christos Faloutsos, and Jia-Yu Pan. Fast random walk with restart and its applications. In *ICDM*, pages 613–622, 2006.

Kristina Toutanova and Colin Cherry. A global model for joint lemmatization and part-of-speech prediction. In *ACL-IJCNLP*, pages 486–494, 2009.

Kristina Toutanova and Mark Johnson. A Bayesian LDA-based model for semi-supervised part-of-speech tagging. *NeurIPS*, 20:1521–1528, 2007.

Kristina Toutanova, Dan Klein, Christopher D. Manning, and Yoram Singer. Feature-rich part-of-speech tagging with a cyclic dependency network. In *NAACL-HLT*, NAACL '03, pages 173–180, USA, 2003. doi: 10.3115/1073445.1073478.

Kristina Toutanvoa and Christopher D Manning. Enriching the knowledge sources used in a maximum entropy part-of-speech tagger. In *Joint SIGDAT conference on Empirical methods in natural language processing and very large corpora*, pages 63–70, 2000.

Elizabeth Closs Traugott. Revisiting subjectification and intersubjectification. *Subjectification, Intersubjectification and Grammaticalization*, 29:71, 2010.

Marcos V. Treviso, Christopher D. Shulby, and Sandra M. Aluisio. Evaluating word embeddings for sentence boundary detection in speech transcripts, 2017.

Marcos Vinícius Treviso, Christopher Shulby, and Sandra Maria Aluísio. Sentence segmentation in narrative transcripts from neuropsychological tests using recurrent convolutional neural networks. *arXiv preprint arXiv:1610.00211*, 2016.

Rocco Tripodi and Roberto Navigli. Game theory meets embeddings: a unified framework for word sense disambiguation. In *EMNLP-IJCNLP*, pages 88–99, 2019.

Yao-Hung Hubert Tsai, Shaojie Bai, Paul Pu Liang, J Zico Kolter, Louis-Philippe Morency, and Ruslan Salakhutdinov. Multimodal transformer for unaligned multimodal language sequences. In *ACL*, volume 2019, page 6558, 2019.

S. C. Tseng. Processing spoken mandarin corpora. *Traitement Automatique Des Langues*, 45(2):89–108, 2004.

Yoshimasa Tsuruoka and Jun'ichi Tsujii. Bidirectional inference with the easiest-first strategy for tagging sequence data. In *HLT-EMNLP*, HLT '05, pages 467–474, USA, 2005. doi: 10.3115/1220575.1220634.

Yulia Tsvetkov, Leonid Boytsov, Anatole Gershman, Eric Nyberg, and Chris Dyer. Metaphor detection with cross-lingual model transfer. In *ACL*, pages 248–258, 2014.

Geng Tu, Taiyu Niu, Ruifeng Xu, Bin Bin Liang, and Erik Cambria. AdaCLF: An adaptive curriculum learning framework for emotional support conversation. *IEEE Intelligent Systems*, 39(4):5–11, 2024.

Zhaopeng Tu, Zhengdong Lu, Yang Liu, Xiaohua Liu, and Hang Li. Modeling coverage for neural machine translation. In *ACL*, pages 76—85, 2016.

Mohammad Tubishat, Norisma Idris, and Mohammad Abushariah. Explicit aspects extraction in sentiment analysis using optimal rules combination. *Future Generation Computer Systems*, 114:448–480, 2021.

Peter D Turney and Patrick Pantel. From frequency to meaning: Vector space models of semantics. *Journal of Artificial Intelligence Research*, 37:141–188, 2010.

Peter Tyrer, Geoffrey M Reed, and Mike J Crawford. Classification, assessment, prevalence, and effect of personality disorder. *The Lancet*, 385(9969):717–726, 2015.

Simone Ueberwasser and Elisabeth Stark. What's up, switzerland? a corpus-based research project in a multilingual country. *Linguistik online*, 84(5), 2017.

Jan Ulrich, Gabriel Murray, and Giuseppe Carenini. A publicly available annotated corpus for supervised email summarization. In *AAAI08 email workshop*, pages 1–6, 2008.

Alexandra Uma, Dina Almanea, and Massimo Poesio. Scaling and disagreements: Bias, noise, and ambiguity. *Frontiers in Artificial Intelligence*, 5:1–11, 2022.

Alexandra N Uma, Tommaso Fornaciari, Dirk Hovy, Silviu Paun, Barbara Plank, and Massimo Poesio. Learning from disagreement: A survey. *Journal of Artificial Intelligence Research*, 72:1385–1470, 2021.

Chirayu Upadhyay, Hasan Abu-Rasheed, Christian Weber, and Madjid Fathi. Explainable job-posting recommendations using knowledge graphs and named entity recognition. In *2021 IEEE International Conference on Systems, Man, and Cybernetics*, pages 3291–3296, 2021.

Olga Uryupina, Ron Artstein, Antonella Bristot, Federica Cavicchio, Francesca Delogu, Kepa J Rodriguez, and Massimo Poesio. Annotating a broad range of anaphoric phenomena, in a variety of genres: the ARRAU corpus. *Natural Language Engineering*, 26(1):95–128, 2020.

Paul E. Utgoff, Neil C. Berkman, and Jeffery A. Clouse. Decision tree induction based on efficient tree restructuring. *Machine Learning*, 29(1):5–44, 1997. doi: 10.1023/A:1007413323501.

Özlem Uzuner, Brett R South, Shuying Shen, and Scott L DuVall. 2010 i2b2/va challenge on concepts, assertions, and relations in clinical text. *Journal of the American Medical Informatics Association*, 18(5):552–556, 2011.

Ozlem Uzuner, Andreea Bodnari, Shuying Shen, Tyler Forbush, John Pestian, and Brett R South. Evaluating the state of the art in coreference resolution for electronic medical records. *Journal of the American Medical Informatics Association*, 19 (5):786–791, 2012.

Ana Valdivia, Victoria Luzón, Erik Cambria, and Francisco Herrera. Consensus vote models for detecting and filtering neutrality in sentiment analysis. *Information Fusion*, 44:126–135, 2018.

Hans van Halteren. A default first order family weight determination procedure for WPDV models. In *Fourth Conference on Computational Natural Language Learning and the Second Learning Language in Logic Workshop*, 2000.

Hans Van Halteren. Chunking with WPDV models. In *Fourth Conference on Computational Natural Language Learning and the Second Learning Language in Logic Workshop*, 2000.

Hans van Halteren, Walter Daelemans, and Jakub Zavrel. Improving accuracy in word class tagging through the combination of machine learning systems. *Comput. Linguist.*, 27(2):199–229, 2001.

Cynthia Van Hee, Els Lefever, and Véronique Hoste. SemEval-2018 task 3: Irony detection in English tweets. In *12th International Workshop on Semantic Evaluation*, pages 39–50, 2018.

Colette M Van Kerckvoorde. *An Introduction to Middle Dutch*. De Gruyter Mouton, 2019.

Emiel van Miltenburg, Wei-Ting Lu, Emiel Krahmer, Albert Gatt, Guanyi Chen, Lin Li, and Kees van Deemter. Gradations of error severity in automatic image descriptions. In *13th International Conference on Natural Language Generation*, pages 398–411, 2020.

Ashish Vaswani, Noam Shazeer, Niki Parmar, Jakob Uszkoreit, Llion Jones, Aidan N Gomez, Łukasz Kaiser, and Illia Polosukhin. Attention is all you need. *NeurIPS*, 30, 2017.

Leonid Velikovich, Sasha Blair-Goldensohn, Kerry Hannan, and Ryan McDonald. The viability of web-derived polarity lexicons. In *NAACL-HLT*, pages 777–785, 2010.

Ben Verhoeven, Walter Daelemans, and Barbara Plank. Twisty: a multilingual Twitter stylometry corpus for gender and personality profiling. In *Tenth International Conference on Language Resources and Evaluation (LREC'16)*, pages 1632–1637, 2016.

Kanishk Verma and Brian Davis. Implicit aspect-based opinion mining and analysis of airline industry based on user-generated reviews. *SN Computer Science*, 2(4): 1–9, 2021.

Loïc Vial, Benjamin Lecouteux, and Didier Schwab. Sense vocabulary compression through the semantic knowledge of WordNet for neural word sense disambiguation. In *10th Global WordNet Conference*, pages 108–117, 2019.

Felipe Viegas, Mário S Alvim, Sérgio Canuto, Thierson Rosa, Marcos André Gonçalves, and Leonardo Rocha. Exploiting semantic relationships for unsupervised expansion of sentiment lexicons. *Information Systems*, 94:101606, 2020.

Renata Vieira and Massimo Poesio. An empirically-based system for processing definite descriptions. *Computational Linguistics*, 26(4):539–593, 2000.

Supriti Vijay and Aman Priyanshu. NERDA-Con: Extending NER models for continual learning–integrating distinct tasks and updating distribution shifts. *arXiv preprint arXiv:2206.14607*, 2022.

Marc Vilain, John D Burger, John Aberdeen, Dennis Connolly, and Lynette Hirschman. A model-theoretic coreference scoring scheme. In *MUC-6*, pages 45–52, 1995.

Julio Villena, Janine García-Morera, Miguel García-Cumbreras, Eugenio Martínez-Cámara, Maria Martín-Valdivia, and L. López. Overview of TASS 2015. In *TASS 2015: Workshop on Sentiment Analysis at SEPLN co-located with 31st SEPLN Conference*, pages 13–21, 09 2015.

Alessandro Vinciarelli and Gelareh Mohammadi. A survey of personality computing. *IEEE Transactions on Affective Computing*, 5(3):273–291, 2014.

Oriol Vinyals, Meire Fortunato, and Navdeep Jaitly. Pointer networks. In *NeurIPS*, pages 2692–2700, 2015a.

Oriol Vinyals, Meire Fortunato, and Navdeep Jaitly. Pointer networks. In *28th International Conference on Neural Information Processing Systems-Volume 2*, pages 2692–2700, 2015b.

Duy-Tin Vo and Yue Zhang. Target-dependent Twitter sentiment classification with rich automatic features. In *IJCAI*, 2015.

Luis Von Ahn. Games with a purpose. *Computer*, 39(6):92–94, 2006.

Luis von Ahn, Mihir Kedia, and Manuel Blum. Verbosity: a game for collecting common-sense facts. In *2006 Conference on Human Factors in Computing Systems, CHI*, pages 75–78, 2006.

Denny Vrandečić. Ontology evaluation. In *Handbook on Ontologies*, pages 293–313. Springer, 2009.

Denny Vrandecic and Markus Krötzsch. Wikidata: a free collaborative knowledgebase. *Communications of the ACM*, 57(10):78–85, 2014.

Tung Vuong, Salvatore Andolina, Giulio Jacucci, and Tuukka Ruotsalo. Spoken conversational context improves query auto-completion in web search. *ACM Transactions on Information Systems (TOIS)*, 39(3):1–32, 2021.

Lennart Wachowiak and Dagmar Gromann. Systematic analysis of image schemas in natural language through explainable multilingual neural language processing. In *COLING*, pages 5571–5581, 2022.

Christopher Walker, Stephanie Strassel, Julie Medero, and Kazuaki Maeda. ACE 2005 multilingual training corpus. *Linguistic Data Consortium, Philadelphia*, 57: 45, 2006.

Marilyn Walker and Steve Whittaker. Mixed initiative in dialogue: an investigation into discourse segmentation. In *ACL*, pages 70–78, 1990.

Byron C Wallace, Laura Kertz, Eugene Charniak, et al. Humans require context to infer ironic intent (so computers probably do, too). In *ACL*, pages 512–516, 2014.

Hai Wan, Yufei Yang, Jianfeng Du, Yanan Liu, Kunxun Qi, and Jeff Z Pan. Target-aspect-sentiment joint detection for aspect-based sentiment analysis. In *AAAI*, pages 9122–9129, 2020.

Mengting Wan and Julian McAuley. Modeling ambiguity, subjectivity, and diverging viewpoints in opinion question answering systems. In *ICDM*, pages 489–498, 2016.

Qian Wan, Luona Wei, Xinhai Chen, and Jie Liu. A region-based hypergraph network for joint entity-relation extraction. *Knowledge-Based Systems*, 228:107298, 2021.

Xiaojun Wan. Co-training for cross-lingual sentiment classification. In *ACL-IJCNLP*, pages 235–243, 2009.

Bailin Wang and Wei Lu. Neural segmental hypergraphs for overlapping mention recognition. In *EMNLP*, pages 204–214, 2018.

Feixiang Wang, Man Lan, and Wenting Wang. Towards a one-stop solution to both aspect extraction and sentiment analysis tasks with neural multi-task learning. In *IJCNN*, pages 1–8, 2018a.

Hongwei Wang, Fuzheng Zhang, Jialin Wang, Miao Zhao, Wenjie Li, Xing Xie, and Minyi Guo. Exploring high-order user preference on the knowledge graph for recommender systems. *ACM Transactions on Information Systems*, 37(3):1–26, 2019a.

Longyue Wang, Zhaopeng Tu, Xing Wang, and Shuming Shi. One model to learn both: Zero pronoun prediction and translation. In *EMNLP-IJCNLP*, pages 921–930, 2019b.

Ming Wang and Yinglin Wang. A synset relation-enhanced framework with a try-again mechanism for word sense disambiguation. In *EMNLP*, pages 6229–6240, 2020.

Peilu Wang, Yao Qian, Frank K. Soong, Lei He, and Hai Zhao. Part-of-speech tagging with bidirectional long short-term memory recurrent neural network. *CoRR*, abs/1510.06168, 2015a.

Pidong Wang and Hwee Tou Ng. A beam-search decoder for normalization of social media text with application to machine translation. In *NAACL-HLT*, pages 471–481, 2013.

Shan Wang and Francis Bond. Building the Chinese open Wordnet (COW): Starting from core synsets. In *11th Workshop on Asian Language Resources*, pages 10–18, 2013.

Sida Wang and Christopher Manning. Fast dropout training. In *ICML*, pages 118–126, 2013.

Sida I Wang and Christopher D Manning. Baselines and bigrams: Simple, good sentiment and topic classification. In *ACL*, pages 90–94, 2012.

Wei Wang, Ling He, Yenchun Jim Wu, and Mark Goh. Signaling persuasion in crowdfunding entrepreneurial narratives: the subjectivity vs objectivity debate. *Computers in Human Behavior*, 114:106576, 2021.

Wenguan Wang and Yi Yang. Towards data-and knowledge-driven artificial intelligence: A survey on neuro-symbolic computing. *arXiv preprint arXiv:2210.15889*, 2022.

Wenya Wang, Sinno Jialin Pan, Daniel Dahlmeier, and Xiaokui Xiao. Coupled multi-layer attentions for co-extraction of aspect and opinion terms. In *AAAI*, 2017.

Xiaochen Wang, Wenzheng Feng, Jie Tang, and Qingyang Zhong. Course concept extraction in MOOC via explicit/implicit representation. In *Third IEEE International Conference on Data Science in Cyberspace*, pages 339–345, 2018b.

Yanan Wang, Qi Liu, Chuan Qin, Tong Xu, Yijun Wang, Enhong Chen, and Hui Xiong. Exploiting topic-based adversarial neural network for cross-domain keyphrase extraction. In *ICDM*, pages 597–606, 2018c.

Yexiang Wang, Yi Guo, and Siqi Zhu. Slot attention with value normalization for multi-domain dialogue state tracking. In *EMNLP*, pages 3019–3028, 2020.

Yong Wang and Ian H Witten. Induction of model trees for predicting continuous classes. In *European Conference on Machine Learning*, 1996.

Yu Wang, Hanghang Tong, Ziye Zhu, and Yun Li. Nested named entity recognition: A survey. *ACM Transactions on Knowledge Discovery from Data*, 16(6):1–29, 2022.

Zhongyuan Wang, Haixun Wang, Ji-Rong Wen, and Yanghua Xiao. An inference approach to basic level of categorization. In *CIKM*, pages 653–662, 2015b.

William Warner and Julia Hirschberg. Detecting hate speech on the world wide web. In *Second Workshop on Language in Social Media*, pages 19–26, 2012.

Bonnie Lynn Webber. Discourse deixis: Reference to discourse segments. In *ACL*, pages 113–122, 1988.

Bonnie Lynn Webber. *A formal approach to discourse anaphora*. Routledge, 2016.

Kellie Webster, Marta Recasens, Vera Axelrod, and Jason Baldridge. Mind the GAP: A balanced corpus of gendered ambiguous pronouns. *Transactions of the Association for Computational Linguistics*, 6:605–617, 2018.

Wei Wei, Zanbo Wang, Xianling Mao, Guangyou Zhou, Pan Zhou, and Sheng Jiang. Position-aware self-attention based neural sequence labeling. *Pattern Recognition*, 110:107636, 2021.

Zhepei Wei, Jianlin Su, Yue Wang, Yuan Tian, and Yi Chang. A novel hierarchical binary tagging framework for joint extraction of entities and relations. *arXiv preprint arXiv:1909.03227*, 2019.

Ralph Weischedel, Martha Palmer, Mitchell Marcus, Eduard Hovy, Sameer Pradhan, Lance Ramshaw, Nianwen Xue, Ann Taylor, Jeff Kaufman, Michelle Franchini, et al. Ontonotes release 5.0. *Linguistic Data Consortium, Philadelphia, PA*, 23, 2013.

Matthew West. *Developing High Quality Data Models*. Morgan Kaufmann Publishers Inc., 340 Pine Street, Sixth FloorSan FranciscoCAUnited States, 2011.

Peter West, Chandra Bhagavatula, Jack Hessel, Jena D. Hwang, Liwei Jiang, Ronan Le Bras, Ximing Lu, Sean Welleck, and Yejin Choi. Symbolic knowledge distillation: from general language models to commonsense models. In *NAACL*, pages 4602–4625, 2022.

David L Wheeler, Tanya Barrett, Dennis A Benson, Stephen H Bryant, Kathi Canese, Vyacheslav Chetvernin, Deanna M Church, Michael DiCuccio, Ron Edgar, Scott Federhen, et al. Database resources of the national center for biotechnology information. *Nucleic Acids Research*, 36(suppl_1):D13–D21, 2007.

Janyce Wiebe. Tracking point of view in narrative. *Computational Linguistics*, 20 (2):233–287, 1994.

Janyce Wiebe. Learning subjective adjectives from corpora. In *AAAI/IAAI*, pages 735–740, 2000.

Janyce Wiebe and Ellen Riloff. Creating subjective and objective sentence classifiers from unannotated texts. In *International conference on intelligent text processing and computational linguistics*, pages 486–497, 2005.

Janyce Wiebe and Ellen Riloff. Finding mutual benefit between subjectivity analysis and information extraction. *IEEE Transactions on Affective Computing*, 2(4): 175–191, 2011.

Janyce Wiebe, Theresa Wilson, and Claire Cardie. Annotating expressions of opinions and emotions in language. *Language Resources and Evaluation*, 39(2): 165–210, 2005.

Janyce Marbury Wiebe. *Recognizing Subjective Sentences: A Computational Investigation of Narrative Text*. State University of New York at Buffalo, 1990.

Anna Wierzbicka. *Semantic Primitives*. Athena um-Verl, 1972.

Anna Wierzbicka. *Semantics: Primes and universals: Primes and universals*. Oxford University Press, UK, 1996.

Jerry S Wiggins. Personality structure. *Annual Review of Psychology*, 19(1):293–350, 1968.

Robert Wilensky. Some problems and proposals for knowledge representation. Technical report, University of California, Berkeley. Institute of cognitive studies, 1986.

Yorick Wilks. Preference semantics. Technical report, Computer Science Department, Stanford University, 1973.

Yorick Wilks. A preferential, pattern-seeking, semantics for natural language inference. *Artificial Intelligence*, 6(1):53–74, 1975.

Theresa Wilson. Annotating subjective content in meetings. In *Sixth International Conference on Language Resources and Evaluation (LREC'08)*, pages 2738–2745, 2008.

Theresa Wilson, Janyce Wiebe, and Rebecca Hwa. Just how mad are you? finding strong and weak opinion clauses. In *AAAI*, volume 4, pages 761–769, 2004.

Theresa Wilson, Janyce Wiebe, and Paul Hoffmann. Recognizing contextual polarity in phrase-level sentiment analysis. In *EMNLP-HLT*, pages 347–354, 2005.

William Winkler. String comparator metrics and enhanced decision rules in the Fellegi-Sunter model of record linkage. *Section on Survey Research Methods*, 01 1990.

Terry Winograd. Understanding natural language. *Cognitive Psychology*, 3(1): 1–191, 1972.

Terry Winograd. Towards a procedural understanding of semantics. *Revue internationale de philosophie*, pages 260–303, 1976.

Sam Wiseman and Karl Stratos. Label-agnostic sequence labeling by copying nearest neighbors. In *ACL*, pages 5363–5369, 2019.

Sam Wiseman, Alexander M. Rush, Stuart Shieber, and Jason Weston. Learning anaphoricity and antecedent ranking features for coreference resolution. In *ACL-IJCNLP*, pages 1416–1426, 2015.

Sam Wiseman, Alexander M. Rush, and Stuart M. Shieber. Learning global features for coreference resolution. In *NAACL-HLT*, pages 994–1004, 2016.

David S Wishart, Yannick D Feunang, An C Guo, Elvis J Lo, Ana Marcu, Jason R Grant, Tanvir Sajed, Daniel Johnson, Carin Li, Zinat Sayeeda, et al. DrugBank 5.0: a major update to the DrugBank database for 2018. *Nucleic Acids Research*, 46(D1):D1074–D1082, 2018.

René Witte and Sabine Bergler. Fuzzy coreference resolution for summarization. In *2003 International Symposium on Reference Resolution and Its Applications to Question Answering and Summarization (ARQAS)*, pages 43–50, 2003.

Martin Wöllmer, Felix Weninger, Tobias Knaup, Björn Schuller, Congkai Sun, Kenji Sagae, and Louis-Philippe Morency. Youtube movie reviews: Sentiment analysis in an audio-visual context. *IEEE Intelligent Systems*, 28(3):46–53, 2013.

Derek F Wong, Lidia S Chao, and Xiaodong Zeng. isentenizer-: Multilingual sentence boundary detection model. *The Scientific World Journal*, 2014, 2014.

Fai Wong and Sam Chao. isentenizer: An incremental sentence boundary classifier. In *6th International Conference on Natural Language Processing and Knowledge Engineering (NLPKE-2010)*, pages 1–7, 2010.

Beverly Park Woolf. Chapter 5 - communication knowledge. In Beverly Park Woolf, editor, *Building Intelligent Interactive Tutors*, pages 136–182. Morgan Kaufmann, San Francisco, 2009.

Britta Wrede and Elizabeth Shriberg. Spotting "hot spots" in meetings: human judgments and prosodic cues. In *EUROSPEECH*, volume 2003, pages 2805–2808, 2003.

Chao Wu, Qingyu Xiong, Hualing Yi, Yang Yu, Qiwu Zhu, Min Gao, and Jie Chen. Multiple-element joint detection for aspect-based sentiment analysis. *Knowledge-Based Systems*, 223:107073, 2021a.

Chien-Sheng Wu, Steven CH Hoi, Richard Socher, and Caiming Xiong. TOD-BERT: Pre-trained natural language understanding for task-oriented dialogue. In *EMNLP*, pages 917–929, 2020a.

Fangzhao Wu, Yongfeng Huang, Yangqiu Song, and Shixia Liu. Towards building a high-quality microblog-specific chinese sentiment lexicon. *Decision Support Systems*, 87:39–49, 2016.

Jialun Wu, Kai He, Rui Mao, Chen Li, and Erik Cambria. Megacare: Knowledge-guided multi-view hypergraph predictive framework for healthcare. *Information Fusion*, 100(101939), 2023.

Minghao Wu, Fei Liu, and Trevor Cohn. Evaluating the utility of hand-crafted features in sequence labelling. In *EMNLP*, pages 2850–2856, 2018.

Shengqiong Wu, Hao Fei, Yafeng Ren, Donghong Ji, and Jingye Li. Learn from syntax: Improving pair-wise aspect and opinion terms extractionwith rich syntactic knowledge. *arXiv preprint arXiv:2105.02520*, 2021b.

Shijie Wu and Ryan Cotterell. Exact hard monotonic attention for character-level transduction. *arXiv preprint arXiv:1905.06319*, 2019.

Wei Wu, Fei Wang, Arianna Yuan, Fei Wu, and Jiwei Li. CorefQA: Coreference resolution as query-based span prediction. In *ACL*, pages 6953–6963, 2020b.

Wen Wu. Implicit acquisition of user personality for augmenting recommender systems. In *22nd International Conference on Intelligent User Interfaces Companion*, pages 201–204, 2017.

Wentao Wu, Hongsong Li, Haixun Wang, and Kenny Qili Zhu. Probase: a probabilistic taxonomy for text understanding. In *ACM SIGMOD International Conference on Management of Data*, pages 481–492, 2012.

Yang Wu, Yanyan Zhao, Xin Lu, Bing Qin, Yin Wu, Jian Sheng, and Jinlong Li. Modeling incongruity between modalities for multimodal sarcasm detection. *IEEE MultiMedia*, 28(2):86–95, 2021c.

Youyou Wu, Michal Kosinski, and David Stillwell. Computer-based personality judgments are more accurate than those made by humans. *National Academy of Sciences*, 112(4):1036–1040, 2015.

Youyou Wu, David Stillwell, H Andrew Schwartz, and Michal Kosinski. Birds of a feather do flock together: Behavior-based personality-assessment method reveals personality similarity among couples and friends. *Psychological Science*, 28(3): 276–284, 2017.

Zhen Wu, Chengcan Ying, Fei Zhao, Zhifang Fan, Xinyu Dai, and Rui Xia. Grid tagging scheme for aspect-oriented fine-grained opinion extraction. *arXiv preprint arXiv:2010.04640*, 2020c.

Yu Xia, Quan Wang, Yajuan Lyu, Yong Zhu, Wenhao Wu, Sujian Li, and Dai Dai. Learn and review: Enhancing continual named entity recognition via reviewing synthetic samples. In *ACL*, pages 2291–2300, 2022.

Pan Xiao, YongQuan Fan, and YaJun Du. A personality-aware followee recommendation model based on text semantics and sentiment analysis. In *Natural Language Processing and Chinese Computing*, pages 503–514, 2018.

Chenyan Xiong, Russell Power, and Jamie Callan. Explicit semantic ranking for academic search via knowledge graph embedding. In Rick Barrett, Rick Cummings, Eugene Agichtein, and Evgeniy Gabrilovich, editors, *26th International Conference on World Wide Web*, pages 1271–1279, 2017.

Shengzhou Xiong, Yihua Tan, and Guoyou Wang. Explore visual concept formation for image classification. In Marina Meila and Tong Zhang, editors, *ICML*, volume 139 of *Machine Learning Research*, pages 11470–11479. PMLR, 2021.

Fangzhi Xu, Qika Lin, Jiawei Han, Tianzhe Zhao, Jun Liu, and Erik Cambria. Are large language models really good logical reasoners? a comprehensive evaluation from deductive, inductive and abductive views. *arXiv preprint arXiv:2306.09841*, 2024.

Hu Xu, Bing Liu, Lei Shu, and Philip S Yu. Double embeddings and CNN-based sequence labeling for aspect extraction. *arXiv preprint arXiv:1805.04601*, 2018.

Hua Xu, Fan Zhang, and Wei Wang. Implicit feature identification in Chinese reviews using explicit topic mining model. *Knowledge-Based Systems*, 76:166–175, 2015a.

Ke Xu, Yunqing Xia, and Chin-Hui Lee. Tweet normalization with syllables. In *ACL-IJCNLP*, pages 920–928, 2015b.

Kelvin Xu, Jimmy Ba, Ryan Kiros, Kyunghyun Cho, Aaron Courville, Ruslan Salakhudinov, Rich Zemel, and Yoshua Bengio. Show, attend and tell: Neural image caption generation with visual attention. In *ICML*, pages 2048–2057, 2015c.

Lu Xu, Hao Li, Wei Lu, and Lidong Bing. Position-aware tagging for aspect sentiment triplet extraction. *arXiv preprint arXiv:2010.02609*, 2020a.

Lu Xu, Yew Ken Chia, and Lidong Bing. Learning span-level interactions for aspect sentiment triplet extraction. In *ACL-IJCNLP*, pages 4755–4766, 2021.

Qiannan Xu, Li Zhu, Tao Dai, Lei Guo, and Sisi Cao. Non-negative matrix factorization for implicit aspect identification. *Journal of Ambient Intelligence and Humanized Computing*, 11:2683–2699, 2020b.

Huong Nguyen Thi Xuan, Anh Cuong Le, et al. Linguistic features for subjectivity classification. In *2012 International Conference on Asian Language Processing*, pages 17–20, 2012.

Ying Xue, Jianshan Sun, Yezheng Liu, Xin Li, and Kun Yuan. Facial expression-enhanced recommendation for virtual fitting rooms. *Decision Support Systems*, 177:114082, 2024.

Zhenzhen Xue, Dawei Yin, and Brian D Davison. Normalizing microtext. In *AAAI*, 2011.

Vikas Yadav and Steven Bethard. A survey on recent advances in named entity recognition from deep learning models. In *COLING*, pages 2145–2158, 2018.

Shahpar Yakhchi, Amin Beheshti, Seyed Mohssen Ghafari, and Mehmet Orgun. Enabling the analysis of personality aspects in recommender systems. *arXiv preprint arXiv:2001.04825*, 2020.

Kosuke Yamada, Ryohei Sasano, and Koichi Takeda. Incorporating textual information on user behavior for personality prediction. In *ACL*, pages 177–182, 2019.

Hang Yan, Junqi Dai, Tuo Ji, Xipeng Qiu, and Zheng Zhang. A unified generative framework for aspect-based sentiment analysis. In *ACL-IJCNLP*, pages 2416–2429, 2021a.

Hang Yan, Tao Gui, Junqi Dai, Qipeng Guo, Zheng Zhang, and Xipeng Qiu. A unified generative framework for various NER subtasks. In *ACL-IJCNLP*, pages 5808–5822, 2021b.

Yukun Yan and Sen Song. Local hypergraph-based nested named entity recognition as query-based sequence labeling. *arXiv preprint arXiv:2204.11467*, 2022.

Zhiheng Yan, Chong Zhang, Jinlan Fu, Qi Zhang, and Zhongyu Wei. A partition filter network for joint entity and relation extraction. In *EMNLP*, pages 185–197, 2021c.

Zhijun Yan, Meiming Xing, Dongsong Zhang, and Baizhang Ma. EXPRS: An extended pagerank method for product feature extraction from online consumer reviews. *Information & Management*, 52(7):850–858, 2015.

Chin Lung Yang, Peter C Gordon, Randall Hendrick, and Jei Tun Wu. Comprehension of referring expressions in chinese. *Language and Cognitive Processes*, 14 (5-6):715–743, 1999.

Dongqiang Yang and David Martin Powers. *Measuring semantic similarity in the taxonomy of WordNet*. Australian Computer Society, Australia, 2005.

Feifan Yang, Xiaojun Quan, Yunyi Yang, and Jianxing Yu. Multi-document transformer for personality detection. In *AAAI*, pages 14221–14229, 2021a.

Feifan Yang, Tao Yang, Xiaojun Quan, and Qinliang Su. Learning to answer psychological questionnaire for personality detection. In *EMNLP*, pages 1131–1142, 2021b.

Hsin-Chang Yang and Zi-Rui Huang. Mining personality traits from social messages for game recommender systems. *Knowledge-Based Systems*, 165:157–168, 2019.

Jie Yang, Shuailong Liang, and Yue Zhang. Design challenges and misconceptions in neural sequence labeling. *arXiv preprint arXiv:1806.04470*, 2018.

Jingyuan Yang, Xinbo Gao, Leida Li, Xiumei Wang, and Jinshan Ding. SOLVER: Scene-object interrelated visual emotion reasoning network. *IEEE Transactions on Image Processing*, 30:8686–8701, 2021c.

Shuoheng Yang, Yuxin Wang, and Xiaowen Chu. A survey of deep learning techniques for neural machine translation. *arXiv preprint arXiv:2002.07526*, 2020a.

Songlin Yang and Kewei Tu. Bottom-up constituency parsing and nested named entity recognition with pointer networks. In *ACL*, pages 2403–2416, 2022.

Tao Yang, Feifan Yang, Haolan Ouyang, and Xiaojun Quan. Psycholinguistic tripartite graph network for personality detection. In *ACL-IJCNLP*, pages 4229–4239, 2021d.

Xi Yang, Jiang Bian, William R. Hogan, and Yonghui Wu. Clinical concept extraction using transformers. *J. Am. Medical Informatics Assoc.*, 27(12):1935–1942, 2020b.

Xiaofeng Yang, Guodong Zhou, Jian Su, and Chew Lim Tan. Coreference resolution using competition learning approach. In *ACL*, pages 176–183, 2003.

Yi Yang and Jacob Eisenstein. A log-linear model for unsupervised text normalization. In *EMNLP*, pages 61–72, 2013.

Yi Yang and Arzoo Katiyar. Simple and effective few-shot named entity recognition with structured nearest neighbor learning. In *EMNLP*, pages 6365–6375, 2020.

Zhilin Yang, Ruslan Salakhutdinov, and William Cohen. Multi-task cross-lingual sequence tagging from scratch. *arXiv preprint arXiv:1603.06270*, 2016a.

Zhilin Yang, Ruslan Salakhutdinov, and William W. Cohen. Transfer learning for sequence tagging with hierarchical recurrent networks. *CoRR*, abs/1703.06345, 2017.

Zichao Yang, Diyi Yang, Chris Dyer, Xiaodong He, Alex Smola, and Eduard Hovy. Hierarchical attention networks for document classification. In *NAACL-HLT*, pages 1480–1489, 2016b.

Zonglin Yang, Xinya Du, Erik Cambria, and Claire Cardie. End-to-end case-based reasoning for commonsense knowledge base completion. In *EACL*, pages 3509–3522, 2023.

Zonglin Yang, Li Dong, Xinya Du, Hao Cheng, Erik Cambria, Xiaodong Liu, Jianfeng Gao, and Furu Wei. Language models as inductive reasoners. In *EACL*, pages 209–225, 2024.

Fanglong Yao, Xian Sun, Hongfeng Yu, Wenkai Zhang, Wei Liang, and Kun Fu. Mimicking the brain's cognition of sarcasm from multidisciplines for Twitter sarcasm detection. *IEEE Transactions on Neural Networks and Learning Systems*, 2021.

Liang Yao, Chengsheng Mao, and Yuan Luo. KG-BERT: BERT for knowledge graph completion. *CoRR*, abs/1909.03193, 2019.

Yuanzhou Yao, Zhao Zhang, Yongjun Xu, and Chao Li. Data augmentation for few-shot knowledge graph completion from hierarchical perspective. In *COLING*, pages 2494–2503, 2022.

Deming Ye, Yankai Lin, Jiaju Du, Zhenghao Liu, Peng Li, Maosong Sun, and Zhiyuan Liu. Coreferential Reasoning Learning for Language Representation. In *EMNLP*, pages 7170–7186, 2020.

Hai Ye and Lu Wang. Semi-supervised learning for neural keyphrase generation. In *EMNLP*, pages 4142–4153, 2018.

Weijie Yeo, Teddy Ferdinan, Przemyslaw Kazienko, Ranjan Satapathy, and Erik Cambria. Self-training large language models through knowledge detection. *arXiv preprint arXiv:2406.11275*, 2024a.

Weijie Yeo, Ranjan Satapathy, and Erik Cambria. Plausible extractive rationalization through semi-supervised entailment signal. In *ACL*, pages 5182–5192, 2024b.

Weijie Yeo, Ranjan Satapathy, Siow Mong Goh, and Erik Cambria. How interpretable are reasoning explanations from prompting large language models? In *NAACL*, pages 2148–2164, 2024c.

Eray Yildiz and A Cüneyd Tantuğ. Morpheus: A neural network for jointly learning contextual lemmatization and morphological tagging. In *Workshop on Computational Research in Phonetics, Phonology, and Morphology*, pages 25–34, 2019.

Qingyu Yin, Weinan Zhang, Yun Zhang, and Ting Liu. A deep neural network for chinese zero pronoun resolution. In *IJCAI*, pages 3322–3328, 2017a.

Qingyu Yin, Yu Zhang, Weinan Zhang, and Ting Liu. Chinese zero pronoun resolution with deep memory network. In *EMNLP*, pages 1309–1318, 2017b.

Qingyu Yin, Yu Zhang, Wei-Nan Zhang, Ting Liu, and William Yang Wang. Deep reinforcement learning for Chinese zero pronoun resolution. In *ACL*, pages 569–578, 2018a.

Qingyu Yin, Yu Zhang, Weinan Zhang, Ting Liu, and William Yang Wang. Zero pronoun resolution with attention-based neural network. In *COLING*, pages 13–23, 2018b.

Qingyu Yin, Weinan Zhang, Yu Zhang, and Ting Liu. Chinese zero pronoun resolution: A collaborative filtering-based approach. *ACM Transactions on Asian and Low-Resource Language Information Processing*, 19(1):1–20, 2019.

Tom Young, Erik Cambria, Iti Chaturvedi, Hao Zhou, Subham Biswas, and Minlie Huang. Augmenting end-to-end dialogue systems with commonsense knowledge. In *AAAI*, pages 4970–4977, 2018.

Bowen Yu, Zhenyu Zhang, Xiaobo Shu, Tingwen Liu, Yubin Wang, Bin Wang, and Sujian Li. Joint extraction of entities and relations based on a novel decomposition strategy. In *ECAI 2020*, pages 2282–2289. IOS Press, 2020.

Guoxin Yu, Jiwei Li, Ling Luo, Yuxian Meng, Xiang Ao, and Qing He. Self question-answering: Aspect-based sentiment analysis by role flipped machine reading comprehension. In *EMNLP*, pages 1331–1342, 2021.

Hong Yu and Vasileios Hatzivassiloglou. Towards answering opinion questions: Separating facts from opinions and identifying the polarity of opinion sentences. In *EMNLP*, pages 129–136, 2003.

Jianfei Yu, Jing Jiang, and Rui Xia. Global inference for aspect and opinion terms co-extraction based on multi-task neural networks. *IEEE/ACM Transactions on Audio, Speech, and Language Processing*, 27(1):168–177, 2018.

Jianfei Yu, Jieming Wang, Rui Xia, and Junjie Li. Targeted multimodal sentiment classification based on coarse-to-fine grained image-target matching. In *IJCAI*, 2022.

Jianxing Yu, Zheng-Jun Zha, Meng Wang, and Tat-Seng Chua. Aspect ranking: Identifying important product aspects from online consumer reviews. In *ACL-HLT*, pages 1496–1505, 2011a.

Jianxing Yu, Zheng-Jun Zha, Meng Wang, Kai Wang, and Tat-Seng Chua. Domain-assisted product aspect hierarchy generation: Towards hierarchical organization of unstructured consumer reviews. In *EMNLP*, pages 140–150, 2011b.

Cuixin Yuan, Junjie Wu, Hong Li, and Lihong Wang. Personality recognition based on user generated content. In *2018 15th International Conference on Service Systems and Service Management (ICSSSM)*, pages 1–6, 2018.

Dayu Yuan, Julian Richardson, Ryan Doherty, Colin Evans, and Eric Altendorf. Semi-supervised word sense disambiguation with neural models. In *COLING*, pages 1374–1385, 2016.

Tan Yue, Rui Mao, Heng Wang, Zonghai Hu, and Erik Cambria. KnowleNet: Knowledge fusion network for multimodal sarcasm detection. *Information Fusion*, 100:101921, 2023.

Yessi Yunitasari, Aina Musdholifah, and Anny Kartika Sari. Sarcasm detection for sentiment analysis in indonesian tweets. *Indonesian Journal of Computing and Cybernetics Systems*, 13(1):53–62, 2019.

Amir Zadeh, Minghai Chen, Soujanya Poria, Erik Cambria, and Louis-Philippe Morency. Tensor fusion network for multimodal sentiment analysis. In *EMNLP*, pages 1103–1114, 2017.

Nurulhuda Zainuddin, Ali Selamat, and Roliana Ibrahim. Improving Twitter aspect-based sentiment analysis using hybrid approach. In *Intelligent Information and Database Systems*, pages 151–160, 2016.

Nasser Zalmout and Nizar Habash. Joint diacritization, lemmatization, normalization, and fine-grained morphological tagging. *arXiv preprint arXiv:1910.02267*, 2019.

Nasser Zalmout and Nizar Habash. Utilizing subword entities in character-level sequence-to-sequence lemmatization models. In *COLING*, pages 4676–4682, 2020.

Omnia Zayed, John Philip McCrae, and Paul Buitelaar. Figure me out: A gold standard dataset for metaphor interpretation. In *LREC*, pages 5810–5819, 2020.

Eugene B. Zechmeister, Andrea M. Chronis, William L. Cull, Catherine A. D'Anna, and Noreen A. Healy. Growth of a functionally important lexicon. *Journal of Reading Behavior*, 27(2):201–212, 1995.

Amir Zeldes. The GUM corpus: Creating multilayer resources in the classroom. *Language Resources and Evaluation*, 51(3):581–612, 2017.

Daniel Zeman, Jan Hajic, Martin Popel, Martin Potthast, Milan Straka, Filip Ginter, Joakim Nivre, and Slav Petrov. CoNLL 2018 shared task: Multilingual parsing from raw text to universal dependencies. In *CoNLL 2018 Shared Task: Multilingual parsing from raw text to universal dependencies*, pages 1–21, 2018.

Feifei Zhai, Saloni Potdar, Bing Xiang, and Bowen Zhou. Neural models for sequence chunking. *arXiv preprint arXiv:1701.04027*, 2017.

Bowen Zhang, Xu Huang, Zhichao Huang, Hu Huang, Baoquan Zhang, Xianghua Fu, and Liwen Jing. Sentiment interpretable logic tensor network for aspect-term sentiment analysis. In *COLING*, pages 6705–6714, 2022.

Chen Zhang, Qiuchi Li, Dawei Song, and Benyou Wang. A multi-task learning framework for opinion triplet extraction. *arXiv preprint arXiv:2010.01512*, 2020a.

Congle Zhang, Tyler Baldwin, Howard Ho, Benny Kimelfeld, and Yunyao Li. Adaptive parser-centric text normalization. In *ACL*, pages 1159–1168, 2013.

Hongming Zhang, Yan Song, Yangqiu Song, and Dong Yu. Knowledge-aware pronoun coreference resolution. In *ACL*, pages 867–876, 2019.

Hongming Zhang, Daniel Khashabi, Yangqiu Song, and Dan Roth. Transomcs: From linguistic graphs to commonsense knowledge. In *IJCAI*, pages 4004–4010, 2020b.

Hongming Zhang, Xin Liu, Haojie Pan, Yangqiu Song, and Cane Wing-Ki Leung. ASER: A large-scale eventuality knowledge graph. In *Web Conference 2020, WWW*, pages 201–211, 2020c.

Jun Zhang, Yan Yang, Chencai Chen, Liang He, and Zhou Yu. KERS: A knowledge-enhanced framework for recommendation dialog systems with multiple subgoals. In *EMNLP*, pages 1092–1101, 2021a.

Meishan Zhang, Yue Zhang, and Duy-Tin Vo. Neural networks for open domain targeted sentiment. In *EMNLP*, pages 612–621, 2015a.

Meishan Zhang, Yue Zhang, and Guohong Fu. End-to-end neural relation extraction with global optimization. In *EMNLP*, pages 1730–1740, 2017a.

Qi Zhang, Yang Wang, Yeyun Gong, and Xuanjing Huang. Keyphrase extraction using deep recurrent neural networks on twitter. In Jian Su, Xavier Carreras, and Kevin Duh, editors, *EMNLP*, pages 836–845, 2016.

Rui Zhang, Cícero Nogueira dos Santos, Michihiro Yasunaga, Bing Xiang, and Dragomir Radev. Neural coreference resolution with deep biaffine attention by joint mention detection and mention clustering. In *ACL*, pages 102–107, 2018.

Tong Zhang, Fred Damerau, and David E Johnson. Text chunking using regularized winnow. In *ACL*, pages 539–546, 2001.

Wei Zhang, Clement Yu, and Weiyi Meng. Opinion retrieval from blogs. In *CIKM*, pages 831–840, 2007.

Wen Zhang, Taketoshi Yoshida, Xijin Tang, and Tu-Bao Ho. Improving effectiveness of mutual information for substantival multiword expression extraction. *Expert Systems with Applications*, 36(8):10919–10930, 2009.

Wenhao Zhang, Hua Xu, and Wei Wan. Weakness finder: Find product weakness from Chinese reviews by using aspects based sentiment analysis. *Expert Systems with Applications*, 39(11):10283–10291, 2012.

Wenxuan Zhang, Yang Deng, Xin Li, Yifei Yuan, Lidong Bing, and Wai Lam. Aspect sentiment quad prediction as paraphrase generation. In *EMNLP*, pages 9209–9219, 2021b.

Wenxuan Zhang, Xin Li, Yang Deng, Lidong Bing, and Wai Lam. Towards generative aspect-based sentiment analysis. In *ACL-IJCNLP*, pages 504–510, 2021c.

Xiang Zhang, Junbo Zhao, and Yann LeCun. Character-level convolutional networks for text classification. *NeurIPS*, 28, 2015b.

Xulang Zhang, Rui Mao, and Erik Cambria. A survey on syntactic processing techniques. *Artificial Intelligence Review*, 56:5645–5728, 2023a.

Xulang Zhang, Rui Mao, Kai He, and Erik Cambria. Neurosymbolic sentiment analysis with dynamic word sense disambiguation. In *EMNLP Findings*, pages 8772–8783, 2023b.

Xulang Zhang, Rui Mao, and Erik Cambria. Granular syntax processing with multi-task and curriculum learning. *Cognitive Computation*, 16, 2024a.

Xulang Zhang, Rui Mao, and Erik Cambria. Multilingual emotion recognition: Discovering the variations of lexical semantics between languages. In *IJCNN*, 2024b.

Xulang Zhang, Rui Mao, and Erik Cambria. SenticVec: Toward robust and human-centric neurosymbolic sentiment analysis. In *ACL Findings*, pages 4851–4863, 2024c.

Yuhao Zhang, Victor Zhong, Danqi Chen, Gabor Angeli, and Christopher D. Manning. Position-aware attention and supervised data improve slot filling. In *EMNLP*, pages 35–45, 2017b.

Ziming Zhang, Ze-Nian Li, and Mark S Drew. AdaMKL: A novel biconvex multiple kernel learning approach. In *2010 20th International Conference on Pattern Recognition*, pages 2126–2129, 2010.

Han Zhao, Zhengdong Lu, and Pascal Poupart. Self-adaptive hierarchical sentence model. In *IJCAI*, pages 4069–4076, 2015.

He Zhao, Longtao Huang, Rong Zhang, Quan Lu, and Hui Xue. SpanMlt: A span-based multi-task learning framework for pair-wise aspect and opinion terms extraction. In *ACL*, pages 3239–3248, Online, July 2020a.

He Zhao, Longtao Huang, Rong Zhang, Quan Lu, and Hui Xue. SpanMlt: A span-based multi-task learning framework for pair-wise aspect and opinion terms extraction. In *ACL*, pages 3239–3248, 2020b.

Jialiang Zhao and Qi Gao. Annotation and detection of emotion in text-based dialogue systems with cnn. *arXiv preprint arXiv:1710.00987*, 2017.

Jieyu Zhao, Tianlu Wang, Mark Yatskar, Vicente Ordonez, and Kai-Wei Chang. Gender bias in coreference resolution: Evaluation and debiasing methods. In *NAACL-HLT*, pages 15–20, 2018.

Lujun Zhao, Xipeng Qiu, Qi Zhang, and Xuanjing Huang. Sequence labeling with deep gated dual path CNN. *IEEE/ACM Transactions on Audio, Speech, and Language Processing*, 27(12):2326–2335, 2019.

Qiyun Zhao, Hao Wang, Pin Lv, and Chen Zhang. A bootstrapping based refinement framework for mining opinion words and targets. In *CIKM*, pages 1995–1998, 2014.

Shanheng Zhao and Hwee Tou Ng. Identification and resolution of Chinese zero pronouns: A machine learning approach. In *EMNLP-CoNLL*, pages 541–550, 2007.

Suncong Zheng, Feng Wang, Hongyun Bao, Yuexing Hao, Peng Zhou, and Bo Xu. Joint extraction of entities and relations based on a novel tagging scheme. In *ACL*, pages 1227–1236, 2017.

Yaowei Zheng, Richong Zhang, Suyuchen Wang, Samuel Mensah, and Yongyi Mao. Anchored model transfer and soft instance transfer for cross-task cross-domain learning: A study through aspect-level sentiment classification. In *Web Conference*, pages 2754–2760, 2020.

Peixiang Zhong, Di Wang, and Chunyan Miao. Knowledge-enriched transformer for emotion detection in textual conversations. In *EMNLP-IJCNLP*, pages 165–176, 2019.

Xiaoshi Zhong and Erik Cambria. Time expression recognition and normalization: A survey. *Artificial Intelligence Review*, 56:9115–9140, 2023.

Zexuan Zhong and Danqi Chen. A frustratingly easy approach for entity and relation extraction. In *NAACL-HLT*, pages 50–61, 2021.

Zhi Zhong and Hwee Tou Ng. It makes sense: A wide-coverage word sense disambiguation system for free text. In *ACL*, pages 78–83, 2010.

Deyu Zhou, Zhikai Zhang, Min-Ling Zhang, and Yulan He. Weakly supervised POS tagging without disambiguation. *ACM Transactions on Asian and Low-Resource Language Information Processing*, 17(4):1–19, 2018.

GuoDong Zhou and Jian Su. Error-driven HMM-based chunk tagger with context-dependent lexicon. In *EMNLP-VLC*, pages 71–79, 2000.

Houquan Zhou, Yu Zhang, Zhenghua Li, and Min Zhang. Is POS tagging necessary or even helpful for neural dependency parsing?, 2020a.

Kun Zhou, Wayne Xin Zhao, Shuqing Bian, Yuanhang Zhou, Ji-Rong Wen, and Jingsong Yu. Improving conversational recommender systems via knowledge graph based semantic fusion. In *SIGKDD*, pages 1006–1014, 2020b.

Nina Zhou, Xuancong Wang, and AiTi Aw. Dynamic boundary detection for speech translation. In *APSIPA*, pages 651–656, 2017.

Ran Zhou, Xin Li, Ruidan He, Lidong Bing, Erik Cambria, Luo Si, and Chunyan Miao. MELM: Data augmentation with masked entity language modeling for low-resource NER. In *ACL*, pages 2251–2262, 2022.

Yan Zhou, Fuqing Zhu, Pu Song, Jizhong Han, Tao Guo, and Songlin Hu. An adaptive hybrid framework for cross-domain aspect-based sentiment analysis. In *AAAI*, pages 14630–14637, 2021a.

Zhengyu Zhou, In Gyu Choi, Yongliang He, Vikas Yadav, and Chin-Hui Lee. Using paralinguistic information to disambiguate user intentions for distinguishing phrase structure and sarcasm in spoken dialog systems. In *2021 IEEE Spoken Language Technology Workshop (SLT)*, pages 1020–1027, 2021b.

Zhi-Hua Zhou and Xu-Ying Liu. Training cost-sensitive neural networks with methods addressing the class imbalance problem. *IEEE Transactions on Knowledge and Data Engineering*, 18(1):63–77, 2005.

Luyao Zhu, Wei Li, Rui Mao, Vlad Pandelea, and Erik Cambria. PAED: Zero-shot persona attribute extraction in dialogues. In *ACL*, pages 9771–9787, 2023.

Luyao Zhu, Wei Li, Rui Mao, and Erik Cambria. HIPPL: Hierarchical intent-inferring pointer network with pseudo labeling for consistent persona-driven dialogue generation. *IEEE Computational Intelligence Magazine*, 2024a.

Luyao Zhu, Rui Mao, Erik Cambria, and Bernard J. Jansen. Neurosymbolic AI for personalized sentiment analysis. In *International Conference on Human-Computer Interaction (HCII)*, Washington DC, USA, 2024b.

Yin Zhuang, Zhen Liu, Ting-Ting Liu, Chih-Chieh Hung, and Yan-Jie Chai. Implicit sentiment analysis based on multi-feature neural network model. *Soft Computing*, 26(2):635–644, 2022.

Anne Zimmerman, Joel Janhonen, and Emily Beer. Human/AI relationships: challenges, downsides, and impacts on human/human relationships. *AI and Ethics*, 2023.

GPSR Compliance

The European Union's (EU) General Product Safety Regulation (GPSR) is a set of rules that requires consumer products to be safe and our obligations to ensure this.

If you have any concerns about our products, you can contact us on ProductSafety@springernature.com

In case Publisher is established outside the EU, the EU authorized representative is:

Springer Nature Customer Service Center GmbH
Europaplatz 3
69115 Heidelberg, Germany

The manufacturer's authorised representative in the EU is Springer
Nature Customer Service Centre GmbH, Europaplatz 3, 69115 Heidelberg,
Germany. If you have any concerns regarding our products, please
contact ProductSafety@springernature.com

Printed and bound by CPI Group (UK) Ltd, Croydon, CR0 4YY
24/04/2026
02096316-0010